Advances in Intelligent Systems and Computing

Volume 217

Series Editor

J. Kacprzyk, Warsaw, Poland

For further volumes:
http://www.springer.com/series/11156

Sigeru Omatu · José Neves
Juan M. Corchado Rodríguez
Juan F. De Paz Santana
Sara Rodríguez González
Editors

Distributed Computing and Artificial Intelligence

10th International Conference

Editors
Sigeru Omatu
Graduate School of Engineering
Osaka Prefecture University
Osaka
Japan

José Neves
Department of Computing Science
University of Minho
Braga Codex
Portugal

Juan M. Corchado Rodríguez
Department of Computing Science and Control
Faculty of Science
University of Salamanca
Salamanca
Spain

Juan F. De Paz Santana
Department of Computing Science and Control
Faculty of Science
University of Salamanca
Salamanca
Spain

Sara Rodríguez González
Department of Computing Science and Control
Faculty of Science
University of Salamanca
Salamanca
Spain

ISSN 2194-5357 ISSN 2194-5365 (electronic)
ISBN 978-3-319-00550-8 ISBN 978-3-319-00551-5 (eBook)
DOI 10.1007/978-3-319-00551-5
Springer Cham Heidelberg New York Dordrecht London

Library of Congress Control Number: 2013937503

Printed on acid-free paper

Springer is part of Springer Science+Business Media (www.springer.com)

Preface

The International Symposium on Distributed Computing and Artificial Intelligence 2013 (DCAI 2013) is a forum to present applications of innovative techniques for solving complex problems. Artificial intelligence is changing our society. Its application in distributed environments, such as the internet, electronic commerce, environment monitoring, mobile communications, wireless devices, distributed computing, to mention only a few, is continuously increasing, becoming an element of high added value with social and economic potential, in industry, quality of life and research. These technologies are changing constantly as a result of the large research and technical effort being undertaken in both universities and businesses. The exchange of ideas between scientists and technicians from both the academic and industry sector is essential to facilitate the development of systems that can meet the ever-increasing demands of today's society. The present edition brings together past experience, current work and promising future trends associated with distributed computing, artificial intelligence and their application in order to provide efficient solutions to real problems.

This conference is a stimulating and productive forum where the scientific community can work towards future cooperation in Distributed Computing and Artificial Intelligence areas. Nowadays it is continuing to grow and prosper in its role as one of the premier conferences devoted to the quickly changing landscape of distributed computing, artificial intelligence and the application of AI to distributed systems. This year's technical program will present both high quality and diversity, with contributions in well-established and evolving areas of research. Specifically, 101 papers were submitted from over 19 different countries (Colombia, Czech Republic, Finland, France, Germany, Greece, Italy, Japan, Libya, Mauritius, Mexico, Morocco, Poland, Portugal, Romania, Slovakia, Spain, Thailand, Tunisia), representing a truly "wide area network" of research activity. The DCAI'13 technical program has selected 75 papers. As in past editions, it will be special issues in journals such as Journal of Artificial Intelligence (IJAI), the International Journal of Imaging and Robotics (IJIR) and the International Journal of Interactive Multimedia and Artificial Intelligence (IJIMAI). These special issues will cover extended versions of the most highly regarded works.

This symposium is organized by the Bioinformatics, Intelligent System and Educational Technology Research Group (http://bisite.usal.es/) of the University of Salamanca. The present edition was be held in Salamanca, Spain, from 22nd to 24th May 2013.

We thank the sponsors (Telefónica Digital, Indra, INSA, Ingeniería de Software Avanzado S.A., IBM, JCyL, IEEE Systems Man and Cybernetics Society Spain, AEPIA *Asociación Española para la Inteligencia Artificial*, APPIA *Associação Portuguesa Para a Inteligência Artificial*, CNRS *Centre national de la recherche scientifique*), the Local Organization members and the Program Committee members for their hard work, which was essential for the success of DCAI'13.

Salamanca
May 2013

Sigeru Omatu
José Neves
Juan M. Corchado Rodríguez
Juan F. De Paz Santana
Sara Rodríguez González (Eds.)

Organization

General Chairs

Ajith Abraham	Norwegian University of Science and Technology
Juan M. Corchado	University of Salamanca (Spain)
James Llinas	State University of New York (USA)
José M. Molina	Universidad Carlos III de Madrid (Spain)
Andre Ponce de Leon	University of Sao Paulo at Sao Carlos (Brazil)

Scientific Chair

Sigeru Omatu (Chairman)	Osaka Institute of Technology (Japan)
José Neves (Chairman)	Universidade do Minho, Braga (Portugal)

Organizing Committee

Jesús García Herrero	University Carlos III of Madrid (Spain)
Juan M. Corchado (Chairman)	University of Salamanca(Spain)
Sara Rodríguez (Co-Chairman)	University of Salamanca (Spain)
Juan F. De Paz (Co-Chairman)	University of Salamanca (Spain)
Javier Bajo	Polytechnic University of Madrid (Spain)
Dante I. Tapia	University of Salamanca (Spain)
Fernando de la Prieta Pintado	University of Salamanca (Spain)
Davinia Carolina Zato Domínguez	University of Salamanca (Spain)
Gabriel Villarrubia González	University of Salamanca (Spain)
Alejandro Sánchez Yuste	University of Salamanca (Spain)
Antonio Juan Sánchez Martín	University of Salamanca (Spain)
Cristian I. Pinzón	University of Salamanca (Spain)
Rosa Cano	University of Salamanca (Spain)
Emilio S. Corchado	University of Salamanca (Spain)

Eugenio Aguirre	University of Granada (Spain)
Manuel P. Rubio	University of Salamanca (Spain)
Belén Pérez Lancho	University of Salamanca (Spain)
Angélica González Arrieta	University of Salamanca (Spain)
Vivian F. López	University of Salamanca (Spain)
Ana de Luís	University of Salamanca (Spain)
Ana B. Gil	University of Salamanca (Spain)
Mª Dolores Muñoz Vicente	University of Salamanca (Spain)

Scientific Committee

Zbigniew Pasek	IMSE/University of Windsor (Canada)
Adriana Giret	Politechnich University of Valencia (Spain)
Alberto Fernández	University Rey Juan Carlos (Spain)
Álvaro Herrero	University of Burgos (Spain)
Ana Carolina Lorena	Federal University of ABC (Brazil)
Andre Coelho	University of Fortaliza (Brazil)
Ângelo Costa	University of Minho (Portugal)
Antonio Moreno	University Rovira y Virgili(Spain)
Antonio Pereira	Instituto Politécnico de Leiria (Portugal)
Araceli Sanchís	University Carlos III of Madrid (Spain)
Ayako HIRAMATSU	Osaka Sangyo University (JAPAN)
B. Cristina Pelayo García-Bustelo	University of Oviedo (Spain)
Bianca Innocenti	University of Girona (Spain)
Bogdan Gabrys	Bournemouth University (UK)
Bruno Baruque	University of Burgos (Spain)
Carina González	University of La Laguna (Spain)
Carlos Carrascosa	Politechnich University of Valencia (Spain)
Carmen Benavides	University of Leon (Spain)
Daniel Glez-Peña	University of Vigo (Spain)
David Griol Barres	University Carlos III of Madrid (Spain)
Davide Carneiro	University of Minho (Portugal)
Dídac Busquets	University of Girona (Spain)
Dongshik KANG	Ryukyu University (JAPAN)
Eladio Sanz	Unversity of Salamanca (Spain)
Eleni Mangina	University College Dublin (Ireland)
Emilio Corchado	University of Burgos (Spain)
Eugenio Aguirre	University of Granada (Spain)
Eugénio Oliveira	University of Porto (Portugal)
Evelio J. González	University of La Laguna (Spain)
Faraón Llorens Largo	University of Alicante (Spain)
Fernando Díaz	Univesity of Valladolid (Spain)
Fidel Aznar Gregori	University of Alicante (Spain)
Florentino Fdez-Riverola	University of de Vigo (Spain)

Francisco Pujol López	Polytechnic University of Alicante (Spain)
Grzegorz Bocewicz -Koszalin	University of technology (Poland)
Helder Coelho	University of Lisbon (Portugal)
Ichiro Satoh	Thammasat University (Japan)
Ivan López Arévalo	Lab. of Information Technology Cinvestav-(Mexico)
Jamal Dargham	University of Malaysia, Saba, (Malaysia)
Javier Carbó	University Carlos III of Madrid (Spain)
Javier Martínez Elicegui	Telefónica I+D (Spain)
Jesús García Herrero	University Carlos III of Madrid (Spain)
Joao Gama	University of Porto (Portugal)
Johan Lilius	Åbo Akademi University (Finland)
José R. Villar	University of Oviedo (Spain)
Juan A. Botia	University of Murcia (Spain)
Juan Pavón	Complutense University of Madrid (Spain)
José M. Molina	University Carlos III of Madrid (Spain)
José R. Méndez	University of Vigo (Spain)
José V. Álvarez-Bravo	University of Valladolid (Spain)
Joseph Giampapa	Carnegie Mellon (USA)
Juan Manuel Cueva Lovelle	University of Oviedo (Spain)
Juan Gómez Romero	University Carlos III of Madrid (Spain)
Kazutoshi Fujikawa	Nara Institute of Science and Technology (Japan)
Lourdes Borrajo	University of Vigo (Spain)
Luis Alonso	University of Salamanca (Spain)
Luis Correia	University of Libon(Portugal)
Luis F. Castillo	Autonomous University of Manizales (Colombia)
Luís Lima	Polytechnic of Porto, (Portugal)
Manuel González-Bedia	University of Zaragoza (Spain)
Manuel Pegalajar Cuéllar	University of Granada (Spain)
Manuel Resinas	University of Sevilla (Spain)
Marcilio Souto	Federal University of Rio Grande do Norte (Brazil)
Margarida Cardoso	ISCTE (Portugal)
Maria del Mar Pujol López	University of Alicante (Spain)
Michael Zaki	University Rostok (Germany)
Michifumi Yoshioka	Osaka Prefecture University, (JAPAN)
Masanori Akiyoshi	Osaka University (JAPAN)
Masaru Teranishi	Hiroshima Institute of Technology (Japan)
Masatake Akutagawa	University of Tokushima (JAPAN)
Michifumi Yoshioka	Osaka Prefecture University, (JAPAN)
Miguel Ángel Patricio	University Carlos III of Madrid (Spain)
Miguel Delgado	University of Granada(Spain)
Miguel Molina	University of Granada(Spain)
Miguel Rebollo	University of Vigo (Spain)
Mohd Saberi Mohamad	University of Technology Malaysia (MALAYSIA)
Noki Mori	Osaka Prefecture University (Japan)

Nora Muda	National University of Malaysia (Malaysia)
Norihisa Komoda	Osaka University (Japan)
Paulo Novais	Polytechnic University of Minho (Portugal)
Pawel Pawlewski	Poznan University of Technology (Poland)
Peter Fobrig	University Rostok (Germany)
Ichiro Satoh	Thammasat University (Japan)
Rafael Corchuelo	Catholic University of Sevilla (Spain)
Ramón Rizo	University of Alicante (Spain)
René Leistikow	University Rostok (Germany)
Rodrigo Mello	University of Sao Paulo (Brazil)
Rubén Fuentes	Complutense University of Madrid (Spain)
Rui Camacho	University of Porto (Portugal)
Sebastian Bader	University of Rostock (Germany)
Seiichi Ozawa	Kobe University, JAPAN
Seiji Ito	Hiroshima Institute of Technology (Japan)
Seyedsaeid Mirkamali	University of Mysore (India)
Shanmugasundaram Hariharan	B.S.A. Rahman University Vandalur (India)
Silvana Aciar	University of Girona (Spain)
Shiro Tamaki	University of the Ryukyus (Japan)
Sofya V. Zhukova	St. Petersburg State University (Russia)
Toru Yamamoto	Hiroshima University (JAPAN)
Takuya Maekawa	Osaka University (JAPAN)
Vicente Botti	Politechnich University of Valencia (Spain)
Vicente Julián	Politechnich University of Valencia (Spain)
Victor J. Sosa-Sosa	Laboratory of Inf. Technology (LTI) (México)
Viet-Hoang VU	Hanoi University of Technology (Vietnam)
Waldo Fajardo	University of Granada(Spain)
Worawan Diaz Carballo	National Institute of Informatics (Thailand)
Yi Fang	Purdue University (USA)
Yusuke NOJIMA	Osaka Prefecture University (JAPAN)
Yutaka MAEDA	Kansai University (JAPAN)
Zbigniew Banaszak	University of Koszalin (Poland)

Contents

Image Processing, Tracking, Robotic, Control and Industrial Systems

Multi-agent Systems

Distributed Computing, Grid Computing, Cloud Computing

Bioinformatics, Biomedical Systems, e-health, Ambient Assisting Living

Data mining, Information Extraction, Semantic, Knowledge Representation

Artificial Intelligence Applications

Erratum

Improving the Performance of NEAT Related Algorithm via Complexity Reduction in Search Space

Heman Mohabeer and K.M.S. Soyjaudah

Department of Electrical and Electronics Engineering,
University of Mauritius,
Reduit, Mauritius
heman.mohabeer@ieee.org, ssoyjaudah@uom.ac.mu

Abstract. In this paper, we focus on the learning aspect of NEAT and its variants in an attempt to solve benchmark problems through fewer generations. In NEAT, genetic algorithm is the key technique that is used to complexify artificial neural network. Crossover value, being the parameter that dictates the evolution of NEAT is reduced. Reducing crossover rate aids in allowing the algorithm to learn. This is because lesser interchange among genes ensures that patterns of genes carrying valuable information is not split or strayed during mating of two chromosomes. By tweaking the crossover parameter and with some minor modification, it is shown that the performance of NEAT can be improved. This enables NEAT algorithm to evolve slowly and retain information even while undergoing complexification. Thus, the learning process in NEAT is greatly enhanced as compared to evolution

Keywords: Crossover, NEAT, Learning, Evolution, Genetic algorithm, artificial neural network.

1 Introduction

Neuroevolution of augmented topologies (NEAT) is the offspring developed from the fusion of genetic algorithms (GA) and artificial neural network (ANN). In 2004, K.Stanley et al. [1] postulated NEAT which has since then emerged as an exciting prospect in the development of artificial intelligence. When applied to reinforcement learning task, NEAT has outpaced other conventional method [2], [3], [4] initially developed to cater for this task. The major change in NEAT when applied to reinforcement learning task is its ability to learn and complexify, thus enabling it to perform while requiring no supervision.

So far, NEAT has not yet been applied to major applications except in benchmarks problems to enable one to compare the performance of NEAT to other methods such as simulated annealing and hill climbing [5], [6]. Instead, several variants [7], [8] based on NEAT have been created. These variants

S. Omatu et al. (Eds.): *Distrib. Computing & Artificial Intelligence*, AISC 217, pp. 1–7.
DOI: 10.1007/978-3-319-00551-5_1

enhanced the behavior of NEAT while solving both reinforcement and sparse reinforcement tasks. Feature selection NEAT (FS-NEAT) [9] is one such variant that helps the algorithm to choose its own input from a set of data which enables it to perform with fewer inputs and provides better performance. This is achieved by learning the network's inputs, topology, and weights simultaneously. Inputs are one of the major parameters affecting the behaviour of a machine learning system. FS- NEAT allows important data to be selectively chosen from a set input so as to trigger the minimum effect on the complexity of the network. Doing so reduce time to find the desired solution and greatly broadens the set of tasks to which they can be practically applied.

Until now, GA has been a major attribute in the performance of Neuroevolution. Its ability to emulate evolution has made it a vital component of ANN. The two main parameters that contribute to evolution are crossover and mutation. Mutations, in NEAT can modify both connection weights and network structures [1], [9]. Connection weights mutate as in any neuroevolution system; structural mutations, which allow complexity to increase, add either a new connection or node to the network. Through mutation, genomes of varying sizes are created, sometimes with completely different connections specified at the same positions. Crossover is solely possible if the system is able to identify the genes matchup between any individuals in a population. This is the main reason for assigning an innovation number to every gene so as to have their chronological representation in the system. When crossing over, the genes in both genomes with the same innovation numbers are lined up. Genes that do not match are inherited from the more fit parent, or if they are equally fit, from both parents randomly.

Crossover value in NEAT has been input as per reported theoretical value originally used in GA. Initially, it was postulated that the crossover rate for maximum efficiency in GA was around 0.7 to 0.9. In this paper we present the effect of varying the crossover value bearing in mind that GA exclusively evolves ANN and does not directly solve the particular problem to which NEAT is applied. It is shown that hypothetical value for crossover rate only works best when GA is used to solve problems in which it is applied autonomously. In this paper, the aspect of learning, which is a slow process, has been taken into consideration. We propose a new crossover rate for NEAT that outperforms the current rate used for every variant of NEAT. Section one provides insights about crossover as one of the major parameter in NEAT. Section two describes the simulations and results while section three describes the discussions. Finally, we conclude and provide ground for future works in section four.

1.1 Crossover

In biological world, mating is important to ensure the continual survival of a species. It also helps to create diversity among the species since traits from each parent are transferred in the resulting offspring. In this way the offspring is said to carry a portion of both parents. Normally, in real world the genes are evenly distributed and the offspring may carry a genotype from parent one while another genotype from parent two. The selection of genotype is a random process. No genotype of a parent is predominant on the other parent. When homologous chromosomes are lined up during meiosis, they can, in a very precise way, exchange genetic material [10]. Variation in the genome is greatly increased by two processes occurring during meiosis: independent assortment and crossing over. Both these processes are extremely complex, but in themselves are not subject to selection. They rearrange the genetic material randomly, resulting in new combinations of the material. As this reshuffling occurs at the level of the genotype, it is not subject to natural selection until the new combinations have been expressed in the phenotype [10]. In this way, it is almost impossible to predict the precise trait of the offspring.

Fortunately, in artificial world, it is possible to control the behavior of the genotype during the mating process and thus embark only those which are fittest into the next generation. To achieve this feat, each gene is assigned a fitness value. This value is calculated based on its position relative to the solution in the state space. In this way the next generation of offspring can be cultivated with genes that have higher fitness value and this continues until the solution is reached. The transfer of genes with higher fitness from parent to offspring contributes to the learning process in the ANN. Single handedly GA cannot learn. It follows the concept of survive or die. However when applied to other algorithms such as neural network as in the case of NEAT, it hugely contributes to the learning process.

1.2 Learning in NEAT

Machine learning, which has always been an area of interest in AI, is concerned with the study of computer algorithms that improve automatically through experience [11]. Reinforcement learning (RL) [12], [13] deals with the problem of how an agent, situated in an environment and interacting with it, can learn a policy, a mapping of states to actions, in order to maximize the total amount of reward it receives over time, by inferring on the immediate feedback returned in the form of scalar rewards as a consequence of its actions. Neural network has the ability to deal with reinforcement learning. It has been successfully been applied to task such as task such as XOR, pole balancing and the double pole balancing. The rate of learning in NN is quite slow and the system requires large amount of training to be able to provide efficient solution. This is mostly dependant on the topology and complexity of the system. In the case of NEAT, learning has not really been the prime concern. Instead, focus was made on the complexification of neat in order to achieve a state whereby the system was able to solve a given reinforcement problem. The complexification of NEAT in this paper was computed in terms of number of generations achieved in order to solve the problem.

1.3 Methodology

The methodology adopted in this paper was quite straightforward. It was argued that crossover had an impact on learning and thus the ability to solve a reinforcement problem in fewer complexifications. The NEAT algorithm was tweaked so that only the crossover rate acted as a variable. In this way, the other parameters such as mutation and speciation were kept at bay. The crossover rate determines the percentage of offspring which will be generated from crossover in the offspring at each generation. If you do not crossover enough, there is not enough sharing of genes. If you crossover too much, good sections of individuals get split up a lot. According to this statement a balance has to be searched for in order to get the optimal workability of the system. The benchmark problems that were used for experimentation purposes were the XOR, pole balancing and the double pole balancing. Since it comprises of reinforcement learning task, learning is an integral phase along with the process of complexification. Simulations were carried out using the NEAT algorithms from [14]. The program was slightly modified to tailor it to a more systematic approach. Figure 1 shows a small portion of the code. The part which is circled highlights the parameters that have been varied from a range of 0.1 to 0.9. A new feature, which consists of monitoring the learning rate of the neural network, has also been added. The main purport is to provide evidence on the influence of crossover rate on the learning aspect of the neural network.

```
Crossover. Percentage =0.8; % percentage governs that way in which new population will be composed from old population. Exception: species with just one individual can only use mutation

Crossover.probability_interspecies=0.0; % if crossover has been selected, this probability governs the intra/interspecies parent composition being used for the crossover

Crossover.probability_multipoint=0.6;%standard-crossover in which matching connection genes are inherited randomly from both parents. In the (1-crossover.probability_multipoint) cases, weights of the new connection genes are the mean of the corresponding parent genes
```

Fig. 1 Sample code displaying the crossover parameters in NEAT.

2 Simulations and Results

It is argued that crossover has an impact on learning and thus the ability to solve a reinforcement problem in fewer complexifications. The NEAT algorithm has been tweaked so that only the crossover rate acted as a variable. In this way, the other parameters such as mutation and speciation have been kept at bay. The crossover rate determines the percentage of offspring which will be generated from crossover in the offspring at each generation. The benchmark problems that were used for experimentation purposes were the XOR, pole balancing and the double pole balancing. Since it comprises reinforcement learning task, learning is an integral phase along with the process of complexification. Simulations were carried out using the NEAT algorithms from [14]. The algorithm was slightly modified to tailor it to a more systematic approach. The first problem to be tackled was the XOR problem. The reason for this choice was that among the three benchmark problem, XOR was the one that invited a minimalistic degree of complexity in NEAT. At each stage where the crossover rate were modified, fifty simulations were carried out to ensure that the result obtained were consistent. It was noted that the best results was obtained from a crossover rate of 0.3. The same concept was applied for the pole balancing and the double pole balancing. Table 1 shows the number of generations achieved by NEAT for solving each problem using a crossover of 0.3 as compared to the theoretical value used while designing NEAT. The initial number of generation is the number achieved while using theoretical value of crossover rate. The new number of generation provides novel values achieved when the crossover rate was reduced to 0.3.

Table 1 Depicts the number of generations to solve the stated reinforcement learning

	Initial number of generations	New number of generations
XOR	67	7
Pole balancing	250	75
Double pole balancing	286	114

3 Discussions and Future Works

This paper provides evidence that a balance between evolution and learning contributes to a significant improvement NEAT. Based on the simulations performed, it has been shown that learning and crossover has a close relationship. By reducing the crossover rate it was observed that the benchmark problems could be solved more quickly and with minimal amount of complexity. Probability of crossover or crossover rate is the parameter that affects the rate at which the crossover operator is applied. A higher crossover rate introduces new strings more quickly into the population, that is, the population quickly evolves. Normally, in the case of GA, it should make the system more efficient and converge toward the solution more quickly. However in NEAT the converse is observed. This is because quick evolution prevents the system from learning thus the ability of NN in NEAT is stripped off its main ability. Theoretically, a low crossover rate may cause stagnation due to the lower exploration rate. However, a rate of 0.3 in a properly designed NEAT algorithm ensures the balance between learning and exploring the state space. Exploration is much slower but the leap in the state space from the current position toward the optimal solution is much more significant. Consider the following argument: useful information carried by a chromosome may consist of a bundle of genes. These genes are individually assigned with a higher fitness value compared to others in the same chromosome. When performing crossover using high rates, the bundle may be broken and thus their information capacity may diminish. However, while reducing crossover rate the bundle is preserved and this reflects on their learning capabilities therefore converging toward the results in a more efficient manner. Moreover, the work presented in this paper has a significant impact since lots of applications have been developed with NEAT as the core algorithm therefore, variants of NEAT such as RBF-NEAT and FS- NEAT is also likely to have their performance improved significantly.

Reinforcement learning task is mostly concerned situations where the solution is achieved through actions induced from response to external environment. Being an intelligent agent, the learning ability of NN is predominant on the exploration skills provided by GA since GA is only secondary toward solving the problem in NEAT. GA only helps to optimize NN via its topology and weight optimization. Alone GA is unable to learn, nonetheless, like genetic algorithms, reinforcement learning is an unsupervised learning problem. However, unlike genetic algorithms, agents can learn during their lifetimes. It is also noted that since the benchmark problems are solved in fewer generations, this ensures that the complexity of the system is nominal. This results in faster computational processing and lesser time to search for solution. It is expected that the proposed change will have an impact on all the variants of NEAT such as the compositional pattern producing network and the real time NEAT. A study could also be conducted on the relationship of crossover and learning in the context of intelligent agents such as the one described in this paper.

Acknowledgment. The financial contribution of the Tertiary Education Commission is gratefully acknowledged.

References

1. Stanley, K.O., Miikkulainen, R.: Efficient Evolution of Neural Network Topologies. In: Proceedings of the 2002 Congress on Evolutionary Computation (CEC 2002), Honolulu, Hawaii (2002)
2. Gomez, F., Miikkulainen, R.: Solving non-Markovian control tasks with neuroevolution. In: Proceedings of the 16th International Joint Conference on Artificial Intelligence. Morgan Kaufmann, Denver (1999)
3. Moriarty, D.E.: Symbiotic Evolution of Neural Networks in Sequential Decision Tasks. PhD thesis, Department of Computer Sciences, The University of Texas at Austin, Technical Report UT-AI97-257 (1997)
4. Zheng, Z.: A benchmark for classifier learning. Technical Report TR474, Basser Department of Computer Science, University of Sydney, N.S.W. Australia (2006), Anonymous ftp from `http://ftp.cs.su.oz.au.in/pub/tr`

5. Granville, V., Krivanek, M., Rasson, J.P.: Simulated annealing: A proof of convergence. IEEE Transactions on Pattern Analysis and Machine Intelligence 16(6), 652–656 (1994)
6. Cano, A., Gomez, M., Moral, S.: Application of a hill-climbing algorithm to exact and approximate inference in credal networks. In: 4th International Symposium on Imprecise Probabilities and Their Applications, Pittsburgh, Pennsylvania (2005)
7. Stanley, K.O.: Exploiting Regularity without Development. In: Proceedings of the 2006 AAAI Fall Symposium on Developmental Systems. AAAI Press, Menlo Park (2006)
8. Stanley, K.O., Miikkulainen, R.: Competitive coevolution through evolutionary complexification. Journal of Artificial Intelligence Research 21, 63–100 (2004)
9. Whiteson, S., Stanley, K.O., Miikkulainen, R.: Automatic Feature Selection in Neuroevolution. In: Proceedings of the Genetic and Evolutionary Computation Conference Workshop on Self-Organization (GECCO 2004), Seattle, WA (2004)
10. `http://amazingdiscoveries.org/Cdeception_gene_chromosome_sexual_reproduction.hmtl` (accessed December 17, 2011)
11. Moriarty, D.E., Miikkulainen, R.: Efficient reinforcement learning through symbiotic evolution. Machine Learning 22, 11–32 (1996)
12. Baird, L.C.: Residual Algorithms: Reinforcement Learning with Function Approximation. In: Prieditis, A., Russell, S. (eds.) Proceedings of the Twelfth International Conference on Machine Learning, July 9-12 (1995)
13. Sutton, R.S., Barto, A.G.: Reinforcement Learning: An Introduction. MIT Press, Cambridge (1998)
14. `http://www.cs.utexas.edu/users/ai-lab/?neatmatlab` (accessed on October 25, 2010)

Intelligent Application to Reduce Transit Accidents in a City Using Cultural Algorithms

Fernando Maldonado, Alberto Ochoa*, Julio Arreola, Daniel Azpeitia, Ariel De la Torre, Diego Canales, and Saúl González

Autonomous University of Ciudad Juárez
alberto.ochoa@uacj.mx

Abstract. Ciudad Juárez, a large city located along the Mexico-United States border with a population over a million people in 87 km^2 has recently experienced a history of violence and insecurity related directly to organized crime: assaults, kidnappings, multi-homicides, burglary, between others. However the second leading cause of death in the city is associated with traffic accidents: 1,377 deaths in 2011 alone. For this reason, citizens have actively pursued specific programs that would decrease the overwhelming statistics: 3,897 deaths from 2008 to 2012. The reason of the following project is to provide drivers with a technological tool with indicators and sufficient information based off statistics compiled by the *Centro de Investigaciones Sociales* (Social Research Centre) at Autonomous University of Ciudad Juárez and other public sources. Then drivers would have more information on possible traffic accidents before they happened. This research tries to combine a Mobile Device based on Cultural Algorithms and Data Mining to determine the danger of suffering a traffic accident in a given part of the city during a specific time.

Keywords: Data Mining, Cultural Algorithms and Mobile Devices.

1 Introduction

According to literature is feasible to use an application to determine the possibility that a traffic accident will occur using data associated with most frequented areas. Then the users will be aware of alternate routes to their destination, among other advantages. Using technological tools, data analysis and mathematical models would then decrease drivers' chances of suffering a traffic accident. The aim of this project is to create a model that uses these tools in a mobile geographic information system (SIGMA) on levels of traffic accidents in specific areas of Ciudad Juárez, which is presented in detail below, this research include the analysis of Cultural Algorithms to determine the danger that a traffic accident would occur during a specific time. We realize an exhaustive analysis of other

* Corresponding author.

S. Omatu et al. (Eds.): *Distrib. Computing & Artificial Intelligence*, AISC 217, pp. 9–17.
DOI: 10.1007/978-3-319-00551-5_2

similar research, the only similar context is explain in [6], where the authors calculated the insecurity of a vehicular group which requires product delivery in different places with random scheduling, but this research does not considers real on-time statistics and the perspective to suggest a different route to travel or stay for a determinate period of time.

2 Project Development

We developed this research project into three sections, which are: modules of application development, implementation of the server and the intelligent module associated with Cultural Algorithms and Data Mining. We will use the operating system Android since it has a considerable market segment while there has been a lot of development using inexpensive equipment, especially in developing countries. Android is free software, so any developer can download the SDK (development kit) that contains your API [2]. However we consider some minor drawbacks: there are Android users that had complained about the fragmentation of the platform due to the different versions. This research's scope is to prevent traffic accidents in Ciudad Juárez, where there were 27 deaths per 100,000 people in 2011.

3 Components of the Application

The first step is to get the coordinates where the user is located, and then sent them to the server, which calculates the number of traffic incidents within the closest radio specified in the configuration of the mobile device, the results are processed to determine a numerical index which will then be represented with a color for better end-user visualization. All this information is obtained from a transit databases and analyzed with Data Mining. After the values are sent to the mobile device and interpreted to construct an URL which is then sent to the Google Maps API. Then a map will be ultimately display with the area indicators.

The figure 1 shown below represents the operation of the SIGMA, which is divided into two parts, the mobile application and Web services hosted on the server. The database, where the incidences data is stored, has 3 tables:

Type_accident: Where the different types of traffic accidents are characterized.
Neighborhood: Here is the geo-reference of the neighborhood.
Accidents: It lists the incidents raised in the different regions of Ciudad Juárez, which are also expressed in decimal degrees geo-referenced to facilitate further calculations, and a narrative guide associated with features of the accident.

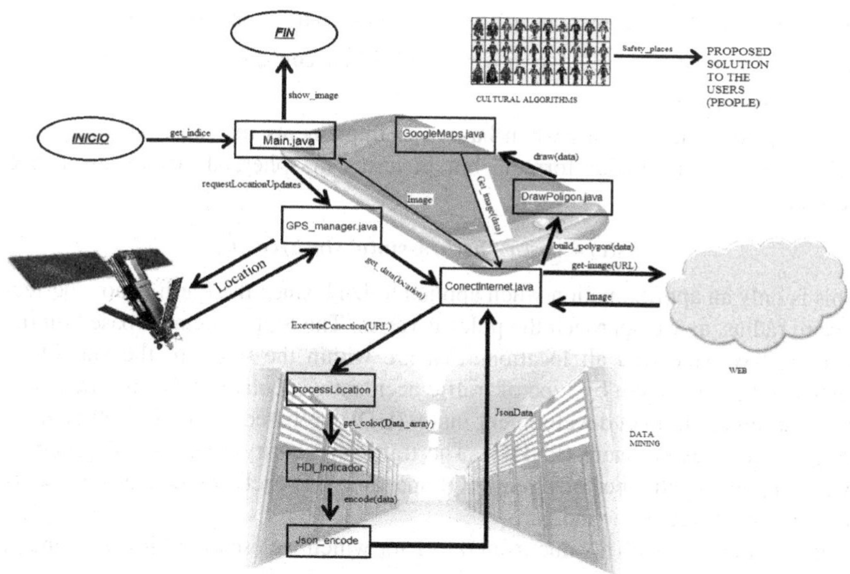

Fig. 1 Functional diagram of the SIGM including hybrid methodology

The structure of the database has a table called accidents, where each record contains the geographic coordinates expressed in decimal degrees, which determine the specific position of the accidents [1] and is visualized I Figure 2. To locate nearby points within a radius requires four parameters: latitude, longitude, and distance equatorial radius, the latter determines the maximum distance the search radius, these parameters are part of a method based on havesine formula, which is used to calculate great circle distances between two points on a sphere. Haversine function that is different from haversine formula is given by the function of semiversine where:

$$\boldsymbol{semiversin(\theta) = haversin\ (\theta) = \frac{versin(\theta)}{2} = sin^2(\frac{\theta}{2})}$$

Haversine formula for any two points on a sphere:
d is the distance between two points (along a great circle of the sphere),
R is the radius of the sphere,
φ 1 is the latitude of point 1,
φ 2 is the latitude of point 2, and
Δ λ is the length difference

$$\mathrm{haversine}\left(\frac{\mathrm{d}}{\mathrm{R}}\right) = \mathrm{haversin}\ (\varphi 1 - \varphi 2) + \cos(\varphi 1)\cos(\varphi 2)\mathrm{haversine}(\Delta\,\lambda)$$

Solving haversine formula can calculate the distance, either by applying the function inverse haversine or by using the arcsine where:

$$\mathbf{h = haversine}\left(\frac{\mathbf{h}}{\mathbf{R}}\right)$$

$$\mathrm{d = R\,haversine^{-1}(h) = 2R\,arcsin}\,(\sqrt{\mathrm{h}})$$

With this formula we can construct the following instructions, taking into account the role of semiversine we have: $\mathbf{h} = \sin^2\left(\frac{\varphi 1 - \varphi 2}{2}\right) + \cos(\varphi 1).\cos(\varphi 2).\sin^2\left(\frac{\Delta\lambda}{2}\right)$
$d = 2R \arcsin(\sqrt{h})$

For performance reasons when haversine implements the function using SQL statements used a similar formula but in terms of spherical cosine law called cosine.

$$cos(c) = cos(a)\, cos(b) + sin(a)\, sin(b)\, cos(C)$$

This is only an approximation when applied to land, since the Earth is no a perfect sphere radius, as we approach the poles it varies. The SQL statement based on the formula haversine find all locations that are within the range of the variable $ radius and groups them by type of traffic accident. Calculate of the uncertainty is modeled on the human development index (HDI) prepared by the United Nations Program for Development (PNDU). The traffic insecurity index is composed of two components Number of Events (Q) and time (H) each component represents half the total value of the index:

Each expressed with a value from 0 to 1 for which we use the following general formula:

Component index = (Value – minimum) / (Maximum – minimum)

The component Q is calculated by the number of events in the area, determined by the position to calculate the maximum and minimum, accidents are grouped by neighborhood. According to the CIS-ICSA at Autonomous University of Ciudad Juárez, Cuatro Siglos Avenue has the most accidents, and thus is the maximum, with 47 deaths. Whenever an avenue exceeds this number, there would be a new maximum; which means that the maximum is dynamic depending on the data entering the database. To calculate the H component data are grouped in ranges of time, for example suppose that areas of 300 m radius 13 deaths are grouped in the following form and consult the index at 4:38 pm, as is shown in Table 1.

Table 1 Example of Neighborhood number of homicides range by hour

Amount	Time Range
5	**5:00 a 6:00 pm.**
3	**4:00 a 5:00pm.**
3	**11:00 a 12:00pm.**
2	**9:00 a 10:00am.**

The time when we consulted put us in the range of 4:00 to 5:00 pm. And the calculation of the H component would be: $\boldsymbol{H} = \frac{3-2}{5-2} = .333$

Using data from the previous example would be the component Q: $\boldsymbol{Q} = \frac{13-2}{53-2} = .215$

The index would be as follows $\boldsymbol{indice} = \frac{1}{2}(.215) + \frac{1}{2}(.333) = .274$

Having calculated the numerical index, a color is assigned depending on the range in which is positioned according to the table 2 divided into eight classes.

Table 2 Ranges of colors according to the numeric index

0 - .125		.56 - .625	
.126 - .25		**.626 - .75**	
.26 - .375		**.751 - .875**	
.376 - .5		**.876 - 1**	

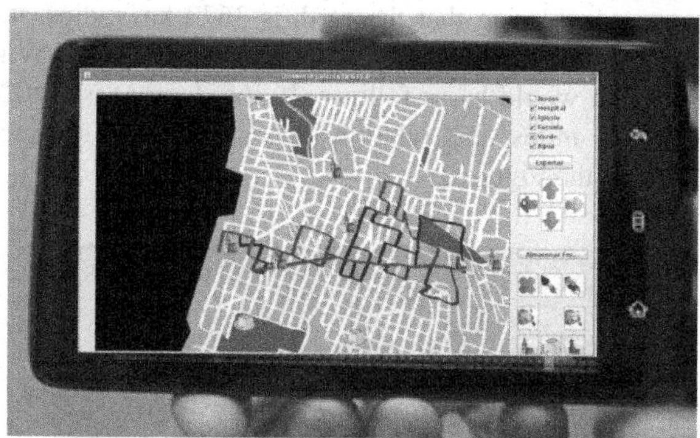

Fig. 2 Intelligent Tool recommend another two routes after obtain data from a transit accident

4 Implementation of the Intelligent Application

There are several aspects that need to be taken into account when designing mobile applications: limited screen size, different resolution and screen sizes across devices. Therefore, the designer has to develop the interface uniformly so that it suits most devices. This module explains how to work with different layouts provided for the Android API. The programming interface we will work on is XML. There are several ways to obtain the geographical position of the device, however we would thoroughly use GPS and access points (Wi-Fi) nearby; both perform similarly but differ in accuracy, speed and resource consumption. Data Server Communication is the most important module because it allows communication with the server, allowing you to send the GPS position obtained by receiving the processed image and map of our location, thus showing the outcome of your application that is the indicator of insecurity. To communicate to a server requires a HTTP client which sends parameters and establishes a connection using TCP / IP. The HTTP client can access any other server or service as this is able to get response from the server and interpret by a stream of data. The Android SDK has two classes with which we can achieve this, HttpClient and HTTPPOST. With the class HttpClient is done to connect to a remote server, it needs HTTPPOST class will have the URI or URL of the remote server. This method receives a URL as a parameter and using classes HTTPPOST HttpClient and the result is obtained and received from the server, in this specific case is only text, which can be JSON or XML format. Here, the server responds with a JSON object that would give the indicator is then used to create the map. For the construction of the polygon that indicates the rate of incidents in a certain radius of the current position is not possible to create it using the GPS coordinates that yields, as these are specified in "degrees" and requires the unit to convert to meters. For this you need to know how an arc equals the terrestrial sphere, which depends on the place on earth where it is located and the address where you are, the simplest case is to measure an arc in Equator, considering that the earth is 3670 km radius, the perimeter of serious Equator radio 2, which would be equal to 40.024 miles. With this you can get a relationship that would be as follows. If 360 degrees is 40.024 miles then a degree is 111,000.18 miles, this relationship can add and subtract yards to the position, as shown in the Figure 3.

Fig. 3 Acquisition of polygon map with the position of the Android application

For the preparation of graphics, we use a class supported with Cultural Algorithm proposed which facilitates the manipulation of data to express visually, using different types of graphs: as in the Figure 4.

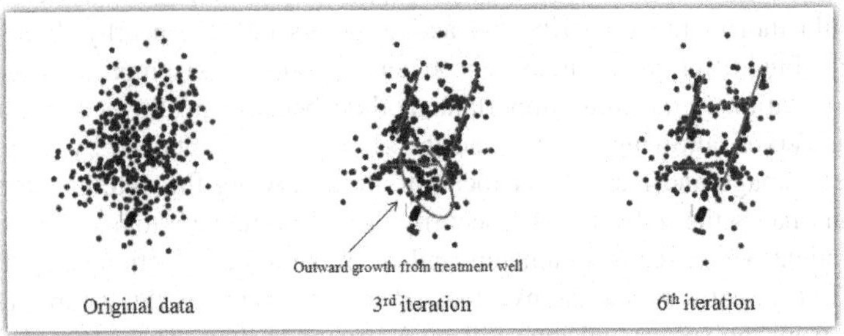

Fig. 4 Statistics Graphics related with deaths in traffic accidents

To implement the model, the application is installed in operating system devices with Android 2.2 or higher, which tests the system in different areas of the city based on the previously research related with Cultural Algorithms on Urban Transport [5], by answering a questionnaire of seven questions to all users after a scholar semester have elapsed since installing the application. The questions' purpose is to raise awareness of the performance, functionality and usability of the system, the demonstration of this application is shown in figure 5. To understand in a proper manner the functionality of this Intelligent Tool, we proposed evaluate our hybrid approach and compare with only data mining analysis and random select activities to protect in the city, we analyze this information based on the unit named "époques", which is a time variable to determine if a change exists in the proposed solution according at a different situation of traffic accidents.

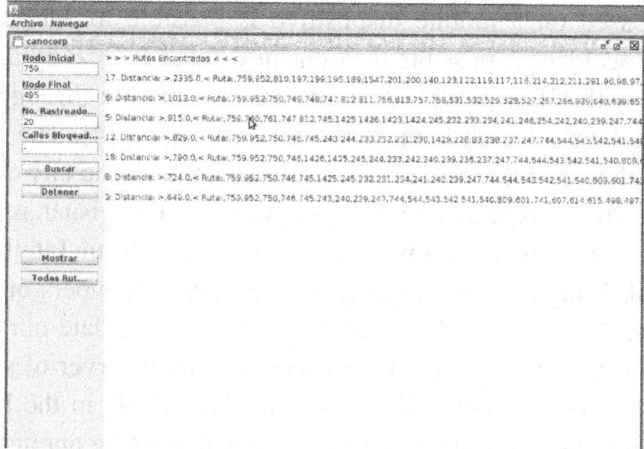

Fig. 5 Implementation and use of Hybrid Intelligent Application based on Cultural Algorithms and Data Mining

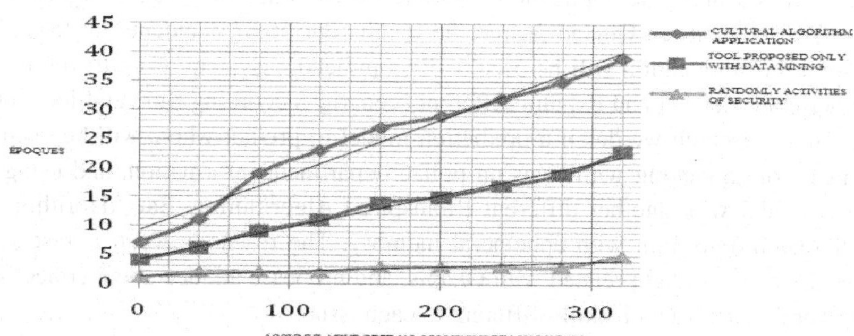

Fig. 6 Solutions proposed to the problem of traffic accidents: (blue) our hybrid approach; (red) only using data mining analysis and (green) using randomly actions to improve the safety of the users

We consider different scenarios to analyze during different time, as is possible see in the Figure 6, and apply a questionnaire to a sample of users to decide search a safety place in the City. Users that receive transit information with traffic accidents (via Data Mining Analysis) try to improve their space of solution. But when solutions are sent using both Cultural Algorithms and Data Mining then it was possible to determine secure solutions by the users and describe the real situation of possible danger. The use of our proposal solution improves 82% the possibilities of delivering a suggestion, to drive away from the possible traffic accidents, against a randomly action and 28% against only using Data Mining analysis. These types of messages would permit in the future a significant decrease in the possibility of suffering a fatal traffic accident.

5 Conclusions

With the use of this innovative application combining Cultural Algorithms and Data Mining based on a mobile dispositive is possible to determine the places where a traffic accident is possible to occur in Ciudad Juárez; by an alert sent to a mobile device with GPS, providing statistical information through a Web server that returns the level of rush hour in the area consulted [4]. The most important contribution is the possible prevention of future deaths in the city caused by traffic accidents. The future research will be to improve the visual representation of traffic problems with real on-time information through an Intelligent Diorama. This design will bring common information to family members or social network close members. Another possible extension will be to update our database using recent data from the local Transit Department's central server of security. Taking into account the 17,500 deaths related to traffic accidents in the last six years in Mexico, we think this innovative technology has promising application in another metropolitan cities in Latin America with similar problems of traffic, such as: Araguaiana, Blumenau, Joinville, Londrina, Manaus, Rondonópolis, Uberlândia. There exists a plausible application for motorcyclist since they face a higher risk of traffic accidents in Ciudad Juárez: 55 deaths out of 67 accidents in 2012. In addition this application will be used as Recommender System when traveling to another countries [7] and explain different scenarios according time and location. In a future research we detail a prediction model to predict where will be occurs an attack of carjacking with basis on police department information, and using a hybrid model with another different Bioinspired Algorithm as Bat Algorithm or Wolf search algorithm with ephemeral memory. The research group is research about a novel paradigm related with Okapis, and proposes an idea about collective behavioral in a herd with skills different to each issue.

References

1. Ignacio, A.F.: Las coordenadas geográficas y la proyección UTM, Universidad de Valladolid, Ingeniería cartográfica, geodésica y fotogrametría (Febrero 2001)
2. Alejandro, A.R.: Estudio del desarrollo de aplicaciones RA para Android. Trabajo de fin de Carrera. Catalunya, España (2011)

3. Orlando, B.J.: Sistema de Información Geográfica Móvil Basado en Comunicaciones Inalámbricas y Visualización de Mapas en Internet. M.C. tesis, Ensenada, Baja California, México (2011)
4. Cáceres, A.: Sistemas de Información Geográfica. Profesorado en Geografía. Instituto Formación Docente P.A.G. (2007)
5. Reyes, L.C., Zezzatti, C.A.O.O., Santillán, C.G., Hernández, P.H., Fuerte, M.V.: A Cultural Algorithm for the Urban Public Transportation. In: Corchado, E., Graña Romay, M., Manhaes Savio, A. (eds.) HAIS 2010, Part II. LNCS (LNAI), vol. 6077, pp. 135–142. Springer, Heidelberg (2010)
6. Glass, S., Vallipuram, M., Portmann, M.: The Insecurity of Time-of-Arrival Distance-Ranging in IEEE 802.11 Wireless Networks. In: ICDS Workshops, pp. 227–233 (2010)
7. Souffiau, W., Maervoet, J., Vansteenwegen, P., Vanden Berghe, G., Van Oudheusden, D.: A Mobile Tourist Decision Support System for Small Footprint Devices. In: Cabestany, J., Sandoval, F., Prieto, A., Corchado, J.M. (eds.) IWANN 2009, Part I. LNCS, vol. 5517, pp. 1248–1255. Springer, Heidelberg (2009)

Synchronization Policies Impact in Distributed Agent-Based Simulation

Omar Rihawi, Yann Secq, and Philippe Mathieu

LIFL (CNRS UMR 8022), Université Lille 1, France
First.LastName@univ-lille1.fr

Abstract. When agents and interactions grow in a situated agent-based simulations, requirements in memory or computation power increase also. To be able to tackle simulation with millions of agents, distributing the simulator on a computer network is promising but raises issues related to time consistency and synchronization between machines. This paper study the cost in performances of several synchronization policies and their impact on macroscopic properties of simulations. To that aims, we study three different time management mechanisms and evaluate them on two multi-agent applications.

Keywords: Distributed simulations, large scale multi-agent system, situated agent-based simulations, synchronization policies.

1 Introduction

Multi-agent systems (MAS) are made of autonomous entities, that interact in an environment to achieve their own goals [11], producing emergent properties at the macroscopic level. To simulate millions of agents, we need a huge memory and computation. The goal in this kind of applications is not to observe individual interactions (microscopic level), but to observe population behaviours (macroscopic level). In some cases, if some agents fail or cannot interact as fast as other agents, it should not be critical to the global simulation outcome.

To achieve such scalability, one can distribute computations over a computer network. Nevertheless, in a distributed setting, problematics linked to time management and synchronization appear and have to be addressed. This paper presents a first study of synchronization costs in performances and the impact of synchronization policies on the preservation of emergent macroscopic properties.

The first section details the notion of simulated time and introduce three main synchronization policies. The second section introduces the platform that we have developed to experiment synchronization issues. The third and last section details experimentations made on two applications, prey-predator and capture the flag, to benchmark the impact of synchronization policies on simulation outcomes.

S. Omatu et al. (Eds.): *Distrib. Computing & Artificial Intelligence,* AISC 217, pp. 19–26.
DOI: 10.1007/978-3-319-00551-5_3

2 Time and Synchronization in MAS Simulation

The word *time* is often defined as a nonspatial continuum in which events occur in irreversible succession from the past through the present to the future (time flow notion [3]). In simulations, several notions of time are involved: user time (the real time) and simulated time, which is a set of small durations used to produce evolutions within a simulation. Simulated time has been defined by Lamport [7] through a logical clock that induce a partial ordering of events, and has been refined in distributed context as *Logical Virtual Time* (LVT) by Jefferson [5].

In MAS simulations, a common implementation to enable the simulation dynamic is to query all agents for their current action and to apply this set of actions. This round of talk defines as a simulation tick or *Time Step* (TS). In centralized MAS, there is only one TS that organize interactions between agents in a given period. Whereas, in a distributed MAS, there is one TS per machine which could not be the same on all. The main problem in such systems is time management between machines [12]. In order to guaranty causality, we have to synchronize TS between all machines. However, several policies of synchronization can be proposed. In distributed systems, there are mainly two approaches: *conservative* (or *synchronous*) and *optimistic* (or *asynchronous*) synchronization [8] [2].

We can define three synchronization policies for distributed MAS simulations: *1)* ***Strong Synchronization (SS):*** in this policy: all machines TS are synchronized. This guaranty that all agents execute the same number of actions. But, the simulator needs more messages to synchronize machines. This conservative approach strictly avoids causality errors but introduce more delays. *2)* ***Flexible Synchronization:*** in this policy, machines are allowed to progress at different pace. As *optimistic* synchronization, which allows machines to advance in TS with detecting causality errors and recovering to previous states by rollback mechanism [4]. This mechanism rolls back by anti-event messages to reconstruct a previous state. But, if there are too much rollbacks, communication cost raises. *Time Window (TW)* synchronization is another flexible policy that can be proposed. In TW, machines can progress in TS until they reach the biggest possible difference W, between the slowest and fastest machines. With this window permission W, machines can avoid some delay of SS but in the same time, it can affect macroscopic behaviours outcome by allowing some agents in different TS to interact. These interactions can be ignored if their impact do not affect the simulation. This is a strong hypothesis that is studied in section 4. *3)* ***No Synchronization (NS):*** last policy simply drops all synchronization between machines. It is like TW policy with ∞ window size. This policy exploits the speed of all machines, but it is obviously not fitted for all applications.

To summarise, this paper studies whether synchronization constraints can be relaxed without impacting macroscopic behaviours emergence.

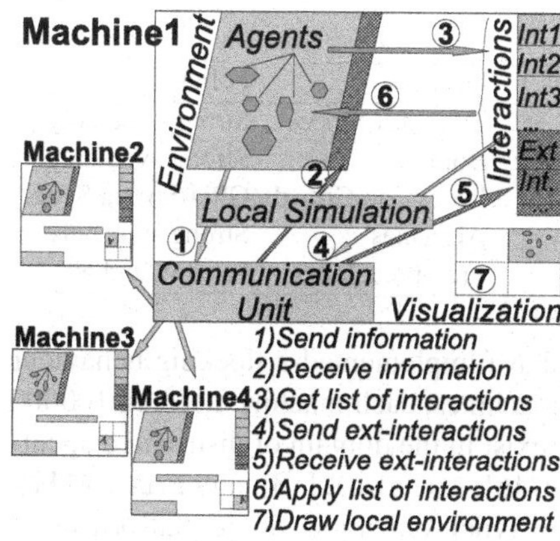

Fig. 1 Architecture of our testbed

Fig. 2 Five million agents on 50 machines

3 Testbed to Benchmark Synchronization Policies

To evaluate synchronization policies, we have developed a distributed simulator. Each machine handles a subset of the environment with its agents. The simulator can be run with one of these policies: *strong*, *time window* or *no* synchronization.

Figure 1 shows 4 machines that execute a distributed agent-based simulation. Each machine has: *local simulation*, *communication unit* and *part of the environment* with its agents. *Local simulation* is a top-manager layer, which manages all tasks: from interactions between agents, receiving information from neighbourhood machines and local visualization. *Communication unit* manages exchange messages between machines and informs local simulation about machines TS. In each TS, machines follow 7 main steps: 1) it sends information to all neighbours about the environment state near them. 2) it waits for new information from others to inform local agents about the neighbours' environments. 3) each machine asks local agents about their next desired actions. 4) it sends the external interactions, which are interactions between agents from different machines, to others. 5) each machine receives interactions from others. 6) possible interactions are applied. 7) finally, local visualization and machines go to next TS. These synchronization policies have similar communication protocols with small differences. *Strong Synchronization* (SS) has in each communication state a notification, which requires an acknowledgement from others. Especially for the last state of communication, each machine should be synced for next TS and suspended until all others are ready. In case of *Time Window* policy, each machine sends a notification of its current TS, and checks if it has permission to switch to next TS. If the difference between its TS and slowest machine TS is less than *W* number of steps (*W* is user defined) then it can progress to next TS. Whereas, *No Synchronization* (NS) policy, machines can progress to next TS without any notification.

Table 1 Comparison between platforms

Platform	NbOfAgents	NbOfMachines	Model	Policy
`Repast`	68 billions	32000 cores	Triangles Model	Strong sync only
`DMASON`	10 millions	64	Boids	Strong sync only
`FLAMEGPU`	11000	GPU	Pedestrian Crowds	GPU-Strong Sync
`AglobeX`	6500	22 cores	Airplanes	Strong sync only
Our testbed	100 millions	200 machines	Prey-predator	SS, TW and NS

To evaluate our testbed scalability, we have implemented a flocking behaviour similar to Reynolds [10] and run it on 50 machines, each machine holding 100000 agents (Figure 2). Many platforms already exist in the domain of distributed agent-based simulation: `Repast` [9], `FLAMEGPU` [6], `AglobeX` [13], and `DMASON` [1]. But all these platforms rely only on strong synchronization. Some platforms distribute the simulation on a network using a middleware, which provides a shared memory between Java Virtual Machines. So, it is free from all distribution considerations. Others are working with master/slave approaches with only strong synchronization mechanism through the server. So, no platform can be considered as a testbed for our work. Table 1 shows a comparison between all these platforms.

4 Experimentations

In this section, we describe two applications that have been implemented and benchmarked to quantify the impact of the three proposed synchronization policies. It seems obvious that time inconsistency have not the same effect in all applications. For example, the boids application can run without synchronization and still produce the emerging flocking behaviour. Thus, we want to determine with the following experimentations the impact of synchronization policies on simulations, more precisely on the expected macroscopic behaviour. To study synchronization impacts, we have implemented one application that is extremely affected by synchronization issues, while the other is not:

1) The prey-predator (PP) model is a classical multi-agent application that involve two kind of agents: preys and predators. Both kind reproduces themselves at a given pace, but predators seek and eat preys. If a predator do not find preys quickly enough, it dies of starvation. This application illustrates population co-evolution in a simplified ecosystem. An example of such model is the *Wolf-Sheep-Grass* simulation proposed by Wilensky [14].

2) Capture the flag (CTF) model has been developed to illustrate the fact that if a simulation outcome relies on timing issues (population growth speed in this case), thus synchronization policies can introduce a bias. To achieve this goal, we propose the use of a simplified capture the flag application with two competing populations. For each population, we have two kind of agents: flag agents that produce new attackers at a given rate, and attacker agent that protect their flags or attack the other population. When two agents attack each other, they both die.

4.1 Synchronization Performance

We have executed experimentations on a LAN network where all machines have similar hardware. Most experimentations have run until 2 millions TS and the only parameter modified is W, the time window size, $W = 0$ *Strong Synchronization* (SS), then 10, 100, 1000, 10000, 100000 until ∞ or *No Synchronization* (NS). Table 2 shows that *No Synchronization* (NS) always gives the maximum speed, because it is free from all synchronization issues. But, in some application like CTF, the simulation outcome is biased.

Table 2 Prey-predator on 4 machines: *No Synchronization* always performs better

Time Steps	2000	4000	6000	8000	10000
SS execution time (hours)	1	2.5	3.5	4.7	6
NS execution time (hours)	0.8	1.4	2	2.7	3.4

A comparative test of both models stability (PP and CTF) shows that, PP model stays stable for more than 2 Millions TSs. For all experimentations, the co-evolution of PP has been preserved until we reach 2 millions TSs, even without any synchronization. Thus, this application is stable with respect to synchronization policies, stable meaning here that all types of agent still exist in the model. Whereas, CTF model has been unstable after W is bigger than N a number of steps ($N \geq 0$ depends on the initial configuration).

In next sections, we study invalid, internal and external interactions on the stable model to see how these interactions are changed when we switch synchronization policies. Then, we study CTF instability in details by exploring initial configuration biases.

4.2 Interactions Effects on Prey-Predator (PP) Model

External-Interactions (EI) are interactions happening between agents that belong to different machines. Figure 3 shows EI between agents which are executed in PP model (with 5000 agents of wolf, sheep and grass), with size of TW: $W = 0$ or SS, 2, 100 and until NS or ∞ (all tests until TS = 100000). Results show that for SS we have a lot of EI, this is normal because information about agents can be sent and received by other machines. However, NS or TW significantly reduce EI, even with small window size. It can be explained because the information about agents is always sent to others machines, but it is not received from others within the corresponding time step. Thus, there is not a lot of EI because they are dropped because of differences between time steps. **Invalid-Time-Step Interactions** happens between two agents coming from different machines that are not in the same TS. Figures 4 shows that invalid interactions percentage increase with time window size (W). However, this

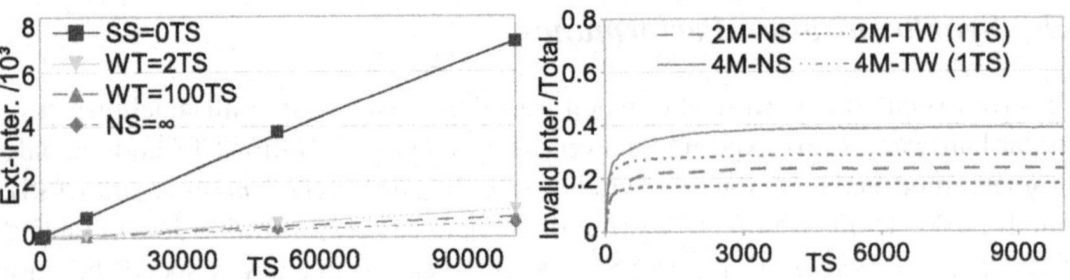

Fig. 3 External interactions in PP model with different synchronization polices

Fig. 4 Invalid interactions between 2 & 4 machines in PP model

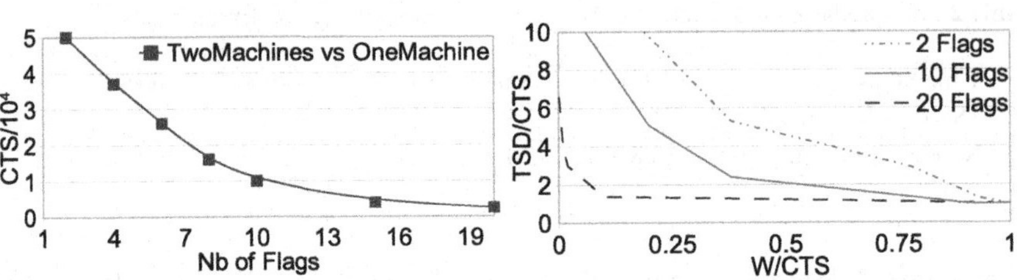

Fig. 5 CTS vs Number of flags in CTF model

Fig. 6 *TSD* for different configurations of CTF Model with different size of TW (*W*)

percentage is less than 40% of total interactions. In case of 4 machines, we have doubled the percentage of two machines, which is because agents can swap between 4 machines more often than in two. Again, NS gives more invalid interactions than TW, because there are more differences between machines TS so information are dropped.

4.3 Capture the Flag Model Instability

As previous experimentations, simulations run on two machines for 2 millions TS and with a TW evolving between 0 to ∞. The first configuration is defined with one flag per machine. This configuration, nearly like the prey predator model, is stable with respect to synchronization issues. If we choose another configuration, like 20 flags per machine, we get different results. For TW size $W > 10000$ the model is no more stable and all flags from one population are captured.

We have defined another initial configuration to evaluate synchronization policies when machines loads are not the same. We have used three machines: the first contains all flags from the first population and two other machines contain half flags of the second population in each. The aim is to have more load on one machine, so the model becomes unstable: the second population should always win because its attacker production will be higher. The convergence should be faster with more flexible synchronization policies, and fastest without synchronization.

We define the *Critical Time Step* (CTS) as the necessary TS to completely destroy the model if no synchronization has been used. Figure 5 shows that CTS will be decreased by increasing the number of flags, from that $CTS = \frac{\alpha_1}{F}$, α_1 is constant which depends on the initial configuration and machines which have been used. We define the *Time-Step to Destroy* (TSD) as the necessary TS to completely destroy the model if W has been used ($TSD = CTS$ *if and only if* $W = \infty$ or NS). Figure 6 visualizes results after scaling each configuration to its CTS (as each one has its own CTS). This figure shows that the TSD decrease if W increase, then and according to the figure: $TSD/CTS = \frac{\alpha_2}{W/CTS}$, α_2 is constant. If we replace $CTS = \frac{\alpha_1}{F}$, then $TSD = \frac{\alpha}{W \times F^2}$, α is constant. That means, TSD decrease if the number of flags increase or W has been increased too.

Figure 6 also shows that for all configurations the model stay stable for a small TW. However, and in all configurations, we can divide each curve into two main parts. The first with W from 0 to 30% of CTS, the model stay stable for a long TSs. That mean, we can give permissions of advancing in TS between different machines until 30% of its CTS, and in this part W strongly affect all curves. The second part with W bigger than 30% of CTS, the TSD decreases slowly according to W.

5 Conclusion

To simulate billions of interacting agents, we have to distribute the simulator. A safe approach consists in splitting the environment into smaller parts and using a strong synchronization policy, but it implies a high cost in message exchanges and execution time. This paper has explored a relaxation of this constraint to speed up execution time and has identified applications where this relaxation do not degrade simulations outcome.

We have studied three synchronization policies for distributed multi-agent simulations: *Strong Synchronization*, *Time-Window Synchronization* and *No Synchronization*. Experimentations show that some applications, like prey-predator model, stay stable with any synchronization policy. Whereas in others models, like capture the flags, it can be strongly affected by changing these policies. We have studied how interactions are changed when we switch the synchronization policies on prey-predator model and we have explored in details the instability of capture the flags model when a biased-initial configuration is used.

Experimentations presented in this paper are a first step, we have to experiment other kind of applications to illustrate synchronization policies impacts and see if results presented in this paper are suitable for other applications. For example, emergency scenario of town simulation can be calculated faster with *No Synchronization* policy than *Strong Synchronization*.

References

[1] Cordasco, G., De Chiara, R., Mancuso, A., Mazzeo, D., Scarano, V., Spagnuolo, C.: A framework for distributing agent-based simulations. In: Alexander, M., et al. (eds.) Euro-Par 2011, Part I. LNCS, vol. 7155, pp. 460–470. Springer, Heidelberg (2012)

[2] Fujimoto, R.M.: Parallel and distributed simulation systems. Wiley (2000)

[3] Gold, T.: Why time flows: The physics of past & future. Daedalus 132(2), 37–40 (2003)

[4] Gupta, B., Rahimi, S., Yang, Y.: A novel roll-back mechanism for performance enhancement of asynchronous checkpointing and recovery. Informatica 31, 1–13 (2007)

[5] Jefferson, D.R.: Virtual time. ACM Trans. Program. Lang. Syst. 7, 404–425 (1985)

[6] Karmakharm, T., Richmond, P., Romano, D.: Agent-based large scale simulation of pedestrians with adaptive realistic navigation vector fields. In: Theory and Practice of Computer Graphics, pp. 67–74 (2010)

[7] Lamport, L.: Ti clocks, and the ordering of events in a distributed system. Commun. ACM 21, 558–565 (1978)

[8] Logan, B., Theodoropoulos, G.: The distributed simulation of multiagent systems. Proceedings of the IEEE 89(2), 174–185 (2001)

[9] Minson, R., Theodoropoulos, G.K.: Distributing repast agent-based simulations with hla. In: Euro. Simulation Interoperability Workshop, pp. 04–046 (2004)

[10] Reynolds, C.: Steering behaviors for autonomous characters (1999)

[11] Russell, S.J., Norvig, P.: Artificial intelligence: a modern approach. Prentice-Hall (1996)

[12] Scerri, D., Drogoul, A., Hickmott, S., Padgham, L.: An architecture for modular distributed simulation with agent-based models. In: AAMAS 2010 Proceedings, pp. 541–548 (2010)

[13] Šišlák, D., Volf, P., Jakob, M., Pěchouček, M.: Distributed platform for large-scale agent-based simulations. In: Dignum, F., Bradshaw, J., Silverman, B., van Doesburg, W. (eds.) Agents for Games and Simulations. LNCS (LNAI), vol. 5920, pp. 16–32. Springer, Heidelberg (2009)

[14] Wilensky, U.: Netlogo wolf-sheep predation model (1997)

Parallel Hoeffding Decision Tree for Streaming Data

Piotr Cal and Michał Woźniak

Wrocław University of Technology,
Department of Systems and Computer Networks,
Wybrzeże Wyspiańskiego 27, 50-370 Wrocław, Poland
{piotr.cal,michal.wozniak}@pwr.wroc.pl

Abstract. Decision trees are well known, widely used algorithm for building efficient classifiers. We propose the modification of the Parallel Hoeffding Tree algorithm that could deal with large streaming data. The proposed method were evaluated on the basis of computer experiment which were carried on few real datasets. The algorithm uses parallel approach and the Hoeffding inequality for better performance with large streaming data. The paper present the analysis of Hoeffding tree and its issues.

Keywords: machine learning, supervised learning, decision tree, parallel decision tree, pattern recognition.

1 Introduction

Nowadays decision trees have a numerous algorithms for creating a very good classifiers. They are used in recognition, data mining, decision support systems and many others. Main advantages of decision trees are readability (especially with small number of attribute), not large complexity, noise resistance and mostly high accuracy. They are often used in expert systems where data are represented as a pair of attributes vector and output value (class). In decision trees, some measures of information gain are used in order to choose the best attribute [9]. They are several good methods to compute information gain so that no special attribute type was treated as better or worse.

Very popular and basic algorithms are ID3 [11] and its extension to C4.5 [12], which could handle with continuous attributes and missing values. Those and other methods were not previously designed in the direction of large data computing, which became very important issue for the practical use. To find solution of this problem, the modification of algorithms was done by the parallelization of known methods or by the invention new parallel algorithms.

It is not necessary to have all data from a large dataset to split the node. Sometimes a part of data is just enough as shown by Domingos and Hulten [2]. They use Hoeffding inequality [3] to evaluate the amount of data needed

S. Omatu et al. (Eds.): *Distrib. Computing & Artificial Intelligence,* AISC 217, pp. 27–35.
DOI: 10.1007/978-3-319-00551-5_4

to split the node. The trees which use this technique are called Hoeffding trees and they work better with non-continuous attributes. Hence, a modification of the tree algorithm is needed, for example such as the modification provided by Agrawal and Jin [5]. They use few methods from algorithm called SPIES [6]. Also, the improvment of hoeffding tree is made by adding the option nodes [10], [4].

The parallelization of algorithms can be done in different ways. A method called PDT (Parallel Decision Trees) was proposed by Kufrin [8]. The algorithm works with master-slave architecture and splits the data among slaves which generate statistics. In [13], few techniques of parallelization was analyzed for LDT and C4.5 algorithms. They use feature, node and data based parallelization and show speedup for few popular datasets. Another method which is dedicated to the streaming data is SPDT proposed by [1] . Each processor constructs histograms and send information to a master processor which splits a node.

Currently the data counted in millions are common. Sometimes the number of data is too large to create satisfied classifiers in satisfied time. We can use weak classifiers in multiple classifier system [7] but it could couse deacreas in accuracy. Additionaly, sometimes we want to rebuild classifiers to keep high accuracy with following data, because the concept of this data has changed. Hoeffding inequality gives us important information. It shows that only part of the data is needed to make sure, that chosen attribute is the best for all data sets. Using parallelization with Hoeffding inequality, we can build accurate decision trees very fast.

In addition the power of single processor unit stops increasing because of physical barrier. But still, computation power is growing by using multiprocessors architecture. Creating a parallel solutions is very important to adjust the efficiency of algorithms with devices possibilities.

The paper proposes Parallel Hoeffding Decision Tree algorithm. The main goals were the evaluation of efficiency of this solution. Especially, the paper focuses on the analysis of time and the influence of Hoeffding inequality. The proposed method was evaluated on several real datasets.

2 Algorithm

In this section we describe main features of this parallel solution. Proposed algorithm constitutes the extension of the Quinlan algorithm ID3 [11]. In order to choose the best attribute on the given node, this method uses entropy to compute an information gain.

Although, the complexity of computation is not high, the algorithm could run too slow for large streaming data. For this reason we use Hoeffding inequality. It will allow us to take only a part of the data with security that the chosen attribute will be the same as for the whole data [2].

Suppose r is a random variable whose range is R (for an information gain it is $\log_c$, where c is the number of classes). After its n independence observation, we have its average value $\bar{r}$. Then, with Hoeffding dependency, we know that the real value of the $\bar{r}$ is less or equal $\bar{r} - \epsilon$ with $1 - \delta$ probability, where ϵ can be computed from equation (1).

$$\epsilon = \sqrt{\frac{R^2 \ln(\frac{1}{\delta})}{2n}} \tag{1}$$

The advantage of this solution is independence from data distribution. But the size of data subset could be bigger than some other limitation methods which relay on data distribution.

Bearing this in mind, our algorithms have to compute information gain for all the valid attributes of this node using the part of the data. After this process, the values of information gain for the best attribute (X_a) and the second best one (X_b) are stored and compared (2).

$$\Delta G = Gain(S, X_a) - Gain(S, X_b) \geq 0 \tag{2}$$

Next we determine the value of δ, and then Hoeffding dependency guarantees that X_a is the best attribute to split the node with probability $1 - \delta$, if size of the subset is n and $\Delta G \leq \epsilon$ (1 and 2). If this does not happen the node has to wait for the additional data. After next data flow the process of choosing attribute is repeated. Nodes which satisfy condition are called Hoeffding nodes and algorithm performs split operation to build deeper nodes if it is needed.

We construct the parallel version of attribute processing. The algorithm needs to get a gain value from all available attributes to split node. Every attribute can be processed independently from another. We use the very common master slave architecture. The main processor (workstation, core) called "master" assigns a tasks to another processors called "slaves". In our case these tasks are based on a computation of information gain for a specified attribute. The master is responsible for assigning and collecting results from slaves. After collecting any results, the master checks if the gain is actually the best one or the second best one for the node and the master stores it. After collecting results for all available attributes, the master checks Hoeffding inequality and then decides whether to split node or to wait for more data. The main master procedures are presented in pseudocode (1).The "receive results" function simply collects results from one slave, updates best gain and second best gain if necessary and put free processor to the set of available processors. The function "do task" is run for a dedicated slave and computes an information gain.

All slave tasks are to find the information gain for a given attribute. The master does not have to send all features from samples but only a relevant one. This spares a network recurses which are very important for large datasets. If an attribute is discrete, a slave has to simply compute the information gain. But if the attribute is continuous, a slave creates fixed number of intervals and treats value in the one interval as one discrete value. This operation is

Algorithm 1. parallel attribute processing (master)

```
Require: {a_0, a_1, ..., a_n} ∈ A – set of available attributes
  {p_0, p_1, ..., p_k} ∈ P – set of available processors (slaves)
  S – set of samples (instances)
  g – gain
  ATTRIBUTE_PROCESSING (P,A,S)
  while A ≠ ∅ do
        if P ≠ ∅ then
              take p_i ∈ P, a_j ∈ A
              do parallel(p_i) DO_TASK (p_i, a_j, S)
              P = P − p_i, A = A − a_j
        else
              RECEIVE_RESULTS (p, a, g)
        end if
  end while
  while P ≠ k do
        RECEIVE_RESULTS (p, a, g)
  end while
```

needed because Hoeffding Tree works better with discrete attributes. After processing, information about intervals is also send to the master node. Algorithm creates one additional interval because when streaming data flow, the value which does not match any intervals could appear.

The proposed algorithm can be configured by user to build strongly overtrained deep trees and trees adapted to the specific data. Created node became Hoeffding node if satisfy Hoeffding inequality or became a temporary leaf. Temporary leaves can be treated like a normal internal nodes (and build without Hoeffding inequality) or like normal leaf. Because algorithm works for streaming data, if more samples appear on the temporary leaf, it will be reconsidered as a potential Hoeffding node again.

The application was created using C++ programing language. We use MPI (message passing interface) library which is designed for the message passing communication. In contrast to the the shared memory communication which can work in the shared memory architecture, message-passing can work also on the distributed memory architecture. This way, we can run our algorithm on one multi-core workstation and on any computer cluster which supports MPI.

3 Experiment

The main goal of experiment is to show the speedup of proposed method and analysis of issues related to the Hoeffding Tree on real data sets.

3.1 Data Sets

We use few popular data sets from UCI repository to evaluate the performance of algorithm. Table (1) presents names, number of classes, size and number of attributes for the each of them.

Table 1 Datasets description

name	classes	size	attributes	
			continuous	discrete
abalone	27	4177	7	1
isolet	28	6238	617	0
census	2	100000	7	33
connect-4	3	67557	0	42
covtype	7	581012	10	44

3.2 Methodology

All tests are carried out on the cluster of workstation (COW) with 9 computers connected with Ethernet network for the distributed memory architecture and on the computer with 8-core CPU with 3.3 GHz each for shared memory architecture. On the single workstation in COW we run virtual machine with designated one core 3.1 GHz and Ubuntu operating system. Besides, the application is designed for steaming data, we run it with all available data at once.

3.3 Evaluation Method

The time of building tree is computed for different number of slaves from 1 to 8. We configure the application to build a full overtrained tree and a parametrized tree (parameters are selected for datasets to avoid very deep tree). We run application five times for all slaves configurations and memory architecture each and calculate average. The reference time is a time for configuration with one slave because the master node controls tree building process and cannot compute information gain. We use speedup term but it is different to some extend because our reference time is not from sequential algorithm but from one slave configuration and the same memory architecture. However, the time of sequential version can be compared to the time of one slave configuration in shared memory architecture. In addition, tests carried out on the distributed memory architecture (COW) use the Fast Ethernet network (100 Mb/s)and network overhead is much more bigger than in the networks designed for computer clusters.

3.4 Results

Table 2 and table 3 show the distribution of nodes in full overtrained and parametrized trees. The number of Hoeffding nodes is very small and for datasets with the small amount of training instances, it equals zero. The presented accuracy is conducted on training datasets so it only shows that tree is built correctly. For overtrained configuration, the number of Hoeffding nodes is greater because it has simply more internal nodes. But when we

Table 2 Distribution of nodes in full overtrained trees for different datasets. $\delta = 0.1$

dataset	instances	nodes	internal nodes	hoeffding nodes	leaves	accuracy
abalone	4177	4106	868	0	3238	0,64
isolet	6238	3232	646	0	2586	1
census	100000	58900	6345	207	52555	0.99
connect-4	67557	41612	13870	174	27742	1
covtype	581012	1867263	858175	202	1009088	0.92

Table 3 Distribution of nodes in parametrized trees for different datasets. $\delta = 0.1$

dataset	instances	nodes	internal nodes	hoeffding nodes	leaves	accuracy
abalone	4177	844	171	0	673	0.4
isolet	6238	711	142	0	569	0.81
census	100000	17439	1368	9	16071	0.96
connect-4	67557	4390	1463	58	2927	0.81
covtype	581012	76387	29040	133	47347	0.83

Table 4 Location of Hoeffding nodes and sample flow. HN - Hoeffding nodes, IN - internal nodes, $\delta = 0.1$

	census			connect-4			covtype		
deep	HN	IN	sample	HN	IN	sample	HN	IN	sample
1	0	1	100000	0	1	67557	1	1	581012
2	0	10	62486	2	3	67557	2	4	581012
3	1	111	55224	0	8	63020	5	8	518997
4	3	251	39416	4	19	53217	4	17	504311
5	2	279	27320	4	42	44936	9	35	485617
deeper	3			48			112		

look on the parametrized configuration, the ratio of Hoeffding nodes to the internal nodes is bigger for covtype and connect-4 datasets. From a practical point of view, we want to have Hoeffding nodes on the higher level in trees where a lot of data come.

Table 4 shows where Hoeffding nodes are located in trees for three datasets and parametrized configuration. For census dataset the first Hoeffding node shows only at level three. This weak result could be caused by class distribution in this dataset. Census has two classes and the majority of instances are from one class. This causes that information gain provided by the attribute could be very similar to another. In this case we need a much more instances to meet Hoeffding inequality.

Better results are shown for connect-4 and covtype datasets. Especially for covtype, we have a good number of Hoeffding nodes in higher levels of tree. Due to these nodes, we do not have to use all data to choose the best attribute to split tree. Thus it could reduce computation time. Because application is designed for streaming data, once we establish Hoeffding nodes, we do not have to check more data if all its ancestors are Hoeffding nodes too. Time for

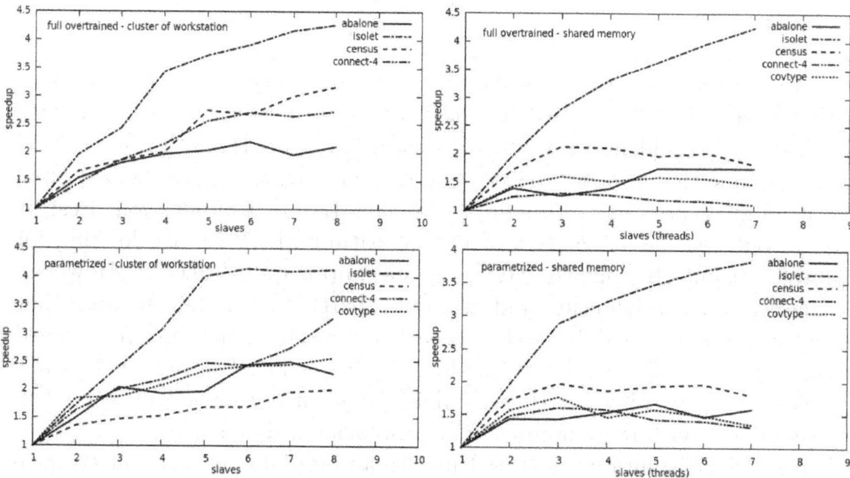

Fig. 1 Speedup for parametrized and full overtrained tree for distributed and shared memory architecture. Reference time is one slave configuration. $\delta = 0.1$

updating the tree is reduced but we should mention that this manner could spare time only if concept drift does not occur.

Fig 1 shows speedup (acceleration) of the algorithm according to the number of slaves. In both architecture the best results are achieved for isolet dataset. Isolet has above 600 attributes and all of them are continuous. Continuous attributes need more computer power because they firstly must be converted to discrete ones. In the shared memory architecture the speedup is often insignificant especially for more then three threads. The master node cannot send tasks and data as fast as they are calculated by slaves. The solution of this problem could be fixed by using shared memory communication because it prevents algorithm from copying data to the slaves. In distributed memory architecture, the speedup is better for all datasets. Also the best results are achieved for isolet data. Because speedup is measured to the reference time of one slave configuration we do not see the difference between the parametrized and the overtrained tree configuration. However, the time of full overtrained tree is always at least three times greater than from parametrized version.

In the cluster of workstation tests, we can see that the decreasing of speedup growth is irregular. The reason for this situation is that used network is not only for processors communication and some noise traffic could be generated by operating system.

4 Remarks

The application increases the performance of building tree in two ways. First, thanks to Hoeffding nodes we do not have to use all data to split the tree in a node or rebuild nodes if their ancestors are also Hoeffding nodes and we can assume that concept drift does not occur. Next, the parallelization of algorithm gives possibility to gain the acceleration of computation. Besides that we only parallelize a part of the algorithm where we are looking for the best attribute to split the node, the speedup is quite effective. However, the master slave architecture and attribute searching parallelism cause that this approach is not scalable well. In our tests, when we use very high speed communication (shared memory architecture) the increase of speedup was not effective after using 3-4 threads. The situation is more optimistic when we use dataset with large number of continuous attributes.

The provided solution is created for streaming data. If we want to have a good classifier, we need to improve it as long as we get new data. Our solution with Hoeffding inequality allow systems to collect a large set of data and build a tree possibly similar to tree built on the basis of infinite data set (with a given probability).

Our algorithm uses MPI library which gives it possibility to work with shared and distributed memory architecture. However, the performance is highly depending on communication between the master node and slaves. We can assume that our parallel approach is better for the shared memory architecture because in many modern machines the communication between processors (threads) is faster than in the distributed memory architecture. Adding the parallel processing of nodes could increase scalability and performance which will be next improvements in our application.

References

1. Ben-Haim, Y., Tom-Tov, E.: A streaming parallel decision tree algorithm. J. Mach. Learn. Res. 11, 849–872 (2010)
2. Domingos, P., Hulten, G.: Mining high-speed data streams. In: Proceedings of the Sixth ACM SIGKDD International Conference on Knowledge Discovery and Data Mining, New York, NY, USA, pp. 71–80 (2000)
3. Hoeffding, W.: Probability inequalities for sums of bounded random variables. Journal of the American Statistical Association, 13–30 (1963)
4. Ikonomovska, E., Gama, J., Zenko, B., Dzeroski, S.: Speeding-up hoeffding-based regression trees with options. In: ICML, pp. 537–544. Omnipress (2011)
5. Jin, R., Agrawal, G.: Communication and memory efficient parallel decision tree construction. In: The 3rd SIAM International Conference on Data Mining (2003)
6. Jin, R., Agrawal, G.: Efficient decision tree construction on streaming data. In: Proceedings of the Ninth ACM SIGKDD International Conference on Knowledge Discovery and Data Mining, New York, NY, USA, pp. 571–576 (2003)

7. Kacprzak, T., Walkowiak, K., Wozniak, M.: Optimization of overlay distributed computing systems for multiple classifier system - heuristic approach. Logic Journal of the IGPL 20(4), 677–688 (2012)
8. Kufrin, R.: Decision trees on parallel processors. In: Parallel Processing for Artificial Intelligence 3. Elsevier Science, pp. 279–306. Elsevier (1995)
9. Mitchell, T.M.: Machine Learning, 1st edn. McGraw-Hill, Inc., New York (1997)
10. Pfahringer, B., Holmes, G., Kirkby, R.: New options for hoeffding trees. In: Orgun, M.A., Thornton, J. (eds.) AI 2007. LNCS (LNAI), vol. 4830, pp. 90–99. Springer, Heidelberg (2007)
11. Quinlan, J.R.: Induction of decision trees. Mach. Learn. (1986)
12. Quinlan, J.R.: C4.5: Programs for Machine Learning. Morgan Kaufmann (1993)
13. Yildiz, O.T., Dikmen, O.: Parallel univariate decision trees. Pattern Recogn. Lett. 28(7), 825–832 (2007)

Stochastic Decentralized Routing of Unsplittable Vehicle Flows Using Constraint Optimization

Maksims Fiosins*

Clausthal University of Technology, Julius-Albert Str. 4,
D-38678 Clausthal-Zellerfeld, Germany
maksims.fiosins@tu-clausthal.de

Abstract. A decentralized solution to the unsplittable flow problem (UFP) in a transport network is considered, where each flow uses only one route from source to sink and the flows cannot be separated into parts in intermediate nodes. The flow costs in each edge depend on the combination of the assigned flows as well as on external random variables. The distributions of the random variables are unknown, only samples are available. In order to use the information available in the samples more effectively, several resamples are constructed from the original samples. The nodes agree on the resamples in a decentralized way using a cooperative resampling scheme. A decentralized asynchronous solution algorithm for the flow routing problem in these conditions is proposed, which is based on the ADOPT algorithm for asynchronous distributed constraint optimization (DCOP). An example illustrating the proposed approach is presented.

1 Introduction

The network flow routing is important for different applications, such as traffic management, logistics, and telecommunications. In the present paper we focus on a special case of the flow routing problem, namely that of unsplittable flows. This means that, in contrast to the usual formulation, the flows in the considered network cannot be split; each flow should use only one route from the source to the sink.

The considered problem corresponds to the routing of several vehicle flows in an urban street network. This network includes sources and sinks, and a set of the flows of vehicles that are moving from the sources to the sinks (each flow corresponds to a specific origin-destination pair). The vehicles and intersection controllers are equipped with communication devices that allow them to transmit information and establish short-term peer-to-peer communication. No centralized regulation in the considered system is performed; the intersection controllers decide in a decentralized manner how each vehicle flow is to be routed. For this purpose, they communicate with neighboring intersections and agree on the routing policy [7].

* Supported by the Lower Saxony University of Technology (NTH) project "Planning and Decision Making for Autonomous Actors in Traffic" (PLANETS).

S. Omatu et al. (Eds.): *Distrib. Computing & Artificial Intelligence,* AISC 217, pp. 37–44.
DOI: 10.1007/978-3-319-00551-5_5

Approaching an intersection, a vehicle sends its source and destination to the intersection controller. Based on this information, the intersection controller recommends the direction in which the vehicle should proceed. Note that the vehicle flows are supposed to be unsplittable, which means that to the vehicles with the same source-destination combination the same direction at each intersection will be recommended. This limitation is usually introduced in order to simplify the routing process and make the controllers' decisions clearer for traffic participants.

The general problem is to reduce the total travel time in the network. The travel time along a street depends on the number of vehicles situated in this street. We suppose that this dependence is not linear and corresponds to the fundamental traffic diagram. Also external random factors (incidents, weather etc.) influence the travel time in the streets. The distributions of the random variables are unknown; only sample populations are available.

We investigate a decentralized solution to the considered problem. For this purpose, we formulate the problem in the form of distributed constraint optimization (DCOP). However, standard algorithms for DCOP, such as DPOP [15] or ADOPT [13], cannot be used directly. An additional difficulcy is maintaining random variables.

The contribution of this paper as follows. First, we modify the ADOPT [13] algorithm to obtain an asynchronous distributed method for the problem of unsplittable network flow routing. And second, we add the capability to work with random variables, represented by samples, by combining the collaborative sampling approach of E[DPOP] with resampling when the distributions are partially known.

The paper is organized as follows. In Section 2 the state of the art is considered. Section 3 presents a formal problem statement. In section 4 we describe our solution approach. In section 5 we illustrate our approach with a numerical example. The last section contains conclusion and final remarks.

2 State of the Art

Classical maximum network flow (MNF) problem has been extensively examined in the literature. A standard Ford-Fulkerson algorithm [8] or network simplex algorithm [10] can be used to solve it effectively. An important type of MNF problems is the multi-commodity flow problem, where multiple flow demands between different source and sink nodes exist. For the case of more than two commodities no polynomial-time algorithms are known. Simple efficient algorithms solve this problem for restricts classes of graphs or in the case of fractional flows [11].

If it is required that each of the flows uses one route only, the problem becomes an unsplittable flow problem (UFP). This problem has been of interest during the last years due to its complexity and practical importance. A number of polynomial algorithms for the restricted versions of the problem have been developed [5].

Introducing stochastic components into UFP exacerbates this problem. A dynamic network flow model for a network that changes over time was introduced by Ford and Fulkerson [8] and extended by Glockner and Nemhauser for random arc capacities [9]; in this case, a multistage stochastic linear program can be used. These models are used to introduce stochastic components in inventory models, bottleneck transportation problem [1], etc.

In the case of solvers' distribution, the UFP can be solved using DCOP methods. There are two main approaches for obtaining the exact solution to DCOP: dynamic programming (DPOP) [15] and deep search (ADOPT) [13]. DPOP requires a linear number of messages, but exponential message size. ADOPT operates with linear sized messages, but the number of messages is exponential. In this paper, we will modify the ADOPT algorithm, because the memory size and one-message transmission time are limited for technical reasons. In addition, ADOPT allows asynchronous solutions, which are stable when communication delays or failures occur. Many applications of DCOP were developed, for example, for EDP, agent plan coordination, security in logistics operations [14], etc.

Stochastic DCOP (SDCOP) [12] is a special type of DCOP that incorporates the additional influence of random variables. When the distributions of random variables are unknown, the resampling approach can be effective [2], [3]. During the resampling procedure, the random samples are formed several times ("resamples"), the estimator of interest is calculated for each resample ("resampling estimators"), and the resampling estimators are then joined to produce a solution [6]. Various approaches for SDCOP were developed, for example independent utility distributions [4] or the E[DPOP] algorithm based on collaborative sampling [12]. We use this approach to introduce stochastic components into the ADOPT algorithm.

3 Problem Formulation

We consider a part of a city, represented by a transport network, a type of directed graph $G = (V, E)$ that contains exactly one node (source) of in-degree 0, exactly one node (sink) of out-degree 0 and no loops.

Let $E^+(v) \subset E$ be a set of edges, leaving the vertex $v \in V$ and $E^-(v) \subset E$ a set of edges, entering the vertex $v \in V$. Let $N^+(v) \subset V$ be an out-neighborhood of the vertex $v \in V$: $N^+(v) = \{u \in V : e(v,u) \in E^+(v)\}$, $N^-(v) \subset V$ be an in-neighborhood of the vertex $v \in V$: $N^-(v) = \{u \in V : e(u,v) \in E^-(v)\}$. For simplicity of the following description we suppose that there is maximally one edge between each pair of neighbor vertices, which implies $|E^+(v)| = |N^+(v)|$ and $|E^-(v)| = |N^-(v)|$ $\forall v \in V$. We write $n^+(v,i)$ to refer the i-th out-neighbor vertex of v, $e^+(v,i)$ to refer the i-th out edge of v, $i = 1, 2, \ldots, |N^+(v)|$.

Suppose $U = \{u_1, u_2, \ldots, u_m\}$ is the set of unsplittable flows that should be routed in this network. We represent the flow routing as follows. Let $U(e) \subset U$ be a subset of the flows, actually routed through the edge $e \in E$. Denote $U^-(v) = \cup_{e \in E^-(v)} U(e)$ a set of the flows entering the vertex v and $U^+(v) = \cup_{e \in E^+(v)} U(e)$ a set of the flows leaving the vertex v. For all vertices $v \in V$, excluding source and sink, we require

$$U^-(v) = U^+(v), \quad (1)$$

so we will usually write $U(v)$ to refer the flows in the intermediate vertex $v \in V$. As well, $U^+(source) = U^-(sink) = U$, $U^-(source) = \emptyset$, $U^+(sink) = \emptyset$.

A routing of the flows at the vertex $v \in V$ is a decision, according to which the controller of the vertex v assigns the flows from $U(v)$ to the edges from $E^+(v)$ (and accordingly to the vertices from $V^+(v)$). In fact, such decision is an ordered partition of the set $U(v)$ to $|N^+(v)|$ non-overlapping subsets. Let $d(v,u) \subset U(v)$ be a set of the flows, which are routed to the out-neighbor $u \in N^+(v)$ of the vertex v (through the edge $e(v,u) \in E^+(v)$). We require $\cup_{u \in N^+(v)} d(v,u) = U(v)$, $\cap_{u \in N^+(v)} d(v,u) = \emptyset$. Define $d(v) = < d(v,u) | u \in N^+(v) >$.

Let $D(v,U(v))$ be a set of all partitions, which are allowed for current flow set $U(v)$ of the vertex $v \in V$. So the partition $d(v)$ is allowed, if $d(v) \in D(v,U(v))$. If $D(v,U(v)) = \emptyset$, this means that the flow set is not allowed for the vertex v; the vertex will have infinity local cost in this case.

The random factors such as weather, incidents, activities in the city etc. influence the traffic conditions. For this purpose we introduce a set of random variables $Y = \{Y_1, Y_2, \ldots, Y_k\}$. The distributions of the variables Y are unknown, only samples H_i of (historical) observations are available. We denote the sets $Y_{v,u} \subset Y$, which have an influence in the edges $e(v,u) \in E$.

The cost of the flow at the edge $e(v,u) \in E$ is defined by a cost function $f_{v,u}(U(e(v,u)), Y_{v,u})$, which depends on the flows $U(e(v,u)) \subset U$ assigned to the edge $e(v,u) \in E$ as well as the random variables $Y_{v,u}$. The flow cost for a whole network is a sum of the costs over all edges:

$$F(U) = \sum_{e(v,u) \in E} f_{v,u}(U(e(v,u)), Y_{v,u}). \tag{2}$$

In order to satisfy the constraint (1) we rewrite (2) in respect to the vertices. Each vertex $v \in V$ forms a partition $d(v) \in D(v,U(v))$ for its flows $U(v)$, where the set of flows $d(v,u)$ is sent to the neighbor $u \in N^+(v)$ through the edge $e(v,u) \in E^+(v)$. The cost for the edge $e(v,u)$ is $f_{v,u}(d(v,u), Y_{v,u})$. So we have

$$F(U) = \sum_{v \in V} \sum_{u \in N^+(v)} f_{v,u}(d(v,u), Y_{v,u}), \qquad d(v) \in D(v,U(v)), \tag{3}$$

where for each $v \in V$ except a source

$$U(v) = \cup_{u \in N^-(v)} d(u,v). \tag{4}$$

Our goal is to minimize the expectation of (3)

$$E_Y[F(U)] \to \min \tag{5}$$

with respect to constraints (4) by calculating the partitions $d(v) \in D(v,U(v))$.

Our goal is to describe decentralized approach to this problem solution. An additional requirement to the algorithm is that it should be asynchronous and reliable to the possible failures in communication. We developed a method for the flows routing in a transport network, which is in fact a modification of ADOPT algorithm.

4 Proposed Algorithm

Suppose that each vertex $v \in V$ is controlled by an agent $a(v) \in A$, where A is a set of agents controlling the intersections. The costs $f_{v,u}(\cdot)$ are available to the agent

$a(v)$ only locally. The communication is possible with neighbor agents, namely the agent $a(v)$ sends messages and receive answers from the agents $a(u)$, $u \in N^+(v)$.

First, a Deep First Search (DFS) tree T from the initial network G is produced. The procedure of DFS tree construction is rather standard and can be executed in linear time depending on the number of vertices. We call the edges that do not belong to T as back edges. Define $P(v) \in V$ a parent of a vertex $v \in V$ in the DFS tree.

For the case of small samples it is more attractive to perform the resampling procedure, which allows to use data in different combinations [2], [3]. However as our algorithm is decentralized, it is important that the same samples are used in calculations. We use a collaborative resampling approach for this purpose.

The algorithm uses r resample sets. Each resample set q, $q = 1,2,\dots,r$ contains the resamples with replacement of a length n_j for the variables $Y_j \in Y$. It produces r solutions, one for each resample set, which are later joined to a single final solution.

Denote $H_j^{*q}(v)$ a q-th resample of random variable Y_j, used by the node v, $H^*(v)$ the information over all random variables and all resamples. The node v forms the sets of the samples $H_{v,u}^{*q}(v)$ for each edge $e(v,u) \in E$, which correspond to variables $Y_{v,u}$ and uses them in the cost calculation.

The source node performs the resampling for the outgoing edges by sampling the variables $\bigcup_{u \in N^+(source)} Y_{source,u}$, stores this information in $H^{*q}(source,u)$ and sends the samples to its child nodes. By receiving the samples from its parent, the child node u updates its local sample information $H^*(u)$, sends it to its children as well as informs its parent in DFS tree about the actual samples. By receiving the sample information from its children, if the node detects some changes in the samples, it sends the sample information to its children as well as parent in the DFS tree.

Each agent $a(v)$ maintains a set $CurrentFlows^{*q}(v)$, $q = 1,\dots,r$, which contains pairs of neighbor edges together with corresponding sets of the flows sent between them; the q-th resampling set was used for their calculation:

$$CurrentFlows^{*q}(v) = \{(u_1,w_1,d^{*q}(u_1,w_1)),(u_2,w_2,d^{*q}(u_2,w_2)),\dots\},$$

where $w_i \in N^+(u_i)$; u_i and w_i can be possibly equal to v and $d^{*q}(u_i,w_i)$ is a flow decision according to the q-th resampling realization of the random variables. This means that $CurrentFlows^{*q}(v)$ contains a partial solution of the flow routing problem, known to the agent $a(v)$.

The elements $(u,v,d^{*q}(u,v)) \in CurrentFlows^{*q}(v)$, $u \in N^-(v)$ define a set of known incoming flows for the node v. We refer this set as $U(v,CurrentFlows^{*q}(v))$.

$$U(v,CurrentFlows^{*q}(v)) = \bigcup_{(u,v,d^{*q}(u,v)) \in CurrentFlows^{*q}(v)} d^{*q}(u,v). \tag{6}$$

Now let us define an average local cost of a node v with respect to random factors Y. The random factors for the edge $e(u,v)$ are represented by the resampling sets $H_{u,v}^{*q}(v)$. The q-th local cost, $q = 1,2,\dots,r$ is defined as an average cost of the partition $d(v)$ calculated by the q-th resampling set:

$$\delta^{*q}(v,d(v)) = \sum_{u \in N^+(v)} \operatorname*{avg}_{H_{v,u}^{*q}(v)} f_{v,u}(d(v,u),H_{v,u}^{*q}(v)). \tag{7}$$

Further a lower bound variable is defined for each possible partition $d(v)$ is

$$LB^{*q}(v,d(v)) = \delta^{*q}(v,d(v)) + \sum_{u \in N^+(v)} lb^{*q}(v,u,d(v,u)), \tag{8}$$

where $lb^{*q}(v,u,d(v))$ is a field of the node v, which contains a cost information (LB field) from the child node u if a set of the flows $d(v,u)$ is sent there and q-th resampling subset is used for calculations. This information is transmitted from child to parent using the COST messages.

The corresponding lower bound value over all partitions is defined as:

$$LB^{*q}(v) = \min_{d(v) \in D(v)} LB^{*q}(v,d(v)). \tag{9}$$

The FLOWS message (analog of VALUE message in the ADOPT algorithm) informs the child nodes about the partition assignments. The agent $a(v)$ stores a set $d^*(v) = \{d^{*q}\}$, $q = 1,2,\ldots,r$ of the partitions, which correspond each resampling set. If the agent decides to change (or first selects) some of the partitions $d^{*q}(v) \in D(v)$, it sends a FLOWS message to all its child nodes $u \in N^+(v)$. The FLOWS message contains a set of flows, sent from the parent node v to the child node u according to each of the partitions $d^{*q}(v)$, so $d^{*q}(v,u)$.

The child agent answers to the FLOWS message by the COST message. Note however, that the agent sends all COST messages to its parent $P(v)$ in the DFS tree only. Other parent nodes will receive this information indirectly. The COST message contains the fields $LB^{*q}(u)$ as well as $CurrentFlows^{*q}(u)$ of the child node u. By receiving the COST message from the child node u, the parent node v saves the values $LB^{*q}(u)$ in the fields $lb^{*q}(u,d^{*q}(v))$, which correspond to its current partitions $d^{*q}(v)$, if it did not change them since the FLOWS message was sent.

5 Experimental Results

We consider the flows $U = \{u_1,u_2,\ldots,u_7\}$ in a simple transportation network. We suppose that there are three random factors Y_1, Y_2 and Y_3, which take values from the set $\{0,1\}$. The probabilities of these values are unknown, only samples are available.

The cost functions $f_{v,u}(U(e(v,u)),Y_{v,u})$ are defined for each node. Each function contains a cost of sending each subset $U(e(v,u)) \subset U$ of the flows. In general, it should be defined for all $2^7 - 1 = 127$ combinations of possible flows; however the edges can contain less flows that reduces dimension. Another dimension for the cost function are possible values of random variables Y_i. Table 1 presents a cost function $f_e()$ definition of the edge e, which depends on the random variables Y_1 and Y_3.

Now we set a number of the resamples $r = 10$ and the size of each resample $n = 20$. The nodes agree about the used resamples and the set of solutions for each resample set was calculated. Table 2 presents the obtained solution.

Some of the results are the same, which correspond to the same (or similar) agreed samples. They can be just averaged, and the average corresponds to the resampling point estimator of the solution. As well it is possible to construct the distribution of the solution, construct its resampling confidence interval etc.

Table 1 Cost function $f_e(U(e),\{Y_1,Y_3\})$ for the edge e

	$Y_1=0;Y_3=0$			$Y_1=0;Y_3=1$			$Y_1=1;Y_3=0$			$Y_1=1;Y_3=1$		
$U(e)$	{1}	{2}	{1,2}	{1}	{2}	{1,2}	{1}	{2}	{1,2}	{1}	{2}	{1,2}
$f_e()$	5	7	11	6	9	13	7	8	14	12	15	23

Table 2 Solution structure for $r=10$ resamples

q	1	2	3	4	5	6	7	8	9	10
$F^{*q}(U)$	103.2	94.3	106.8	106.8	103.2	103.2	96.9	106.8	103.2	94.3

Table 3 demonstrates complexity of the algorithm. We make simple transformations of the graph in order to assess a number of messages depending on the number of the vertices as well as on the number of the flows.

Table 3 The number of messages for various number of vertices $|V|$ and flows $|U|$ in the network

$\|V\|$	9	9	10	10	11	11	12	12
$\|U\|$	6	7	7	8	8	9	8	9
# of messages	$2.3*10^3$	$6.8*10^3$	$1.3*10^4$	$6.7*10^4$	$3.1*10^5$	$9.4*10^5$	$8.3*10^5$	$2.7*10^6$

These results show that the complexity of the algorithm is exponential depending on the number of vertices $|V|$ as well as the number of flows $|U|$. However the results are acceptable for considered number of vertices and flows.

6 Conclusions

We developed a distributed asynchronous algorithm for unsplittable flows routing in the transport network. For this purpose, the ADOPT algorithm for DCOP was modified. We added an opportunity to work with random variables (factors) represented by samples; the vertices can agree about the used samples in decentralized way by applying collaborative resampling approach.

The results show that the algorithm solves the considered problem and correspond the requirements. So it can be used effectively for decentralized vehicle flow routing. However the number of messages is exponential depending on the number of vertices and the number of flows.

In the future work we are going to develop an stochastic approximation of this algorithm in order to avoid exponential number of messages and get an acceptable solution. Another important direction of the research is true parallel implementation of the algorithm in order to assess its efficiency on parallel platform (GPU).

References

1. Ahuja, R.K., Magnanti, T.L., Orlin, J.B.: Network Flows: Theory, Algorithms, and Applications. Prentice-Hall, Englewood Cliffs (1993)
2. Andronov, A., Fiosina, J., Fiosins, M.: Statistical estimation for a failure model with damage accumulation in a case of small samples. J. of Stat. Planning and Inf. 139, 1685–1692 (2009)
3. Andronov, A., Fiosins, M.: Applications of resampling approach to statistical problems of logical systems. Acta et Comm. Univ. Tartuensis de Mathematica 8, 63–72 (2004)
4. Atlas, J., Decker, K.: Task scheduling using constraint optimization with uncertainty. In: Proc. of CAPS 2007, pp. 25–27 (2007)
5. Azar, Y., Regev, O.: Strongly polynomial algorithms for the unsplittable flow problem. In: Aardal, K., Gerards, B. (eds.) IPCO 2001. LNCS, vol. 2081, pp. 15–29. Springer, Heidelberg (2001)
6. Fiosins, M., Fiosina, J., Müller, J.P.: Change point analysis for intelligent agents in city traffic. In: Cao, L., Bazzan, A.L.C., Symeonidis, A.L., Gorodetsky, V.I., Weiss, G., Yu, P.S. (eds.) ADMI 2011. LNCS (LNAI), vol. 7103, pp. 195–210. Springer, Heidelberg (2012)
7. Fiosins, M., Fiosina, J., Müller, J.P., Görmer, J.: Reconciling strategic and tactical decision making in agent-oriented simulation of vehicles in urban traffic. In: Proc. of the 4th Int. ICST Conf. Simulation Tools and Techniques, pp. 144–151. ACM Digital Library (2011)
8. Ford, L., Fulkerson, D.: Flows in Networks. Princeton University Press (1962)
9. Glockner, G., Nemhauser, G.: A dynamic network flow problem with uncertain arc capacities: Formulation and problem structure. Operations Research 48(2), 233–242 (2000)
10. Goldberg, A.V., Grigoriadis, M.D., Tarjan, R.E.: Use of dynamic trees in a network simplex algorithm for the maximum flow problem. Mathematical Programming 50, 277–290 (1991)
11. Karakostas, G.: Faster approximation schemes for fractional multicommodity flow problems. ACM Trans. Algorithms 4(1), 13:1–13:17 (2008)
12. Léauté, T., Faltings, B.: E[DPOP]: Distributed Constraint Optimization under Stochastic Uncertainty using Collaborative Sampling. In: Proc. of the IJCAI 2009, pp. 87–101 (2009)
13. Modi, P.J., Shen, W., Tambe, M., Yokoo, M.: Adopt: Asynchronous distributed constraint optimization with quality guarantees. Artificial Intelligence Journal 161, 149–180 (2005)
14. Ottens, B., Faltings, B.: Coordinating agent plans through distributed constraint optimization. In: Proc. of MASPLAN 2008 (2008)
15. Petcu, A., Faltings, B.: Dpop: A scalable method for multiagent constraint optimization. In: Proc. of the 19th Int. Joint Conf. on Artificial Intelligence, pp. 266–271 (2005)

Extending Semantic Web Tools for Improving Smart Spaces Interoperability and Usability

Natalia Díaz Rodríguez[1], Johan Lilius[1], Manuel Pegalajar Cuéllar[2], and Miguel Delgado Calvo-Flores[2]

[1] Turku Centre for Computer Science (TUCS), Department of Information Technologies, Åbo Akademi University, Turku, Finland
{ndiaz,jolilius}@abo.fi

[2] Department of Computer Science and Artificial Intelligence, University of Granada, Spain
{manupc,mdelgado}@decsai.ugr.es

Abstract. This paper explores the main challenges to be tackled for more accessible, and easy to use, Smart Spaces. We propose to use Semantic Web principles of interoperability and flexibility to build an end-user graphical model for rapid prototyping of Smart Spaces applications. This approach is implemented as a visual rule-based system that can be mapped into SPARQL queries. In addition, we add support to represent imprecise and fuzzy knowledge. Our approach is exemplified in the experimental section using a context-aware test-bed scenario.

Keywords: Smart Space, Fuzzy Ontology, Interoperability, End-user Application Development.

1 Introduction

Smart Spaces were built in relation with the vision in which computers work on behalf of users, they have more autonomy, and they are able to handle unanticipated situations. This implies the use of Artificial Intelligence (AI), agents and machine learning. One of the main goals of Smart Spaces is to achieve device interoperability through standard machine readable information to allow easy construction of mash-ups of heterogeneous applications. A *Smart Space* (SS) is an abstraction of physical ubiquitous space that allows devices to join and leave the space as well as sharing information. New devices are constantly being introduced, often involving different information formats or coming from diverse sources. Lack of interoperability between systems and devices easily becomes a problem, resulting in them not being used efficiently. In order to have devices that interoperate seamlessly, their respective data and functionality must be easily integrated and accessible.

SSs are considered to be context-aware systems; therefore, a key requirement for realizing them, is to give computers the ability to understand their situational conditions [4]. To achieve this, contextual information should be represented in adequate ways for machine processing and reasoning. Semantic technologies suite well this purpose because ontologies allow, independently, to share knowledge, minimizing redundancy. Furthermore, semantic languages can act as metalanguages to define

S. Omatu et al. (Eds.): *Distrib. Computing & Artificial Intelligence,* AISC 217, pp. 45–52.
DOI: 10.1007/978-3-319-00551-5_6

other special purpose languages (e.g., policy languages), which is a key advantage for better interoperability than that one of tools that share no roots of constructs.

We focus on providing the end-user the possibility of rapidly prototype the behaviour of a SS, abstracting away technical details. When presenting a semantic SS to the user, the interface model, even if simplified, must adhere to the Semantic Web (SW) formal models. Domain Specific Languages (DSL) demonstrate support in this abstraction for integration of metamodels using ontologies (e.g. in [16]). Concerning end-user frameworks and GUIs for developing SS applications, we can find works that simplify the tasks to the user when creating their own services, through simple rules.

The survey of programming environments for novice programmers [9] shows how to lower the barriers to programming, which is one of our main aims; let non expert users to take part in the configuration of a Smart Space. Some good examples of end-user visual editors that simplify the tasks to the user when creating their own services or applications, through simple rules, are *If This Then That* [1] for online social services, *Twine* [2] for applications based on sensor interaction or Valpas [12] intelligent environment for assisted living.

A graphical tool to prototype context-aware applications is *iCap* [14], where coding is not required, but rather using window controls and IF-THEN rules. Their public is developers that want to rapidly test and iterate ubicomp applications. Other work in this line is a reconfiguration framework focused on tackling system variability and policy definition in runtime [6]. They use *PervML* as DSL, Feature Modelling techniques and Model Driven Engineering (MDE) principles such as code generation. End-user interaction evaluation is pending.

In [17] a rule-based framework with puzzle-like pieces allows to specify domain-specific situations with service invocations and state changes. Its use in a PDA showed quicker progress in users with less provided help. In *SiteView* [1], a tangible interface allows to control simple conjunctive rules of a "world-in-miniature" to help users create and view the effects of the rules.

A more complex system, aimed at end-users, is [15]. It is based on a pervasive navigation environment which uses spatial and resource annotations from the users' pictures. Pervasive Maps allow to model, explore and interact with complex pervasive environments.

In [5], an ECA (Event-Condition-Action) rule system allows end-users to control and program their SS in a drag & drop environment with wildcards filters and textual expressions, allowing in this way complex rules to be formulated. Finally, the dataflow rule language in [3] shows increased expressiveness in young non-programmers. However, previous frameworks lack underlying semantic capabilities and support for fuzzy rule expressions through linguistic labels. There does not exist a GUI for visualizing both fuzzy ontologies and fuzzy rules. Thus, our proposal takes the interaction with context-aware Smart Spaces to a higher level, allowing easier prototyping of the SS's behaviour by a) providing semantics to enhance the

[1] `http://ifttt.com`

[2] `http://supermechanical.com/twine/`

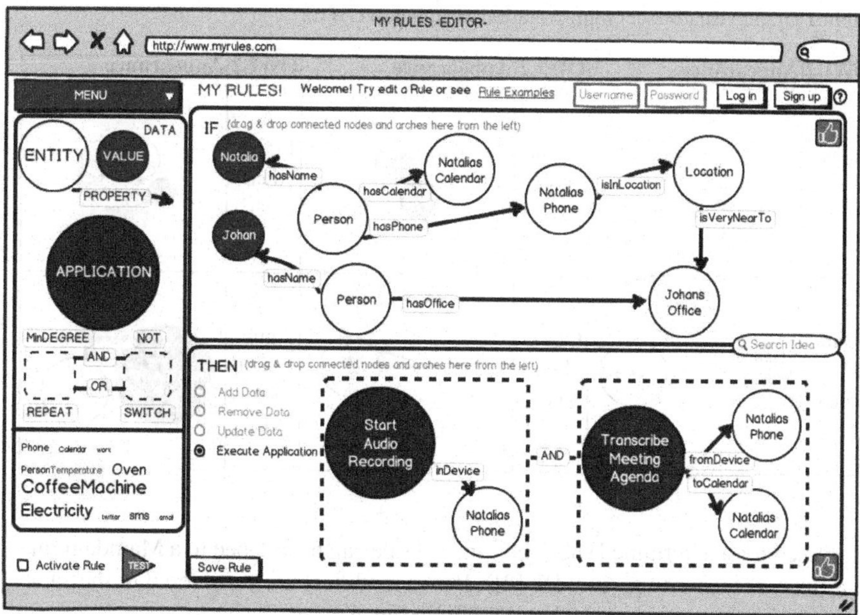

Fig. 1 User interface mock-up and example of semantic rule construction

context-awareness, and b) permitting imprecise every-day life expressions. Next section elaborates more on our proposal and Section 3 discusses the approach and suggests future directions.

2 A Framework for Rapid Application Development of Context-Aware Smart Spaces

Our contribution consist of a graphical interface for representing, visualizing and interacting with SSs information. It allows any end-user to model his own applications without knowledge of programming. Data gathering is possible by aggregation of different ontologies and datasets. The interface is based on simple IF-THEN rules applied to graph-based data. A node can be of two types, representing an OWL class (*Entity*, large and white) or a data property value (small and purple). An arc can represent a data property or object property, depending on the type of the destination node (*destNode*={Class or Value}). The THEN clause of the rule serves to 1) add, remove or update information in form of arcs and nodes (representing RDF triples) from the knowledge base, or 2) execute a registered browser-based application (with associated service grounding), possibly using concrete and well defined individuals or properties described in the IF clause or Linked Data. Registered web or Linked Data services are represented in large grey nodes. Subgraphs in IF and THEN clauses can be connected with logical operators and included into loops expressed in the rule's consequent. A minimum degree of satisfiability can be

Table 1 Graph visual model representation mapping to OWL2

OWL2	Appearance	OWL2	Appearance	OWL2	Appearance
Class	CLASS	Data type	Numeric Text	Individual	Location Notolos Calendar Notolo Person
Data Property	Value dataProperty CLASS	Object Property	CLASS A objectPrope CLASS B	Application/ Service	APPLICATION requiredParameter CLASS

expressed for a determined subgraph, since a rule can be mapped to a Mamdani rule in a fuzzy reasoner (e.g. *fuzzyDL* [2]). Fuzzy modifiers are considered in the same way as crisp properties (e.g. *isVeryNearTo* in Fig. 1).

The graph-based and "puzzle"-like pieces to edit rules with take inspiration from the successful *Scratch* framework [11]. Variable bindings are correct, by construction of the user interface, through letting the user allocate pieces only in the positions in which corresponding data ranges and domains are allowed.

This intuitive way of expressing a rule's condition, by dragging and joining compatible (data type-wise) nodes and arcs, can be easily translated into SPARQL query patterns (e.g., conditions in the *WHERE* field) and allow fast formulation of mash-up applications. Table 1 summarizes the mapping applied to transform end-user visual model representations into OWL2 entities.

A great power of visual languages is their ability of categorizations of certain primitives, and the graphical properties, to carry semantic information. Furthermore, elements of their syntax can intrinsically carry semantic information. To develop an effective visual language, i.e., a language that can be easily and readily interpreted and manipulated by the human reader, we followed guidelines for visual language design [7, 10]. These can be summarized as *morphology as types, properties of graphical elements, matching semantics to syntax, extrinsic imposition of structure and pragmatics for diagrams*. In our UI, we attach meaning to the components of the language both naturally (by exploiting intrinsic graphical properties such as keeping the underlying RDF graph structure) and intuitively [7] (taking consideration of human cognition, e.g. using colour to distinguish literals from classes). E.g., we considered the primary properties of graphical objects [8] to design our language's construct symbols.

2.1 *End-User Graphical Model Mapping to SPARQL*

Through structuring the edition of applications as simple IF-THEN rule statements, and by using an underlying graph-based graphical structure, an end-user can model semantic behaviour, by means of classes, individuals and relationships. The RDF store, which reflects its content on the left side of the UI, shows only legal relationships and properties associated to each entity. Simple SPARQL queries can extract the required data to be presented in each view, each moment the user hovers a specific entity or menu. E.g., given a class, show its object properties associated. For example, to get the object properties of the class *GenericUser*, the following query would return *hasCalendar, worksForProject, performsActivity*, etc.

```
SELECT DISTINCT ?pred
WHERE { ?pred rdfs:domain ha:GenericUser.
        ?pred rdfs:range ?object.
        ?object a owl:Class.}
```

The mapping that transforms a graphical rule into a SPARQL query is below:

```
Initialize counter for ClassNode variables, n to 0.
Initialize processedNodes dictionary to empty.
<-IF CLAUSE MAPPING->
For each ClassNode in IFClause of the Rule:
    For each Arc leaving from ClassNode:
        If destNode is a Datatype: // Data Property Triple
            Add patterns (?indiv_n a ClassName) and
                 (?indiv_n dataProp destNodeDataValue) to WHERE
            Add originNode and its index n to processedNodes
            Increment variable index n
        Else: // The Triple represents an Object Property
            If originNode is processed, obtain its index x
                If destNode is processed, get its index z
                    Add pattern (?indiv_x objectProp ?indiv_z) to WHERE
                Else:
                    Add pattern (?indiv_x objectProp ?indiv_n) to WHERE
                    Add destNode and its index n to processedNodes
                    Increment variable index n
            Else:
                If destNode is processed, get its index y
                    Add pattern(?indiv_n objectProp ?indiv_y) to WHERE
                    Add originNode and its index n to processedNodes
                    Increment variable index n
                Else:
                    Add pattern (?indiv_n objectProp ?indiv_n+1) to
                        WHERE
                    Add originNode and destNode to processedNodes
                    Increment variable index n by 2
<-THEN CLAUSE MAPPING->
If THENClause.type is APP: // Execute external App
    For each ClassNode in THENClause:
        If ClassNode is processed, obtain its index w &
            add '?indiv_w' to SELECT
        Else: "ERROR: Class Nodes in APP parameters need to be defined
            in IFClause". Exit
    QueryResult = Run SPARQL Query with {SELECT, WHERE}
    Execute set of AppNodes with QueryResult as parameters
Else:
    If THENClause.type is ADD: // Add triples
        For each Arc marked toAdd, add pattern to INSERT
    Else:
        If THENClause.type is REMOVE: // Remove triples
            For each Arc marked toDelete, add pattern to DELETE
    Run SPARQL query including {SELECT, WHERE, INSERT, DELETE}
```

The algorithm "parses" first the IF, followed by the THEN clause in the graphical model, to finally run a SPARQL query with the parameters collected in some of the array structures for SELECT, INSERT, DELETE and WHERE. Each *(origNode, Arc, destNode)* structure in the visual model corresponds to a triple pattern (subject, predicate, object). A counter n keeps track of node indexes to keep unique naming for each variable associated in the SPARQL query. Every arc is processed and, depending on the type of its destination node (line 6 & 11), the pattern is modelled as a)an individual's data property or b)an object property pattern. Generated patterns are added to the WHERE field of the query. For arcs and nodes in the THEN clause, the visual model's equivalent triple patterns are added to the INSERT or DELETE fields of the SPARQL query, respectively, since these are triples that must be marked as *toAdd, toRemove* or *toUpdate*. Finally, in THEN clause, some application (grey) nodes may require input parameters, that can reuse information from entities declared in the IF clause.

Rule Example Scenario: To study the viability of the ubiquitous model, we propose a location and context-aware scenario where positioning sensors are available through, e.g., each person's phone. We developed a Human Activity ontology that models different kind of users, their interactions and the activities they perform on the environment. Let us suppose the end-user wants to create a rule which allows her to start recording audio of the weekly meeting with his supervisor, automatically when she gets into his room: "If *Natalia* enters the room of his supervisor *Johan*, start audio-recording the meeting agenda in her phone's calendar". The aim would be keeping track, for future reference, of the agenda points and brainstorming ideas discussed, on the user's calendar. First of all, the user would select, from the GUI left menu the needed entities, the datatype values for identifying the individuals *Johan* and *Natalia*, and the relations which connect these with each other. The query produced by our algorithm is below. Although 4 lines longer, it is equivalent to a straightforward query written by somebody with knowledge of SPARQL:

```
SELECT  ?calendar1 ?phone2
WHERE{ ?user0 a ha:User.
       ?user0 ha:hasName "Natalia"^^xsd:string.
       ?user0 ha:hasCalendar ?calendar1.
       ?user0 ha:hasPhone ?phone2.
       ?user0 ha:isInLocation ?location3.
       ?phone2 ha:isInLocation ?location3.
       ?location3 ha:isNear ?office4.
       ?user5 a ha:User.
       ?user5 ha:hasName "Johan"^^xsd:string.
       ?user5 ha:hasOffice ?office4.}
```

3 Discussion and Future Work

This paper focuses on providing ordinary end-users with an accessible and functional SS vision through a tool that allows to exploit the potential of SW technologies, without requiring technical knowledge and supporting everyday life tasks.

Our visual model mock-up proposal, can as well serve as an educative interface for teaching basic SW technologies and logic programming ideas intuitively. However, our main contribution is a general purpose visual language based on a semantic metamodel, that supports query federation and (imprecise) rule composition for rapid development of mash-up applications. Our end-user model pushes the devised evolution of the SW from a data modelling to a computational medium [13] by bringing the advantages of the SW closer to any non expert user. Our contribution follows visual language design guidelines [7] for an intuitive, *well matched*, visual language, i.e., its representation clearly captures the key features of the represented artefact (in our case RDF triples), in addition to simplify various desired reasoning tasks (i.e., hiding namespaces and query languages). The applications of use range from assisted living and health care to home automation or industry processes.

Future work will complement the prototype framework to support complete fuzzy reasoning, develop the graphical model (and its usability) and propose an activity model to represent a higher level human behaviour, in which the end-user can control daily activities, save and exchange rules. The proposed architecture, its support for imprecise rules and fuzzy reasoning, show the path for dealing with current issues on SSs' usability. Since it is clear that having a *Scratch* for real life problem modelling would be of great use, future works, aiming at tackling the mentioned issues, will overcome and better model a context-aware SW accessible not only by machines but also by any human.

Acknowledgement. The research work presented in this paper is funded by TUCS (Turku Centre for Computer Science).

References

1. Beckmann, C.: Siteview: Tangibly programming active environments with predictive visualization. Intel Research Tech. Report, pp. 167–168 (2003)
2. Bobillo, F., Straccia, U.: *fuzzyDL*: An expressive fuzzy description logic reasoner. In: 2008 International Conference on Fuzzy Systems (FUZZ 2008), pp. 923–930. IEEE Computer Society (2008)
3. Bolós, A.C., Tomás, P.P., Martínez, J.J., Agües, J.A.M.: Evaluating user comprehension of dataflows in reactive rules for event-driven AmI environments. In: Proceedings of the V International Symposium on Ubiquitous Computing and Ambient Intelligence, UCAmI (2011)
4. Chen, H., Finin, T., Joshi, A.: Semantic web in a pervasive context-aware architecture. In: Artificial Intelligence in Mobile System (AIMS 2003), In conjunction with UBICOMP, pp. 33–40 (2003)
5. García-Herranz, M., Haya, P., Alamán, X.: Towards a ubiquitous end-user programming system for smart spaces 16(12), 1633–1649 (June 2010)
6. Giner, P., Cetina, C., Fons, J., Pelechano, V.: A framework for the reconfiguration of ubicomp systems. In: Corchado, J.M., Tapia, D.I., Bravo, J. (eds.) UCAMI 2008. ASC, vol. 51, pp. 1–10. Springer, Heidelberg (2009)
7. Gurr, C.: Computational diagrammatics: diagrams and structure. In: Besnard, D., Gacek, C., Jones, C.B. (eds.) Structure for Dependability: Computer-Based Systems from an Interdisciplinary Perspective. Springer, London (2006)

8. Gurr, C.: Visualizing a logic of dependability arguments. In: Cox, P., Fish, A., Howse, J. (eds.) Visual Languages and Logic Workshop (VLL 2007), vol. 274, pp. 97–109 (2007); Workshop within IEEE Symposium on Visual Languages and Human Centric Computing VL/HCC 2007
9. Kelleher, C., Pausch, R.: Lowering the barriers to programming: A taxonomy of programming environments and languages for novice programmers. ACM Comput. Surv. 37(2), 83–137 (2005)
10. Moody, D.: The physics of notations: Toward a scientific basis for constructing visual notations in software engineering. IEEE Trans. Softw. Eng. 35(6), 756–779 (2009)
11. Resnick, M., Maloney, J., Monroy-Hernandez, A., Rusk, N., Eastmond, E., Brennan, K., Millner, A., Rosenbaum, E., Silver, J., Silverman, B., Kafai, Y.B.: Scratch: programming for all. Communications of the ACM 52(11), 60–67 (2009)
12. Rex, A.: Design of a caregiver programmable assistive intelligent environment. Aalto University (2011)
13. Rodriguez, M.A., Bollen, J.: Modeling computations in a semantic network. Computing Research Repository (CoRR). ACM, abs/0706.0022 (2007)
14. Sohn, T.Y., Dey, A.K.: iCAP: An informal tool for interactive prototyping of context-aware applications. In: Extended Abstracts of CHI, pp. 974–975 (2003)
15. Vanderhulst, G., Luyten, K., Coninx, K.: Pervasive maps: Explore and interact with pervasive environments. In: 2010 IEEE International Conference on Pervasive Computing and Communications (PerCom), March 29-April 2, pp. 227–234 (2010)
16. Walter, T., Ebert, J.: Combining dSLs and ontologies using metamodel integration. In: Taha, W.M. (ed.) DSL 2009. LNCS, vol. 5658, pp. 148–169. Springer, Heidelberg (2009)
17. Zhang, T., Brügge, B.: Empowering the user to build smart home applications. In: Proceedings of 2nd International Conference on Smart Homes and Health Telematic (ICOST 2004), Singapore. Palviainen, Marko Series. Marko Palviainen (2004)

OPTICS-Based Clustering of Emails Represented by Quantitative Profiles

Vladimír Špitalský and Marian Grendár

Slovanet a.s., Záhradnícka 151, 821 08 Bratislava, Slovakia
{vladimir.spitalsky,marian.grendar}@slovanet.net

Abstract. OPTICS (Ordering Points To Identify the Clustering Structure) is an algorithm for finding density-based clusters in data. We introduce an adaptive dynamical clustering algorithm based on OPTICS. The algorithm is applied to clustering emails which are represented by quantitative profiles. Performance of the algorithm is assessed on public email corpuses TREC and CEAS.

1 Introduction

Email classification and spam filtering is a well developed area of research, which lays on an intersection of data mining, machine learning and artificial intelligence; cf. e.g. [1]. Though classification and categorization of emails have attracted considerable interest, in the literature much less attention is paid to email clustering, i.e. unsupervised learning. The main use of clustering in spam filtering is for detection of email campaigns; cf. e.g. [2, 3] or [4] for a use in semi-supervised filtering. Usually emails enter a clustering algorithm after a pre-processing, and are represented in terms of word frequencies (the so-called bag-of-words representation) or by heuristic rules. Emails are commonly clustered by the K-means algorithm. The bag-of-words and heuristic rules representations of emails suffer from several deficiencies such as the language dependence, high computational costs, vulnerability, sensitivity to concept drift, high number of heuristic rules, necessity to update the rules, etc. Choice of the clustering algorithm is usually guided by its simplicity of use. The more advanced clustering algorithms – such as the density-based methods – are usually not considered in the email clustering context, as they require more elaborate tuning.

In [5, 6] a quantitative profile (QP) representation of emails has been proposed, and demonstrated its excellent performance in email classification and spam filtering. QP represents an email by a fixed-dimension vector of numbers that characterize the email; e.g. lengths of lines. In the present communication, the QP representation is explored in the email clustering context. It turns out that clustering by the density-based methods, such as DBSCAN and OPTICS, is well-suited for the QP representation. This motivates the development of an OPTICS-based algorithm AD-OPTICS, an adaptive dynamical clustering algorithm. The main features of

S. Omatu et al. (Eds.): *Distrib. Computing & Artificial Intelligence*, AISC 217, pp. 53–60.
DOI: 10.1007/978-3-319-00551-5_7

AD-OPTICS are: 1) multi-view clustering, 2) tuning by quantiles, 3) multi-step dynamical clustering. AD-OPTICS, when applied to categorization of QP-represented emails, gives a language independent, simple-to-use clustering algorithm, able to uncover nontrivial clusters with high homogeneity. AD-OPTICS with QPs is intended for updating a training corpus of emails, as well as a support in the weighted email classification.

The paper is organized as follows. First we recall the quantitative profile representation of emails. Then the adaptive dynamical OPTICS-based clustering algorithm (AD-OPTICS) is presented in detail. Results of a performance study of AD-OPTICS in email clustering are gathered in Section 4. The concluding section summarizes the main advantages of the algorithm.

2 Quantitative Profile Representation of Emails

The Quantitative Profile (QP) approach to spam filtering and email categorization [5, 6] is motivated by the email shape analysis proposed by Sroufe, Phithakkitnukoon, Dantu, and Cangussu in [7]. Quantitative profile of an email is a p-dimensional vector of numbers that characterize the email. For instance, vector of lengths of the first p lines of an email forms the line profile of the email. The vector of occurrences of letters from an alphabet constitutes the character profile. The line profile is an example of the binary profile, which assumes a set of special characters and the profile is obtained by counting occurrences of letters between two characters from the special set. Besides new instances of the binary profile, several other classes of quantitative profiles were considered in [6], such as the grouped character profile, d-gram grouped character profile, as well as the size profile. A formal definition of some of the profiles is recalled below (cf. [6] for further details).

First, note that in the QP approach, an email is represented as a realization of a vector random variable generated by a hierarchical data generating process. The length n of an email is an integer-valued random variable, with the probability distribution F_n. Given the length, the email is represented by a random vector $X_1^n = (X_1, \dots, X_n)$ from the probability distribution $F_{X_1^n|n}$ with the support in $\mathscr{A}^n$, where $\mathscr{A} = \{a_1, \dots, a_m\}$ is a finite set (alphabet) of size m.

The grouped character profile (CPG) is a character profile with the alphabet $\mathscr{A}'$, where $\mathscr{A}'$ is a partition of $\mathscr{A}$. Here we consider two partitions $\mathscr{A}'$ based on the Unicode character categories. The CPG5 partition consists of the following five groups of characters: uppercase Lu, lowercase Ll, decimal digits Nd, controls Cc with spaces Zs, and the rest. In CPG13, the controls Cc and spaces Zs are separated and the rest is further divided to punctuation Pd, Ps, Pe, Po, symbols Sm, Sc, Sk and others.

The d-gram grouped character profile (dCPG) is based on the alphabet $(\mathscr{A}')^d$, where d is specified in advance. The dCPG profile consists of the counts of occurrences of the d-grams in an email.

Finally, the size profile comprises the information on the size of an email (in bytes), the size of selected header fields and also their CPGs, and the number and the size of email parts of selected content types, with their CPGs.

In [6], performance of the Random Forest-based classifiers with several classes of QP was assessed on the public TREC and CEAS email corpuses as well as on private corpuses. The performance was compared with that of the optimized SpamAssassin and Bogofilter. In the email categorization task the QP-based Random Forest classifiers attained comparable and even better performance than the other considered filters. Besides the good performance, there are also other benefits of representing emails by quantitative profiles. Quantitative profiles are in general easy to compute. Thanks to the QP-representation the resulting classifier is language independent. The Random Forest classifier adds to it robustness to outlying emails and high scalability.

Motivated by the favorable performance of QP-based filters in the classification task, we consider their use for email clustering. The clustering is performed by a novel adaptive, dynamical, OPTICS-based clustering algorithm, which is described below.

3 Adaptive Dynamical OPTICS-Based Clustering Algorithm AD-OPTICS

Density-based clustering is an approach to clustering which utilizes density of points in forming clusters. DBSCAN and OPTICS are two of the most popular density-based clustering methods. Next we briefly describe these algorithms; for a survey cf. [8].

The DBSCAN (Density-Based Spatial Clustering of Applications with Noise) clustering algorithm was proposed in [9]. It divides objects into clusters according to the given distance function d and two parameters: the radius ε and the number of points $minPts$. An object x is said to be *core* if its (closed) ε-neighborhood $U_d(x,\varepsilon)$ contains at least $minPts$ objects. The clusters are defined in such a way that they are unions of ε-neighborhoods of core points and two core points belong to the same cluster if and only if one of them is density reachable from the other one; recall that an object y is *density reachable* from x provided there are core points $x_0 = x, x_1, \ldots, x_k$ such that $x_{i+1} \in U_d(x_i, \varepsilon)$ for every i and $y \in U_d(x_k, \varepsilon)$. Objects x such that $U_d(x,\varepsilon)$ does not contain any core point are called *noise objects*; these objects do not belong to any cluster. For the details see [9].

One of the main problems with DBSCAN is the choice of the radius ε; small ε means that many objects are noise and large ε means that essentially different clusters can be glued together. To overcome these difficulties, in [10] the authors proposed OPTICS (Ordering Points To Identify the Clustering Structure) algorithm. Analogously as DBSCAN, also OPTICS depends on the distance d and two parameters $\varepsilon_{\max}$ and $minPts$. But unlike DBSCAN, OPTICS is not a clustering algorithm. Its purpose is to linearly order all objects in such a way that closest objects (according to the distance d) become neighbors in the ordering. This is achieved by defining the so-called *core-distance* $cd(x)$ and *reachability-distance* $rd(x)$ for every object x. OPTICS guarantees that, for every $\varepsilon \leq \varepsilon_{\max}$, if $rd(x) \leq \varepsilon$ then x belongs to the same ε-DBSCAN cluster as its predecessor. Thus, for any given $\varepsilon \leq \varepsilon_{\max}$, the ε-DBSCAN clusters correspond to the maximal intervals in the OPTICS ordering such that $rd(x) \leq \varepsilon$ for every but the first object of the interval. For other algorithms extracting (hierarchical) clusterings from OPTICS, see [10, 11, 12].

Concerning the choice of $\varepsilon_{\max}$, if it is too small, OPTICS cannot extract information about the clustering structure. On the other hand, with growing $\varepsilon_{\max}$ the runtime complexity of OPTICS grows dramatically, cf. e.g. [13]. Considerable effort was devoted to the choice of the density threshold for DBSCAN and OPTICS, cf. e.g. [13, 14, 15]

In the proposed algorithm AD-OPTICS we address the following problems.

- First, we want the algorithm to run in a reasonable time even for large databases. Since most of the runtime is spent by OPTICS, in AD-OPTICS the value of the crucial OPTICS parameter $\varepsilon_{\max}$ is chosen in an adaptive way.
- The second problem associated with OPTICS is the issue of cluster extraction. Several methods are proposed in the literature; cf. e.g. [10, 11, 12]. We use the flat DBSCAN clusters, where the cuts are determined in a novel way: by quantiles of reachability distance.
- Due to the time constraint employed to set the value of $\varepsilon_{\max}$, a part of the clustering structure can be lost. To solve this problem, the left-aside objects are sent into a new round of OPTICS.
- The final problem relates to clustering of complex objects, i.e. objects with multiple views, for which several metrics can be applied. AD-OPTICS uses a simple strategy: in each round, a metric is selected. For other multi-view density-based clustering methods, cf. e.g. [16].

Input parameters of AD-OPTICS are: the minimal cluster size *minClSize*, the minimal number *minSizeToCluster* of left-aside objects which can be further clustered and maximal time *opticsTime* for running one instance of OPTICS. The main loop of the algorithm can be summarized as follows.

1. Objects, which were left aside in the previous round, comprise the set D.
2. Select a metric d.
3. Determine $\varepsilon_{\max}$ by estimated runtime of OPTICS and *minPts* relative to the given minimal cluster size.
4. Run OPTICS on D with d, $\varepsilon_{\max}$ and *minPts*.
5. Set ε-levels $(\varepsilon_i)_i$ by the quantiles of reachability.
6. Extract ε_i-DBSCAN clusters in the bottom-up way.
7. a) If there are no clusters, go to Step 2.
   b) If the number of unclustered objects is less than *minSizeToCluster*, terminate.
   c) Otherwise, go to Step 1.

The OPTICS parameter $\varepsilon_{\max}$ is selected such that the expected OPTICS runtime is below a given time threshold *opticsTime*. An estimate of the expected runtime is based on an assumption that the expected time needed for running OPTICS is given as a product of the number of objects and the average time needed for a range query with radius $\varepsilon_{\max}$.

To extract clusters from obtained OPTICS ordering, we use several DBSCAN "cuts" with ε's corresponding to quantiles of reachability distance. Notice that if ε is α-quantile of reachability distance then the ratio of clustered objects is approximately α. The obtained DBSCAN clusterings are processed from bottom to top; a cluster is accepted if it has sufficient size.

Of course, there are many other conceivable possibilities for cluster extraction. For instance, one can use a cluster validity index to accept a cluster. Furthermore, the OPTICS reachability plot can be used for cluster extraction in various other ways, cf. e.g. [10, 11, 12]. One of the advantages of the proposed simple approach is that instead of running OPTICS on the entire dataset, the present approach allows to run OPTICS on a sample of data. Afterwards, with the OPTICS-selected ε-thresholds much simpler DBSCAN can be run on the entire dataset, saving both runtime and memory.

4 Performance Study

We have applied AD-OPTICS to email clustering. To this end we have used two public corpuses TREC07 and CEAS08 of ham and spam emails. TREC07 comprises 25 220 hams and 50 199 spams, CEAS08 consists of 27 126 hams and 110 579 spams.

The algorithm was implemented in Java and the results were processed in R [17]. The implementation of OPTICS algorithm was based on that from WEKA 3.6.5 and it was backed by the Vantage Point Tree [18]. The Java library Mime4J was used for parsing emails and obtaining size profiles. Let us note that for the other four profiles (two binary and two d-gram ones) we have used no email parsing at all, with one small exception: in the two body-profiles we skipped email headers.

Emails were represented by six quantitative profiles: two size profiles (SP-CPG13 and SP-CPG5), two binary profiles (brackets profile BRP applied to the whole email and BRP applied to email body) and two d-gram grouped character profiles (3CPG5-whole and 2CPG13-body). For more detailed specification of the profiles cf. Section 2 and [6]. To every profile L_1-distance was applied. The minimal cluster size *minClSize* was set to 100 and the minimal number of objects for further clustering *minSizeToCluster* was set to 5000. The time allowed for one OPTICS run was set to 20 minutes.

Next we summarize the results obtained by AD-OPTICS on the two considered corpuses. Since the corpuses are labeled it is possible to quantify how well a clustering splits hams and spams. For every cluster we determine its *majority label*, that is, the most frequent label of emails in the cluster, and we define the *purity of a cluster* to be the percentage of emails with the majority label. The weighted (by size) average of cluster's purities is the so-called *purity of a clustering*. Formally, if $\mathscr{C}$ is a clustering and $\mathscr{L}$ is a labeling of emails, then the purity of $\mathscr{C}$ is

$$\text{purity}(\mathscr{C}) = \frac{1}{n} \sum_{C \in \mathscr{C}} \max_{L \in \mathscr{L}} |C \cap L| ,$$

where n is the size of the corpus.

Table 1 Number of clusters according to purity 100%, 90–100%, 0-90%, with size proportions in parenthesis

corpus	clusters	100%	90–100%	0–90%	purity
TREC07	431	331 (72.4%)	58 (12.7%)	42 (14.9%)	94.3%
CEAS08	724	638 (83.0%)	55 (8.5%)	31 (8.5%)	97.5%

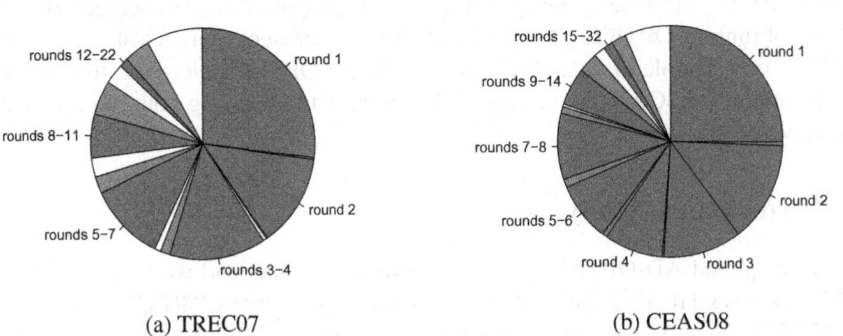

(a) TREC07 (b) CEAS08

Fig. 1 Purity 100% (blue), 90–100% (light blue) and 0–90% (white) of clusters according to rounds. Later rounds are aggregated

The results are summarized by Table 1 and Figure 1. The number of produced clusters is acceptable. We can see that vast majority of emails was split into pure clusters. Among them roughly 25% were found in the first round of the algorithm. First few rounds produced almost exclusively 100% pure clusters. Naturally, the number of pure clusters, as well as purity of clusters, decreased with round.

Purity of clusterings obtained from AD-OPTICS is in fact the success rate of classification conditioned upon known majority labels of clusters. Hence it can be compared with success rate of supervised classification. For instance, for the TREC corpus, SpamAssassin with optimized weights has success rate 94.6%, and BogoFilter has 98.4%, cf. the supplement of [5]. From this point of view, the purity of obtained clusterings is excellent.

5 Conclusions

We have proposed an adaptive dynamical clustering OPTICS-based algorithm AD-OPTICS. The algorithm chooses the key OPTICS parameter ε_{max} in an adaptive manner and does not require laborious tuning. Clusters are extracted from OPTICS in a novel way: by quantiles of reachablity distance. Objects are clustered in a dynamic way: those which were left out of clusters, enter the new round of clustering. AD-OPTICS can naturally handle complex objects with multi-view representations. AD-OPTICS is simple to configure. Indeed, the only input parameters to the algorithm are the maximum time which a single round of the algorithm could take, the minimal cluster size, and the minimal number of the left-aside (unclustered) emails. Finally, the presented approach makes possible running OPTICS on a sample of data, and then, using the OPTICS-suggested ε-thresholds, apply DBSCAN clustering to the whole dataset.

The algorithm was applied to email clustering, with emails represented by several Quantitative Profiles (QP). The AD-OPTICS configuration parameters are independent of the particular QP used for email representation. On the two public corpuses, AD-OPTICS produced excellent results. Vast majority of emails were clustered into homogeneous clusters. Moreover, along "trivial" clusters (those, for which one can easily construct a regular expression for filtering them out), also highly non-trivial and still homogeneous clusters were obtained.

The AD-OPTICS application to clustering QP-represented emails is easy to extend in several directions: emails can be represented by any QP and distance of emails can be measured by different metrics. When predicted labels of emails are available, they can be seamlessly integrated into the clustering process. Finally, thanks to the QP representation, the clustering is language-independent.

Acknowledgement. Stimulating feedback from Peter Farkaš, Ján Gallo, Peter Kočík, Erik Lehotský, Jana Majerová, Dušan Slivoň, Jana Škutová and Stanislav Záriš is gratefully acknowledged. This paper was prepared as a part of the "SPAMIA" project, MŠ SR 3709/2010-11, supported by the Ministry of Education, Science, Research and Sport of the Slovak Republic, under the heading of the state budget support for research and development.

References

1. Almeida, T.A., Yamakami, A.: Advances in spam filtering techniques. In: Elizondo, D.A., Solanas, A., Martinez, A. (eds.) Computational Intelligence for Privacy and Security. SCI, vol. 394, pp. 199–214. Springer, Heidelberg (2012)
2. Haider, P., Scheffer, T.: Bayesian clustering for email campaign detection. In: ICML 2009, pp. 385–392. ACM, New York (2009)
3. Qian, F., Pathak, A., Charlie Hu, Y., Morley Mao, Z., Xie, Y.: A case for unsupervised-learning-based spam filtering. In: SIGMETRICS 2010, pp. 367–368. ACM, New York (2010)
4. Whissell, J.S., Clarke, C.L.A.: Clustering for semi-supervised spam filtering. In: CEAS 2011, pp. 125–134. ACM, New York (2011)
5. Grendár, M., Škutová, J., Špitalský, V.: Spam filtering by quantitative profiles. Intnl. J. Comp. Sci. Issues 9, 265–271 (2012)
6. Grendár, M., Škutová, J., Špitalský, V.: Email categorization and spam fitering by random forest with new classes of quantitative profiles. In: Compstat 2012, pp. 283–294. ISI/IASC (2012)
7. Sroufe, P., Phithakkitnukoon, S., Dantu, R., Cangussu, J.: Email shape analysis. In: Kant, K., Pemmaraju, S.V., Sivalingam, K.M., Wu, J. (eds.) ICDCN 2010. LNCS, vol. 5935, pp. 18–29. Springer, Heidelberg (2010)
8. Kriegel, H.P., Kröger, P., Sander, J., Zimek, A.: Density-based clustering. WIREs DMKD 1(3), 231–240 (2011)

9. Ester, M., Kriegel, H.P., Sander, J., Xu, X.: A Density-Based Algorithm for Discovering Clusters in Large Spatial Databases with Noise. In: Proc. 2nd Int. Conf. on KDDM, pp. 226–231. AAAI Press (1996)
10. Ankerst, M., Breunig, M., Kriegel, H.P., Sander, J.: OPTICS: Ordering Points to Identify the Clustering Structure. In: Proc. ACM SIGMOD 1999, pp. 49–60. Springer, Heidelberg (1999)
11. Sander, J., Qin, X., Lu, Z., Niu, N., Kovarsky, A.: Automatic Extraction of Clusters from Hierarchical Clustering Representations. In: Proc. 7th Pacific-Asia Conference on KDDM, pp. 75–87 (2003)
12. Brecheisen, S., Kriegel, H.P., Kröger, P., Pfeifle, M.: Visually Mining Through Cluster Hierarchies. In: Proceedings of the 4th SIAM International Conference on Data Mining, pp. 400–411. SIAM (2004)
13. Achtert, E., Böhm, C., Kröger, P.: DeLi-Clu: Boosting Robustness, Completeness, Usability, and Efficiency of Hierarchical Clustering by a Closest Pair Ranking. In: Ng, W.-K., Kitsuregawa, M., Li, J., Chang, K. (eds.) PAKDD 2006. LNCS (LNAI), vol. 3918, pp. 119–128. Springer, Heidelberg (2006)
14. Gorawski, M., Malczok, R.: AEC Algorithm: A Heuristic Approach to Calculating Density-Based Clustering *Eps* Parameter. In: Yakhno, T., Neuhold, E.J. (eds.) ADVIS 2006. LNCS, vol. 4243, pp. 90–99. Springer, Heidelberg (2006)
15. Cassisi, C., Ferro, A., Giugno, R., Pigola, G., Pulvirenti, A.: Enhancing density-based clustering: Parameter reduction and outlier detection. Info. Sys. 38, 317–330 (2013)
16. Achtert, E., Kriegel, H.P., Pryakhin, A., Schubert, M.: Hierarchical Density-Based Clustering for Multi-Represented Objects. In: MCD 2005. ICDM (2005)
17. R Development Core Team: R: A language and environment for statistical computing. R Foundation for Statistical Computing, Vienna, Austria (2010) ISBN 3-900051-07-0, `http://www.R-project.org`
18. Yianilos, P.N.: Data structures and algorithms for nearest neighbor search in general metric spaces. In: Proceedings of the Fourth Annual ACM-SIAM Symposium on Discrete Algorithms, pp. 311–321. SIAM (1993)

Optimal Saving and Prudence in a Possibilistic Framework

Ana Maria Lucia Casademunt[1] and Irina Georgescu[2]

[1] Cordoba University, Faculty of Business Administration, Department of Sociology, Cordoba, Spain
alucia@etea.com

[2] Universidad Loyola–Andalucia, Department of Quantitative Methods, Cordoba, Spain and Academy of Economic Studies, Department of Economic Cybernetics, Bucharest, Romania
irina.georgescu@csie.ase.ro

Abstract. In this paper we study the optimal saving problem in the framework of possibility theory. The notion of possibilistic precautionary saving is introduced as a measure of the way the presence of risk (represented by a fuzzy number) influences a consumer in establishing the level of optimal saving. The equivalence between the prudence condition (in the sense of Kimball) and a positive possibilistic precautionary saving is proved. Some relations between possibilistic risk aversion, prudence and possibilistic precautionary saving are established.

Keywords: precautionary saving, prudence, possibility theory.

1 Introduction

The effect of risk on saving was studied for the first time by Leland [9], Sandmo [12] and Drèze and Modigliani [4]. They showed that if the third derivative of the utility function is positive, then the precautionary saving is positive. Kimball introduced in [8] the notion of "prudence" and established its relation with optimal saving.

This paper aims to approach optimal saving and prudence in the context of Zadeh's possibility theory [13]. The first contribution of this paper is a model of optimal saving, similar to the one in [8] or [6], p. 95. The notion of possibilistic precautionary saving (associated with a weighting function f, a fuzzy number A representing the risk and a utility function representing the consumer) is introduced and necessary and sufficient conditions for its positivity are established. The second contribution refers to some relations between the degree of absolute prudence [8], possibilistic risk aversion [7] and possibilistic precautionary saving. Among others, the possibilistic precautionary premium is defined as a possibilistic measure of precautionary motive. This notion is analogous to (probabilistic) precautionary premium of [8].

We will survey the content of the paper. In Section 2 is recalled, according to [2], [3], [5] the definition of fuzzy numbers and some associated indicators: possibilistic expected utility, possibilistic expected value and possibilistic variance. The equivalence between the concavity (resp. convexity) of a continuous utility function with a possibilistic Jensen–type inequality is proved.

S. Omatu et al. (Eds.): *Distrib. Computing & Artificial Intelligence,* AISC 217, pp. 61–68.
DOI: 10.1007/978-3-319-00551-5_8

In Section 3 the possibilistic two–period model of precautionary saving is studied. The consumer is represented by two utility functions u and v and the risk, present in the second period, is described by a fuzzy number. The expected lifetime utility of the model is defined with the help of the notion of possibilistic expected utility. The main introduced notion is possibilistic precautionary saving. It measures the changes on optimal saving produced by the presence of risk in the second period. If this indicator has a positive value than by adding the risk the consumer will choose a greater level of optimal saving. The main result of the section characterizes the positivity of possibilistic precautionary saving by the prudence condition $v''' \geq 0$.

Section 3 begins by recalling the Arrow–Pratt index [1], [10], the degree of absolute prudence [8] and possibilistic risk premium [7]. A result of the section characterizes the property of possibilistic risk premium to be decreasing in wealth by the comparison between prudence and absolute risk aversion (prudence is larger than absolute risk aversion). Then the notion of possibilistic precautionary premium is introduced and some of its properties which establish relations between prudence, possibilistic risk aversion and possibilistic precautionary saving are proved.

2 Possibilistic Expected Utility

Let X be a non–empty set. A fuzzy subset of X (shortly, fuzzy set) is a function $A : X \to [0,1]$. A fuzzy set A is normal if $A(x) = 1$ for some $x \in X$. The support of A is defined by $supp(A) = \{x \in \mathbf{R} | A(x) > 0\}$.

Assume $X = \mathbf{R}$. For $\gamma \in [0,1]$, the γ–level set $[A]^\gamma$ is defined by

$$[A]^\gamma = \begin{cases} \{x \in \mathbf{R} | A(x) \geq \gamma\} \text{ if } \gamma > 0 \\ cl(supp(A)) \text{ if } \gamma = 0 \end{cases}$$

($cl(supp(A))$ is the topological closure of $supp(A)$.)

The fuzzy set A is fuzzy convex if $[A]^\gamma$ is a convex subset of $\mathbf{R}$ for all $\gamma \in [0,1]$.

A fuzzy subset A of $\mathbf{R}$ is a *fuzzy number* if it is normal, fuzzy convex, continuous and with bounded support. If A, B are fuzzy numbers and $\lambda \in \mathbf{R}$ then the fuzzy numbers $A + B$ and λA are defined by

$$(A+B)(x) = \sup_{y+z=x} \min(A(y), B(z))$$

$$(\lambda A)(x) = \sup_{\lambda y = x} A(y)$$

A non–negative and monotone increasing function $f : [0,1] \to \mathbf{R}$ is a *weighting function* if it satisfies the normality condition $\int_0^1 f(\gamma) d\gamma = 1$.

Let f be a weighting function and $u : \mathbf{R} \to \mathbf{R}$ a continuous utility function. Assume that A is a fuzzy number whose level sets have the form $[A]^\gamma = [a_1(\gamma), a_2(\gamma)]$ for any $\gamma \in [0,1]$.

The possibilistic expected utility $E(f,u(A))$ is defined by:

(1) $E(f,u(A)) = \frac{1}{2}\int_0^1 [u(a_1(\gamma)) + u(a_2(\gamma))]f(\gamma)d\gamma$

If u is the identity function of $\mathbf{R}$ then $E(f,u(A))$ is the possibilistic expected value [3]:

(2) $E(f,A) = \frac{1}{2}\int_0^1 [a_1(\gamma) + a_2(\gamma)]f(\gamma)d\gamma$

If $u(x) = (x - E(f,A))^2$ for any $x \in \mathbf{R}$ then $E(f,u(A))$ is the possibilistic variance [3]:

(3) $Var(f,A) = \frac{1}{2}\int_0^1 [(a_1(\gamma) - E(f,A))^2 + (a_2(\gamma) - E(f,A))^2]f(\gamma)d\gamma$

When $f(\gamma) = 2\gamma$, $\gamma \in [0,1]$, $E(f,A)$ and $Var(f,A)$ are the notions introduced by Carlsson and Fullér in [2].

Proposition 1. *[7] Let g,h be two utility functions and $a,b \in \mathbf{R}$. If $u = ag + bh$ then $E(f,u(A)) = aE(f,g(A)) + bE(f,h(A))$*

Proposition 2. *If u is a continuous utility function then the following assertions are equivalent:*
(i) u is concave;
(ii) $E(f,u(A)) \le u(E(f,A))$ for any fuzzy number A.

Proof. (i) $\Rightarrow$ (ii) Let A be a fuzzy number such that $[A]^\gamma = [a_1(\gamma), a_2(\gamma)]$ for $\gamma \in [0,1]$. Since u is concave, for any $\gamma \in [0,1]$ the following inequality holds:
$\frac{u(a_1(\gamma))+u(a_2(\gamma))}{2} \le u(\frac{a_1(\gamma)+a_2(\gamma)}{2})$
Taking into account that $f \ge 0$ and applying Jensen inequality it follows
$E(f,u(A)) = \int_0^1 \frac{u(a_1(\gamma))+u(a_2(\gamma))}{2} f(\gamma)d\gamma \le$
$\le \int_0^1 u(\frac{a_1(\gamma)+a_2(\gamma)}{2}) f(\gamma)d\gamma \le u(\int_0^1 \frac{a_1(\gamma)+a_2(\gamma)}{2} f(\gamma)d\gamma) = u(E(f,A))$

(ii) $\Rightarrow$ (i) Let $a,b \in \mathbf{R}$, $a < b$. We consider the fuzzy number A for which $a_1(\gamma) = a$ and $a_2(\gamma) = b$ for any $\gamma \in [0,1]$. Then $E(f,u(A)) = \frac{u(a)+u(b)}{2}$ and $u(E(f,A)) = u(\frac{a+b}{2})$. By hypothesis, we will have $\frac{u(a)+u(b)}{2} \le u(\frac{a+b}{2})$. This inequality holds for any $a,b \in \mathbf{R}$ and u is continuous. By [11], Ex. 3, p. 67, it follows that u is concave. □

Corollary 1. *If u is a continuous utility function then the following are equivalent:*
(i) u is convex
(ii) $u(E(f,A)) \le E(f,u(A))$ for any fuzzy number A.

3 A Possibilistic Model of Precautionary Saving

The probabilistic two–period model of precautionary saving from [6], p. 95 is characterized by the following data:

- $u(y)$ and $v(y)$ are the utility functions of the consumer for period 0, resp. 1
- for period 0 there exists a sure income y_0 and for period 1 an uncertain income given by a random variable $\tilde{y}$

• s is the level of saving for period 0

Assume that u, v have the class C^3 and $u' > 0$, $v' > 0$, $u'' < 0$, $v'' < 0$. The expected lifetime utility of the model is
(1) $V(s) = u(y_0 - s) + M(v((1+r)s + \tilde{y}))$
where r is the rate of interest for saving.

The consumer's problem is to choose that value of s for which the maximum of $V(s)$ is attained.

The possibilistic model of optimal saving that we are going to build further starts from the same data, except for the fact that $\tilde{y}$ will be replaced by a fuzzy number.

We fix a weighting function f and a fuzzy number A whose level sets are $[A]^\gamma = [a_1(\gamma), a_2(\gamma)]$, $\gamma \in [0,1]$.

The (possibilistic) expected lifetime utility $W(s)$ of our model will be defined using the notions of possibilistic expected utility from the previous section:

(2) $W(s) = u(y_0 - s) + E(f, v((1+r)s + A))$

Relation (2) can be written:

(3) $W(s) = u(y_0 - s) + \frac{1}{2}\int_0^1 [v((1+r)s + a_1(\gamma)) + v((1+r)s + a_2(\gamma))] f(\gamma) d\gamma$

By derivation, from (3) one obtains:

(4) $W'(s) = -u'(y_0 - s) + \frac{1+r}{2}\int_0^1 [v'((1+r)s + a_1(\gamma)) + v'((1+r)s + a_2(\gamma))] f(\gamma) d\gamma$

which can be written:

(5) $W'(s) = -u'(y_0 - s) + (1+r)E(f, v'((1+r)s + A))$

Deriving it one more time it follows

(6) $W''(s) = u''(y_0 - s) + \frac{(1+r)^2}{2}\int_0^1 [v''((1+r)s + a_1(\gamma)) + v''((1+r)s + a_2(\gamma))] f(\gamma) d\gamma$

One considers the following optimization problem:
(7) $\max_s W(s)$

Proposition 3. *(i) W is a strictly concave function.*
(ii) The optimal solution $s^* = s^*(A)$ of problem (7) is given by $W'(s^*) = 0$.

Proof. (i) By hypothesis, $u'' < 0$, $v'' < 0$, thus by formula (6) it follows $W''(s) < 0$ for any $s \in \mathbf{R}$.
(ii) follows from (i). □

By Proposition 3 (ii) and (5), it follows that the optimal solution s^* is determined by the following equality:

(8) $u'(y_0 - s^*) = (1+r)E(f, v'((1+r)s^* + A))$

We consider now the optimal saving model in which in period 1 we don't have uncertainty any more: the uncertain income A is replaced by the sure income $E(f, A)$. The lifetime utility of this model is:

(9) $W_1(s) = u(y_0 - s) + v((1+r)s + E(f,A))$

and the optimization problem becomes

(10) $\max_s W_1(s) = W_1(s_1^*)$

In this case one has

(11) $W_1'(s) = -u'(y_0 - s) + (1+r)v'((1+r)s + E(f,A))$

The optimal solution $s_1^* = s_1^*(E(f,A))$ of problem (10) is given by $W_1'(s^*) = 0$, which, by (11), is written:

(12) $u'(y_0 - s_1^*) = (1+r)v'((1+r)s_1^* + E(f,A))$

The difference $s^* - s_1^*$ will be called *possibilistic precautionary saving* (associated with y_0, r and A). This indicator measures the way the presence of the possibilistic risk A causes changes in consumer's decision to establish the optimal saving.

The following proposition is the main result of the paper. The key–element of its proof is the application of Proposition 2.

Proposition 4. *The following assertions are equivalent:*
(i) $s^*(A) - s_1^*(A) \geq 0$ *for any fuzzy number* A*;*
(ii) $v'''(x) \geq 0$ *for any* $x \in \mathbf{R}$.

Proof. Let A be a fuzzy number. From (11) and (8) one obtains, by denoting $s^* = s^*(A)$:

$$\begin{aligned} W_1'(s^*) &= -u'(y_0 - s^*) + (1+r)v'((1+r)s^* + E(f,A)) \\ &= (1+r)[v'((1+r)s^* + E(f,A)) - E(f, v'((1+r)s^* + A))] \end{aligned}$$

Since W_1' is a strictly decreasing function one has

$s^*(A) \geq s_1^*(A)$ iff $W_1'(s^*(A)) \leq W_1'(s_1^*(A)) = 0$

Taking into account the value of $W_1'(s^*)$ computed above one obtains:

$s^*(A) \geq s_1^*(A)$ iff $v'((1+r)s^*(A) + E(f,A)) \leq E(f, v'((1+r)s^*(A) + A))$

But

$v'((1+r)s^*(A) + E(f,A)) = v'(E(f,(1+r)s^*(A) + A))$ thus

$s^*(A) \geq s_1^*(A)$ iff $v'(E(f,(1+r)s^*(A) + A)) \leq E(f, v'((1+r)s^*(A) + A))$

The previous inequality holds for any fuzzy number A, thus, by Corollary 1, the following equivalences follow:

- $s^*(A) \geq s_1^*(A)$ for any fuzzy number A
- v' is convex
- $v'''(x) \geq 0$ for any $x \in \mathbf{R}$ □

Condition (1) of Proposition 4 (=the positivity of possibilistic precautionary saving) expresses the fact that the presence of risk leads to the increase of optimal saving, and condition (2) is the well-known property of "prudence" introduced by Kimball in [8]. Thereby Proposition 4 can be regarded as a possibilistic version of Kimball's result by which the positivity of probabilistic precautionary saving is equivalent with prudence [8]. Since condition (2) is present both in Kimball's result and in Proposition 4, we conclude that the positivity of possibilistic precautionary saving is equivalent with the positivity of probabilistic precautionary saving.

4 Prudence and Possibilistic Risk Aversion

We consider an agent with the utility function u of class C^2 and $u' > 0$, $u'' < 0$. The *Arrow–Pratt index* r_u is defined by [1], [10]:

(1) $r_u(x) = -\frac{u''(x)}{u'(x)}, x \in \mathbf{R}$.

If u has the class C^3 then the *degree of absolute prudence* P_u was defined by Kimball in [8]:

(2) $P_u(x) = -\frac{u'''(x)}{u''(x)}, x \in \mathbf{R}$

One notices that $P_u \geq 0$ iff $u''' \geq 0$. If $g = -u'$ then $P_u = r_g$.

r_u is a measure of risk aversion and P_u is a measure of the agent's prudence in front of risk, In the above mentioned papers, r_u and P_u are indicators for analyzing probabilistic risk. By [7], the Arrow–Pratt index is an efficient instrument for the study of risk represented by fuzzy numbers.

We fix a weighting function f, a utility function u, a fuzzy number A and a real number x. We define the *possibilistic risk premium* $\pi(x,A,u)$ as the unique solution of the equation:

(3) $E(f,u(x+A)) = u(x+E(f,A) - \pi(x,A,u))$

Proposition 5. *[7]* $\pi(x,A,u) \approx \frac{1}{2} r_u(x+E(f,A))Var(f,A)$

Let u_1, u_2 be the utility functions of two agents such that $u_1' > 0$, $u_2' > 0$, $u_1'' < 0$, $u_2'' < 0$. We denote $r_1 = r_{u_1}$ and $r_2 = r_{u_2}$.

Proposition 6. *[7] The following assertions are equivalent:*

(a) $r_1(x) \geq r_2(x)$ *for any* $x \in \mathbf{R}$

(b) For any $x \in \mathbf{R}$ *and for any fuzzy number A,* $\pi(x,A,u_1) \geq \pi(x,A,u_2)$.

The above result is the possibilistic analogue of Pratt theorem [10]. It shows how using the Arrow–Pratt index one can compare the aversions to possibilistic risk of the two agents.

The following proposition establishes a connection between the possibilistic risk aversion and prudence.

Proposition 7. *The following assertions are equivalent:*

(i) For any fuzzy number A, the possibilistic risk premium $\pi(x,A,u)$ *is decreasing in wealth:* $x_1 \leq x_2$ *implies* $\pi(x_2,A,u) \leq \pi(x_1,A,u)$;

(ii) For all $x \in \mathbf{R}$, $P_u(x) \geq r_u(x)$ *(prudence is larger than absolute risk aversion).*

Proof. Let A be a fuzzy number with $[A]^\gamma = [a_1(\gamma), a_2(\gamma)]$, $\gamma \in [0,1]$.

From [3] it follows:

$u(x+E(f,A) - \pi(x,A,u)) = \frac{1}{2}\int_0^1 [u(x+a_1(\gamma)) + u(x+a_2(\gamma))]f(\gamma)d\gamma$

Deriving and applying again (3) for $g = -u'$ one obtains:

$(1-\pi'(x,A,u))u'(x+E(f,A) - \pi(x,A,u)) = -\frac{1}{2}\int_0^1 [g(x+a_1(\gamma))+$
$g(x+a_2(\gamma))]f(\gamma)d\gamma =$
$= -E(f,g(x+A)) = -g(x+E(f,A) - \pi(x,A,g))$

From these equalities it follows:
$\pi'(x,A,u) = \frac{g(x+E(f,A)-\pi(x,A,g))-g(x+E(f,A)-\pi(x,A,u))}{u'(x+E(f,A)-\pi(x,A,u))}$

By hypothesis, $u' > 0$ and g is strictly increasing, thus the following assertions are equivalent:
• $\pi(x,A,u)$ is decreasing in x
• For all x, $\pi'(x,A,u) \leq 0$
• For all x, $g(x+E(f,A)-\pi(x,A,g)) \leq g(x+E(f,A)-\pi(x,A,u))$
• For all x, $\pi(x,A,g) \geq \pi(x,A,u)$

By these equivalences and Proposition 6, the following assertions are also equivalent:
• For any fuzzy number A, $\pi(x,A,u)$ is decreasing in x
• For any fuzzy number A and $x \in \mathbf{R}$, $\pi(x,A,g) \geq \pi(x,A,u)$
• For any $x \in \mathbf{R}$, $r_g(x) \geq r_u(x)$

Since $P_u(x) = r_g(x)$ the equivalence of assertions (i) and (ii) follows. □

The (probabilistic) *precautionary premium* was introduced in [8] as "a measure of the strength of precautionary saving motive". We define now a similar notion in a possibilistic context.

Let v be a utility function of class C^3 with $v' > 0$, $v'' < 0$ and $v''' > 0$. The *possibilistic precautionary premium* $\phi(x,A,v)$ associated with wealth x, a fuzzy number A representing the risk and the utility function v is the unique solution of the equation:
(4) $E(f,v'(x+A)) = v'(x+E(f,A)-\phi(x,A,v))$

One notices that $\phi(x,A,v) = \pi(x,A,-v')$ therefore we can apply to $\phi(x,A,v)$ all the results valid for possibilistic risk premium. In particular, Propositions 5, 6 and 7 lead to

Proposition 8. $\phi(x,A,v) \approx \frac{1}{2}P_v(x+E(f,A))Var(f,A)$

Proposition 9. *Let v_1, v_2 be two utility functions of class C^3 with $v_1' > 0$, $v_2' > 0$, $v_1'' < 0$, $v_2'' < 0$, $v_1''' > 0$, $v_2''' > 0$. The following assertions are equivalent:*
(a) $P_{v_1}(x) \geq P_{v_2}(x)$ for any $x \in \mathbf{R}$
(b) For any $x \in \mathbf{R}$ and for any fuzzy number A, $\phi(x,A,v_1) \geq \phi(x,A,v_2)$

Proposition 10. *Let v be a utility function of class C^4 with $v' > 0$, $v'' < 0$, $v''' > 0$, $v^{iv} < 0$. The following assertions are equivalent:*
(i) For any fuzzy number A, the possibilistic precautionary premium $\phi(x,A,v)$ is decreasing in x;
(ii) $-\frac{v^{iv}(x)}{v'''(x)} \geq -\frac{v'''(x)}{v''(x)}$ for any $x \in \mathbf{R}$

We return to the model of precautionary saving of Section 3, assuming that v has the class C^3 and $v' > 0$, $v'' < 0$, $v''' > 0$. By the optimum condition (8) of Section 3 and equation (4) from above, it follows:
$u'(y_0 - s^*) = (1+r)E(f,v'((1+r)s^* + A))$

$$= (1+r)v'((1+r)s^* + E(f,A) - \phi((1+r)s^*,A,v))$$

We consider the case $r = 0$ and $u = v$. Then $u'(y_0 - s^*) = u'(s^* + E(f,A) - \phi(s^*,A,v))$, from where, taking into account that u' is injective it follows:

$s^* = \frac{1}{2}(y_0 + \phi(s^*,A,v) - E(f,A))$

5 Concluding Remarks

The possibilistic approach of the optimal saving problem is founded on the hypothesis that risk situations are represented by fuzzy numbers, and consumers are described by their utility functions. The formulation of the optimal saving problem uses the notion of possibilistic expected utility from [7].

This paper contains the following contributions:

- the characterization of the concavity of continuous utility functions by a possibilistic Jensen–type inequality;
- the definition of the notion of possibilistic precautionary saving and the characterization of prudence condition (in the sense of Kimball) by the positivity of possibilistic precautionary saving;
- the relation between possibilistic risk aversion and prudence;
- the definition of possibilistic precautionary premium as "strength" of possibilistic precautionary saving, its approximate calculation and its use to compare the degrees of absolute prudence of two consumers.

References

1. Arrow, K.J.: Essays in the Theory of Risk Bearing. North Holland, Amsterdam (1970)
2. Carlsson, C., Fullér, R.: On possibilistic mean value and variance of fuzzy numbers. Fuzzy Sets Syst. 122, 315–326 (2001)
3. Carlsson, C., Fullér, R.: Possibility for Decision. Springer (2011)
4. Drèze, J., Modigliani, F.: Consumption decision under uncertainty. J. Economic Theory 5, 308–355 (1972)
5. Dubois, D., Prade, H.: Possibility Theory. Plenum Press, New York (1988)
6. Eeckhoudt, L., Gollier, C., Schlesinger, H.: Economic and Financial Decisions under Risk. Princeton University Press (2005)
7. Georgescu, I.: Possibility Theory and the Risk. Springer (2012)
8. Kimball, M.S.: Precautionary saving in the small and in the large. Econometrica 58, 53–73 (1990)
9. Leland, H.: Saving and uncertainty: the precautionary demand for saving. Quarterly J. Economics 82, 465–473 (1968)
10. Pratt, J.: Risk aversion in the small and in the large. Econometrica 32, 122–130 (1964)
11. Rudin, W.: Analiză Reală şi Complexă. Ediţia a III-a, Theta, Bucharest (1999)
12. Sandmo, A.: The effect of uncertainty on saving decision. Rev. Economic Studies 37, 353–360 (1970)
13. Zadeh, L.A.: Fuzzy sets as a basis for a theory of possibility. Fuzzy Sets Syst. 1, 3–28 (1978)

Semantic Annotation and Retrieval of Services in the Cloud

Miguel Ángel Rodríguez-García[1], Rafael Valencia-García[1],
Francisco García-Sánchez[1], José Javier Samper-Zapater[2], and Isidoro Gil-Leiva[1]

[1] Departamento de Informática y Sistemas,
Universidad de Murcia. Campus de Espinardo 30100 Murcia, Spain
{miguelangel.rodriguez,valencia,frgarcia,isgil}@um.es
http://www.um.es

[2] Departament d'Informàtica,
Escola Tècnica Superior d'Enginyeria, Universitat de València, Avda. de la Universidad,
s/n, 46100 Burjassot, Valencia, Spain
jose.j.samper@uv.es
http://www.uv.es

Abstract. Recently, the economy has taken a downturn, which has forced many companies to reduce their costs in IT. This fact has, conversely, benefited the adoption of innovative computing models such as cloud computing, which allow businesses to reduce their fixed IT costs through outsourcing. As the number of cloud services available on the Internet grows, it is more and more difficult for companies to find those that can meet their needs. Under these circumstances, enabling a semantically-enriched search engine for cloud solutions can be a major breakthrough. In this paper, we present a fully-fledged platform based on semantics that (1) assist in generating a semantic description of cloud services, and (2) provide a cloud-focused search tool that makes use of such semantic descriptions to get accurate results from keyword-based searches. The proposed platform has been tested in the ICT domain with promising results.

1 Introduction

The future Internet will be based on services and this new trend will have significant impact on domains such as e-Science, education and e-Commerce. Consequently, the Web is evolving from a mere repository of information to a new platform for business transactions and information interchange. Large organizations are increasingly exposing their business processes through Web services technology for the large-scale development of software, as well as for the sharing of their services within and outside the organization. New paradigms for software and services engineering, such as Software-as-a-Service (SaaS) and the cloud computing model, promise to create new levels of efficiency through large-scale sharing of functionality and computing resources.

S. Omatu et al. (Eds.): *Distrib. Computing & Artificial Intelligence*, AISC 217, pp. 69–77.
DOI: 10.1007/978-3-319-00551-5_9

Cloud computing is a technological paradigm that permits to offer computing services over the Internet [Zhang et al., 2010]. In the current socio-economic climate, the affordability of cloud computing has gained popularity among today's innovations. Under these circumstances, more and more cloud services become available. Consequently, it is becoming increasingly difficult for service consumers to find and access those cloud services that fulfill their requirements. Semantic approaches have proven to be very effective in improving search processes [Vidoni et al., 2011]. However, providing semantic descriptions for all the cloud solutions currently available on the Internet is a very time-consuming task. Natural language processing (NLP) tools can help in automating the translation of the existent cloud-related natural language descriptions into semantically equivalent ones. In this paper, we present a semantic-based platform to annotate and retrieve services in the cloud.

In last decade several semantic annotation systems have been developed. However, as of today there is still not a standard approach for semantic annotation [Uren et al., 2006]. For this reason, semantic annotation systems have been classified based on some parameters such as 'standard format', 'ontology support', 'document evolution' and 'automation' [Uren et al., 2006]. Concerning 'standard formats', several formats are recommended by the World Wide Web Consortium to build ontologies. The most extended formats in the context of semantic annotation are RDF, RDF Schema, and OWL. The two former formats are used by the following approaches Armadillo [Chapman et al., 2005], CREAM [Handschuh and Staab, 2003]. OWL, on the other hand, is supported by others tools such as CERNO [Kiyavitskaya et al., 2009], EVONTO [Tissaoui et al., 2011], and KIM [Popov et al., 2003]. The application proposed here is also based on OWL.

Additionally, one property that is often desired in the scope of semantic annotation is multiple ontologies support, which allows to expand the knowledge to cover different domains. There are several tools such as KIM, CREAM or Armadillo that have been developed to support the use of multiple ontologies. In contrast, CERNO, S-CREAM [Handschuh et al, 2002] or EVONTO do not include this feature.

In the annotation context, there are a number of constraints related to computational cost guiding the way to process multiples ontologies, as follows: (i) the ontologies that are to be used must be merged, or (ii) annotations have to explicitly declare to which ontology they refer. Given performance and computational costs constraints, it is more appropriate to have several mid-size ontologies than a big merged ontology. In fact, some techniques have been proposed that split huge ontologies into several modules to make them more manageable for computers [Cuenca-Grau et al., 2007].

A further interesting property of ontology-based systems is that of ontology evolution. It refers to the process of changing the ontologies over time by, for example, adding or modifying new classes or individuals, or removing knowledge and ensuring the consistency of the annotations against the ontologies that are being modified. EVONTO, KIM, S-CREAM and CREAM implement an ontology evolution approach. Other semantic annotation tools such as CERNO or Armadillo do not cover this feature. In our work, support for both multiple ontology and ontology evolution is provided.

Almost all the current semantic annotation tools provide support for document evolution. For example, while Armadillo, CREAM, KIM and EVONTO update the annotations if a change is made in one or more documents, S-CREAM and CERNO do not.

Three kinds of semantic annotation systems can be distinguished: manual, fully automated and semi-automated. Manually annotating documents with semantic content is a very time-consuming task [Cravegna et al., 2002]. Therefore, the tendency is toward providing semi-automated tools within the current ontology-based annotation systems. Examples of this trend are CERNO and S-CREAM. There are also some fully-automated tools such as Armadillo, KIM and EVONTO.

The rest of the paper is organized as follows. The components that take part in the platform and their overall architecture are described in Section 2. In Section 3, a case use scenario in the information and communications technologies (ICT) domain and its evaluation is shown. Finally, conclusions and future work are put forward in Section 4.

2 Platform Architecture

The focus of the work described here is the development of a fully-fledged application for the semantically-enhanced search of services in the cloud. The architecture of the proposed approach is shown in Fig. 1. The approach is based on three main modules: (i) the semantic annotation module, (ii) the semantic indexing module, and (iii) the semantic search engine. In a nutshell the system works as follows: First, natural language descriptions of the services in the cloud are semantically represented and annotated. Then, from these annotations a semantic index is created using the classic vector space model. Finally, a semantic search engine permits to retrieve the matching services from keyword-based searches. Next, these components are described in detail.

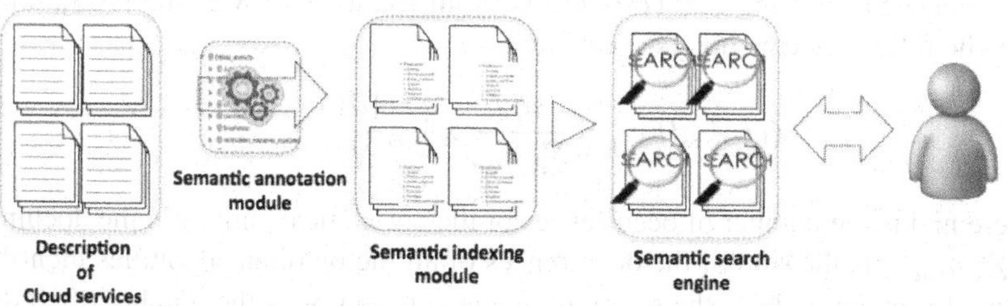

Fig. 1 System architecture

2.1 Semantic Annotation Module

This tool receives both domain ontologies and a natural language description of cloud services as inputs. Then, using a set of natural language processing (NLP) tools, it obtains a semantic annotation for the analyzed cloud services descriptions in accordance with the domain ontologies and Wordnet. This module is based on the methodology presented in [Valencia-García et al., 2008] and is composed of two main phases: the NLP phase and the semantic annotation phase.

The main aim of the NLP stage is the extraction of the morphosyntactic structure of each sentence. For this purpose, a set of NLP software tools, including a sentence detection component, a tokenizer, a set of POS (Part-Of-Speech) taggers, a set of lemmatizers and a set of syntactic parsers, have been developed. The GATE framework[1] has been employed to build some of the components required for the NLP phase. GATE is an infrastructure for developing and deploying software components that process human language.

During the second phase, the cloud services descriptions are annotated with the classes and individuals of the domain ontologies by following the process described next. First, the most important linguistic expressions are identified by means of statistical approaches based on the syntactic structure of the text. Then, for each linguistic expression, the system tries to determine whether such expression is an individual of a class of the domain ontology.

The outcome of the semantic annotation module is a list of semantic annotations defined in terms of the ontology. The classes and individuals in the annotations represent terms that have been extracted from the cloud services descriptions.

2.2 *Semantic Indexing Module*

In this module, the system retrieves all the annotated knowledge from the previous module and tries to create fully-filled annotations with this knowledge. This step is based on the work presented in [Castells et al., 2007]. Each annotation of each document is stored in a database and has a weight assigned. The annotation weight reflects how relevant the ontological entity is for the document meaning. Weights are calculated by using the TF-IDF algorithm [Salton and McGill, 1983], which uses the following equation (see equation 1).

$$(tf - idf)_{i,d} = \frac{n_{i,d}}{\sum_k n_{k,d}} \times \log \frac{|D|}{N_i} \tag{1}$$

where ni,d is the number of occurrences of the ontological entity i in the document d, $\sum_k n_{k,d}$ is the sum of the occurrences of all the ontological entities identified in the document d, |D| is the set of all documents and Ni is the number of all documents annotated with i.

In this scenario, the cloud services descriptions are the documents to be analyzed. For each description, an index is calculated based on the adaptation of the classic vector space model presented in [Castells et al., 2007]. Each service is represented as a vector in which each dimension corresponds to a separate ontological concept of the domain ontology. The value of each ontological concept dimension is calculated as follows (see equation 2).

---

1 http://gate.ac.uk/

$$(v_1, v_2, \dots, v_n)_d \; where \; v_i = \sum_{j=1}^{n} \frac{tf - idf_{j,d}}{e^{dist(i,j)}} \quad (2)$$

where dist(i,j) is the semantic distance between the concept i and concept j in the domain ontology. This distance is calculated by using the taxonomic (subclass_of) relationships of concepts in the domain ontology. So, the distance between a concept and itself is 0, the distance between a concept and its taxonomic parent or child is 1 and so on.

The outcome of the semantic indexing module is a list of semantic concepts sorted according to equation 2. Each assigned value represents both the relevance of the corresponding concept in all the analyzed descriptions and its relationships with other concepts in the domain ontology.

2.3 Semantic Search Engine

This module is responsible for finding services in the cloud from a keywords-based query. This process takes advantage of the semantic content and annotations previously gathered by the system.

First, users introduce a series of keywords and the system identifies which concepts in the domain ontology are referred by them. As it has been explained in the previous section, each service is represented as a vector in which each dimension corresponds to a separate concept of the domain ontology. Then, the semantic search engine calculates a similarity value between the query q and each service s. In order to do that, the cosine similarity is used (see equation 3):

$$sim(q, s) = cos\theta = \frac{q \times s}{|q| \times |s|} \quad (3)$$

A ranking of the most relevant cloud services that are related to the topics referenced in the query is then defined by using the similarity function showed in equation 3. The 's' vector is the one calculated by equation 2 for each service description. The 'q' vector, on the other hand, is the one created from the concepts extracted from the search engine query. The θ symbol represents the angle that separates both vectors, and describes the similitude grade between two documents.

3 Case Use: Annotation and Retrieval of ICT Services in the Cloud

The platform described in the previous section has been implemented and tested in the ICT domain. For this, in the first place, an ontology of the ICT domain has been developed. Next, around 100 different services with their description in natural language have been selected to be annotated by the system.

3.1 ICT Ontology

In this work, ontologies that semantically describe the functional properties of ICT applications have been studied. A representative example within this area is shown in [Lasheras et al., 2009], where an OWL (Web Ontology Language) ontology for requirements specification documents is developed and used for modeling reusable security requirements. The semantic description of the functionality of software components has been addressed in [Happel et al., 2006]. Here, the KOntoR system allows semantic descriptions of components to be stored in a knowledge base and semantic queries to be run on it (using the SPARQL language). The OWL ontology-based DESWAP system is presented in [Hartig et al., 2008]. In the context of this project, a knowledge base with comprehensive se-mantic descriptions of software and its functionalities was developed. Thus, by taking into account the shortcomings of developing a new ontology from scratch, the ontologies developed under the scope of the DESWAP project have been reused in this work to represent the features and functional properties of software projects.

3.2 Evaluation

During a first stage, representatives of an ICT organization are required to input a set of interesting services in the cloud with their descriptions. Then they are semantically annotated and stored in the ontology repository. The Sesame RDF repository, backed up by a MySQL database, has been used to implement the ontology repository.

Once the semantic indexes have been created, the experiment starts. This experimental evaluation aims at elucidating whether the semantic search engine module of the proposed platform is useful. Ten topic-based queries were issued. For each query, a set of cloud services was manually selected. At the same time, the semantic search engine was asked to perform the same task, in an automatic way. These results were then compared to those produced by the manual selection.

The average time taken for the human expert to complete each search throughout the cloud services repository, which contains 106 services, was 180,98 seconds. In contrast, the tool proposed in this paper executed each query at a rate of 0,78 seconds.

The final results of the experiment are shown in Table 1. The system obtained the best scores for queries of the topic "Databases", with a precision of 0.92, a recall of 0.89, and a F1 measure of 0.90. In general, the system obtains better results in precision (88% on average) than the results of recall (82% on average). Hence, these results are promising.

Table 1 Precision, recall and F1 of the experiment

Topics	Precision	Recall	F1
J2EE	0,89	0,81	0,85
application server	0,82	0,76	0,79
Databases	0,92	0,89	0,90
Enterprise information systems	0,9	0,83	0,86

4 Discussion and Conclusion

Semantic annotation and retrieval of cloud services is a challenging task and addresses the issue of finding the service or services with the functionality that meets the users' needs. In this paper, a semantic platform for the annotation of cloud services from their natural language descriptions and their retrieval from key-word based searches has been proposed. The system presented here automatically annotates different cloud services from their natural language description, which can be available in a number of document formats such as XML, HTML or PDF. Besides, the proposed platform has been implemented taking into account a multiontology environment (with OWL 2 ontologies) to be able to cope with several domains. Moreover, it supports the evolution of the source documents, thus maintaining the coherence between the natural language descriptions and the annotations, which are stored using a semantic Web-based model.

An experiment has been carried out with the objective of checking whether the system is useful for semantically annotating and retrieving services in the cloud. The results of the experiment are promising. However, they do not reflect the actual potential of the approach, since the experiment has been performed at a very small scale. Thus, a more complete and thorough validation of the system is planned by applying the system to a larger set of services and by using statistical methods for analyzing the results obtained.

Several issues remain open for future work. So far, the services have been analyzed by exploring their natural language descriptions. It could be beneficial to also use semantic information about their functionality by using ontologies that can describe these services as shown in [Ortegón-Cortázar et al., 2012]. Additionally, we are currently working on upgrading this system and converting it into a recommendation system in which users could set their preferences and the system would return only those services that are relevant to them in a particular domain. Finally, we plan to study the possibility of offering a search service also including an opinion mining engine, such as the one presented in [Peñalver-Martínez et al., 2011], which permits to obtain the sentimental classification of the services in order to provide information about their non-functional properties.

Acknowledgements. This work has been supported by the Spanish Ministry for Science and Innovation and the European Commission (FEDER / ERDF) through project SeCloud (TIN2010-18650).

References

1. Castells, P., et al.: An adaptation of the vector-space model for ontology-based information retrieval. IEEE Transactions on Knowledge and Data Engineering 19(2), 261–272 (2007)
2. Sam, C., et al.: Armadillo: Integrating knowledge for the semantic web. In: Proceedings of the Dagstuhl Seminar in Machine Learning for the Semantic Web (2005)
3. Ciravegna, F., Dingli, A., Petrelli, D., Wilks, Y.: User-system cooperation in document annotation based on information extraction. In: Gómez-Pérez, A., Benjamins, V.R. (eds.) EKAW 2002. LNCS (LNAI), vol. 2473, pp. 122–137. Springer, Heidelberg (2002)
4. Cuenca-Grau, B., et al.: Extracting Modules from Ontologies: A Logic-based Approach. In: Proc. of the 3rd OWL Experiences and Directions Workshop. CEUR, vol. 258 (2007)
5. Handschuh, S., Staab, S., Ciravegna, F.: S-CREAM – Semi-automatic CREAtion of Metadata. In: Gómez-Pérez, A., Benjamins, V.R. (eds.) EKAW 2002. LNCS (LNAI), vol. 2473, pp. 358–372. Springer, Heidelberg (2002)
6. Handschuh, S., Staab, S.: Cream: Creating metadata for the semantic web. Comput. Networks 42(5), 579–598 (2003)
7. Happel, H.-J., et al.: KOntoR: An Ontology-enabled Approach to Software Reuse. In: Proc. of the 18th Int. Conf. on Software Engineering and Knowledge Engineering, SEKE, San Francisco (2006)
8. Hartig, O., et al.: Designing Component-Based Semantic Web Applications with DESWAP. In: Proceedings of the Poster and Demonstration Session at the 7th International Semantic Web Conference (ISWC), Karlsruhe, Germany (2008)
9. Kiyavitskaya, N., et al.: Cerno: Light-weight tool support for semantic annotation of textual documents. Data Knowl. Eng. 68(12), 1470–1492 (2009)
10. Lasheras, J., et al.: Modelling Reusable Security Requirements based on an Ontology Framework. Journal of Research and Practice in Information Technology 41(2), 119–133 (2009)
11. Ortegón-Cortázar, G., et al.: Adding semantics to cloud computing to enhance service discovery and access. In: do Nascimento, R.P.C. (ed.) Proceedings of the 6th Euro. American Conference on Telematics and Information Systems (EATIS 2012), pp. 231–236. ACM, New York (2012), http://doi.acm.org/10.1145/2261605.2261639, doi:10.1145/2261605.2261639
12. Popov, B., Kiryakov, A., Kirilov, A., Manov, D., Ognyanoff, D., Goranov, M.: KIM – Semantic Annotation Platform. In: Fensel, D., Sycara, K., Mylopoulos, J. (eds.) ISWC 2003. LNCS, vol. 2870, pp. 834–849. Springer, Heidelberg (2003)
13. Peñalver-Martínez, I., Valencia-García, R., García-Sánchez, F.: Ontology-Guided Approach to Feature-Based Opinion Mining. In: Muñoz, R., Montoyo, A., Métais, E. (eds.) NLDB 2011. LNCS, vol. 6716, pp. 193–200. Springer, Heidelberg (2011)
14. Salton, G., McGill, M.J.: Introduction to modern information retrieval. McGraw-Hill (1983) ISBN 0070544840

15. Tissaoui, A., et al.: EVONTO: Joint evolution of ontologies and semantic annotations (short paper). In: Dietz, J. (ed.). Dans: International Conference on Knowledge Engineering and Ontology Development (KEOD 2011). INSTICC - Institute for Systems and Technologies of Information, Control and Communication, pp. 1–6 (2011)
16. Uren, V., et al.: Semantic Annotation for Knowledge Management: Requirements and a Survey of the State of the Art. Journal of Web Semantics 4(1), 14–28 (2006)
17. Valencia-García, R., et al.: A knowledge acquisition methodology to ontology construction for information retrieval from medical documents. Expert Systems 25(3), 314–334 (2008)
18. Vidoni, R., et al.: An intelligent framework to manage robotic autonomous agents. Expert Systems with Applications 38(6), 7430–7439 (2011)
19. Zhang, Q., et al.: Cloud computing: state-of-the-art and research challenges. J. Internet Serv. Appl. 1(1), 7–18 (2010)

Simulating a Team Behaviour of Affective Agents Using Robocode

António Rebelo[1], Fábio Catalão[1], João Alves[1], Goreti Marreiros[3,4], Cesar Analide[1,2], Paulo Novais[1,2], and José Neves[1,2]

[1] Universidade do Minho, Campus de Gualtar, 4700 Braga, Portugal
{pg19827,pg19832,pg20688}@alunos.uminho.pt
[2] CCTC – Centro de Ciências e Tecnologias da Computação
{analide,pjon,jneves}@di.uminho.pt
[3] GECAD – Knowledge Engineering and Decision Support Group, Porto, Portugal
mgt@isep.ipp.pt
[4] Institute of Engineering – Polytechnic of Porto, Porto, Portugal

Abstract. The study of the impact of emotion and affect in decision making processes involved in a working team stands for a multi-disciplinary issue (e.g. with insights from disciplines such as Psychology, Neuroscience, Philosophy and Computer Science). On the one hand, and in order to create such an environment we look at a team of affective agents to play into a battlefield, which present different emotional profiles (e.g. personality and mood).On the other hand, to attain cooperation, a voting mechanism and a decision-making process was implemented, being Robocode used as the simulation environment. Indeed, the results so far obtained are quite satisfying; the agent team performs quite well in the battlefield and undertakes different behaviours depending on the skirmish conditions.

1 Introduction

Traditionally, emotions and affects have been ignored in classic decision making methods [1]. However, in the last years, researchers of distinct areas (e.g. Psychology, Neuroscience and Philosophy) have begun to explore the role of the affect as a positive influence on human decision-making. Currently, the representation of human emotions in artificial environments is a common issue in Artificial Intelligence.

In 2003, Ortony [2] discussed the main characteristics that an agent must have to be considered believable. There, it was defended that agents should have consistent motivational and behaviours states. In order to ponder this option, it is reinforced that agents need not only a robust model of emotions but also have to implement a proper model of personality, which will contribute to give them coherence, consistence and some degree of predictability.

In this work it is proposed the development of dissimilar affective robots and the simulation of their behaviour and performance in a battlefield environment.

S. Omatu et al. (Eds.): *Distrib. Computing & Artificial Intelligence,* AISC 217, pp. 79–86.
DOI: 10.1007/978-3-319-00551-5_10

Undeniably, it will be shaped robots with different emotional profiles and analysed their behaviour, either when act per se or when are part of a team. In order to create robots that may in a consistent way express emotions felt during the course of a battle, and to make the system more similar to human perception, some insights lent from the field of psychology will be considered [2,17,19]. As a simulation environment it will be used Robocode, whose objective is to code a robot to beat others in a battlefield [3]. It provides a simple setting, which allows for an easy understanding of robot coding concepts. Robocode is also a very flexible platform, which admits the use of Artificial Intelligence related methodologies and strategies for problem solving in teamwork [4].

This paper comprises in section 2 a brief description of the psychological concepts involved in this work, and in section 3 an overview of Robocode environment and the main movements allowed to robots in a battlefield. Section 4 and 5 presents and discuss our approach to create affective robots in Robocode. Finally, some conclusions are presented in section 6.

2 Background

In this section it will be given a brief description of the main psychological concepts that will be incorporated into the affective.

2.1 Affect

It is often found in the literature the use of alternative terms such as emotion, affect and mood. Here it is adopted the definition of Forgas [5], that recognises affect as the most generic and used term to refer to mood and emotion. Emotion is normally referred to as an intense experience of short duration (seconds to minutes), with a specific origin, and in general, any person is conscious of the situation. On the other hand, moods have a propensity to be less intensive, longer lasting (hours or even days) and in general remain unacquainted. Moods may be caused by an intense or recurrent emotion, or yet by environmental aspects.

The psychology literature is full of examples on how emotions affect the decision-making process [6,7,8]. The frequently changing emotional states of an individual influence their behaviour and their interactions with those around him/her, which in the present context are other group members. For example, the phenomenon of emotional contagion is the tendency to express and feel emotions that are similar to and influenced by others. This phenomenon may be analysed as the modal mood of the group, in terms of a particularly salient emotion that one of the group members is feeling [9].

We propose to incorporate emotions into our system using the OCC model of Ortony, Clore and Collins [10], an archetypal that is widely used for emotion simulation of embodied agents [11-15]. In OCC, agent's concerns are divided into goals (i.e. desired states of the world), standards (i.e. ideas on how people should act) and preferences (i.e. likes and dislikes), and distributed across twenty-two properly representable emotion categories or "types". To reduce the complexity in his original model, Ortony proposed a simplified one with 12 (twelve) emotional categories divided into 6 (six) positive (i.e. joy, hope, relief, pride, gratitude, and love) and 6 (six) negative categories (i.e. distress, fear, disappointment, remorse, anger, and hate) [2]. We expect this reduced model to be adequate for our purposes.

It is also possible to find other approaches to infer agents moods, most of them related to the set of agent experimented emotions. Definitely, in this work it will be used the approach proposed by Albert of Mehrabian, that ponders that agent's mood is calculated according to 3 (three) variables, namely Pleasure (P), Arousal (A) and Dominance (D) [16].

2.2 Personality

The differences in personality manifest themselves in different ways in all aspects of psychological life (e.g. affect, behaviour, motivation, perception, cognition). Moreover, it may be stated that personality has a key role in the conduct of a particular agent. Agent's individual differences and personalities will interfere and influence aspects of their psychology, such as the way it perceives emotions, feels affection, behaviours, motivations, and cognition [17][18]. Despite the high degree of disagreement around the best way to represent an agent personality, there is some support in favour of the Five Factor Model (FFM) [19], which is the personality model more common in computer applications, and according to it, the individual differences are captured in the form of traits, i.e., Openness, Conscientiousness, Extraversion, Agreeableness and Neuroticism [4]. For these reasons, FFM was chosen as the model to be used in this work.

3 Robocode Environment

The ROBOCODE is an event driven environment (e.g. robot bumps into a wall and robot hitting another robot). There are 5 (five) main devices which allow for the robot control, i.e., movement (forward, backward), tank-body rotation, gun-rotation, radar-rotation. A battle in Robocode is composed by several rounds; at the beginning is assigned to each robot a fixed level of energy. During the battle, the robots' energy level may decrease and/or increase (e.g. hitting other robot increases the energy level, bump a wall or getting hit by a shell decreases its energy level). Once the energy level gets to zero, the robot is simply dismissed; the same happens if other robot hits its peers. To consider a round finished, only one robot may remain in the battlefield.

4 Affective Team

4.1 *Robots Affective Model*

The emotional system is built on two important factors, specifically the emotions and the mood of the robot. These are two distinct components in terms of intensity and duration. The mood of a robot is of low intensity and long duration, while emotions have high intensity but are specific to a particular event, being therefore brief or transitory. The agent personality is paramount to establish its initial mood. In Figure 1, one may see the general approach pursue for the creation of the affective model, in which the environment stands for the framework in which our robot will be immersed, i.e., the physical environment and the other robots that are in the same somatic setting.

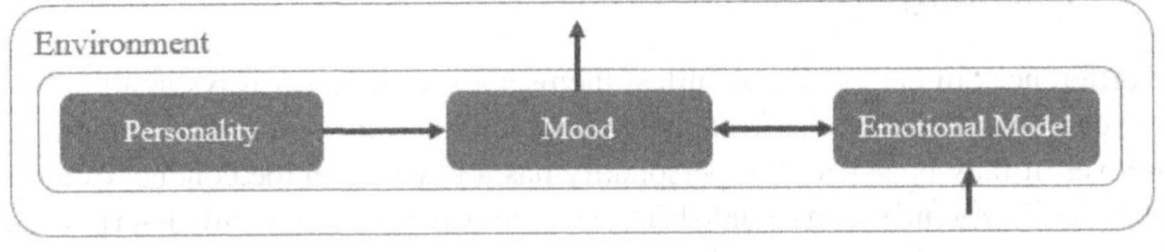

Fig. 1 The affective model

4.1.1 Modelling Emotions

Each emotion has an intensity value; however each robot feels emotions with a distinctive strength, which is contingent to several factors. The intensity of the emotion is given in terms of the equation [18]:

$$Intensity_{emotion} = \frac{\sqrt{(P)^2 + (A)^2 + (D)^2}}{\sqrt{3}}$$

These emotions are triggered through their own actions and also through interactions with other robots or even events. As it was mentioned before, the emotions considered in this work are identified in the revised version of the OCC model, i.e., joy, hope, relief, pride, gratitude, love, distress, fear, disappointment, remorse, anger, and hate [2]. Over time, the intensity of emotions decreases or decays, leading to a state of irrelevance; emotion decay is obtained by counting the number of rounds passed since the emotion was felt. The emotion intensity in a given robot is obtained by using the rule [18]:

$$I_{emotion\ felt\ (r)} = I * e^{-r} * n \quad (2)$$

i.e., the intensity of an emotion felt by a robot varies according to the number of counted *rounds* (*r*) since the emotion was triggered, the *Intensity* of emotion (*I*) and the value of *neuroticism* (*n*) of the robot in question.

4.1.2 Modelling Mood

Emotions change the current mood depending on its intensity and the personality of a robot. The mood of a robot is modelled based on the work of Mehrabian, in terms of *Pleasure* (*P*), *Arousal* (*A*) and *Dominance* (*D*), which forms the *PAD* space [11]. The *P* dimension is related to the emotional state of the robot, being positive or negative; the *A* scale denotes its level of physical and mental activity; the *D* one indicates its feeling of being in control. Mehrabian defined 8 (eight) types of mood based on *PAD* values, which were adopted to represent the mood of robots. Mehrabian also established a relationship between the *OCEAN* model of 5 (five) dimensions with the three-dimensional *PAD* space. Through this relationship it is possible to set the mood of the original robots according to their personality, i.e., the *OCEAN* = (*O*, *C*, *E*, *A*, *N*) model is used to obtain the robots personality values. Therefore, the initial mood of a robot may be calculated as follows [17]:

$$P = (O, C, E, A, N),\ where\ O, C, E, A, N \in [-1,1]$$

$$Mood = (P, A, D),\ where\ P, A, D \in [-1,1]$$

$$P = 0.59 \times A + 0.19 \times N + 0.21 \times E$$

$$A = -0.57 \times N + 0.30 \times A + 0.15 \times O$$

$$D = 0.60 \times E - 0.32 \times A + 0.25 \times O$$

Once the initial mood of a robot is computed, it is necessary to define how it is going to evolve. In one`s case the mood swings are due to the occurrence of events and the actions performed by a robot in combat; the emotions generated may be positive or negative, and follow the pattern presented in the OCC model. Over time, and verifying the absence of emotions, mood stabilizes at its initial values. In the present setting it was decided to use an adaptation of the simplified version of the OCC Model .The emotions Love and Hate have been removed due to their non-application in the context of this work.

4.2 Esteem

Due to the conduct of the robots on the ground and the events triggered by them, a robot goes through diverse emotional states, which may be positive, offering a good practice, or negative, resulting in discomfort. Having this in mind, it becomes obvious that a robot will have a good indebtedness with respect to a counterpart that causes good practices and little regard for one that only produces bad experiences.

On the other hand each robot team has a value of esteem for their equals in combat. This assessment is central, once it affects many of the actions of the robots. As an example, let us look at the selection of the next enemy to target. If one of the opponents has a very low esteem, it entertains a high probability of being chosen to be targeted by the voting robot, i.e., the values of esteem have a significant weight in the decision making process and in voting.

4.3 Decision, Votes and Leader

Each robot has its own personality and mood. In this way, they have different desires and goals. In order to achieve some kind of cooperation, it was necessary to create a leader and a decision mechanism able to focus on two key aspects of the battle: the next target of the opposing teams and the movement that should be adopted.

The leader has the responsibility of holding a referendum and responding to it by publishing the results. The leader should ask for the votes of the other members of the team, even of those that have been lethally wounded, which is the case when too many rounds have passed since the last referendum, or if the target of the team has been slaughtered. Upon the death of the team leader, the element with more energy will announce itself as the new leader and resume the voting mechanism. The entire team votes, giving a preference value to each of the antagonists to defeat and a preference value to each of the possible engagements to use. When the robot evaluates the scores of its enemies, it takes into account the following variables, i.e., its mood, the esteem for that opponent, the opponent's remaining energy and the distance to it.

During the analysis of the votes, the leader may disregard some opinions or give added weight to others. In order to accomplish this practice, the leader takes into account the esteem of the teammate who voted. If the teammate is having a good performance in combat, then the level of appreciation of the leader is high, and so the leader values its opinion. Otherwise, if the teammate is failing too many shots and hitting its own teammates, then the leader has a low level of esteem to it and its opinion is devalued against that of others. When the results of the voting process are treated, the team leader may either accept them and pronounce the results or, if the results are going in a different direction of its own interpretation of the situation on the ground, the team leader may proceed in order to reach its purposes. The probability of the group decision being manipulated by the leader is directly proportional to its value of thoughtfulness and inversely proportional to its value of agreeableness.

4.4 Movements

In 2009, Nakagawa work look at a reformation of the motion method to control affective nuances in robots [20]. In order to revise diverse types of motions without changing their meanings, this method uses three parameters: velocity and extension of motion and a basic posture. Following Nakagawa idea, the mood of the robots along the battle is shaped. Robots, when subjected to successive negative emotions tend to do more unwise decisions and to reduce teamwork. When robots are subjected to successive positive emotions they tended to be more relaxed and to increase teamwork.

The robots developed in this project may implement one of the following types of engagement: *defensive*, *opportunistic* and *tactical*. The *defensive* robot has as its main goal to keep itself at a secure distance from its opponents. The *opportunist* robot changes its self-confident attitude when is struck by a bullet or when someone crashes with it. This is a robot that times its attacks based on the level of security it feels. If it does not take damage in the last rounds, it feels confident and takes the opportunity to get closer to its enemies, improving its chances of striking. However, if an enemy manages to decrease its energy, the *opportunistic* robot retreats and starts to avoid its nemeses, hoping that its interest on it will fall and may look at another robot. A *tactical* robot is an entity that performs its actions based on a defined line of behaviour. This robot plans its navigation before the encounter, and executes it disregarding any outside interference. It relies on tactic and perception of the battlefield as it keeps itself near the borderlines of the conflict.

5 Simulation Analysis

According to the simulation results, a set of conclusions may be made based on the robots behavior. It can be said that emotions influenced the outcome of the simulations. The emotions shaped the mood of the robots along the battle, which in turn shaped the actions and votes consummated. Successive negative emotions felt by the robots make them more likely to do unwise decisions, bringing down teamwork, which is a typical sign of emotion instability. On the other hand, robots feeling consecutive positive emotions proved to be more relaxed and to cooperate as a team. Simulations have shown that personality played an important role as well, as robots with confident and cool personality would tolerate negative situations and stick to the plan, while others with a more neurotic personality would panic and stray away from the plan. This is consistent with the literature [21], as emotions have different intensities and durations depending on the individual personality. Based on the environment in which the simulation takes place, these types of behavior are in concordance with real life situations. Teams are able to act accordingly to the robots emotions and personality.

6 Conclusions and Future Work

In this work it were presented robots capable of feeling emotions and act accordingly, namely changing its behaviour. It stands for an approach to create an affective team of robots to be used in a battlefield; it was used the *PAD* mood space, which is able to support *OCEAN* and *OCC* models. Each robot of the team has its own personality and mood. In order to achieve cooperation, a voting mechanism and a decision-making process was implemented. The results are quite satisfying as the team nurses a very good performance on the battlefield and assumes diverse conducts that are contingent on the battle conditions. As future work, it is intended to create other types of movement that a team may endorse, as well as to develop a more multifaceted system to choose the governance.

References

1. Marreiros, G., Ramos, C., Neves, J.: Emotion and Group Decision Making in Artificial Intelligence. In: Cognitive, Emotive and Ethical Aspects of Decision-Making in Humans and in AI, vol. IV, pp. 41–46 (2005) ISBN 1-894613-86-4
2. Ortony, A.: On making believable emotional agents believable. In: Traple, R.P. (ed.) Emotions in Humans and Artefacts. MIT Press, Cambridge (2003)
3. Hartness, K.: Robocode: using games to teach artificial intelligence. J. Comput. Sci. Coll. 19(4), 287–291 (2004)
4. Howard, P., Howard, J.: The BIG FIVE Quickstart: An introduction to the five factor model of personality for human resource professionals. Center for Applied Cognitive Studies, Charlotte (2004)
5. Forgas, J.: Mood and judgment: The affect infusion model (AIM). Psychological Bulletin 117, 39–66 (1995)
6. Loewenstein, G., Lerner, J.S.: The role of affect in decision making. In: Handbook of Affective Sciences. Oxford University Press (2003)
7. Schwarz, N.: Emotion, cognition, and decision making. Cognition and Emotion 14(4), 433–440 (2000)
8. Barsade, S.: The Ripple Effect: Emotional Contagion and Its Influence on Group Behavior. Administrative Science Quarterly 47, 644–675 (2002)
9. Neumann, R., Strack, F.: Mood contagion: The automatic transfer of mood between persons. Journal of Personality and Social Psychology 79, 211–223 (2000)
10. Ortony, A., Clore, G.L., Collins, A.: The cognitive structure of emotions. Cambridge University Press, Cambridge (1988)
11. Gratch, J., Marsella, S.: Evaluating a computational model of emotion. Journal of Autonomous Agents and Multiagent Systems 11(1), 23–43 (2006)
12. Mourão, D., Paiva, A.: EToy: Building an affective physical interface. In: 2nd Workshop on Attitude, Personality and Emotions in User-Adapted Interaction (2001)
13. Lim, M.Y., Dias, J., Aylett, R., Paiva, A.: Improving Adaptiveness in Autonomous Characters. In: Prendinger, H., Lester, J.C., Ishizuka, M. (eds.) IVA 2008. LNCS (LNAI), vol. 5208, pp. 348–355. Springer, Heidelberg (2008)
14. Bída, M., Brom, C.: Towards a platform for the education in emotion modeling based on virtual environments. In: 3rd Workshop on Emotion and Computing, pp. 45–52 (2008)
15. Marreiros, G., Santos, R., Ramos, C., Neves, J.: Context-aware emotion-based model for group decision making. IEEE Intelligent Systems Magazine 25(2), 31–39 (2010)
16. Ryckman, R.: Theories of Personality. Thomson/Wadsworth (2004)
17. Mehrabian: Analysis of the Big-Five Personality Factors in Terms of the PAD Temperament Model. Australian Journal of Psychology 46 (1996)
18. Santos, R., Marreiros, G., Ramos, C., Neves, J., Bulas-Cruz, J.: Personality, Emotion, and Mood in Agent-Based Group Decision Making. IEEE Intelligent Systems 26(6), 58–66 (2011)
19. McCrae, R.R., John, P.O.: An introduction to the five-factor model and its applications. Journal of Personality 60, 175–215 (1992)
20. Nakagawa, K., Shinozawa, K., Ishiguro, H., Akimoto, T., Hagita, N.: Motion modification method to control affective nuances for robots. In: Int. Conf. Intelligent Robots and Systems, pp. 5003–5008 (2009)
21. Haifang, L., Haipeng, H., JunJie, C.: A new layered model of affect. In: Int. Asia Conf. on Informatics in Control, Automation and Robotics (CAR), pp. 261–264 (2010)

Qualitative Acceleration Model: Representation, Reasoning and Application

Ester Martinez-Martin[1], Maria Teresa Escrig[2], and Angel P. del Pobil[1]

[1] Universitat Jaume-I, Castellón, Spain
{emartine,pobil}@icc.uji.es
[2] Cognitive Robot, S.L. Parque Científico, Tecnológico y Empresarial (ESPAITEC), Universitat Jaume-I, Castellón, Spain
mtescrig@c-robots.com

Abstract. On the way to autonomous service robots, spatial reasoning plays a main role since it properly deals with problems involving uncertainty. In particular, we are interested in knowing people's pose to avoid collisions. With that aim, in this paper, we present a qualitative acceleration model for robotic applications including representation, reasoning and a practical application.

1 Introduction

Recent research is interested in building autonomous systems helping people in their daily tasks, specially when they are tedious and/or repetitive. These *common* tasks can involve poorly defined situations. On this matter, humans have a remarkable capability to solve them without any measurements and/or any computations. Familiar examples are parking a car, cooking a meal, or summarizing a story. That is, people make decisions mostly based on perceptual information, rather than accurate measurements [14]. Thus, qualitative reasoning properly fits this problem since it works on representation formalisms close to human conceptual schemata for reasoning about the surrounding physical environment (e.g. [13]).

Focusing on safety in Robotics, a keypoint is detecting and following all-embracing elements in order to avoid collisions, especially when they are human beings. Although some devices have been developed to avoid collisions, they considerably restrict the system's autonomy and flexibility. In addition, they present imprecise data and, for that reason, a qualitative model is required. In this paper, we present a new qualitative model of acceleration combined with orientation, which provides surrounding element's pose, allowing the system to properly avoid potential collisions.

In qualitative spatial reasoning, a particular aspect of the physical world, a magnitude, is considered. That is, a system of qualitative relationships between entities which cover that aspect of the world to some degree is developed. Examples of that can be found in many disciplines (e.g. geography [12], psychology [7], ecology [1],

S. Omatu et al. (Eds.): *Distrib. Computing & Artificial Intelligence,* AISC 217, pp. 87–94.
DOI: 10.1007/978-3-319-00551-5_11

biology [6], robotics [9] and Artificial Intelligence [3]). Actually, a qualitative representation of a magnitude results from an abstraction process and it has been defined as that representation that *makes only as many distinctions as necessary to identify objects, events, situations, etc. in a given context for that magnitude* in [11]. The way to define those distinctions depends on two different aspects:

1. **the level of *granularity***. In this context, *granularity* refers to a matter of precision in the sense of the amount of information which is included in the representation
2. **the distinction between *comparing* and *naming* magnitudes** (as stated in [2]). This distinction refers to the usual comparison between *absolute* and *relative*. From a spatial point of view, this controversy corresponds to the way of representing the relationships among objects (see Fig. 1). From the distinction between *absolute* and *relative* pointed out by Levinson [8], an object *b* is *any compared relationship* to another object *a* from the same Point of View (*PV*) when *comparing* magnitudes are considered. It is worth noting that the comparison depends on the orientation of both objects *with respect to* (wrt) the *PV*, since objects *a* and *b* can be at any orientation wrt the *PV*. On the other hand, *naming* magnitudes divides the magnitude of any concept into intervals (sharply or overlapped separated, depending on the context) such that qualitative labels are assigned to each interval. Note that the result of reasoning with this kind of regions can provide imprecision. This imprecision will be solved by providing disjunction in the result. That is, if an object can be found in several qualitative regions, q_i or q_{i+1} or ... or q_n, then all possibilities are listed as follows $\{q_i, q_{i+1}, \ldots, q_n\}$ by indicating this situation

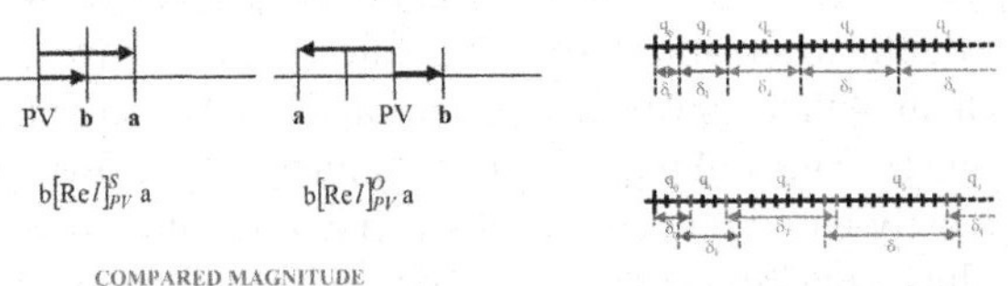

Fig. 1 An example of *compared* distances as represented in [4] (left) and an example of structure relations in *naming* magnitudes with sharply and overlapped separated qualitative areas (right)

From that starting point, and on the way to develop intelligent abilities to solve some service robotics problems, in this paper, we present a qualitative naming acceleration model including its qualitative representation, the reasoning process and a real robotic application. With that aim, the structure of this paper is as follows: the proposed *qualitative acceleration* model is analyzed in Section 2, while a practical application is described in Section 3. Finally, some conclusions and future work are presented in Section 4.

2 Qualitative Acceleration Model

The acceleration is the physical concept that measures how an object's speed or direction changes over time. Physically, it can be defined as:

$$Acceleration = \frac{Velocity}{Time} = \frac{Space}{Time^2} \tag{1}$$

2.1 Representation

The first issue to be solved concerns to how *acceleration* is represented. So, from the previous acceleration definition and focusing on developing its qualitative naming model, the acceleration representation will consist of three elements:

1. the **number of objects** implied in each relation (i.e. *arity*). From the physical definition of acceleration, the relationships to be defined imply two objects such that an object acts as reference and the other one is referred.
2. the **set of acceleration relations between objects**. It depends on the considered level of *granularity*. In a formal way, this set of relations is expressed by means of the definition of a *Reference System (RS)* composed of:
   - a *set of qualitative symbols* in increasing order represented by $Q = \{q_0, q_1, ..., q_n\}$, where q_0 is the qualitative symbol closest to the *Reference Object (RO)* and q_n is the one furthest away, going to infinity. Here, by cognitive considerations, the acceptance areas have been chosen in increasing size
   - the *structure relations*, $\Delta r = \{\delta_0, \delta_1, ..., \delta_n\}$, describe the acceptance areas for each qualitative symbol q_i. So, δ_0 corresponds to the acceptance area of qualitative symbol q_0; δ_1 to the acceptance area of symbol q_1 and so on. These acceptance areas are quantitatively defined by means of a set of close or open intervals delimited by two extreme points: the initial point of the interval j, δ_j^i, and the ending point of the interval j, δ_j^e, such that the structure relations are rewritten as:

   $$\begin{cases} \Delta r = \left\{ \left[\delta_0^i, \delta_0^e\right[, \left[\delta_1^i, \delta_1^e\right[, \ldots, \left[\delta_n^i, \delta_n^e\right[\right\} \text{ if open intervals are considered} \\ \Delta r = \left\{ \left[\delta_0^i, \delta_0^e\right], \left[\delta_1^i, \delta_1^e\right], \ldots, \left[\delta_n^i, \delta_n^e\right] \right\} \text{ otherwise} \end{cases}$$

   Therefore, the acceptance area of a particular acceleration entity, $AcAr(entity)$, is δ_j if its value is between the initial and ending points of δ_j, that is, $\delta_j^i \leq value(entity) \leq \delta_j^e$
3. the **operations**. The number of operations associated to a representation corresponds to the possible change in the PV. In this case, as only two objects are implied in the acceleration relationships, only one operation can be defined: *inverse*.

Note that a particularity of acceleration is that the values for the intervals in which the workspace is divided into, can be both positives and negatives and this feature has to be considered when the reasoning process is designed.

2.2 The Basic Step of the Inference Process

The *Basic Step of the Inference Process (**BSIP**)* for the acceleration concept can be defined as: given two acceleration relationships between three spatio-temporal entities *a*, *b* and *c*, we want to find the acceleration relationship between the two entities which is not initially given. However, it is important to take into account that the relative movement of the implied objects can be at any direction. For that reason, the BSIP is studied integrating the acceleration concept with a qualitative orientation model. Note that, for this case, the qualitative orientational model of Freksa and Zimmerman [5] has been redefined as depicted in Fig. 2 since the RO is always on the object *b*. In that way, it is possible to reason with the extreme angles of the defined structure relations for the *Orientation Reference System (**ORS**)*.

Given that qualitative areas are defined by intervals, we use the two operations to add and subtract qualitative intervals presented in our previous work [10]. In particular, the functions to be performed are *qualitative_sum* (obtains the sum of two qualitative intervals), *qualitative_difference* (provides the subtraction of two qualitative intervals), *Find_UB_qualitative_sum* (obtains the qualitative interval corresponding to the Upper Bound (UB) of the qualitative sum of two qualitative intervals) and *Find_LB_qualitative_difference* (provides the qualitative interval corresponding to the Lower Bound (LB) of the qualitative subtraction of two qualitative intervals). In addition, five new functions are defined: *pythagorean_theorem_LB* and *pythagorean_theorem_UB* that obtain the qualitative interval respectively corresponding to the lower and upper bounds when the Pythagorean theorem is applied; *intermediate_orientation* provides the orientations existing between the two ones given as input (e.g. *intermediate_orientation(right, straight-front)* will be *front-right*); *open_interval*, from an orientation defined with a closed interval and another with an open interval, returns that corresponding to an open interval; and, *all_orientation_relationships* returns all the defined qualitative orientations.

Therefore, the BSIP for acceleration has been solved as follows (see Fig. 3): when any acceleration relationship is zero, both acceleration and orientation will be equal to the other involved relationship. When the two acceleration relationships have the same orientation, the resulting relationship has the same orientation and its value corresponds to the qualitative sum of both relationships. On the contrary, if the relationships have an opposite orientation, the resulting relationship will be obtained as their qualitative difference and its orientation will be equal to that of the highest acceleration value (in absolute values). On the other hand, in the case both

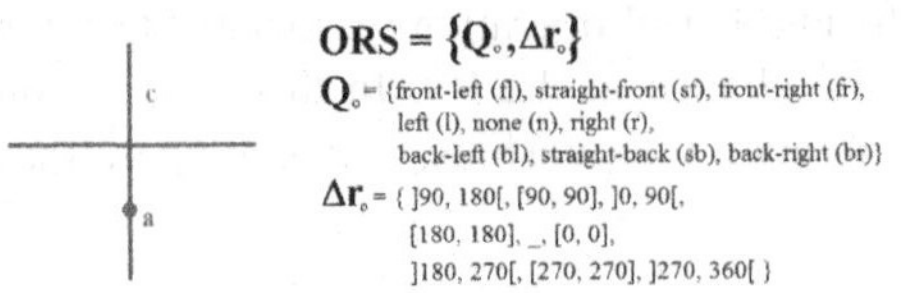

c
a
ORS = {Q$_o$,Δr$_o$}
Q$_o$ = {front-left (fl), straight-front (sf), front-right (fr), left (l), none (n), right (r), back-left (bl), straight-back (sb), back-right (br)}
Δr$_o$ = {]90, 180[, [90, 90],]0, 90[, [180, 180], _, [0, 0],]180, 270[, [270, 270],]270, 360[}

Fig. 2 Redefinition of the Orientation Reference System (ORS) of Freksa and Zimmerman [5] by means of its set of qualitative symbols (Q_O) and its structure relations (Δ_O)

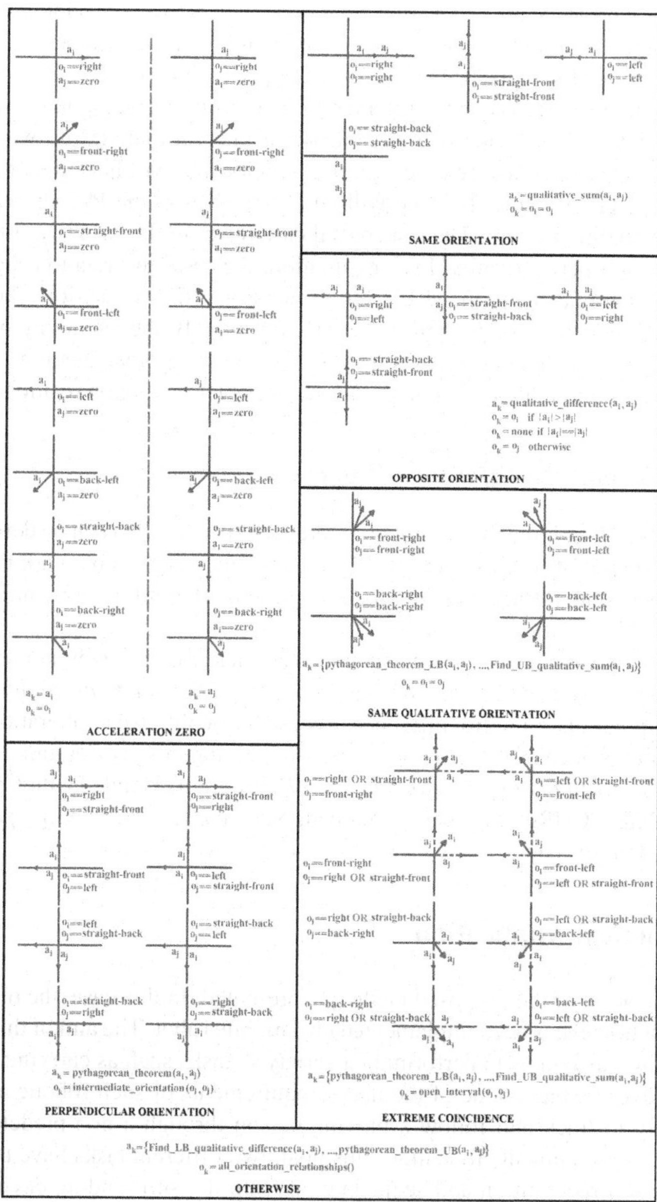

Fig. 3 Graphical resolution of the BSIP for acceleration, where a_i and a_j represents the acceleration relationships given as input with their corresponding orientation relationships (o_i and o_j); and a_k and o_k are the resulting acceleration and orientation relationships

relationships have the same orientation but it corresponds to an open interval, the resulting relationship has the same orientation, although its value will be a disjunction of acceleration relationships from the result of applying the Pythagorean theorem to the UB qualitative sum. When the orientation relationships corresponds to an open and a close interval such that one extreme of an interval matches up with an extreme of the other interval, then the resulting relationship will have the orientation of the open interval, while its value will be obtained from the Pythagorean theorem and the qualitative sum. The last special case refers to the case two orientation relationships are perpendicular. In that situation, the resulting relationship results from the Pythagorean theorem, whereas its orientation is the orientation relationship between the orientations of the initial relationships. Finally, the remaining situations are solved by means of qualitative difference and the Pythagorean theorem. With regard to its orientation, it corresponds to all the possible orientation relationships.

2.3 *The Complete Inference Process*

From the BSIP definition, the Complete Inference Process (CIP) can be defined. So, mainly, it consists of repeating the BSIP as many times as possible with the initial information and the information provided by some BSIP until no more information can be inferred.

As knowledge about relationships between entities is often given in the form of *constraints*, the CIP can be formalized as a *Constraint Satisfaction Problem (CSP)*. So, the computation of the full inference process for qualitative acceleration can be viewed as an instance of the CSP. So, on way of solving this acceleration inference process, a *Constraint Logic Programming (CLP)* program extended with *Constraint Handling Rules* (CHRs) has been developed. Nevertheless, for lack of space, it is not provided in this paper.

3 A Practical Application

A real application of the proposed method is presented. In this case, the qualitative acceleration model has been implemented on a mobile robot. The aim of this system is to assist human beings in performing a variety of tasks such as carrying person's tools or delivering parts. One of the major requirements of such robotic assistants is the ability to track and follow a moving person through a non-predetermined, unstructured environment. To achieve this goal, two different tasks have to be performed: person recognition and segmentation from the surrounding environment, and motion control to follow the person using the recognition results. In particular, we have focused on developing the qualitative reasoning method to achieve the second task.

For that, an indoor pan-tilt-zoom (PTZ) camera was mounted on a Pioneer 3-DX mobile platform without restricting its autonomy and flexibility. The core of the PTZ system is a Canon VC-C4 analog colour camera with a resolution of $320x240$ pixels, which is integrated with the mobile platform hardware.

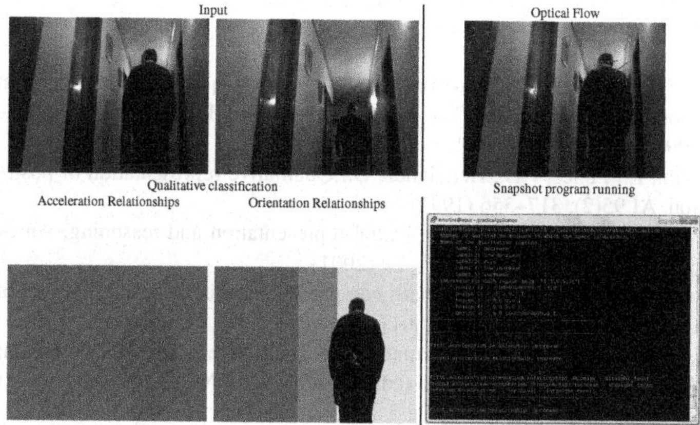

Fig. 4 Results obtained with the real robot when the qualitative acceleration relationships are labelled as $Q = \{\text{decrease, low\_decrease, zero, low\_increase, increase}\}$. On the other hand, orientation relationships correspond to the modified Freksa and Zimmermann's approach such that fl is coded by red, sf by green, and fr by yellow.

So, on the one hand, the system knows both its acceleration and its orientation through the information gathered from its motors. On the other hand, an image processing based on optical flow provides an estimation of the acceleration and orientation relationships corresponding to the person to be followed. Therefore, from these two relationships (the one obtained by the robotic system itself and the other corresponding to the person from image processing), the system is able to determine the required acceleration-orientation relationship that allows it to know the required trajectory change to properly follow and assist that person. An example of the obtained results can be seen in Fig. 4.

4 Conclusions and Future Work

In this paper, we have developed a qualitative model for physical acceleration such that acceleration and orientation are combined, the basic step of the inference process is expressed in terms of qualitative sums and differences, and, given that knowledge about relationships between entities is often provided in the form of *constraints*, the complete inference process is formalized as a *Constraint Satisfaction Problem (CSP)*.

The qualitative acceleration model defined in these terms allows to automatically estimating people's pose around the system and, therefore, avoiding collisions. This results in safer and more accurate robotic systems, in spite of sensor imprecision.

As future work we will investigate the development of new qualitative models based on intervals of aspects such as: time, weight, body sensations, etc. to provide robots with intelligent abilities to solve service robotics problems.

References

1. Cioaca, E., Linnebank, F., Bredeweg, B., Salles, P.: A qualitative reasoning model of algal bloom in the danube delta biosphere reserve (ddbr). Ecol. Informatics 4(5-6), 282–298 (2009)
2. Clementini, E., Felice, P.D., Hernández, D.: Qualitative representation of positional information. AI 95(2), 317–356 (1997)
3. Cohn, A., Hazarika, S.: Qualitative spatial representation and reasoning: An overview. Fundamenta Informaticae 46(1-2), 1–29 (2001)
4. Escrig, M., Toledo, F.: Reasoning with compared distances at different levels of granularity. In: CAEPIA, Gijón, Spain (2001)
5. Freksa, C.: Using orientation information for qualitative spatial reasoning. In: Frank, A.U., Formentini, U., Campari, I. (eds.) GIS 1992. LNCS, vol. 639, pp. 162–178. Springer, Heidelberg (1992)
6. King, R., Garrett, S., Coghill, G.: On the use of qualitative reasoning to simulate and identify metabolic pathways. Bioinformatics 21(9), 2017–2026 (2005)
7. Knauff, M., Strube, G., Jola, C., Rauh, R., Schlieder, C.: The psychological validity of qualitative spatial reasoning in one dimension. Spatial Cognition & Computation 4(2), 167–188 (2004)
8. Levinson, S.: Space in Language and Cognition. Explorations in Cognitive Diversity. Cambridge University Press, UK (2003)
9. Liu, H., Brown, D., Coghill, G.: Fuzzy qualitative robot kinematics. IEEE Trans. on Fuzzy Systems 16(3), 808–822 (2008)
10. Martínez-Martín, E., Escrig, M.T., del Pobil, A.P.: A General Framework for Naming Qualitative Models Based on Intervals. In: Omatu, S., Paz Santana, J.F., González, S.R., Molina, J.M., Bernardos, A.M., Rodríguez, J.M.C. (eds.) Distributed Computing and Artificial Intelligence. AISC, vol. 151, pp. 681–688. Springer, Heidelberg (2012)
11. Renz, J., Nebel, B.: Qualitative Spatial Reasoning Using Constraint Calculi, pp. 161–215. Springer, Berlin (2007)
12. van de Weghe, N., Cohn, A., de Tré, G., de Maeyer, P.: A qualitative trajectory calculus as a basis for representing moving objects in geographical information systems. Control and Cybernetics 35(1), 97–119 (2006)
13. Westphal, M., Wölfl, S.: Qualitative csp, finite csp, and sat: Comparing methods for qualitative constraint-based reasoning. In: IJCAI, pp. 628–633 (2009)
14. Zadeh, L.: A new direction in ai. toward a computational theory of perceptions. AI Magazine 22(1), 73–84 (2001)

Framework of Optimization Methodology with Use of an Intelligent Hybrid Transport Management System Based on Hopfield Network and Travelling Salesman Problem

Natalia Kubiak and Agnieszka Stachowiak

Poznan University of Technology, Faculty of Engineering Management
Strzelecka 11, 60-965 Poznań, Poland
natalia.kubiak@tlen.pl, agnieszka.stachowiak@put.poznan.pl

Abstract. A medium size (and bigger) company has to have a good (i.e. adjusted to the companies profile and its environment for example economic situation or seasonality of the demand) informatics system. The reason for it is that large amount of date has to be collected on time, transformed and transmitted to allow managers to make adequate and made on time decision. Time factor in this situation is very important and without a good logistic system it is difficult or even impossible to deal with contemporary market. In this paper a framework of a hybrid intelligent system is presented, which will optimize the lead time needed to the transport goods from the company to the receiver. Consequently the time, when the final product will appear on the market will be optimized and thus the product can be both due the client. Transport is often associated with the Travelling Salesman Problem and the method of its approximated solution will be considered.

Keywords: Management, transport, TSP, hybrid system, Hopfield network, expert system.

1 Introduction

1.1 *Time as a Factor in Transport Process*

One of the main tasks of the management in a company is to offer a product on the market [14], i.e. deliver it on time, at a proper price and conditions, to places, where the customer want to buy it [26] [29]. This task is related directly with transport, which is why logistics management seems to be important. The delivery service is rated in view of time (speed of the order realization), punctuality, reliability and quality of the deliveries [12].

The qualities and time of the deliveries influence the costs of maintenance and exhaustion of the stock. When the delivery time is longer, then the size of stocks

S. Omatu et al. (Eds.): *Distrib. Computing & Artificial Intelligence*, AISC 217, pp. 95–102.
DOI: 10.1007/978-3-319-00551-5_12

must be bigger to avoid running out of stocks while waiting for arrival of the next delivery [8].

The necessity of the transport optimization is also important, because the road, on which the goods have to be transported, is longer, since the manufacturing is moved to distant places in the world [11].

That is why time is often considered as one of the main sources of advantage in the competitive economy. The curtailment of time between the order and the delivery can cause the reduction of stock and the costs of its storage [8]. The easiest way to achieve the reduction of time is faster and more efficient transport process.

1.2 Informatics System in Logistic

The main task of informatics in management is to provide useful information (quick, reliable and not requiring to many resources) to the managers [33] [18]. Such information is often based on other, further generated information [25]. The amount of available information is large, hence management can be effective, only when the company has a good IT system [35]. Thus the informatics system becomes one of the inseparable and integral part of the logistic systems [12], therefore IT systems supporting management functions are one of the most important, abundant and growing PC software group [5].

Authors are working in logistics of companies with different production profile and have seen, used and analyzed various different logistic systems. Consequently a thesis can be formulated, that often the systems applied are not well adjusted to the company. Furthermore the adjustment to the new conditions takes a lot of time, sometimes years and when it is done the conditions change again. This is the reason why a well-adjusted system is needed.

Such a well-adjusted system has to be flexible – it should be able to learn and successfully applied in data management. Neural system is an example of such a system. Additionally it can be applied to solve problems, which could not be solved with use of standard algorithms [8].

The price of the final product is influenced by the costs of supply, transport, production and distribution and this is why they have to be rationalized (see [9]). The system, which is considered, will optimize transport by adjusting it to the current supply situation, so that time needed to transport the goods was as short as possible and the loading surface was maximally used.

From the authors experience in management these two factors – short transport time and optimal utilization of the loading surface are often contradict to each other resulting in trade-off relation. Often the complement of the surface is done due collection of goods, that could be collected with the next transport or by adding new pick-up (or unloading) points. Such processing causes increased demand on the warehouse surface, in which they will be stored, or increase the transport time (in the second situation).

Authors suggest a framework of a system, which will adjusts itself to the actual economic situation without updating or changing the management system with help of a neural network and expert system. Thanks to such a system the company can quickly respond to changes with no outlay. The system will influence the transport time by optimization of the loading management. As a result time and money will be spared. The loading surface and the number of trucks needed to transport the goods will be also optimized by continuous adaptation to the current economic situation.

2 Hybrid System

2.1 Artificial Intelligence

One of the first applications of artificial intelligence (AI) is in data base management systems to support the activity of human [1] [23] and now AI is more often applied in many other fields [35] [8] (see [30]). AI is defined as a field in informatics studies that causes, that machines perform an action, which would require intelligence, if a human would do it [8]. That means AI is a computer program, which actions are similar to human actions [24].

The first application of IT was in database management systems [10], which are used in almost every informatics system [1].

In this article two elements of AI are applied, because one of the elements is a self-learning program which adapts to changes in the reality in a dynamic way without modification of the program code. The other element explains the actions of the system and checks if there are any mistakes made.

This approach to programs (hybrid constructions of different technologies of AI) seems to be more efficient than traditional programs [30].

Using neural networks results in numerous benefits, but the main, according to the authors is that they are comfortable and cheap systems with many elements, that process information in a parallel way and they do not require extra informatics system [4] [31].

The applications of neural network concerns optimization problems, for classifications and forecast (for example request of electric energy or rate of Swiss Francs in short period of time - see i.e. [2]). Despite the large amount of applications, the capabilities of their utilization are not yet explored and it seems, that they will effect on the progress in information's technics [28].

2.2 Travelling Salesman Problem and Neural Network

The path length will be optimized like in Travelling Salesman Problem. The optimization will be used with help of neural network. Hopfield introduced an idea of a function in neuron network and showed, that a search for a point, that will minimalize the energy can be moved beyond an electronic circular. With use of this property a solution w nondeterministic in time NP-hard problems, like TSP, can be found [17].

Often cited articles of Hopfield and Tank [16] are one of the first reports on applications of neural networks for such optimization problems [13] and the application was successful. Their work about the TSP aroused large interest, because of its elegant formulation (see [13] [20] [27]).The Hopfield network is ideal suited for optimizations problem [21].

Hopfield and Tank showed in their article, that the Hopfield method can solve TSP with 30 cities with short computing time and good accuracy [34].

According to the TSP a salesman (navigator) will travel through a set of cities (ports, nodes), visiting every city only once, so that the total path length is minimal [3], so N cities will be considered and the shortest path should be found, so that every city was visited once [7], and in the end salesman comes back to the initial city. The path will be closed and its length will be a sum of all distances between the cities, which it is composed of [15].

Exemplary TSP for N cities has a symmetric NxN matrix A with non-positive real coefficient d_{ij},, where d_{ij} denotes the distance from i-th and j-th city. The original Hopfield and Tank formulation [16] use N^2 output nodes in form of a quadrat matrix, where rows correspond to the cities and columns to their position in the path. So a valid solution will be in form of permutation matrix, i.e. exact one city will be in on position and every city will be visited only one [20].

The H-T formulation has a bad scaling property and only for small amount of parameter combination achieves good result in suitable and stable solution [32].

2.3 *Problem Formulation*

A new factor will be considered, i.e. the time that a truck waits a long time for (un)loading. This is a waste of time (where time is so important) and it depends individually on the infrastructure and processes in company, where the truck is. This is an important aspect and it will be considered in hereby paper and in the hybrid system as well in the logistic factor.

The considered optimization is related with the choice of the shortest path, which a truck has to cross to pick up the goods and it is associates to the Travelling Salesman Problem (TSP). The neural network presented in this work will be a Hopfield network and it will be constructed analogically to the neural network that solves TSP. Significant differences between the two problems (TSP and presented problem) exist. For example we assume, that the network will group the suppliers in order of the pick-ups and each group will identify the order in which one truck will pick-up the goods and in the end will travel to the production company. First of all due the limitations related to the weight and gauge of the shipment (we assume, that every shipment is palatable) a situation can appear, in which not every goods can be pick up by one truck, what means, that

1. In one group of suppliers not every supplier have to appear,
2. In one group can also be only one supplier,

3. Sometimes the designated path is longer than the shortest path, because the logistic factor (the transport time should be to long) or the loading surface (which should be maximal filled) will be optimized,
4. A situation can appear, when from a curtain supplier no goods will be picked up, because there will be no shipment to deliver. In this case the input vector in the corresponding place will be put 0, instead of 1.

Input data for the system are ($i, j = 1, \ldots, N$, where N is the number of suppliers)

- The size of the shipment, i.e. the number of pallet places pal_i
- The weight of the shipment g_i
- Suppliers logistic factors u_i,
- The distance between the ith and jth supplier d_{ij},
- Additionally to each supplier a location factor will be attributed p_i, because the suppliers will be grouped in respect of their location and in this matter the pick-ups will be made among the neighbors (the concept is based on articles published by Choy [7], Kahng [20], Khakmardan [22], Joppe [19]),
- Input vector $x = [x_1, \ldots, x_N]$, which consists of elements from a set $\{0,1\}$. It will determinate if the curtain supplier (and corresponding to him neuron) is active, that means if from the supplier a shipment will be made.

The authors do not wont to limit the framework to one specific type of truck, so in the system maximal values will be introduced pal_{max}, g_{max}. Futhermore the numer of suppliers in the i group will be set as M_i, B is the number of groups. The output a set of vectors is v, $v = [v_1, v_2, \ldots, v_B], v_k = [v_1^k, v_2^k, \ldots, v_M^k], v_M^k = v_{M_k}$.

The systems workflow is as follow (presented in figure 1):

1. Based on the initial data and initial vector the neuron network will establish a group of suppliers, which will be picked-up be the first truck.
2. The expert system will archive the received group and it will change in the vector x the vectors component, which correspond to the designated suppliers and it will set them as zero.
3. The expert system will restart the neuron network, wherein in the new group the previous denoted suppliers will be not considered
4. The system will work until the input vector will be a zero vector.

The energy function is as follow

$$A \sum_{i=1}^{N} \sum_{j=1}^{N} x_i V_{i,j} d_{i,j} - B \sum_{x=1}^{N} \sum_{j=1}^{N} x_i V_{i,j} (pal_i - pal_{max})^2 -$$
$$C \sum_{x=1}^{N} \sum_{j=1}^{N} x_i V_{i,j} (g_i - g_{max})^2 - D \sum_{x=1}^{N} \left(\left| p_{i-1}^k - p_i^k \right| + \left| p_{i+1}^k - p_i^k \right| \right)^2, \quad (1)$$

where A, B, C i D are positive constant value. Additional conditions should be fulfilled $M \leq N$ and $v_1 : pal_{v_1} = \max_{i \in \{1,\ldots,N\}} pal_i$.

The algorithm of Hopfield and Tank [16] is very sensitive to the initial conditions [20] [16], as also to the construction of the energy function [17], therefore in the system the initial point will be set so, that it will be the supplier, which will have the biggest shipment to pick-up (the biggest loading space, which will be used for the shipment), what is denoted in the last condition.

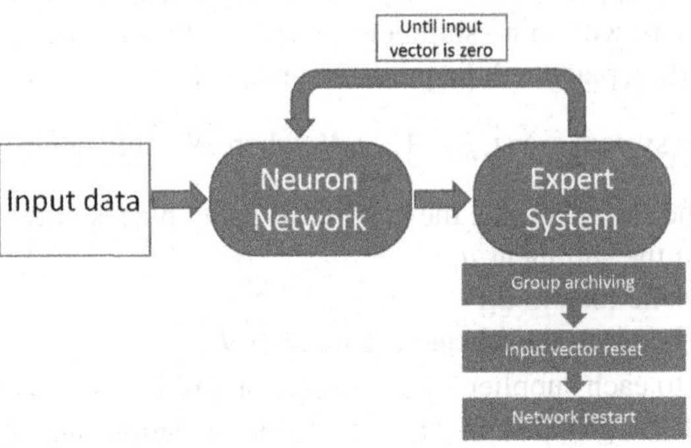

Fig. 1 The scheme of hybrid system [own work]

3 Future Work

The presented approach is the first step in building a big supply software in a company. First of all it has to be tested on real data of a company, to check if the new proposal is more effective, than the system, which is normally used in the company. Because a neural network is used, the optimality is not guaranteed, but a good enough solution in appropriate time is expected. This solution can be also a module unit of a management system of a company. Thanks to the expert system the input data will be matched to neural network. The authors are convinced that such a framework can be also applied in a distribution company, but at first it has to be adapted to it. Only after implementing this framework can be tested, if it is useful and profitable.

Table 1 Comparison of the new solution to programs usually applied by companies

Programme type/Aspect	Standard program	Shell system	New solution
Implementation cost	Average-low	High	Low
Speed	Average	Depends on solution	Fast because of using neural systems
Adaptability	No	Yes	Flexibility (thanks to SE1)
Errors	Few, because of standardization of solutions	Many (system is adjusted to the company)	Average numer of errors
Changes after implementation	Not possible	Possible	Possible

The number of coefficients in the energy function must be so small as possible [6], so after implementing a set of test has to be made with real data for certain company to optimize the structure and number of coefficient.

The results can be different for different companies; therefore final conclusions can be made, when the system will be implemented to different companies.

Usually in neural-expert systems the data is first prepared due the expert system in order to provide an optimal solution. In the denoted situation is not the case. In this framework the data is only grouped at the beginning and then the network looks for the best solution. This art of solution solving has to be tested and the authors want to see, if such algorithm will be good in this problem.

References

1. Bach, M.: Wybrane zagadnienia związane z automatycznym tłumaczeniem zadań wyszukiwania danych sformułowanych w języku polskim na język formalny SQL. In: Grzech, A. (red.) Inżynieria Wiedzy i Systemy Ekspertowe Tom 1. Oficyna Wydawnicza Politechniki Wrocławskiej, pp. 247–256 (2006)
2. Baczyński, D., Bielecki, S., Parol, M., Piotrowski, P., Wasilewski, J.: Sztuczna inteligencja w praktyce, Labolatorium, Oficyna Wydawnicza Politechniki Warszawskiej (2008)
3. Bank, S.I., Avramovic, Z.Z.: Hopfield network in solving travelling salesman problem in navigation. In: Neural Network Applications in Electrical Engineering, NEUREL 2002, pp. 207–210 (2002)
4. Barzykowski, J. (opr.): Współczesna metrologia zagadnienia wybrane. Wydawnictwa Naukowo – Techniczne (2004)
5. Bubnicki, Z.: Teoria i algorytmy sterowania. Wydawnictwo Naukowe PWN (2005)
6. Chao, H., Zheng, N., Yan, G.: Neuron design and stability analysis of neural network for TSP. In: Proceedings of the IEEE International Conference on Systems, Man and Cybernetics, vol. 3, pp. 2453–2458 (1994)
7. Choy, C.S.-T., Siu, W.-C.: New approach for solving the travelling salesman problem using self-organizing learning. In: IEEE International Conference on Neural Networks - Conference Proceedings, January 1, vol. 5, pp. 2632–2635 (1995)
8. Coyle, J.J., Bardi, E.J., Langley Jr., C.J.: Zarządzanie logistyczne. Polskie Wydawnictwo Ekonomiczne (2007)
9. Fijałkowski, J.: Transport wewnętrzny w systemach logistycznych. Wybrane zagadnienia, Oficyna Wydawnicza Politechniki Warszawskiej (2003)
10. Flasiński, M.: Wstęp do sztucznej inteligencji. Wydawnictwo Naukowe PWN (2011)
11. Golińska, P. (red.): Nowoczesne rozwiązania technologiczne w logistyce. Politechnika Poznańska (2010)
12. Gołembska, E. (red. nauk.): Kompendium wiedzy o logistyce Nowe wydanie. Wydawnictwo Naukowe PWN (2010)
13. Goto, A., Kawamura, M.: Solution method using correlated noise for TSP. In: Ishikawa, M., Doya, K., Miyamoto, H., Yamakawa, T. (eds.) ICONIP 2007, Part I. LNCS, vol. 4984, pp. 733–741. Springer, Heidelberg (2008)
14. Grzegorczyk, W.: Strategie marketingowe przedsiębiorstw na rynku Międzynarodowym. Wydawnictwo Uniwersytetu Łódzkiego (2011)
15. Guo, P., Zhu, L., Liu, Z., He, Y.: An ant colony algorithm for solving the sky luminance model parameters. In: Liu, B., Ma, M., Chang, J. (eds.) ICICA 2012. LNCS, vol. 7473, pp. 365–372. Springer, Heidelberg (2012)

16. Hopfield, J.J.: Neurons with graded response have collective computational properties like those of two-state neurons. Proceedings of the National Academy of Sciences of the United States of America 81(10), 3088–3092 (1984)
17. Izumida, M., Murakami, K., Aibara, T.: Analysis of neural network energy functions using standard forms. Systems and Computers in Japan 23(8), 36–45 (1992)
18. Januszewski, A.: Funkcjonalność informatycznych systemów zarządzania, t. 1 Zintegrowane systemy transakcyjne. Wydawnictwo Naukowe PWN/MIKOM (2008)
19. Joppe, A., Cardon Helmut, R.A., Bioch, J.C.: A neural network for solving the travelling salesman problem on the basis of city adjacency in the tour. In: International Joint Conference on Neural Networks, IJCNN, vol. 3, pp. 961–964 (1990)
20. Kahng, A.B.: Traveling salesman heuristics and embedding dimension in the Hopfield model, pp. 513–520 (1989)
21. Kashmiri, S.: The travelling salesman problem and the Hopfield neural network. In: Conference Proceedings - IEEE SOUTHEASTCON 2, pp. 940–943 (1991)
22. Khakmardan, S., Poostchi, H., Akbarzadeh, M.R.: Solving Traveling Salesman problem by a hybrid combination of PSO and Extremal Optimization. In: Proceedings of the International Joint Conference on Neural Networks (IJCNN), January 1, pp. 1501–1507 (2011)
23. Knosola, R., et al.: Zastosowania metod sztucznej inteligencji w inżynierii produkcji. Wydawnictwo Naukowo-Techniczne (2007)
24. Kwiatkowska, A.M.: Systemy wspomagania decyzji Jak korzystać z WIEDZY i informacji. Wydawnictwo Naukowe PWN/MIKOM (2007)
25. Oleński, J.: Ekonomia informacji. Metody, Polskie Wydawnictwo Ekonomiczne (2003)
26. Pomykalski, A.: Zarządzanie i planowanie marketingowe. Wydawnictwo Naukowe PWN (2008)
27. Qu, H., Yi, Z., Xiang, X.: Theoretical analysis and parameter setting of Hopfield neural networks. In: Wang, J., Liao, X.-F., Yi, Z. (eds.) ISNN 2005, Part I. LNCS, vol. 3496, pp. 739–744. Springer, Heidelberg (2005)
28. Osowski, S.: Sieci neuronowe do przetwarzania informacji. Oficyna Wydawnicza Politechniki Warszawskiej (2006)
29. Sławińska, M. (red.): Kompendium wiedzy o handlu. Wydawnictwo Naukowe PWN (2008)
30. Szczerbicki, E., Reidsema, C.: Modelling Design Planning in Concerrent Engineering. In: Intelligent Processing and Manufacturing of Materials, IPMM 1999, vol. 2, pp. 1055–1060 (1999)
31. Tadeusiewicz, R.: Sieci neuronowe. Akademicka Oficyna Wydawnicza RM (1993)
32. Tang, H., Tan, K.C., Yi, Z.: Parameter settings of Hopfield networks applied to traveling salesman problems. In: Tang, H., Tan, K.C., Yi, Z. (eds.) Neural Networks: Computational Models and Applications. SCI, vol. 53, pp. 99–116. Springer, Heidelberg (2007)
33. Wierzbicki, T. (red.): Informatyka w zarządzaniu. Wydawnictwo Naukowe PWN (1986)
34. Yoshiyuki, U.: Solving traveling salesman problem by real space renormalization technique. In: IEEE International Conference on Neural Networks - Conference Proceedings, vol. 7, pp. 4529–4534 (1994)
35. Zieliński, J.S. (red.): Inteligentne systemy w zarządzaniu. Wydawnictwo Naukowe PWN (2000)

Selecting the Shortest Itinerary in a Cloud-Based Distributed Mobility Network

Jelena Fiosina and Maksims Fiosins*

Clausthal University of Technology,
Institute of Informatics,
Julius-Albert Str. 4, D-38678, Clausthal-Zellerfeld, Germany
{Jelena.Fiosina,Maksims.Fiosins}@gmail.com

Abstract. New Internet technologies can considerably enhance contemporary traffic control and management systems (TCMS). Such systems need to process increasing volumes of data available in clouds, and so new algorithms and techniques for statistical data analysis are required. A very important problem for cloud-based TCMS is the selection of the shortest itinerary, which requires route comparison on the basis of historical data and dynamic observations. In the paper we compare two non-overlapping routes in a stochastic graph. The weights of the edges are considered to be independent random variables with unknown distributions. Only historical samples of the weights are available, and some edges may have common samples. Our purpose is to estimate the probability that the weight of the first route is greater than that of the second one. We consider the resampling estimator of the probability in the case of small samples and compare it with the parametric plug-in estimator. The analytical expressions for the expectations and variances of the proposed estimators are derived, which allow theoretical evaluation of the estimators' quality. The experimental results demonstrate that the resampling estimator is a suitable alternative to the parametric plug-in estimator. This problem is very important for a vehicle decision-making procedure to choose route from the available alternatives.

Keywords: traffic control and management, future Internet, stochastic graph, shortest route, resampling, small samples, estimation, simulation.

1 Introduction

Future Internet opportunities open new perspectives on the development of intelligent transport systems (ITS). Technologies such as cloud and grid computing, the Internet of Things concept and ambient intelligence methods allow the development of new applications, to hide the complexity of data and algorithms in the network.

* The research leading to these results has received funding from the European Union Seventh Framework Programme (FP7/2007-2013) under grant agreement No. PIEF-GA-2010-274881.

S. Omatu et al. (Eds.): *Distrib. Computing & Artificial Intelligence*, AISC 217, pp. 103–110.
DOI: 10.1007/978-3-319-00551-5_13

This allows traffic participants to run simple applications on their mobile devices, which provide clear recommendations on how they should act in the current situation. These simple applications are based on the aggregation and processing of large amounts of data, which are collected from different traffic participants and objects. These data are physically distributed and available in virtual clouds. This creates a need for innovative data analysis, processing, and mining techniques, which run in clouds and prepare necessary information for end-user applications.

In this study, we deal with route recommendation systems, which are essential applications in cloud-based ITS. This system includes optimization of the booked itinerary with respect to user preferences, time, fuel consumption, cost, and air pollution to provide better (i.e., quicker, more comfortable, cheaper, and greener) mobility. The recommendations can be made on the basis of static information about the network (traffic lights, public transport schedules, etc.) combined with dynamic information about the current situation and historically stored data about traveling under equivalent conditions. If necessarily, the recommendations of other travelers. can be included. Booking the shortest itinerary is a key aspect in many traffic scenarios with different participants: a dynamic multi-modal journey, a simple private drive through a transport network, or smart city logistical operations. We consider an example of driving through a transport network segment considering the time consumption as the optimization criterion in itinerary comparisons and shortest route selection. In this case, the route recommendation is based on the estimates of the travel time along the route.

For this purpose, an artificial transport network is created, the travel times for alternative routes are estimated, and the best route is selected. Different methods of travel-time forecasting can be used, such as regression models, and neural networks. Most of these are sensitive to outliers or incorrect model selection(e.g. wrong distribution). In these situations, the methods of computational statistics can be effective.

Computational statistics includes a set of methods for non-parametric statistical estimation. The main idea is to use data in different combinations to replace complex statistical inferences by computations. The resampling approach supposes that the available data are used in different combinations to obtain model-free estimators that are robust to outliers. The quality of the estimators obtained is also important.

In the present study, we demonstrate data flows in cloud-based ITS for route recommendations and propose a resampling-based approach for the route comparison in such systems. We derive the properties of the proposed resampling estimators and compare these with traditional plug-in estimators.

The remainder of this study is organized as follows: Section 2 formulates the problem, Sections 3-4 describe the resampling procedure and its properties, Section 5 contains a numerical example, and Section 6 presents the conclusion.

2 Problem Formulation

We consider a cloud-based ITS architecture [8]. In terms of the Internet of Things, the real-world users are represented in the cloud system as virtual agents, which act in the cloud and virtual traffic network. The street network is presented by the virtual transport network, which consists of a digital map as well as the associated ad-hoc network models that allow estimation and forecasting of the important network characteristics for each problem [6]. The virtual agents store the real-time information, which is collected and constantly processed in the cloud. Moreover, the strategies for execution of the cloud application are constantly pre-calculated and checked in the virtual network (e.g., the shortest routes are pre-calculated). When a user runs the cloud application, the pre-calculated strategy is updated with the real-time data and is executed, with respect to the corresponding changes. Data flows and corresponding optimization methods in the cloud-based ITS architecture are presented in Fig. 1.

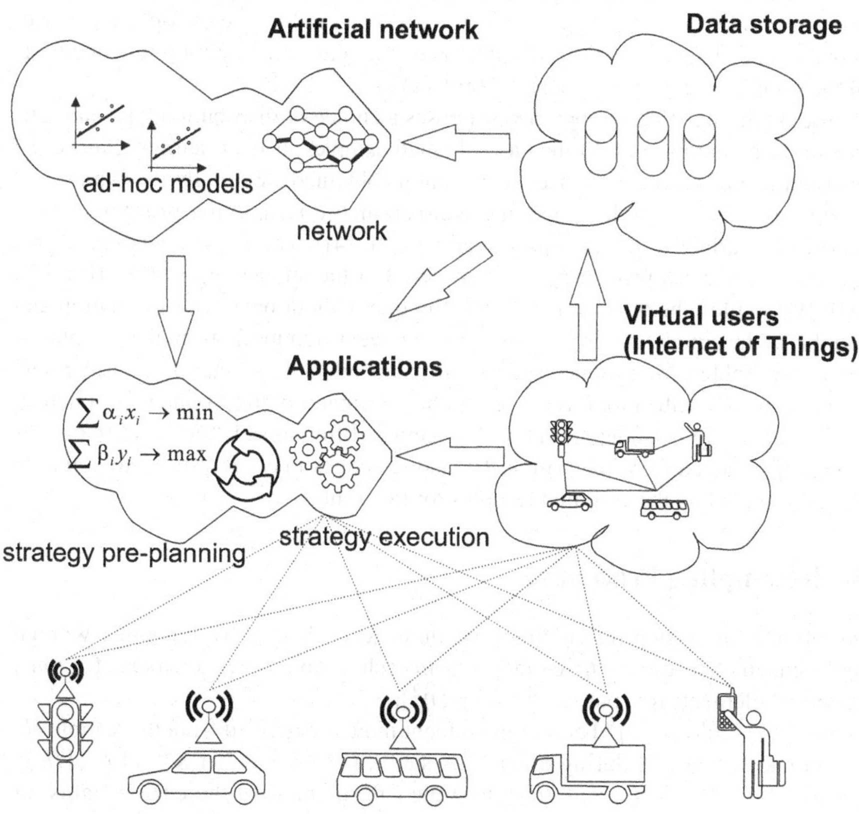

Fig. 1 Data flows and corresponding optimization methods in cloud-based ITS

We consider an application, that provides route recommendations to vehicle drivers. The essential process of this application is the comparison of pre-defined routes. It is based on historical samples of the route segments, which are collected from the virtual users. The candidate routes are compared in the virtual transport network in order to recommend the best route to a user. As the travel times are random, a stochastic comparison should be used. We consider a directed graph $G = (V, E)$ with n edges, $|E| = n$, where each edge $e_i \in E$ has an associated weight X_i (e.g. travel time). We assume that the weights $\{X_1, X_2, \ldots, X_n\}$ are independent random variables (r.v.). A route in the graph is a sequence of edges such that the next edge in the sequence starts from the node, where the previous edge ends. Let us denote a route b as a sequence k^b of edge indices in the initial graph: $k^b = (k_1^b, k_2^b, \ldots, k_{n_b}^b)$, which consists of n_b edges. Hence, a route b is the sequence of edges $r^b = \{e_{k_1^b}, e_{k_2^b}, \ldots, e_{k_{n_b}^b}\}$. The route weight S^b is the sum of the corresponding edge weights, so $S^b = \sum_{i \in k^b} X_i$.

We compare two non-overlapping routes by calculating the probability that the weight of route 1 is greater than that of route 2: $\Theta = P\{S^1 > S^2\}$.

The distributions of the edge weights are unknown, only the samples are available: $H_i = \{H_{i,1}, H_{i,2}, \ldots, H_{i,m_i}\}$, where $i = 1, 2, \ldots, c$, $c \leq n$. Each sample may correspond to one or several edges. An (unknown) cumulative distribution function (cdf) of the sample H_i elements is denoted by $F_i(x)$, $i = 1, 2, \ldots, c$.

The traditional plug-in approach supposes a choice of distribution type and estimation of its parameters. In the case of small samples, it is difficult to choose the distribution law correctly; hence the estimators obtained are usually inaccurate.

Hence, it is preferable to use the non-parametric resampling procedure ([7]), which is a variant of the bootstrap method ([3], [4]). The implementation of this approach to various problems was considered in the studies reported in ([1], [2], [5]). We employ the usual simulation technique without parameter estimation and use this in the simulation process to extract elements randomly from the samples of random variables. We produce a series of independent experiments and accept the average over all realizations as the resampling estimator of the parameter of interest.

Two cases are considered: (1) each edge has different samples, so only one element is extracted from the sample H_i; and (2) edges may correspond to common samples, including the common samples for two routes.

3 Resampling Procedure

We propose an N-step resampling procedure. At each step, we randomly without replacement choose $\eta_i^1 + \eta_i^2$ elements from each sample H_i: η_i^1 elements for route 1, and η_i^2 elements for route 2: $\eta_i = (\eta_i^1, \eta_i^2)$.

Let $J_i^b(l)$, $|J_i^b(l)| = \eta_i^b$ be a set of element indices extracted from the sample H_i, for a route b, $b = 1, 2$, during resampling step l, $i = 1, \ldots, c$. Let $\mathbf{X}^{*l} = \bigcup_{i=1}^{c} \{H_{i,j} : j \in J_i^1(l)\} \cup \bigcup_{i=1}^{c} \{-H_{i,j} : j \in J_i^2(l)\}$ be the l-th resample of the edge weights for both routes, with the weights of route 2 assumed to be negative.

Let $\Psi(\mathbf{x})$ be an indicator function, where $\mathbf{x} = (x_1, x_2, \ldots)$ is a vector of real numbers: $\Psi(\mathbf{x})$ is unity if $\sum\limits_i x_i > 0$; otherwise, it is zero. The average of $\Psi(\mathbf{X}^{*l})$ over all N steps is accepted as the resampling estimator of the probability of interest: $\Theta^* = \frac{1}{N} \sum_{l=1}^{N} \Psi(\mathbf{X}^{*l})$. The resampling procedure is presented as Algorithm 1.

The function $extract(X, n)$ randomly chooses n elements without replacement from the set X. The function $subsample(X, a, n)$ returns n elements from X, starting from position a. These two cases differ with the parameters of the *extract* procedure.

Algorithm 1. Function RESAMPLE

```
function RESAMPLE(H_i, η_i, i = 1,...,c, N)
    for all l ∈ 1,...,N do
        for all i ∈ 1...c do
            X_i^{*l} ← extract(H_i, η_i^1 + η_i^2)
            X1_i^{*l} ← subsample(X_i^{*l}, 1, η_i^1); X2_i^{*l} ← subsample(X_i^{*l}, η_i^1 + 1, η_i^2)
        end for
        X^{*l} = ∪X1_i^{*l} ∪ −X2_i^{*l}; Θ_l ← Ψ(X^{*l})
    end for
    Θ* ← (1/N) Σ_{l=1}^N Θ_l
    return Θ*
end function
```

4 Properties of the Resampling Estimator

The estimator Θ^* is obviously unbiased: $E(\Theta^*) = \Theta$, so we are interested in its variance. Consider the elements extracted at two different steps $l \neq l'$. Moreover, we denote: $\mu = E\,\Psi(\mathbf{X}^{*l})$, $\mu_2 = E\,\Psi(\mathbf{X}^{*l})^2$, $\mu_{11} = E\,\Psi(\mathbf{X}^{*l}) \cdot \Psi(\mathbf{X}^{*l'}), l \neq l'$. Then, the variance is $V(\Theta^*) = E(\Theta^{*2}) - \mu^2 = \{\frac{1}{N}\mu_2 + \frac{N-1}{N}\mu_{11}\} - \mu^2$, for the estimation of which we need the mixed moment μ_{11} depending on the resampling procedure.

Different Samples for Each Edge

In this case, $J_i^b(l)$ consists of one element, denoted as $j_i^b(l)$. This is the index of an element extracted from the sample H_i at step l for route b.

Let $M_i = \{1,2,\ldots,m_i\}$, $U^b : \{i : \eta_i^b \neq \emptyset\}$, $M^b = \prod_{i \in U^b} M_i$ and $\mathbf{j}^b(l) = \{j_i^b(l) : i \in U^b\}$, $\mathbf{j}(l) = (\mathbf{j}^1(l), \mathbf{j}^2(l))$, where $\mathbf{j}^b(l) \in M^b$ and $b = 1,2$.

We use a modification of the ω-pair notation [5]. Let $\omega^b \subset U^b$, $\omega = (\omega^1, \omega^2)$. We assume that two vectors $\mathbf{j}(l)$ and $\mathbf{j}(l')$ produce an ω-pair, if $j_i^b(l) = j_i^b(l')$ for $i \in \omega^b$ and $j_i^b(l) \neq j_i^b(l')$ for $i \notin \omega^b$. In other words, the components of the vectors $\mathbf{j}(l)$ and $\mathbf{j}(l')$ produce the ω-pair if they have the same elements from the samples, whose indices are contained by ω.

Let $A(\omega)$ be an event 'resamples $\mathbf{j}(l)$ and $\mathbf{j}(l')$ for the different steps $l \neq l'$ produce the ω-pair', let $P\{\omega\}$ be the probability of this event, and let $\mu_{11}(\omega)$ be the corresponding mixed moment. The probability of producing the ω-pair is

$$P\{\omega\} = \frac{1}{|M^1||M^2|} \prod_{i \in \bigcup_b \{U^b \setminus \omega^b\}} (m_i - 1).$$

The mixed moment μ_{11} can be calculated with the formula $\mu_{11} = \sum_{\omega \subset U^1 \times U^2} P(\omega)\mu_{11}(\omega)$.

Next, we intend to calculate $\mu_{11}(\omega)$, $\omega \subset U^1 \times U^2$. Let

$$S_l^{dif}(\omega) = \sum_{i \in U^1 \setminus \omega^1} H_{i,j_i^1(l)} - \sum_{i \in U^2 \setminus \omega^2} H_{i,j_i^2(l)},$$
$$S_{ll'}^{com}(\omega) = \sum_{i \in \omega^1} H_{i,j_i^1(l)} - \sum_{i \in \omega^2} H_{i,j_i^2(l)}.$$

Then, $\mu_{11}(\omega)$ can be calculated as

$$\mu_{11}(\omega) = E(\Psi(\mathbf{X}^{*l}) \cdot \Psi(\mathbf{X}^{*l'})|\omega) = \int_{-\infty}^{+\infty} \left(1 - F^d_\omega(-x)\right)^2 dF^c_{\omega(x)},$$

where $F^d_\omega(x)$ is cdf of $S^{dif}_l(\omega)$, $F^c_\omega(x)$ is cdf of $S^{com}_{ll'}(\omega)$ given ω-pair.

Common Samples for Edges

Here, we use the notation of α-pairs ([1], [2], [5]) instead of ω-pairs. Let $J^b_i(l) = \{j^b_{i,1}(l), j^b_{i,2}(l), \ldots, j^b_{i,\eta^b_i}(l)\}, J^b(l) = \{J^b_i(l) : i \in U^b\}, \mathbf{J}(l) = \{J^1(l), J^2(l)\}$, where $J^b_i(l) \subset M^b$, $b = 1,2$, $l = 1,\ldots,N$, $i = 1,2,\ldots,c$.

Let $A^b_i(ll')$ be a set of indices of the common elements, extracted from the sample H_i for route b at steps l and l'. Let $A^{bp}_i(ll')$ be a set of indices of the common elements, extracted from the sample H_i for route b at step l and for route p and at step l'. Let $\bar{A}^{bp}_i(l)$ be a set of indices of the elements from route b at step l, which were in neither route b nor route p at step l', $b,p \in \{1,2\}$ and $b \neq p$:

$A^b_i(ll') = J^b_i(l) \cap J^b_i(l')$, $\bar{A}^{bp}_i(l) = J^b_i(l) \setminus (A^b_i(ll') \cup A^{bp}_i(ll'))$, $A^{bp}_i(ll') = J^b_i(l) \cap J^p_i(l')$, $\bar{A}^{pb}_i(l) = J^p_i(l) \setminus (A^p_i(ll') \cup A^{pb}_i(ll'))$. Let $0 \le \alpha^b_i \le \eta^b_i, 0 \le \alpha^{bp}_i \le \min(\eta^b_i, \eta^p_i)$, $b,p \in \{1,2\}$ and $b \neq p$. Let $\alpha_i = \{\alpha^1_i, \alpha^2_i, \alpha^{12}_i, \alpha^{21}_i\}$, $\alpha = \{\alpha_i\}, i = 1,2,\ldots,c$. Next, we say that $\mathbf{J}(l)$ and $\mathbf{J}(l')$ produce an α-pair, if and only if: $\alpha^1_i = |A^1_i(ll')|$, $\alpha^2_i = |A^2_i(ll')|$, $\alpha^{12}_i = |A^{12}_i(ll')|$, $\alpha^{21}_i = |A^{21}_i(ll')|$. Let $A_{ll'}(\alpha)$ denote the event 'sub-samples $\mathbf{J}(l)$ and $\mathbf{J}(l')$ produce an α-pair', and let $P_{ll'}\{\alpha\}$ be the probability of this event: $P_{ll'}\{\alpha\} = P_{ll'}\{A_{ll'}(\alpha)\}$.

To calculate $\mu_{11}(\alpha)$ we replace ω-pairs with α-pairs. Therefore we need to calculate $P\{\alpha\}$ and $\mu_{11}(\alpha)$. The probability $P\{\alpha\}$ is

$$P\{\alpha\} = \prod_{i \in 1,2,\ldots,c} \frac{\binom{\eta^1_i}{\alpha^1_i}\binom{\eta^2_i}{\alpha^{21}_i}\binom{m_i - \eta^1_i - \eta^2_i}{\eta^1_i - \alpha^1_i - \alpha^{21}_i}}{\binom{m_i}{\eta^1_i}} \times$$

$$\times \binom{\eta^1_i - \alpha^1_i}{\alpha^{12}_i}\binom{\eta^2_i - \alpha^{21}_i}{\alpha^2_i} \frac{\binom{m_i - 2\eta^1_i - \eta^2_i + \alpha^1_i + \alpha^{21}_i}{\eta^2_i - \alpha^{12}_i - \alpha^2_i}}{\binom{m_i - \eta^1_i}{\eta^2_i}},$$

where $\binom{n}{m}$ is a binomial coefficient.

To calculate $\mu_{11}(\alpha)$ we divide each sum into three subsums: $S^{dif}_l(\alpha)$ contains different elements for steps l and l'; $S^{com}_{ll'}(\alpha)$ - the common elements for the same route; $S^{com12}_{ll'}(\alpha)$ - the common elements for different routes. Let $S^{dif}_l(\alpha) = \sum_{i=1}^c \left\{\sum_{j \in \bar{A}^{12}_i(l)} H_{i,j} - \sum_{j \in \bar{A}^{21}_i(l)} H_{i,j}\right\}$, $S^{com}_{ll'}(\alpha) = \sum_{i=1}^c \left\{\sum_{j \in A^1_i(ll')} H_{i,j} - \sum_{j \in A^2_i(ll')} H_{i,j}\right\}$, $S^{com12}_{ll'}(\alpha) = \sum_{i=1}^c \left\{\sum_{j \in A^{12}_i(ll')} H_{i,j} - \sum_{j \in A^{21}_i(ll')} H_{i,j}\right\}$.

As $S^{com}_{ll'}(\alpha) = S^{com}_{l'l}(\alpha)$ and $S^{com12}_{ll'}(\alpha) = -S^{com12}_{l'l}(\alpha)$, $\mu_{11}(\alpha)$ is:

$$\mu_{11}(\alpha) = E\{\Psi(\mathbf{X}^{*l}) \cdot \Psi(\mathbf{X}^{*l'})|\alpha\} = P\left\{\Psi(\mathbf{X}^{*l}) = 1, \Psi(\mathbf{X}^{*l'}) = 1|\alpha\right\} =$$
$$= \int_{-\infty}^{+\infty}\int_{-\infty}^{+\infty}(1 - F_\alpha^d(-x-y)) \times (1 - F_\alpha^d(-x+y))dF_\alpha^c(x)dF_\alpha^{c12}(y),$$

where $F_\alpha^d(x)$ is cdf of $S_l^{dif}(\alpha)$, $F_\alpha^c(x)$ is cdf of $S_{ll'}^{com}(\alpha)$, $F_\alpha^{c12}(x)$ is cdf of $S_{ll'}^{com12}(\alpha)$.

5 Numerical Example

We model a route recommendation in the southern part of the city of Hanover, (Germany), which is shown in Fig. 2 (left), and represented by the graph in Fig. 2 (right). We compare two routes: nodes 9,8,7,5,3,1 (solid) and nodes 9,6,4,2,1 (dashed) for vehicles travelling from 9 to 1.

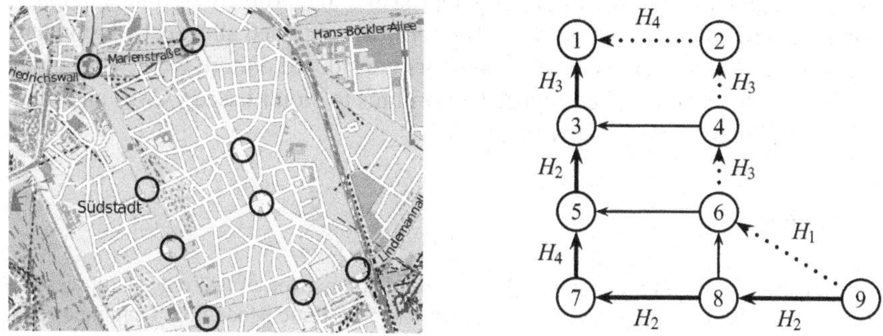

Fig. 2 Street network of the south part of Hanover city and corresponding graph

The cloud-based ITS collects information about travel times for different road segments. We assume that due to technical or organizational limitations, the travel

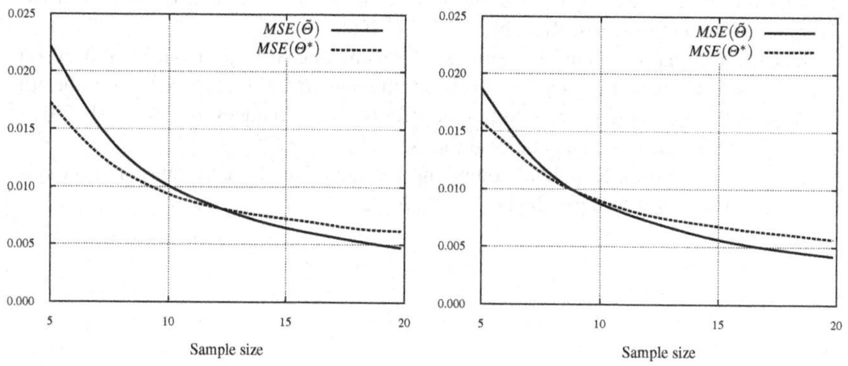

Fig. 3 MSE (vertical axis) of plug-in and res. estimators for $\Theta = 0.5$ (left) and $\Theta = 0.34$ (right)

times on different roads are indistinguishable. The travel times are collected into four samples H_1, H_2, H_3 and H_4, as demonstrated in the graph in Fig. 2 (right).

The traditional methods for route comparison give a biased estimator. As an alternative, we apply the resampling approach. For comparison, we use the mean squared errors of the plug-in $MSE(\tilde{\Theta})$ estimator and the resampling estimator $MSE(\Theta^*) = V(\Theta^*)$ because of $E(\Theta^*) = \Theta$. The experimental results are shown in Fig. 3. We can see that the resampling estimator is effective in most situations.

Conclusion

Cloud applications open new perspectives on intelligent transportation services. Data mining is one of the most important problems for such systems. We demonstrated the application of the resampling approach to the problem of route comparison in route recommendation systems. The formulas obtained allow calculation and comparison of the properties of the estimators considered. Future work will be devoted to the integration of the proposed algorithms to cloud-based TCMS and their validation on large-scale transport networks.

References

1. Afanasyeva, H.: Resampling-approach to a task of comparison of two renewal processes. In: Proc. of the 12th Int. Conf. on Analytical and Stochastic Modelling Techniques and Applications, Riga, pp. 94–100 (2005)
2. Andronov, A., Fioshina, H., Fioshin, M.: Statistical estimation for a failure model with damage accumulation in a case of small samples. Journal of Statistical Planning and Inference 139(5), 1685–1692 (2009), doi:10.1016/j.jspi.2008.05.026
3. Davison, A., Hinkley, D.: Bootstrap Methods and their Application. Cambridge university Press (1997)
4. Efron, B., Tibshirani, R.: Introduction to the Bootstrap. Chapman & Hall (1993)
5. Fioshin, M.: Efficiency of resampling estimators of sequential-parallel systems reliability. In: Proc. of 2nd Int. Conf. on Simulation, Gaming, Training and Business Process Reengineering in Operations, Riga, pp. 112–117 (2000)
6. Fiosins, M., Fiosina, J., Müller, J., Görmer, J.: Reconciling strategic and tactical decision making in agent-oriented simulation of vehicles in urban traffic. In: Proceedings of the 4th International ICST Conference on Simulation Tools and Techniques, pp. 144–151 (2011)
7. Gentle, J.E.: Elements of Computational Statistics. Springer (2002)
8. Li, Z., Chen, C., Wang, K.: Cloud computing for agent-based urban transportation systems. IEEE Intelligent Systems 26(1), 73–79 (2011)

Application of Genetic Algorithms to Determine Closest Targets in Data Envelopment Analysis

Raul Martinez-Moreno, Jose J. Lopez-Espin, Juan Aparicio, and Jesus T. Pastor

Centro de Investigación Operativa
Universidad Miguel Hernández, 03202, Elche, Spain
raul.m.m@hotmail.es, {jlopez,j.aparicio,jtpastor}@umh.es

Abstract. This paper studies the application of a genetic algorithm (GA) for determining closest efficient targets in Data Envelopment Analysis. Traditionally, this problem has been solved in the literature through unsatisfactory methods since all of them are related in some sense to a combinatorial NP-hard problem. This paper presents and studies some algorithms to be used in the creation, crossover and mutation of chromosomes in a GA, in order to obtain an efficient metaheuristic which obtains better solutions.

Keywords: Genetic Algorithms, DEA, Closest Targets.

1 Introduction

Data Envelopment Analysis (DEA) is a non-parametric technique based on mathematical programming for the evaluation of technical efficiency of a set of decision making units (DMUs) in a multi-input multi-output framework[6]. In contrast to other efficiency methodologies, e.g. stochastic frontiers, DEA provides simultaneously both an efficiency score and benchmarking information through efficient targets. Indeed, the efficiency score is obtained from the distance between the assessed DMU and a point on the frontier of the technology, which in turns serves as efficient target for the evaluated unit. In a firm, information on targets may play a relevant role in practice since they indicate keys for inefficient units to improve their performance. Conventional DEA measures of technical efficiency yield targets that are associated with the furthest efficient projection to the evaluated DMU[3]. However, several authors [3, 14] argue that the distance to the efficient projection point should be minimized, instead of maximized, in order to the resulting targets are as similar as possible to the observed inputs and outputs of the evaluated DMU. The argument behind this idea is that closer targets suggest directions of improvement for inputs and outputs that lead to the efficiency with less effort than other alternatives. The determination of closest targets

S. Omatu et al. (Eds.): *Distrib. Computing & Artificial Intelligence*, AISC 217, pp. 111–119.
DOI: 10.1007/978-3-319-00551-5_14

has attracted an increasing interest of researchers in recent DEA literature [1, 2, 13, 14]. Regarding papers that have studied the computational aspects of DEA models associated with the determination of closest targets, we may cite several references [3, 7, 8, 10]. As we will argue in detail in Section 2, all these approaches present some strong points and weaknesses and, consequently, currently there is no approach accepted as the best solution to the problem. From a computational point of view, the problem is difficult enough and this fact justifies the effort to apply new methods in order to overcome it.

In this paper, the approach in [3] is used to be solved through GA with the aim of determining closest targets in DEA. The difficulty of this problem requires us to study the GA by parts, incorporating restrictions while the results are analyzed. The remainder of the paper is organized as follows: In Section 2, a briefly introduction of the main notions associated with Data Envelopment Analysis is presented. In Section 3, the GA is developed. In Section 4 the results of some experiments are summarized, and finally. Section 5 concludes.

2 Data Envelopment Analysis and Closest Targets

In industrial sectors, technologies have been estimated using many different methods over the past 50 years [5]. DEA is a non-parametric technique that allows estimating a piece-wise-linear convex technology, constructed such that no observation of the sample of data lies above it. DEA involves the use of Linear Programming to construct the non-parametric piece-wise surface over the data. Technical efficiency measures associated with the performance of each DMU are then calculated relative to this surface, as a distance to it. Before continuing, we need to define some notation. Assume there are data on m inputs and s outputs for each of n DMUs. For the j-th DMU these are represented by $x_{ij} \geq 0$, $i = 1, \ldots, m$, and $y_{rj} \geq 0$, $r = 1, \ldots, s$, respectively. Regarding the determination of closest targets, a group of researchers [7, 8] focus their work on find all the faces of the polyhedron that defines a technology estimated by DEA. Obviously, the computing time of the suggested algorithms increases significantly as the problem size grows (n+m+s) since this issue is closely related to a combinatorial NP-hard problem. On the other hand, a second group proposes to determine closest targets resorting to Mathematical Programming [3, 10]. In particular, the following Mixed Integer Linear Program is introduced in order to determine the closest targets associate with the Enhanced Russell Graph Meassure[12], a well-known measure of technical efficiency in DEA, for DMU k, k=1,...,n.

$$
\begin{array}{llll}
\max \beta_k - \frac{1}{m}\sum_{i=1}^{m}\frac{t_{ik}^-}{x_{ik}} & & & \\
\text{s.t.} & & & \\
\beta_k + \frac{1}{s}\sum_{r=1}^{s}\frac{t_{rk}^+}{y_{rk}} = 1 & & (c.1) & \\
-\beta_k x_{ik} + \sum_{j=1}^{n}\alpha_{jk}x_{ij} + t_{ik}^- = 0 & \forall i & (c.2) & \\
-\beta_k y_{rk} + \sum_{j=1}^{n}\alpha_{jk}y_{rj} - t_{rk}^+ = 0 & \forall r & (c.3) & \\
-\sum_{i=1}^{m}\nu_{ik}x_{ij} + \sum_{r=1}^{s}\mu_{rk}y_{rj} + d_{jk} = 0 & \forall j & (c.4) & (1) \\
\nu_{ik}, \mu_{rk} \geq 1 & \forall i, r & (c.5, 6) & \\
d_{jk} \leq M b_{jk} & \forall j & (c.7) & \\
\alpha_{jk} \leq M(1 - b_{jk}) & \forall j & (c.8) & \\
b_{jk} = 0, 1 & \forall j & (c.9) & \\
\beta_k \geq 0 & & (c.10) & \\
t_{ik}^-, t_{rk}^+, d_{jk}, \alpha_{jk} \geq 0 & \forall i, r, j & (c.11, 12, 13, 14) &
\end{array}
$$

The idea behind the constraints of the preceding model is the following (see Theorem 1 in [3]): On the one hand, by (c.1)-(c.3), we are considering the set of producible points. On the other hand, with (c.4)-(c.6) we are considering supporting hyperplanes such that all the points of the estimated technology (a polyhedron) lie on or below these hyperplanes. Finally, (c.7) and (c.8) are the key conditions that connect the two previous set of constraints. Specifically, they prevent that the targets generated correspond to interior points of the estimated technology. (c.9) defines bj as a binary variable and (c.10)-(c.14) are the usual non-negative constraints. Thus, as we introduce constraints in the model we get closer to the desired solution. Nevertheless, a drawback of the approach in [3] is that it uses a "big M" to model the key constraints (c.7) and (c.8) and the value of this big positive quantity may be calculated if and only if we previously know all the faces that define the technology. Therefore, this alternative is again associated with a combinatorial NP-hard problem. The same can be claimed with respect to the approach by Jahanshahloo et al.[10].

3 Genetic Algorithms for DEA

Genetic algorithms [11] are used here to obtain a satisfactory solution of Eq.(1). A population of chromosomes representing particular feasible solutions of Eq.(1) is explored. Each chromosome represents a candidate to be the best model and it is composed by β_k, α_{jk}, t_{ik}^-, $t_{rk}^+ \in \mathbb{R}^+$ and $b_{jk} \in \{0,1\}$ being $i = 1, ..., m, j = 1, ..., n, r = 1, ..., s$. An evaluation of each chromosome is calculated by using Eq.(1).

3.1 Defining a Valid Chromosome

A valid chromosome has to satisfy at least the following constraints in equation 1: c.2, c.3, c.8, c.9, c.10, c.11, c.12 and c.14. Four methods to generate the initial population of chromosomes have been tested. Methods 1,2 and 4 are different and 3 is an extension of method 2.

- Method 1. The parameters of the chromosomes are generated randomly.
- Method 2. The process starts obtaining a random β_k and a set of α_{jk} values using algorithm 1. Next the b_{jk} values are generated considering α_{jk} in order to satisfy c.8. The values t_{rk}^{+} and t_{ik}^{-} are deduced from inputs and outputs matrices and the previously generated β_k and α_{jk} using c.2 and c.3. In case of obtaining a valid solution the algorithm ends, if not, algorithm 2 is used.

Algorithm 1. Generate alpha

Require: $X \in \mathbb{R}+^{m,n}$, $Y \in \mathbb{R}+^{s,n}$, DMU k.
Ensure: $\alpha_{10}, ..., \alpha_{n0} \in \mathbb{R}+$, $\forall \alpha_{jk} \geq 0$.

$V^x, V^y \in \mathbb{R}+^n$ such as $V_j^x = \frac{1}{m}\sum_{i=1}^{m} X_{i,j}$ and $V_j^y = \frac{1}{s}\sum_{i=1}^{s} Y_{i,j}$
Sort V^x and V^y in decreasing.
for j:=1,..., n **do**
  if $V_j^x < \lfloor n/2 \rfloor$ **and** $V_j^y >= \lfloor n/2 \rfloor$ **then**
    $\alpha_{jk} \leftarrow 0$
  else if $V_j^x >= \lfloor n/2 \rfloor$ **and** $V_j^y < \lfloor n/2 \rfloor$ **then**
    $\alpha_{jk} \leftarrow$ Generate $0.5 \leq \alpha_{jk} \leq 1$ randomly.
  else if ($V_j^x \geq$ *BoundPoint* **and** $V_j^y \geq$ *BoundPoint*) **or** ($V_j^x <$ *BoundPoint* **and** $V_j^y <$ *BoundPoint*) **then**
    $\alpha_{jk} \leftarrow$ Generate $0 \leq \alpha_{jk} \leq 0.25$ randomly.
  end if
end for
Find the minimum α_{jk} and modify its value in order to satisfy expression 1

Algorithm 2 decreases and increases β_k obtained a closet value to the minimal β_k that satisfy c.11. Two parameters are used: a factor number p and the maximal number of iterations of the process, that also determines the increasing or decreasing value of β_k. To increase β_k, a similar algorithm is used working in a similar way to decrease β_k, but doing the add operation in the first inner loop and using the constraint (c.12) as a condition instead of (c.11) for both inner loops.

- Method 3. This method is an extension of method 2 through adding a third tweak process for the α_{jk} (algorithm 3).
- Method 4. The method 4 consists in an hyper-heuristic (metaheuristic that operates over some other metaheuristic) [4]. A GA has been used to produce sets of α_{jk} and β_k for the initial population, that satisfy c.2, c.3, c.11, c.12 and c.14. The evaluation function in the hyper-heuristic is sum of the negative values of t_{rk}^{+} and t_{ik}^{+}, being the chromosomes with the higher punctuation a better candidate. Moreover β_k and α_{jk} are considered, penalizing values close to 1.

Algorithm 2. Decrease β_k

Require: β_k, $p \in \mathbb{Z}$, $MaxIter \in \mathbb{Z}$.
Ensure: A minimal β_k for the chromosome that satisfy constraint 11: $\forall t_{ik}^- \geq 0$.

```
d ← 0
while d ≤ MaxIter do
   while ∀t⁻_ik ≥ 0 do
      {Decrease β_k while still satisfies (c.11)}
      β_k ← β_k − p/d
      Generate t⁻_ik using expression 1.
   end while
   {At this point (c.11) is not satisfied because of the decrease of β_k. Increase
   β_k value until it is satisfied again}
   repeat
      β_k ← β_k + p/d
      Generate t⁻_ik using expression 1.
   until ∀t⁻_ik ≥ 0
   d ← d + 1
end while
```

Algorithm 3. Generate a chromosome (Method 3)

Require: $X \in \mathbb{R}+^{m,n}$, $Y \in \mathbb{R}+^{s,n}$, $p \in \mathbb{R}$, DMU k, $MaxIter \in \mathbb{Z}$
Ensure: Chromosome c.

```
repeat
   Generate 0 ≤ β_k ≤ 1 randomly.
   Generate α_1k, ..., α_nk using algorithm 1 and t⁻_1k, ..., t⁻_mk and t⁺_1k, ..., t⁺_sk using
   expression 1.
   while ∃t⁻_ik ≤ 0 and ∃t⁺_rk ≤ 0 and number of α_jk ≠ 0 > 2 do
      if ∃t⁻_ik ≤ 0 and ∀t⁺_rk ≥ 0 then
         Increase β_k
      end if
      if ∀t⁻_ik ≥ 0 and ∃t⁺_rk ≤ 0 then
         Decrease β_k
      end if
      if ∃t⁻_ik ≤ 0 or ∃t⁺_rk ≤ 0 then
         Choose α_jk no null randomly and make it equal to zero, decreasing the
         rest α_jk in p and find the minimum α_jk and modify its value in order
         to satisfy expression 1
      end if
   end while
until A valid chromosome is obtained or Iterations ≥ MaxIter or the number
of α_jk no null is lower than n-2
```

3.2 Select the Best Ranking, Crossover and Mutation

In each generation a proportion of the existing population is selected to breed a new generation. A comparison of the evaluations of all the chromosomes in the population is made in each generation, and only part of them (those which are in the best ranking) will survive. The number of chromosomes which survive in each population (called $SurvSize$) is preset. For each two new solutions to be produced ("son" and "daughter"), a pair of "parent" ("father" and "mother") chromosomes is selected from the set of chromosomes.

Due to the number of constraints to satisfy, a traditional crossover method will produce an offspring with a higher rate of non valid chromosomes. With the aim of avoid this, a crossover with different "levels" is introduced in algorithm 4. The defined levels are 1 for β_k, 2 for α_{jk}, 3 for t^-_{ik} and 4 for t^+_{rk}. A level is chosen randomly, values in that level are crossed and those in lower levels deduced to satisfy eq.(1). Finally, in each iteration a chromosome is randomly chosen to be mutated. The procedure is similar to crossover, considering the dependences between variables and deffining different levels of mutation.

4 Experimental Results

Some experiments have been carried out to tune certain parameters of the algorithm. In all of them, $Population = 100$, $SurvSize=\frac{PopSize}{2}$ and $MaxIter = 1000$. The experiments have two objectives: One is to compare the four methods of creating valid chromosomes. The other objective is to study the GA, both varying the problem size. Table 1 shows the result obtained for both objectives. In the first part the averaged percentage of chromosomes and the execution time are shown when the four methods are used. It is possible to see that the number of valid chromosomes obtained using method 3 is higher than those obtained by the rest of methods. Thus, method 3 is used in the generation of valid chromosomes in the GA. In the second part, the solution achieved by the GA and the execution cost are shown.

From the results, it may be concluded that Methods 2 and 4 yield low values for the averaged percentage of valid chromosomes (<11%), while Method 2 generates greater values. However, for Method 2, this percentage decreases as the size of the problem increases (from 50.83% to 13.90%). On the other hand, Method 3 produces the best percentages for each of the four experiments, achieving a value of 100% in the case of m=4, s=2 and n=30. Regarding the GA, relatively high values are observed for some experiments (70% and 84.22%). Nevertheless, in this case, the relation between the percentage of valid chromosomes and the size is unclear.

Algorithm 4. Crossover

Require: Chromosome c_1, Chromosome c_2.
Ensure: Chromosome c_3, Chromosome c_4

```
Generate 0 ≤ γ ≤ 1 randomly.
Generate 0 ≤ φ ≤ 4 randomly.
if φ = 0 then
  {Crossing β_k}
  β_k^3 ← β_k^1 * γ + (1 − γ) * β_k^2
  β_k^4 ← β_k^2 * γ + (1 − γ) * β_k^1
else if φ ≤ 1 then
  {Crossing α_jk}
  for j := 1, ..., n do
    α_jk^3 ← α_jk^1 * γ + (1 − γ) * α_jk^2
    α_jk^4 ← α_jk^2 * γ + (1 − γ) * α_jk^1
  end for
else if φ ≤ 2 then
  {Crossing t_ik^-}
  for i := 1, ..., m do
    t_ik^{-,3} ← t_ik^{-,1} * γ + (1 − γ) * t_ik^{-,2}
    t_ik^{-,4} ← t_ik^{-,2} * γ + (1 − γ) * t_ik^{-,1}
  end for
else if φ ≤ 3 then
  {Crossing t_rk^+}
  for r := 1, ..., s do
    t_rk^{+,3} ← t_rk^{+,1} * γ + (1 − γ) * t_rk^{+,2}
    t_rk^{+,4} ← t_rk^{+,2} * γ + (1 − γ) * t_rk^{+,1}
  end for
end if
Deduce α_jk, b_jk, t_ik^- and t_rk^+ for c1 and c2 in order to satisfy eq.(1)
```

Table 1 Averaged percentage and execution cost (in seconds) of the four methods to generate valid chromosomes and the solution obtained by the GA, its execution cost (in seconds) when varying the problem size

size			Method 1		Method 2		Method 3		Method 4		Genetic		
m	s	n	time	% val.	time	% val.	time	% val.	time	% val.	solution	time	% val.
2	1	15	0.003	0.75	0.008	50.83	26.423	82.08	17.244	10.58	0.8395	0.207	70.00
3	2	25	0.004	0.00	0.010	32.65	6.722	90.05	21.283	0.80	0.8059	0.297	56.90
4	2	30	0.004	0.00	0.022	25.57	0.223	100.00	29.521	1.57	0.7305	0.173	84.22
5	3	40	0.004	0.00	0.019	13.90	13.125	73.90	18.187	0.05	0.7225	0.805	44.55

5 Conclusions and Future Works

It had been used a GA with the aim of solve a DEA problem satisfactorily from a computational point of view. Different population methods have been developed and compared, in order to obtain better results in the GA. Nevertheless it hadn't been tackled the whole problem, remaining for a future work the development of a method considering the constrains 1, 4, 5, 6, 7 and 13 in Eq.(1). Improve the population method and try different heuristics could help to achieve this objective. It is also in mind a deepest study of the relation between the different initial paramethers (*Population*, $MaxIter$, *SurvRate*,...) and the size of the problem with the computing time requiered and the effectiveness of the algorithm as well as the development of a parallel version (with shared and distributed memory) in order to gain computing capability. Finally, a comparasion of the introduced approach with other previous works could be important, especially in terms of accuracy and resources usage.

References

1. Amirteimoori, A., Kordrostami, S.: A Euclidean distance-based measure of efficiency in data envelopment analysis. Optimization 59, 985–996 (2010)
2. Aparicio, J., Pastor, J.T.: On how to properly calculate the Euclidean distance-based measure in DEA. Optimization (2012), doi:10.1080/02331934.2012.655692
3. Aparicio, J., Ruiz, J.L., Sirvent, I.: Closest targets and minimum distance to the Pareto-efficient frontier in DEA. J. Prod. Anal. 28, 209–218 (2007)
4. Burke, E., Hart, E., Kendall, G., Newall, J., Ross, P., Schulenburg, S.: Hyper-heuristics: An Emerging Direction in Modern Search Technology. In: Handbook of Metaheuristics, vol. 16, pp. 457–474 (2003)
5. Coelli, T., Rao, D.S.P., Battese, G.E.: An Introduction to Efficiency and Productivity Analysis. Kluwer Academic Publishers, Boston (1998)
6. Cooper, W.W., Seiford, L.M., Tone, K.: Data envelopment analysis: a comprehensive text with models, applications, references and DEA-solver software. Kluwer Academic Publishers, Boston (2000)
7. Jahanshahloo, G.R., Hosseinzadeh Lotfi, F., Zohrehbandian, M.: Finding the piecewise linear frontier production function in data envelopment analysis. Appl. Math. Compt. 163, 483–488 (2005)
8. Jahanshahloo, G.R., Hosseinzadeh Lotfi, F., Zhiani Rezai, H., Rezai Balf, F.: Finding strong defining hyperplanes of Production Possibility Set. Eur. J. Oper. Res. 177, 42–54 (2007)
9. Jahanshahloo, G.R., Vakili, J., Mirdehghan, S.M.: Using the minimun distance of DMUs from the frontier of the PPS for evaluating group performance of DMUs in DEA. Asia Pac. J. Oper. Res. 29(2), 1250010-1–1250010-25 (2012a)
10. Jahanshahloo, G.R., Vakili, J., Zarepisheh, M.: A linear bilevel programming problem for obtaining the closest targets and minimum distance of a unit from the strong efficient frontier. Asia Pac. J. Oper. Res. 29(2), 1250011-1–1250011-19 (2012b)

11. Mitchell, M.: An Introduction to Genetic Algorithm. MIT Ps (1998)
12. Pastor, J.T., Ruiz, J.L., Sirvent, I.: An Enhanced DEA Russell Graph Efficiency Measure. Eur. J. Oper. Res. 115, 596–607 (1999)
13. Pastor, J.T., Aparicio, J.: The relevance of DEA benchmarking information and the Least-Distance Measure: Comment. Math. Comput. Model. 52, 397–399 (2010)
14. Portela, M.C.A.S., Borges, P.C., Thanassoulis, E.: Finding Closest Targets in Non-Oriented DEA Models: The Case of Convex and Non-Convex Technologies. J. Prod. Anal. 19, 251–269 (2003)

Odor Classification Based on Weakly Responding Sensors

Sigeru Omatu[1], Mitsuaki Yano[1], and Toru Fujinaka[2]

[1] Osaka Institute of Technology Faculty of Engineering, 5-16-1 Omiya Asahi-ku Osaka, 535-8585, Japan
omatu@rsh.oit.ac.jp, yano@elc.oit.ac.jp

[2] Hiroshima University Faculty of Education, 1-1-1 Kagamiyama, Higashi-Hiroshima, 739-8524, Japan
fjnk@hiroshima-u.ac.jp

Abstract. We consider an array sensing system of odors and adopt a layered neural network for classification. Measurement data obtained from fourteen metal oxide semiconductor gas (MOG) sensors are used, where some sensors exhibit relatively weak responses. We propose two methods for enhancing such weak signals to obtain better classification results. One method is to apply scaling to magnify the weak signals as to increase their significance in the classification criteria. The other method also involves magnifying the weak signals. However, predetermined values are assigned in the order of the magnitude of the actual signals. In both methods the group of weak signals is first determined. Then their values are negated prior to scaling, in order to be distinguished from stronger signals. An experiment shows that the accuracy of classifying five kinds of odors is improved from 74% to 85%.

Keywords: feature vector scaling, weak odor signals, odor classification, neural networks.

1 Introduction

Olfaction is one of the five senses, and it is important to achieve the high level of information processing based on the data obtained from odor sensors. We use these five senses to enjoy comfortable human life with communication and mutual understanding. Artificial odor sensing and classification systems through electronic technology are called an electronic nose and they have been developed according to various odor sensing systems and several classification methods [1],[2]. After those papers, there was much development about intelligent electronic nose systems.

Recently, we have developed electronic nose systems to classify the various odors under different densities based on a layered neural network and a competitive neural network of the learning vector quantization method [3], [4],[5],[6].

Although the human olfactory system is not fully understood by physicians, the main components about the anatomy of human olfactory system are known to be the olfactory epithelium, the olfactory bulb, the olfactory cortex, and the higher brain or cerebral cortex as shown in Fig. 1. From our experience, it is difficult to achieve the perfect classification even if we use many sensors since some of them are insensitive

S. Omatu et al. (Eds.): *Distrib. Computing & Artificial Intelligence*, AISC 217, pp. 121–128.
DOI: 10.1007/978-3-319-00551-5_15

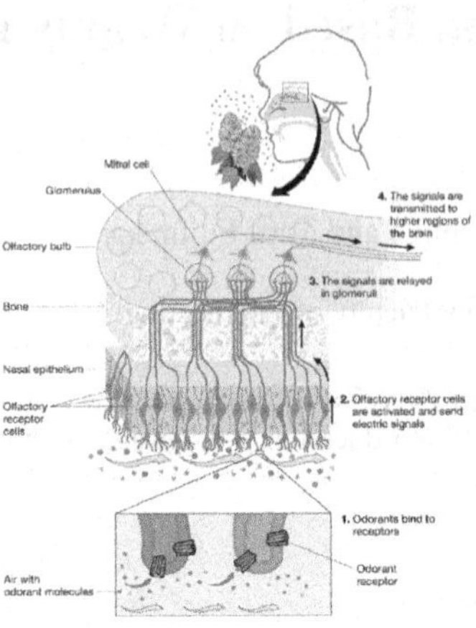

Fig. 1 Olfactory system:EMBO reports (2007)

for some odors. This means that some sensors are actively sensitive and effective for classification. But insensitive sensors may become sensitive for other odors and we cannot eliminate those sensors even if they are insensitive to an odor since they could be useful for the other odors.

To solve the dilemma, we use all sensors regardless of the sensitivity to Specific odor. We extract the features of odors by processing all sensors by reforming the data such that insensitive data become useful. In this paper, we propose two methods to extract the feature vectors by processing the insensitive data.

2 Experimental Data

We carried out two experiments, Experiment I and Experiment II, as shown in Table 1 where variation means fluctuation level for each specimen. In Experiment I, we measured the odors for the same kind of coffees produced in different countries such as Colombia, Guatemala, and Ethiopia and the odors of two kinds of blended coffees namely, Blend Coffee 1 and Blend Coffee 2. In Experiment I, six times of the measurement have been repeated for each coffee. Thus, the total number of features of odors is thirty for the five kinds of coffees. In Experiment II, we have measured odors of three kinds of teas, coffee, and cocoa as shown in Table 1. In Experiment II, we measured six times for each group of specimens as done in Experiment I. Thus, total number of the features in Experiemnt II is also thirty as in Experiment I. The experimental conditions for both Experiment I and II are summarized in Table 2.

The neural network is a layered type based on the error back-propagation method. The numbers of neurons for this experiment are fourteen in the input layer, thirty in the hidden layer, and three in the output layer, namely, 14-30-3 structure. The

Table 1 Experiment I and II

No.	Variation	Specimens
I	large	Three districts coffee (Colombia, Guatemala, Ethiopia), Two blend coffees
II	small	English tea, Green tea, Oolong tea, Coffee, Cocoa

Table 2 Environmental data for Experiment I and II

Subject	Units
Gas flow	1[*l*/min]
Temperature of gas	35[°C]
Odor measurement time	12[min]
Dry air flow time	7[min]

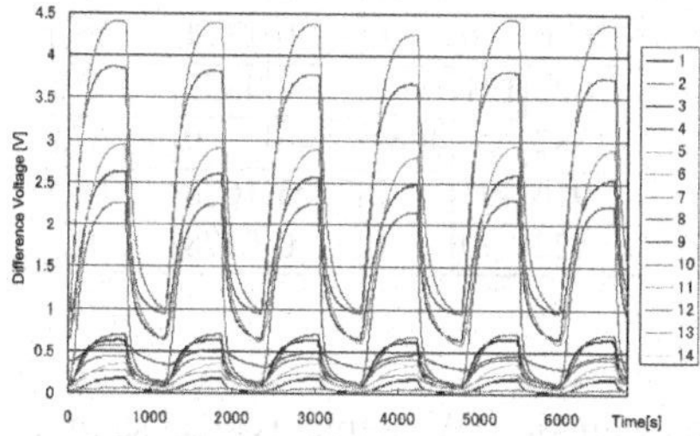

Fig. 2 Sample path of blend coffee 2 of Experiment I

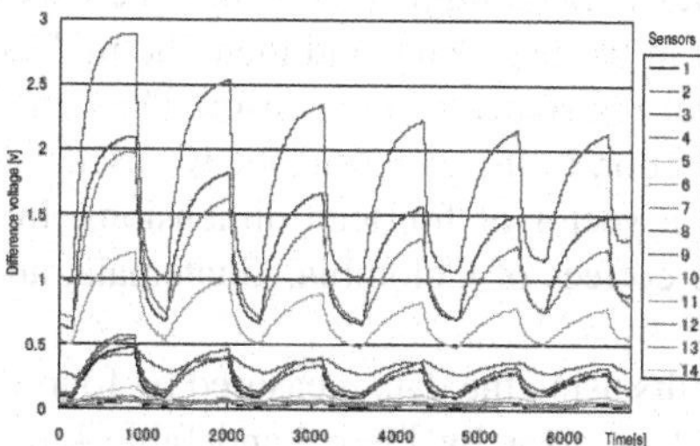

Fig. 3 Sample path of tea of Experiment II

reason for using three neurons in the output layer is that three bits can represent eight numerals where in Experiment I and Experiment II, we use five patterns for classification. We separated the six measurement data for each odor into three training data and three test data. The combination of the data is ${}_6C_3$=20 for each odor and total number of the test data becomes 32×10^5. Among them we have selected

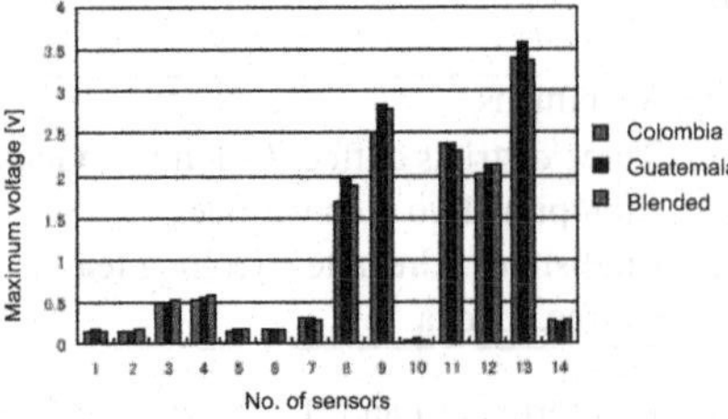

Fig. 4 Feature vectors for coffees in Experiment I

Table 3 Classification results for Experiment I. Here, A is Colombia coffee, B is Guatemala coffee, C is Ethiopia coffee, D is blended coffee 1, and E is blended coffee 2.

	Classification results:60.4 (%)					
Odor data	A	B	C	D	E	Correct (%)
A	4686	0	0	0	1314	78.1
B	1022	121	0	0	4857	2.0
C	1387	0	4067	546	0	67.8
D	956	499	2	3765	1678	62.8
E	522	0	0	0	5478	91.3

six thousand data for the evaluation. We depict typical sample paths of blend coffee 2 of Experiment I and Tea of Experiment II, by Fig. 2 and Fig. 3, respectively.

We select the maximum value of each sample path as a feature of an odor for a sensor. The reason to use the maximum value is that the odor will be accumulated in the sensor for a while in the beginning and then, the molecules will be oxidized by reducing the electrical registance of the sensor. Thus, the maximum value reflects steady state of the odor. In the experiments, we used fourteen sensors. Thus, for each odor, we have a vector of fourteen dimensions. In Fig. 4 we show the feature vectors for three coffees of Colombia, Guatemala, and blend 2 in Experiment I.

The classification results using the data measured by fourteen sensors for Experiment I and Experiment 2 are given by Table 3 and Table 4, respectively.

The results of these experiments are not so good. Especially, for Experiment I it is necessary to improve the classification results. In this case, from Fig. 4 we can see that there are two groups for the feature vectors. One is sensitive for odors and the other is insensitive among fourteen sensors. Thus, if we utilize the data of insensitive sensors, we can expect to improve the classification results since there exist differences of features between odors even if the fluctuation levels are not large.

Table 4 Classification results for Experiment II. Here, A is tea, B is green teas, C is oolong tea, D is coffee, and E is cocoa.

	Classification results:74.3 (%)					
Odor data	A	B	C	D	E	Correct(%)
A	4525	100	513	0	862	75.4
B	101	3967	0	0	1932	66.1
C	983	423	3959	635	0	65.8
D	0	283	0	5717	0	95.3
E	629	1234	2	0	4135	68.9

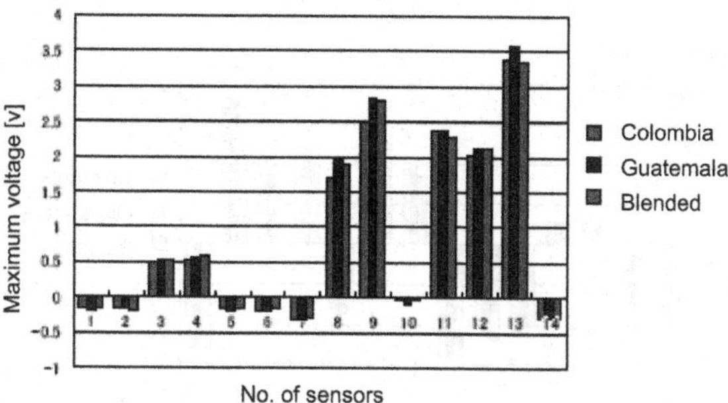

Fig. 5 Feature vectors of the sign reversed version of Experiment I

3 Two Methods for Improving the Classification Results

In what follows, we will propose two methods to use those data in order to improve the classification rate.

3.1 Method 1

Neurons become sensitive near thresholds since the derivative of the sigmoid function becomes maximum at the threshold value. In other words, the outputs of the neurons become insensitive when the absolute value of the net input becomes very large. Thus, it is preferable to move the net input to some range such as [-5,5]. We propose Method 1 as stated in the following. First, we make two groups for the feature vectors such that one is sensitive and the other is insensitive. In Experiment I, lower ranking group (G1) is the sensors 1, 2, 5, 6, 7, and 14 and higher ranking group (G2) is the sensors 3, 4, 8, 9, 11, 12, 13.

Then, reverse the sign of values in G1 (cf. Fig 5) and scale the values with negative sign such that the maximum among G2 is equal to the absolute value of the smallest value of G1 by linear interpolation (cf. Fig 6).

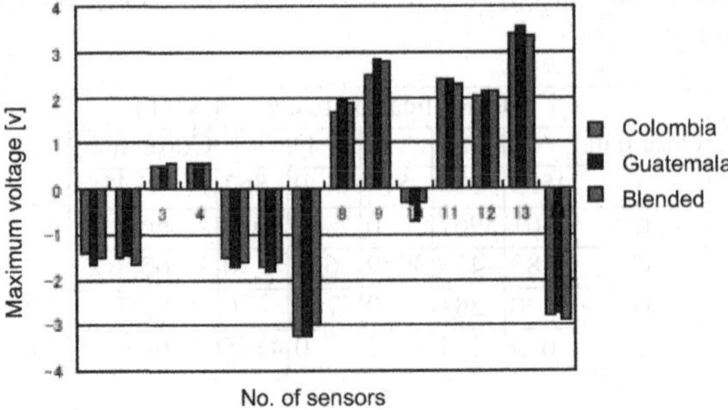

Fig. 6 Feature vectors of linearly interpolated version of Experiment I

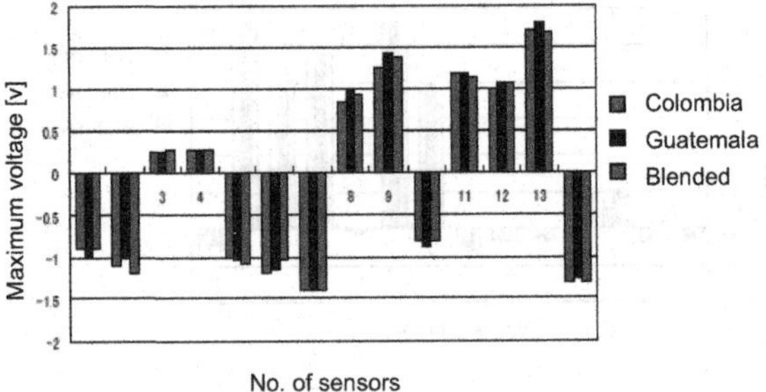

Fig. 7 Feature vectors of ranked version of Experiment I

3.2 Method 2

This method is a ranking of the measurement data. First, separate the data into two groups, G1 and G2 as Method 1 and reverse the sign of values in G1. Then, arrange in decreasing order of absolute value and assign the value 1.4 for the first sensor, 1.3 for the second sensor, etc. in order of decreasing absolute vale. After that, scale the value in G2 in the same way by assigning the values from 1.7, 1.4, 1.3, 1.1, 0.9, 0.3, 0.3. The scaled values are determined by trial and error according to the rule such that G2 values are more important than G1 values (cf. Fig 7)

3.3 Comparion between Method 1 and Method 2

The simulation results for Experiment I and Experiment II by two methods are shown in Table 5-Table 6 and Table 7-Table 8, respectively. From these results, we can see Method 1 could improve the classification results compared with Method 2

Table 5 Classification results by Method 1 for Experiment I

	Classification results:91.0 (%)					
Odor data	A	B	C	D	E	Correct (%)
A	5952	23	4	1	20	99.2
B	51	4793	0	214	942	79.9
C	4	0	5996	0	0	99.9
D	0	440	2	4555	1003	75.9
E	0	0	0	0	6000	100

Table 6 Classification results by Method 2 for Experiment I

	Classification results:87.9 (%)					
Odor data	A	B	C	D	E	Correct (%)
A	5949	51	0	0	0	99.2
B	1255	4445	4	296	0	74.1
C	494	0	4353	1153	0	72.6
D	0	390	0	5610	0	93.5
E	0	0	0	0	6000	100

Table 7 Classification results by Method 1 for Experiment II

	Classification results:85.6 (%)					
Odor data	A	B	C	D	E	Correct (%)
A	4902	0	1098	0	9	81.7
B	81	5768	0	9	142	96.1
C	324	5	5421	248	2	90.4
D	0	0	450	5550	0	92.5
E	1031	94	588	260	4027	67.1

Table 8 Classification results by Method 2 for Experiment II

	Classification results:59.7 (%)					
Odor data	A	B	C	D	E	Correct (%)
A	5178	754	9	0	59	86.3
B	952	3253	0	0	1795	54.2
C	504	1166	556	532	3242	9.3
D	0	347	469	5184	0	86.4
E	601	1644	0	9	3746	62.4

4 Conclusions

In this paper, a new approach to odor classification by using MOG sensors has been discussed. After surveying the odor sensing and classification methods, we have examined two examples, Experiment I and Experiment II to increase the classification accuracy. Basically, we have proposed two methods by using the data measured by insensitive sensors. From these results, using those data is efficient to improve the classification accuracy significantly. We will search new features of odor information for classification in the future.

Acknowledgement. This research has been supported by Grant-in-Aid for Challenging Exploratory Research No. 23360175 of Japan Society for the Promotion of Science and we wish to thank JSPS for their support.

References

1. Milke, J.A.: Application of Neural Networks for discriminating Fire Detectors. In: 10th International Conference on Automatic Fire Detection, AUBE 1995, Duisburg, Germany, pp. 213–222 (1995)
2. Bicego, M.: Odor classification Using similarity-Based Representation. Sensors and Actuators B 110, 225–230 (2005)
3. Omatu, S., Yano, M.: Intelligent Electronic Nose System Independent on Odor Concentration. In: Abraham, A., Corchado, J.M., González, S.R., De Paz Santana, J.F. (eds.) International Symposium on DCAI. AISC, vol. 91, pp. 1–9. Springer, Heidelberg (2011)
4. Omatu, S., Araki, H., Fujinaka, T., Yano, M.: Intelligent Classification of Odor Data Using Neural Networks. In: ADVCOMP 2012, Barcelona, Spain, pp. 1–7 (2012)
5. Omatu, S.: Pattern Analysis for Odor Sensing System, pp. 20–34. IGI Global (2012)
6. Fujinaka, T.; Yoshioka, M., Omatu, S., Kosaka, T.: Intelligent Electronic Nose Systems for Fire Detection Systems Based on Neural Networks. In: The Second International Conference on Advanced Engineering Computing and Applications in Sciences, Valencia, Spain, pp. 73–76 (2008)
7. General Information for TGS sensors, Figaro Engineering (2012), `http://www.figarosensor.com/products/general.pdf`

A Model to Visualize Information in a Complex Streets' Network

Taras Agryzkov[1], José L. Oliver[2], Leandro Tortosa[1], and José F. Vicent[1]

[1] Departamento de Ciencia de la Computación e Inteligencia Artificial, Universidad de Alicante, Spain
taras@taras.es, {tortosa,jvicent}@ua.es

[2] Departamento de Expresión Grafica y Cartografia, Universidad de Alicante, Spain
joseluis.oliver@ua.es

Abstract. This paper discusses a process to graphically view and analyze information obtained from a network of urban streets, using an algorithm that establishes a ranking of importance of the nodes of the network itself. The basis of this process is to quantify the network information obtained by assigning numerical values to each node, representing numerically the information. These values are used to construct a data matrix that allows us to apply a classification algorithm of nodes in a network in order of importance. From this numerical ranking of the nodes, the process finish with the graphical visualization of the network. An example is shown to illustrate the whole process.

1 Introduction

A complex network can be regarded as any collection of nodes, that are interacting as a system and that are connected by directed or undirected edges, that is, a graph. The science of complex networks is characterized by a strong interaction between theory and applications. In this respect complex networks, and specially streets networks, serve as natural models to describe the organization of a system.

The most common approach to convert physical street network into an abstract mathematical graph is mapping street intersections as nodes while mapping streets segments between intersections as links. Because this approach directly maps geographic entities into graph entities with the same dimension, it is call the primal approach. Correspondingly, a dual approach is indirectly mapping streets segments as nodes and intersections as links (see [10, 16, 5]). The abstract dual representation of the street network is more suitable when the subject of analysis focuses on the open spaces that street segments implicit created.

We are here interested in the study of a particular class of complex networks that describe the street pattern of cities. The beginning of these studies can be traced back to the classical works on regional transportation networks based on graph theory [12]. The advent of complex system science and its paradigm [2, 3], jointly with the increasing availability of spatial and time geo-referenced data, has given a new boost to these studies, and several important contributions have appeared recently [4, 18].

S. Omatu et al. (Eds.): *Distrib. Computing & Artificial Intelligence,* AISC 217, pp. 129–136.
DOI: 10.1007/978-3-319-00551-5_16

Masucci et al. (see [15]) study the structural property of the London street network in its dual and primal representation. In [11] the author, by using 40 urban networks in a dual representation, found a small-world structure and a scale-free property for both street length and connectivity degree, and used various centrality indices as indicators of the importance of streets. Lammer et al. (see [13]) developed a comparative analysis of the betweenness distribution in 20 cities in Germany, suggesting a relation with vehicular traffic. Others have focused on centrality in primal and dual representations of streets' networks [7, 8, 9].

The study of complex networks present new methods for detecting the important or relevant nodes in a given network and to distinguish core nodes from peripheral network elements. Several techniques are proposed, ranging from structure preserving model reduction, shortest path trees, network motifs, as well as variations of Google PageRank algorithm (see [6, 14, 17, 1]). The PageRank model uses the structure of the Web to build a Markov chain considering a nonnegative and irreducible matrix M for transition probability called also a primitive matrix. This property guarantees that, according to Frobenius test for primitivity, a stationary vector exists as a solution for the PageRank problem.

This paper discusses a process to graphically view and analyze information related to a network of urban streets, using an algorithm that establishes a ranking of importance of the nodes of the network itself.

2 A Model to Visualize the Information in the Network

The process of displaying information in a graphical way for an urban network that we propose in this paper comprises the following steps:

Step 1 Collecting data.
Step 2 Numerical assignment to each of the network nodes of the data obtained.
Step 3 Applying an algorithm for ranking the nodes.
Step 4 Visualizing the network according to the importance of each node in it.

The data collection process is a field study that consists of collecting the data or information you want to analyze or visualize. Subsequently, these data must be assigned to the nodes of the network so that each node has a set of numerical values associated with the information that is being studied. Then, we apply the algorithm that classifies the nodes and, finally, we proceed with the graphical representation of the result. In this section we describe this process in detail.

2.1 The Abstract Model to Represent Streets' Networks

To represent an abstract model for streets' networks, our starting point is the concept of primal graphs, where intersections are turned into nodes and streets into edges. We see a real example of a streets' network and its primal graph associated in Figure 1.

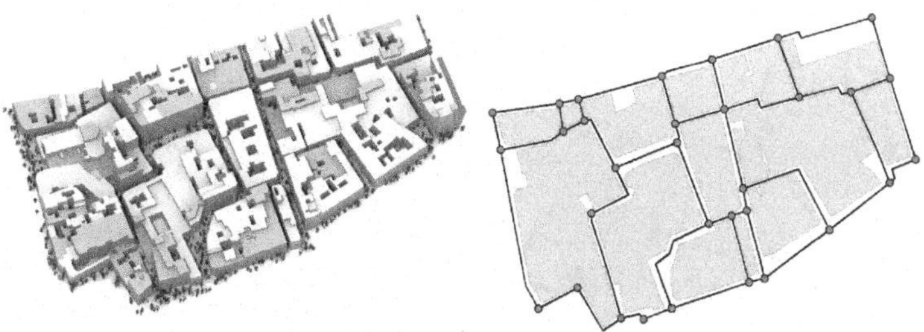

Fig. 1 A primal graph representation from a small part of a city

The model of primal graph that we can see in Figure 1 only represents the urban network, while not allows us to obtain any information on other aspects of the network itself or about its characteristics. If we are representing a city as a network, the primal graph gives us information about the connectivity of the nodes and shows the geometry of it. Therefore, if we need to visualize or understand some relevant information about the network or display some sort of activity, whether commercial or otherwise, we need to include this information somehow.

Our model assigns numerical values to each node, according to the information we are measuring or analyzing. For example, if we are measuring the commercial activity of an urban network, we have to assign numerical values to each node depending on business activity that occurs in their environment. Thus, the process of valuation and allocation of information to numeric values in each of the vertices is essential throughout the process. Following the example of the allocation of business to the nodes of a network, we must consider some details in the allocation process.

- **Nodes and information**. In our evaluation model of the network, a node is a key element because it is the element that contains the most valuable information about the characteristics of the urban network. To every node of the network are assigned values corresponding to quantitative and qualitative allocations adjacent thereto. The values we assign to the node are those who will participate in the state assessment.
- **Allocation criteria of the commercial activity to the nodes.** The approach we take to assign the appropriate allocations to nodes is as follows: to each node we assign only those endowments that pertain to the edges that concur in it. This is a general approach since the idea behind this approach is that each node will have associated commercial allocations that are in its proximity. Often, due to the particular characteristics of the urban framework, it is not as simple as it seems the numerical allocation of endowments to each of the nodes. We see an example to better understand the process of allocating endowments to the nodes.

In Figure 2, we see an example to illustrate the process we follow to establish the allocation criteria of the commercial activity for an urban network. On the left side of the image we have represented a small part of a real urban network and the location

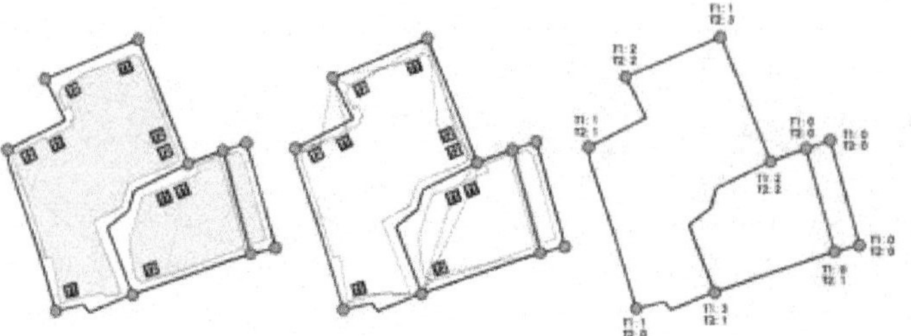

Fig. 2 Three images reflecting the allocation criteria of the commercial activity for the nodes of an urban streets' network

of endowments in this area. The central part of the figure shows the representation of the assignation of the allocations for each node, while in the right part of the image we have already established the commercial activity or corresponding numerical assignment to each of the nodes.

3 The Adapted PageRank Algorithm

Once assigned numerical values of the analyzed information to each node, an algorithm must be run with the aim to classify the nodes by their importance.

The PageRank method [17] was proposed to compute a ranking for every Web page based on the graph of the Web, regardless of their content, based solely on their location in the Web's graph structure. The purpose of the method is obtaining a vector, called *PageRank vector*, which gives the relative importance of the pages.

However, this method is not appropriate when we work with complex urban networks, due to the specific characteristics of these networks. In [1], Agryzkov et al. propose an adaptation of the PageRank model to establish a ranking of nodes in an urban network, taking into account the influence of external information. The central idea behind the algorithm for ranking the nodes is the construction of a data matrix D, which allows us to represent numerically the information of the urban network that we want to analyze and measure.

Let us assume that we have an urban network with n nodes and that $c_1, c_2, \ldots, c_l$ are the l characteristics that we want to quantify at each of the nodes. So, for each node i, d_{ij} represents the value of the characteristic c_j associated to node i. Besides, we construct a vector $\mathbf{v_0}$ with l components according to the importance of each of the characteristics evaluated. With these parameters, the algorithm proposed is:

Algorithm 1 *Algorithm to classify the nodes.*

INPUT: The transition matrix A, the elements d_{ij}, for $i = 1,2,\ldots,n$ and $j = 1,2,\ldots,l$, and the vector $\mathbf{v_0}$.

OUTPUT: The PageRank vector $\mathbf{r}$.

Step 1 *Construct the data matrix D as $D = (d_{ij})$, for $i = 1,2,\ldots,n$ and $j = 1,2,\ldots,l$.*

Step 2 *Compute* $\mathbf{v}$ *by multiplying* $D \cdot \mathbf{v_0} = \mathbf{v}$.
Step 3 *Normalize* $\mathbf{v} \longrightarrow \mathbf{v}^*$.
Step 4 *Construct the matrix* V*, from* $\mathbf{v}^*$.
Step 5 *Construct the matrix* $M' = (1-\alpha)A + \alpha V$*, from* A *and* V.
Step 6 *Compute the eigenvector* $\mathbf{r}$ *associated to the eigenvalue* 1 *for the matrix* M'.

The result of applying this algorithm to a network is a vector $\mathbf{r} = \{r_1, r_2, \ldots, r_n\}$ with n components, where the i-th component represents the ranking of i-th node within the overall network.

Summarizing the key points of this process, we can say that the mean feature of this algorithm is the construction of the matrix D and the vector $\mathbf{v_0}$. Firstly, the matrix D allows us to represent numerically the information we want to study; secondly, the vector $\mathbf{v_0}$ allows us to establish the importance of each of the factors or characteristics that have been measured and take part of D.

In other words, we can say that the algorithm 1 constitutes a model to establish a ranking of nodes in an urban network, with the primary feature that assigns a value to each node according to its significance within the physical network.

4 An Example to Discuss the Model

In this example we develop the model described in the previous sections, considering a part of a real urban network, corresponding to the city of Murcia (Spain). The objective is the visualization and analysis of the city's commercial activity using this model. To perform this task, we have collected data from shops, bars, restaurants, banks, supermarkets that we can find in the city. This information is transferred to vectors associated with each node and, in turn, these data have been used to construct the data matrix D which appears in Algorithm 1.

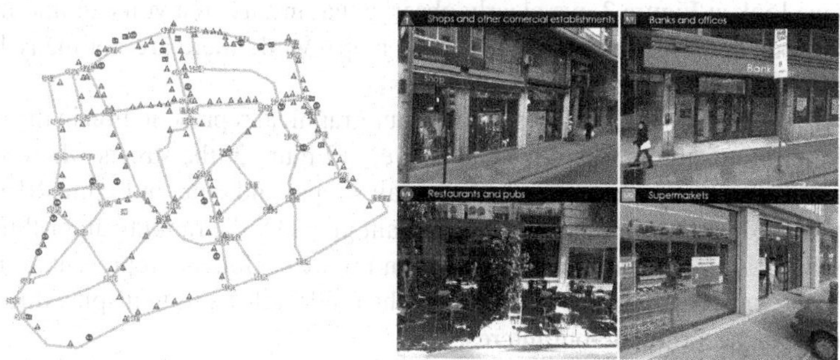

Fig. 3 Assignment of commercial endowments to the nodes in a streets' network

Figure 3 shows how we conduct the process of allocating commercial allocations to each of the nodes. Different colors are used to represent the different types of business that we are considering (bars in red, shops in green, banks in blue and department stores in orange). The idea is to place on the map the different items associated to commercial activity and assign each node those that are in its proximity, with the criteria exposed in Section 2. Once we have all the nodes with its numerical value, we construct the matrix D and execute the algorithm, which gives us a classification of the nodes, according to their importance in the network.

The last step is to represent the ranking of the nodes; to do that task we can use a gradient scale ranging from red (most important nodes) to blue (least important nodes in the network) to visualize each of the nodes.

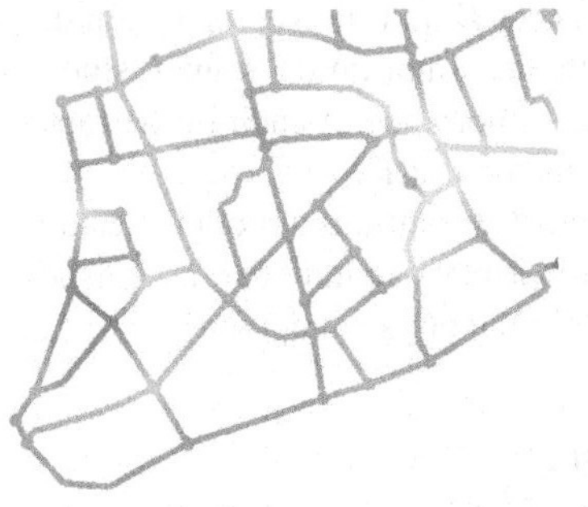

(a) Ranking endowments based on the presence of bars and restaurants.

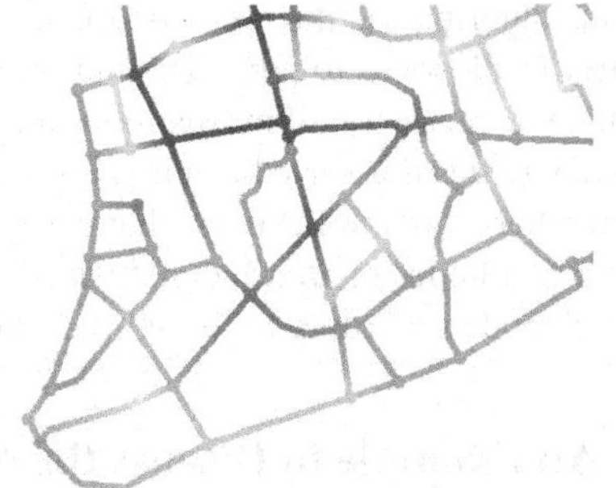

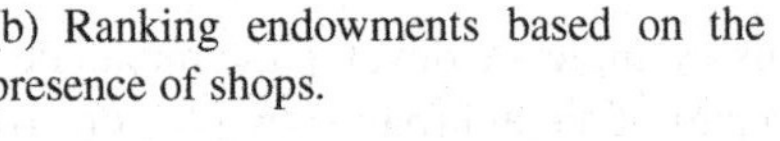

(b) Ranking endowments based on the presence of shops.

Fig. 4 Visualizing the nodes in the example network, according to their importance

Figure 4(a) shows an image of the urban network in which we have represented commercial allocations related to the existence of bars and restaurants. We can see that are colored in red vertices where a certain concentration of such endowments exist. If we look at Figure 3, we clearly observe that in this area you can find many different types of stores (green marks on the map), while there are not many bars, restaurants or offices (blue, red, orange marks).

However, in Figure 4(b) we have focused our graphical representation in the presence of shops and stores of all kinds. Observe, in Figure 3, the streets where there is a higher concentration of green marks (small shops) and note, in Figure 4(b), the areas with a redder tint (vertices most important given by the ranking algorithm).

Note the differences in the images shown in Figure 4. Both are representations of the same network; however, the flexibility of the model allows us to display different characteristics, according to our interests.

It is noteworthy that, although we introduce all the network information in a matrix D (*data matrix*), the vector $\mathbf{v_0}$ we introduce in the algorithm allows us to select the information that we want to view. Thus, in the example shown, the vector is weighted so that the algorithm calculates the classification of nodes giving the highest priority to the endowments of bars and restaurants in Figure 4(a), while to

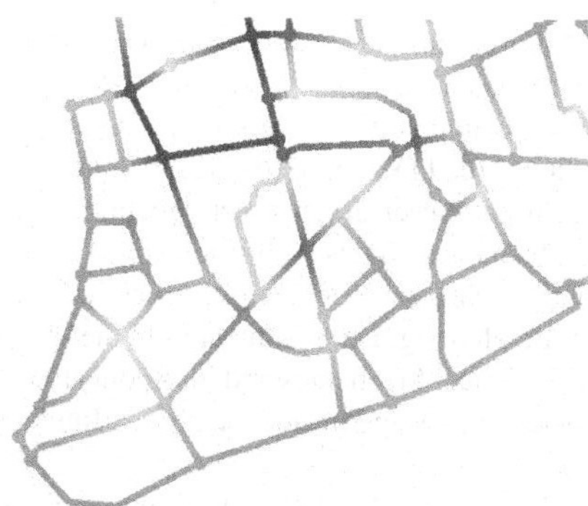

Fig. 5 Visualizing the nodes in the example network, according to their commercial activity

get the image in Figure 4(b) the greatest importance is given to information collected through D to commercial allocations as shops and small businesses.

Similarly, we could generate similar plots to visualize the other commercial endowments as banks, shopping malls, and other. If we weight the vector $\mathbf{v_0}$ assigning equal weight to all types of business, the picture we get is the one shown in Figure 5.

The great advantage of this approach over other models based on classification of nodes (see the bibliography), consists of the introduction of the data matrix D that allows us to quantify the network information for subsequent graphic display. Thus, as a conclusion, we can say that the model allows us to represent the information of a network graphically; thereby, we have a ready access to data on a network in a precise and clear way.

Acknowledgement. This work was partially supported by University of Alicante grants GRE09-02 and GRE10-34 and Generalitat Valenciana grant GV/2011/01.

References

1. Afryzkov, T., Oliver, J.L., Tortosa, L., Vicent, J.: An algorithm for ranking the nodes of an urban network based on concept of PageRank vector. Applied Mathematics and Computation (219), 2186–2193 (2012)
2. Barabasi, A.L.: Emergence of Scaling in Random Networks Science, pp. 286–509 (1999)
3. Barabasi, A.L.: Statistical Mechanics of Complex Networks. Review of Modern Physics 74, 47 (2002)
4. Barthelemy, M.: Spatial Networks. Physics Reports 499, 1–101 (2011)
5. Batty, M.: Cities and Complexity. Understanding Cities with Cellular Automata, Agent Based Models, and Fractals. The MIT Press, Cambridge (2005)
6. Berkhin, P.: A survey on PageRank computing. Internet Mathematics 2(1), 73–120 (2005)
7. Crucitti, P., Latora, V., Porta, S.: Centrality measures in spatial networks of urban streets. Physical Review E 73, 036125 (2006)
8. Crucitti, P., Latora, V., Porta, S.: The network analysis of urban streets: a dual approach. Physica A: Statistical Mechanics and its Applications 369(2), 853–866 (2006)

9. Crucitti, P., Latora, V., Porta, S.: The network analysis of urban streets: a dual approach. Physica A: Statistical Mechanics and its Applications 369(2), 853–866 (2006)
10. Jeong, S.K., Ban, Y.U.: Computational algorithms to evaluate design solutions using Space Syntax. Computer-Aided Design 43(6), 664–676 (2011)
11. Jiang, B.: A topological pattern of urban street networks: Universality and peculiarity. Physica A Statistical Mechanics and its Applications 384, 647–655 (2007)
12. Kansky, K., Danscoine, P.: Measures of network structure. Flux 5(1), 89–121 (1989)
13. Lammer, S., Gehlsen, B., Helbing, D.: Scaling laws in the spatial structure of urban road networks. Physica A Statistical Mechanics and its Applications 363, 89–95 (2006)
14. Langville, A.N., Mayer, C.D.: Deeper inside PageRank. Internet Mathematics 1(3), 335–380 (2005)
15. Masucci, A.P., Smith, D., Crooks, A., Batty, M.: Random planar graphs and the London street network. European Physical Journal B 71, 259–271 (2009)
16. Noh, J.D., Rieger, H.: Random walks on complex networks. Physical Review Letters 92, 118701-1–118701-4 (2004)
17. Page, L., Brin, S., Motwani, R., Winogrand, T.: The pagerank citation ranking: Bringing order to the web. Technical report 1999-66, Stanford InfoLab (1999)
18. Strano, E., Nicosia, V., Latora, V., Porta, S., Barthelemy, M.: Elementary processes governing the evolution of road networks. Scientific Reports 2, 296 (2012)

Case-Based Reasoning Applied to Medical Diagnosis and Treatment

Xiomara Blanco, Sara Rodríguez, Juan M. Corchado, and Carolina Zato

Departamento Informática y Automática, Universidad de Salamanca
Plaza de la Merced s/n, 37008, Salamanca, Spain
{xiopepa,srg,corchado}@usal.es

Abstract. The Case-Based Reasoning (CBR) is an appropriate methodology to apply in diagnosis and treatment. Research in CBR is growing and there are shortcomings, especially in the adaptation mechanism. In this paper, besides presenting a methodological review of the technology applied to the diagnostics and health sector published in recent years, a new proposal is presented to improve the adaptation stage. This proposal is focused on preparing the data to create association rules that help to reduce the number of cases and facilitate learning adaptation rules.

1 Introduction

The case-based reasoning (CBR) is a methodology for reasoning on computers that tries to imitate the behavior of a human expert and learn from the experience of past cases.

The CBR has demonstrated to be a systems development methodology appropriate smart when you want to apply in unstructured domains. Therefore, the choice of this methodology is appropriate for the development of diagnostic support systems in multidisciplinary medical services (1).

Nilsson and Sollenbom analyzed the CBR published between 1999 and 2003 (2), and concluded that hybrid systems using other techniques of Artificial Intelligence (AI) are increasing. This is because the application domain is increasingly complex, the potential use of these systems in the clinical area is high, but much work is still needed. More recent studies of CBR applied to health sciences conclude that there are many opportunities to work in diagnostic support for disabled and elderly (3-5). (5) emphasizes three areas for improvement: (i) reducing the search space in case recovery; (ii) maintaining knowledge always valid, embedding knowledge in cases to assist in reviewing the conclusions; (iii) work adaptation methods that consider allowing local constraints. Another paper about review published in 2011, notes that the CBR has been applied to various medical tasks such as diagnosis, classification, treatment and tutorials (6). For future work these

S. Omatu et al. (Eds.): *Distrib. Computing & Artificial Intelligence*, AISC 217, pp. 137–146.
DOI: 10.1007/978-3-319-00551-5_17

authors (6) suggest using probabilistic and statistical computing due to data systems every time are more and more large, complex and uncertain in clinical environment. Another conclusion is that the automatic adaptation is a weak point and a big challenge, especially in the medical field (6).

This paper focuses on medical CBR systems created between 2008 to 2011. The aim is to report the results of a systematic review of CBR applied to the health sector. We follow the methodology for systematic review of literature for software engineering researchers to Kitchenham (11)..

Another objective is to make a proposal to improve the adaptation stage. We focused on preparing the data to create association rules that help to reduce the number of cases and make learning easier around adaptation rules. This article is organized as follows: section 2 is an overview of the CBR applied to the health sector. Section 3 is shown the steps to implement the revised methodology used in this work. In section 4, the results achieved with the review and new proposals for the adaptation stage are described. In section 5 some conclusions and future work are presented.

2 Case-Based Reasoning Applied to the Health Sector

An early exploration of CBR in the medical field was conducted by Koton (7) and Bareiss (8) in the late 1980s. The CBR is inspired by human reason, i.e. to solve a problem by applying previous experiences adapted to the current situation.

The domain of health sciences offers the scientific community challenges that are difficult to solve with other methods and approaches. In the medical context, the symptoms represent the problem, and diagnosis and treatment are the solution. Aamodt and Plaza (9) have described a life cycle with four main steps (retrieve, reuse, revise and retain) (6).

Learning occurs naturally, if a case is successful the experience is retained. If the case is unsuccessful, the reason for the failure is identified and remembered.

In the 90's many efforts were made to create adaptive algorithms in medicine, but always was specific rules for the CBR built. In the literature there are solutions aimed at:

- Avoid the problem of adaptation.
- Solve the problem: Implementing rules that experts propose

  Adaptation by composition, this is one of his biggest successes in implementing influenza early warning. The solution of multiple cases are combined to produce a new solution, It is efficient if there are few conflicts between different components so a change in one component does not have many effects on others (10).

Adaptation by abstraction of cases was performed in diagnosis and therapy, each medical case contains between 40 and 130 symptoms and syndromes. Cases prototype were abstracted and then adapted, so few restrictions have to control through rules (22).

In Adaptation by abstraction, cases are stored at many levels of abstraction, the adaptation is performed from top to bottom taking the statistical mode. Initially, the solution is adapted to the highest level of abstraction (omitting less relevant details), then the solution is gradually refines and the required details are added. This adaptation can reuse a single case or it can reuse the different cases for different levels of abstraction or refinement of various details of the solution (10).

Unfortunately, the technique is general, but the content of these rules are domain specific. Especially for complex medical tasks generating adaptation rules is often tedious and sometimes even impossible.

However, for therapeutic tasks some adaptation rules can be applied, for example, by substitution adaptation for calculating dose (22).

Views general lines that are included in the case-based reasoning applied to the health sector, the following section summarizes the methodological work done to make a review of the work done in this field to date.

3 Literature Review Methodology

The methodology used is that described by Dr. Barbara Kitchenham in 2007 (11). A systematic review of literature is a secondary study that uses a well-defined methodology to identify, analyze and interpret all available data relating to a specific question. Systematic reviews seek to summarize the existing evidence on a specific topic. You can also identify gaps in current research to suggest areas for future research. A well defined methodology makes it less likely that the results are partial literature.

The steps associated with the planning of the review are: Identify the need for a review. Assign a review. Specify the research question, (it is the most important of any review). Develop a review protocol. Evaluate the review protocol. Perform the review: identify research, selection of primary studies, study quality assessment, data extraction and control and data synthesis. In this point, it is important to identify whether the results of the studies are consistent with each other (ie, homogeneous) or incompatible. Finally, report.

We conducted a systematic review of studies of CBR in medicine between September 2008 and September 2012, We considered only studies that validated their results. Research that were excluded: : CBR applied to tutorials, studies including summaries, systematic reviews and studies of non-human.

4 Results

This section shows the results more significant obtained after the methodological review:

4.1 Results: Review

Out of a total 1018 references were initially identified: 1010 of electronic databases and bibliographies 8 authors and suggestions. See Figure 1 for details. 21 articles were selected.

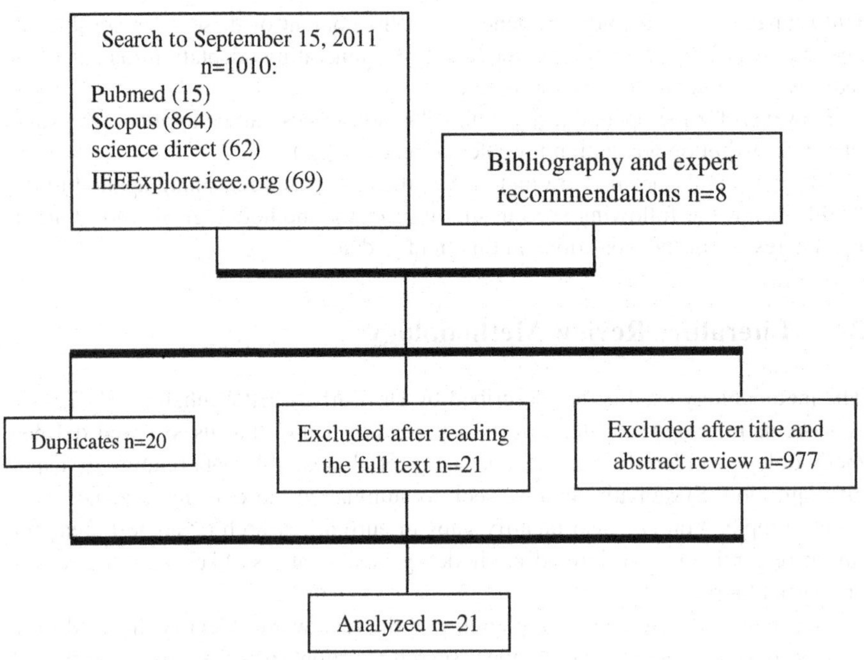

Fig. 1 Table summarizes the flow of study selection

4.2 Results: Trends in Medical CBR

The results of the review are summarized in the following tables and figures.

Table 1 shows the techniques of CBR or artificial intelligence used in each phase of the CBR (retrieval, reuse, revision and retention or learning), and the success rate in system validation.

Table 1 Techniques used for the modules of the CBR and accuracy rate

Ref.	Retrieve	Reuse	Revise	Retain	Analysis	Rate
(1)	Euclidian distance	-	-	-	ten-fold cross-validation	89
(2)	nearest neighbourhood	Dempster–Shafer	Annealing weight	Annealing weight	leave-one-out strategy	81,94
(3)	similarity function	Manual	manual	manual	statistical frequency	93
(4)	Euclidean distance	-	manual	stored case automatically	conditional probability	58
(5)	nearest-neighbor	-	-	Manual	Goodness-of-fit (R2)	81
(6)	Genetic algorithms	-	-	-	3-fold cross validation	92,4
(7)	Global Similarity	-	-	-	statistical frequency	94
(8)	weight set ranked by decision tree	-	-	-	Area under the curve	
(9)	Fuzzy Mathematics and	rules	manual	manual	statistical frequency	65,3
(10)	nearest neighborhood	Manual	-	-	statistical frequency	77,5
(11)	probabilistic	-	-	-	local grading	80
(12)	RMA	hierarchical	manual	manual	5-fold cross-validation	
(13)	nearest neighborhood	-	-	-	statistical frequency	95,58
(14)	nearest neighborhood	Neural net	-	-	correlation	96
(15)	cluster	fuzzy	-	-	statistical frequency	83,15
(16)	Bayesian network	-	-	-	holdout	92.17
(17)	Similarity Measurements	Copy	-	-	correction score	76
(18)	weighted clustering method	A stepwise regression	-	-	statistical frequency	98,9
(19)	similarity measure nearest neighborhood	-	-	-	5-fold cross-validation	94,57
(20)	nearest neighborhood	-	-	-	statistical frequency	
(21)	nearest neighborhood	-	-	-	statistical frequency	

Figure 2 shows a frequency diagram of the application domains of the articles analyzed. It is possible to observe that most were made for medical diagnosis. Figure 3 shows a frequency diagram that indicates what and how many techniques used in the review stage, i.e. what adaptation techniques were used in different CBR.

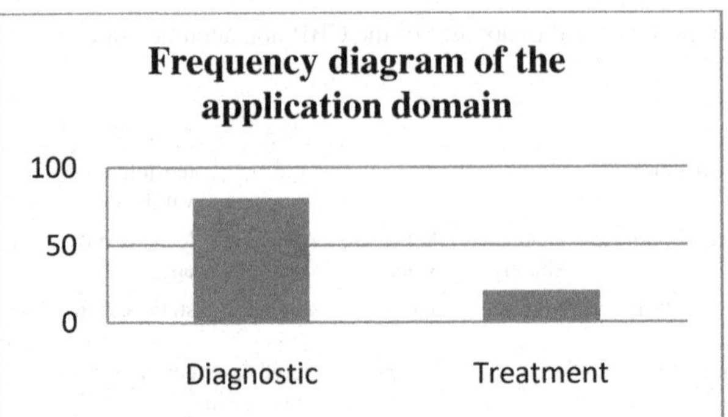

Fig. 2 Frequency diagram of the application domain

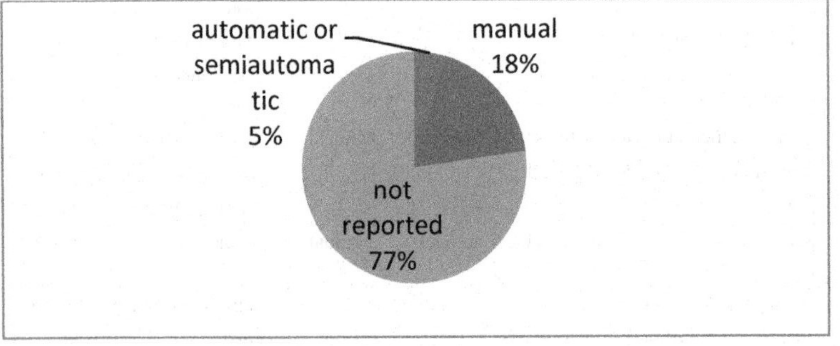

Fig. 3 Percentages of the different techniques used in the revision phase

4.3 *Results: Proposal Medical CBR*

We see that there are shortcomings in the adaptation mechanism, most studies report that it is manually. In this section, we present an adaptation mechanism to apply in medical CBR.

The proposal includes three phases: pre-processing data applying fuzzy sets, this in order to discretize continuous variables and fill empty spaces. This helps us to create better association rules. Then generate association rules for cases and make easier adaptive learning process. Finally matching patterns search through unsupervised neural networks.

4.3.1 Pre Processing Data

Applying fuzzy sets: The domain knowledge is represented by the values of m discrete attributes whose value is discretized in many cases, in many others apply fuzzy, so we reduce the search space (5).

Fill empty spaces: Another problem when working with medical data are incomplete data. This complicates the recovery and adaptation process. One solution is to apply the algorithm proposed by Henry Tirry Petri Myllymäki and (33).

The domain is represented by the values of m discrete attributes $A_1, \dots, A_m$. An attribute A_i has n_i possible values $a_{i1,\dots,}a_{in}$. Cases are coded as a vector:
$C_k = P_k(A_1), \dots, P_k(A_2), \dots P_k(A_m)$

Where $P_k(A_i)$ expresses the probability distribution for attribute values A_i when the case is C_k. $P_k(a_{ij}) = P(A_i = a_{ij} | C_k = 1)$

To calculate these probabilities We can estimated the proportional number of instances of the class C_k in our case are medical diagnostics. Similarly $P_k(a_{ij}) = P(A_i = a_{ij} | C_k = 1)$ can be estimated by the occurrences of values a_{ij} within the class C_k.

Once filled the empty spaces, association rules are constructed. An adaptation problem in medicine is the extreme specificity of individual cases, the generalized abstract prototypes or classes can support the adaptation process.

Each event has a list of features, which usually contains between 40 and 130 symptoms and syndromes, this means that there are many differences between similar cases, the adaptation cannot consider all. Prototypes are created, which contains the most frequently observed characteristics (34).

Structural adaptation is to change the structure of the solution during the adaptation. The proposal aims to implement such adaptation through the construction of association rules.

4.3.2 Construction of Association Rules Using the ART Algorithm

Association Rule Tree Algorithm (ART) builds decision lists that can be viewed as decision trees, makes use of techniques for efficient and scalable association rule mining (35).

Once made the association rules with a certainty factor value above a defined threshold, the association rules are our base cases for training of the neural network that evaluates adaptation.

Considering the algorithm and the suggestions in (36) which is trained back propagation neural network for the adaptation process and the suggestion to use another type of neural networks, our algorithm continues with the recovery by near neighbors and further training of a neural network.

Through near neighbors algorithm compute the similarity between the new case and association rules. After having recovered association rules, we train a neural network to learn adaptation knowledge.

Extracting adaptation knowledge, i.e. the differences between the characteristics of each new case and the characteristics of each retrieved rule and see how these differences affect the diagnosis. The neural network to be used is the Adaptive Resonance Theory that arise in response to two major problems with supervised neural networks: to learn and retain new patterns learned.

4.3.3 Adaptive Resonance Theory

The networks that use unsupervised learning approach has not been used for the design of CBR systems (37). ART stands for Adaptive Resonance Theory English, developed by Stephen Grossberg and Gail Carpenter is a network of three layers: input layer, without performing any pre-processing of the input data. Hidden layer. Output layer neurons is a competitive layer.

The input layer and hidden layer always have the same number of neurons. Each neuron in the input layer is a single neuron in the hidden layer, therefore also spoken of as networks of networks ART two layers instead of three. In our case the entrance are the differences in clinical variables P (aij) and the output are the probabilities of different diagnoses P (Ck).

5 Conclusions

In this paper a research study on primary studies of CBR in the medical field is presented. Moreover, It is demonstrated once again that the CBR is still applied in many medical situations for various tasks such as diagnosis and treatment (4-6).

Research in CBR applied health sector is growing, but most systems are prototypes and not yet available on the market as commercial products, as perceived in (6). The study confirms that using hybrid CBR with AI techniques makes more easily manage the complexities inherent in the data used in the studies analyzed, and better results are obtained (4).

Moreover, we observe that most have very well defined their case retrieval system, but there is a failure in the mechanism of adaptation (see Figure 3). Most studies don't report adaptation technique and others report that they do it manually. As a result of the review we suggest a proposal to improve adaptation phase; we propose solutions to each of the problems reported to the review of CBR. First fuzzy sets and applying Bayes as part of a data preprocessed. After association rules to have fewer cases. Subsequently using neural networks that learn adaptation rules. The next step is to implement each of the phases of this proposal on data from medical diagnosis or treatment, validate results and suggest improvements. Previous studies have evaluated recovery techniques cases, but studies as the present have not been found in the literature. We hope that other studies like this are made to confirm or refute the results of our analysis.

Acknowledgments. This work has been partially supported by the MICINN project TIN 2009-13839-C03-03 (FEDER Support).

References

1. Chuang, C.L.: Case-based reasoning support for liver disease diagnosis. Artif. Intell. Med. 53(1), 15–23 (2011)
2. Petrovic, S., Mishra, N., Sundar, S.: A novel case based reasoning approach to radiotherapy planning. Expert Systems with Applications 38(9), 10759–10769 (2011)

3. Ocampo, E., MacEiras, M., Herrera, S., Maurente, C., Rodríguez, D., Sicilia, M.A.: Comparing Bayesian inference and case-based reasoning as support techniques in the diagnosis of Acute Bacterial Meningitis. Expert Systems with Applications 38(8), 10343–10354 (2011)
4. Ting, S.L., Kwok, S.K., Tsang, A.H.C., Lee, W.B.: A hybrid knowledge-based approach to supporting the medical prescription for general practitioners: Real case in a Hong Kong medical center. Knowledge-Based Systems 24(3), 444–456 (2011)
5. Ahmed, M.U., Begum, S., Funk, P., Xiong, N., von Scheele, B.: A multi-module case-based biofeedback system for stress treatment. Artificial Intelligence in Medicine 51(2), 107–115 (2011)
6. Hsu, K.H., Chiu, C., Chiu, N.H., Lee, P.C., Chiu, W.K., Liu, T.H., et al.: A case-based classifier for hypertension detection. Knowledge-Based Systems 24(1), 33–39 (2011)
7. Agwil, R.O., Shrivastava, D.P.: Integrated Thallassaemia Decision Support System. WSEAS Transactions on Computers 9(8), 857–867 (2010)
8. Huang, M.L., Hung, Y.H., Lee, W.M., Li, R.K., Wang, T.H.: Usage of Case-Based Reasoning, Neural Network and Adaptive Neuro-Fuzzy Inference System Classification Techniques in Breast Cancer Dataset Classification Diagnosis. Journal of Medical Systems (in press)
9. Gu, D.X., Liang, C.Y., Li, X.G., Yang, S.L., Zhang, P.: Intelligent technique for knowledge reuse of dental medical records based on case-based reasoning. Journal of Medical Systems 34(2), 213–222 (2010)
10. Marling, C., Shubrook, J., Schwartz, F.: Toward case-based reasoning for diabetes management: A preliminary clinical study and decision support system prototype. Computational Intelligence 25(3), 165–179 (2009)
11. Toward translational incremental similarity-based reasoning in breast cancer grading. Image Perception, Access and Language IPAL (UMI CNRS 2955, UJF, NUS, I2R), Singapore National University Hospital National University of Singapore Politehnica University of Timisoara, Romania University of Besançon, France Medical Informatics Service, University Hospital of Geneva, Sweden University of Applied Sciences, Western Switzerland, Sierre, Sweden (2009)
12. Rodríguez, S., De Paz, J.F., Bajo, J., Corchado, J.M.: Applying CBR systems to micro array data classification 49, 102–111 (2009)
13. Ahn, H., Kim, K.: Global optimization of case-based reasoning for breast cytology diagnosis. Expert Systems with Applications 36(1), 724–734 (2009)
14. Obot, O.U., Uzoka, F.M.: A framework for application of neuro-case-rule base hybridization in medical diagnosis. Applied Soft Computing 9(1), 245–253 (2009)
15. Fazel Zarandi, M.H., Zarinbal, M., Izadi, M.: Systematic image processing for diagnosing brain tumors: A Type-II fuzzy expert system approach. Applied Soft Computing 11(1), 285–294 (2011)
16. Cruz-Ramírez, N., Acosta-Mesa, H.G., Carrillo-Calvet, H., Barrientos-Martínez, R.E.: Discovering interobserver variability in the cytodiagnosis of breast cancer using decision trees and Bayesian networks. Applied Soft Computing 9(4), 1331–1342 (2009)
17. A Reuse-Based CBR System Evaluation in Critical Medical Scenarios (2009)
18. Fan, C.-Y., Chang, P.-C., Lin, J.-J., Hsieh, J.C.: A hybrid model combining case-based reasoning and fuzzy decision tree for medical data classification. Applied Soft Computing 11(11), 632–644 (2009)
19. Lin, R.H., Chuang, C.L.: A hybrid diagnosis model for determining the types of the liver disease. Computers in Biology and Medicine 40(7), 665–670 (2010)

20. Influenza Forecast: Case-Based Reasoning or Statistics? (2007)
21. Floyd Jr., C.E., Lo, J.Y., Tourassi, G.D.: Case-Based Reasoning Computer Algorithm that Uses Mammographic Findings for Breast Biopsy Decisions. American Journal of Roentgenology 175(5), 1347–1352 (2000)
22. Schmidt, R., Vorobieva, O., Gierl, L.: Case-based Adaptation Problems in Medicine, pp. 267–274 (2003)

Multiple Agents for Data Processing

Ichiro Satoh

National Institute of Informatics
2-1-2 Hitotsubashi, Chiyoda-ku, Tokyo 101-8430, Japan
ichiro@nii.ac.jp

Abstract. This paper proposes a distributed processing framework inspired from data processing. It unique among other data processing for large-scale data, so-called bigdata, because it can locally process data maintained in distributed nodes, including sensor or database nodes with non-powerful computing capabilities connected through low-bandwidth networks. It uses mobile agent technology as a mechanism to distribute and execute data processing tasks to distributed nodes and aggregate their results. The paper outlines the architecture of the framework and evaluates its basic performance.

1 Introduction

MapReduce is a model for processing large data sets. It was originally studied by Google [2] and inspired by the *map* and *reduce* functions commonly used in parallel LISP or functional programming paradigms. Data processing should be provided close to the sources of data, e.g., sensors and database, as much as possible to reduce network traffic. However, existing projects assume that MapReduce is executed in data centers or in-house high performance computing systems, which may be far from the sources of data. This paper proposes a novel implementation of MapReduce to process data at the edges of networks. In fact, it can satisfy the following requirements, which existing MapReduce implementations cannot support.

- The goal of our MapReduce implementation is to directly execute MapReduce processing on ambient computing systems and sensor networks, where each node may have non-powerful processor with the small amount of memory, rather than data centers and high performance server clusters.
- Networks in such systems may be low-band, often disconnected, and dynamic, e.g., wireless sensor networks. Therefore, our implementation should be available in such networks.
- There may be no database or file systems in the target systems. Instead the data that need to be processed are generated or locally maintained in the local storage of nodes.

S. Omatu et al. (Eds.): *Distrib. Computing & Artificial Intelligence,* AISC 217, pp. 147–154.
DOI: 10.1007/978-3-319-00551-5_18

- Every node may be able to support management and/or data processing tasks, but may not initially have any codes for its tasks.

The basic idea behind our implementation is to deploy data processing tasks at the nodes that have the target data at the edge of networks and aggregate the results, rather than to transmit data to servers at the center. It introduces mobile agent technology, where mobile agents are autonomous programs that can travel from computer to computer in a network, at times and to places of their own choosing. The state of the running program is saved, by being transmitted to the destination. The program is resumed at the destination continuing its processing with the saved state. Our implementation of MapReduce defines the management system and data processing tasks as mobile agents and *map* and *reduce* processing in MapReduce are provided by migrating workers, which are implemented as mobile agents, with the results of their processing. It is constructed based on our original mobile agent platform, which is designed for data processing, in particular MapReduce processing.

2 Related Work

The tremendous opportunities to gain new and exciting value from big data are compelling for most organizations, but the challenge of managing and transforming it into insights requires new approaches. MapReduce processing has been used as one of the approaches. It originally supports *Map* and *Reduce* processes [2] The first is to divide a large scale of data into smaller sub-problems and assign them to worker nodes. Each worker node processes the smaller sub-problem. The second is to collect the answers to all the sub-problems and aggregates them as the answer to the original problem it was trying to solve. *Hadoop*, which is one of its one of the most popular implementations of MapReduce, developed and named by Yahoo!. There have been many attempts to improve Hadoop in academic or commercial projects. On the other hand, there have been a few attempts to implement MapReduce itself except for Hadoop. For example, the Phoenix system [6] and the MATE system [5] supported multicore processors with shared memory. Haloop [1] and Twister [3] were designed for MapReduce-based iterative computation. Google's MapReduce, Hadoop, and other existing MapReduce implementations assume their own distributed file systems, e.g., Google file system (GFS) and Hadoop file system (HDFS), or shared memory between processors. For example, Hadoop needs to move target data from the external storage systems to HDFS via networks before its processing. Our MapReduce system does not move data between nodes. Instead, it deploys program codes for defining processing tasks to the nodes that have the data by using the migration of agents corresponding to the tasks and execute the codes with their current local data. In the literature of sensor networks, IoT, and machine-to-machine (M2M), several academic or commercial projects have attempted to support data on the edge, e.g., sensor nodes and embedded computers. For example, Cisco's *Flog Computing* and EMC's computing intend to integrate cloud computing over the Internet and peripheral computers.

3 Mobile Agent-Based MapReduce

This section outlines our mobile agent-based MapReduce processing system and compares between our system and Hadoop, which one of the most typical implementation of MapReduce. The architecture of our MapReduce system is different from existing implementation of MapReduce, including Hadoop.

3.1 *Data Processing at the Edge*

The original MapReduce and its clones are unable to cope cost-effectively if at all with new dynamic data sources and multiple contexts for the large amount of data, which is generated at sensors and devices. More data are generated at the edge of networks, e.g., sensors and devices, than servers, including data centers and cloud computing infrastructure. The transmission of such data from nodes at the edge to server nodes seriously affect the performance of analyzing the data and results in congestion in networks. To solve this problem, data processing tasks are defined as mobile agents and dynamically duplicated and deployed at the nodes that have the target data. Mobile agents also can directly access data from sensors and low-level file systems at their destination nodes. Our approach assumes data at nodes to be independent of one another and can be processed without exchanging data between nodes. Finally, like MapReduce, agents running on nodes carry their results to specified nodes after their processing are done to aggregate the results.

3.2 *Architecture*

The original MapReduce consists of one *master* node and one or more *worker* nodes and Hadoop consists of *job tracker*, *task tracker*, *name*, and *data* nodes, where the first and third corresponds to the master node, the second and forth to data nodes in the original MapReduce. Our MapReduce system has a little different architecture from Google's MapReduce and a far from Hadoop. The system itself is a collection of three kinds of mobile agents, called *Mapper*, *Worker*, and *Reducer*, which should be deployed at appropriate nodes, called *data nodes*, according to the location of the target data. They are still mobile agents so that they can dynamically deployed at nodes according to the location of the target data and available resources to process them.

3.3 *MapReduce Processing*

Our system supports MapReduce processing with mobile agents. Figure 1 shows the basic mechanism of the processing.

- **Map** process: a *Mapper* agent corresponds to the master node of the original MapReduce. The agent makes copies of *Worker* agent and each of the *Worker* agents migrates to one or more data nodes, which locally have the target data. They execute their processing locally at the nodes. When there are multiple

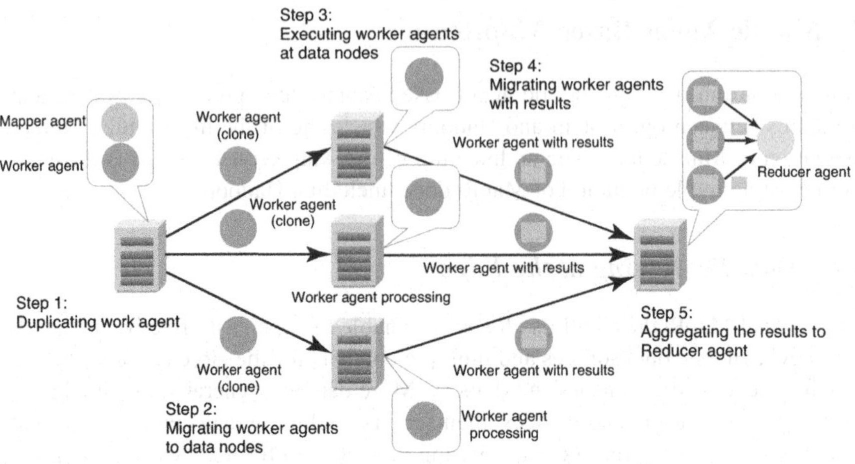

Fig. 1 Mobile agent-based MapReduce processing

Mapper agents in the same time, they can be executed in a specified schedule, e.g., sequential or parallel.

- **Reduce** process: after executing their processing, *Worker* agents migrates to the computer that the *Reducer* agent is running with their results and then send the results to the agent. *Mapper* and *Reducer* agents can be running on the same node.

Note that the amount of the results are by far smaller than the amount of target data. Each *Worker* agent assumes to be executed independently of the others.

4 Design and Implementation

This section describes our mobile agent-based MapReduce system. It consists of two layers; mobile agents and agent runtime systems. The former consists of agents corresponding to job tracker and *map* and *reduce* processing and the latter corresponds to task and data nodes. It was implemented with Java language and operated on the Java virtual machine. In our implementation, the system has been designed independently of MapReduce, because it can support other data processing approaches in the same time. As a result, our approach can be available in other existing mobile agent platforms.

4.1 Agent Runtime System

Each runtime system runs on a computer and is responsible for executing agents at the computer and migrating agents to other computers through networks. The system itself is designed independent of any data processing. Instead, agents running on it support MapReduce processing.

4.1.1 Agent Duplication

Before deploying agents at data nodes, our approach makes one or more copies of task agents. The runtime system can store the state of the agent in heap space in addition to the codes of agents into a bit-stream formed in Java's JAR file format, which can support digital signatures for authentication. The current system basically uses the Java object serialization package for marshaling agents. The package does not support the capturing of stack frames of threads. Instead, when an agent is duplicated, the runtime system issues events to it to invoke their specified methods, which should be executed before it is duplicated and it then suspends their active threads.

4.1.2 Agent Migration

Each runtime system also establishes at most one TCP connection with each of its neighboring systems in a peer-to-peer manner without any centralized management server and exchanges control messages and agents through the connection. When an agent is transferred over a network, the runtime system transmits one or more marshalled agents to the destination data nodes through TCP connections from the source node to the nodes. After arriving at the nodes, they are resumed and activated from the marshalled agents and then their specified methods are invoked to acquire resources and start their processing.

4.1.3 Agent Execution

Each agent can have one or more activities, which are implemented by using the Java thread library. Furthermore, the runtime system maintains the life-cycle of agents. When the life-cycle state of an agent is changed, the runtime system issues certain events to the agent. The system can impose specified time constraints on all method invocations between agents to avoid being blocked forever. Each agent is provided with its own Java class load, so that its namespace is independent of other agents in each runtime system. The identifier of each agent is generated from information consisting of its runtime system's host address and port number, so that each agent has a unique identifier in the whole distributed system. Therefore, even when two agents are defined from different classes whose names are the same, the runtime system disallows agents from loading other agents's classes. To prevent agents from accessing the underlying system and other agents, the runtime system can control all agents under the protection of Java's security manager.

4.2 Mobile Agent

Each agent is defined as a collection of Java objects. It is general-purposed. Instead, we provide agents with a framework for MapReduce processing. Every agent consists of several callback methods to be invoked by the runtime system before or after the life-cycle state of the agent changes, e.g., initialization, execution, arrival, departure, suspension, and termination. It can invoke several fundamental methods used to create a new agent as its child and control the life-cycle of itself and its children, e.g., mobility, duplication, termination. To support existing data processing software for Hadoop, the current implementation can explicitly provide callback methods compatible to Hadoop's classes and interfaces, e.g., `Mapper` and `Reducer`.[1]

Unlike other existing MapReduce implementations, including Hadoop, our system does not have any file system, because nodes in sensor networks and ambient computing systems may lack enrich storage devices. Instead, it provides a tree-structured key value stores (KVSs), where each KVS maps arbitrary string value and arbitrary byte array data and is maintained inside its agent, and directory servers for KVSs in agents. To support *reduce* processing, the root KVS merge KVS of agents into itself. In the current implementation each KVS in each data processing agent is implemented as a hashtable whose keys given as pairs of arbitrary string values and values are byte array data and is carried with its agent between nodes.

5 Current Status

A prototype implementation of this framework was constructed with Sun's Java Developer Kit version 1.6 or later versions. The implementation provided graphical user interfaces to operate the mobile agents. Although the current implementation was not constructed for performance, we evaluated that of several basic operations in a distributed system where eight computers (Intel Core Duo 2 2 GHz with MacOS X 10.7 and J2SE version 6) were connected through a Giga Ethernet.

- The cost of agent duplication was measured as the left of Fig.2, where The agent was simple and consisted of basic callback methods. The cost included that of invocating two callback methods.
- The cost of migrating the same agent between two computers was measured as the right of Fig.2. The cost of agent migration included that of opening TCP-transmission, marshaling the agents, migrating the agents from their source computers to their destination computers, unmarshaling the agents, and verifying security.

We constructed word counting from texts.[2] Fig.3 shows the basic structure of our word counting by using mobile agent-based MapReduce with the screenshots of the word counting.

---

[1] It does not support fully compatible methods, since it does not intend to use existing programs for Hadoop.

[2] Word counting is one of the most typical examples of Hadoop.

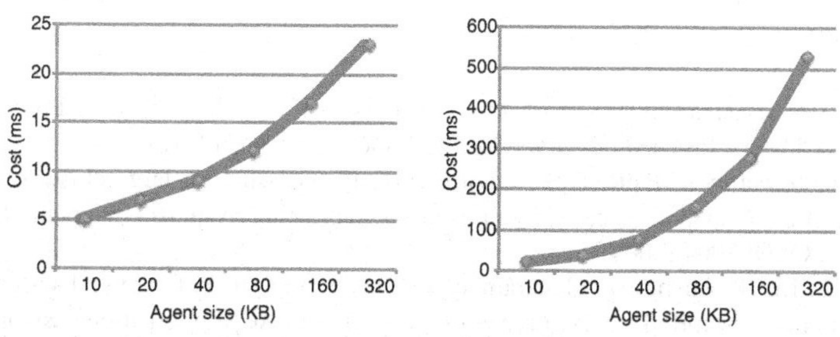

Fig. 2 Mobile agent-based MapReduce processing

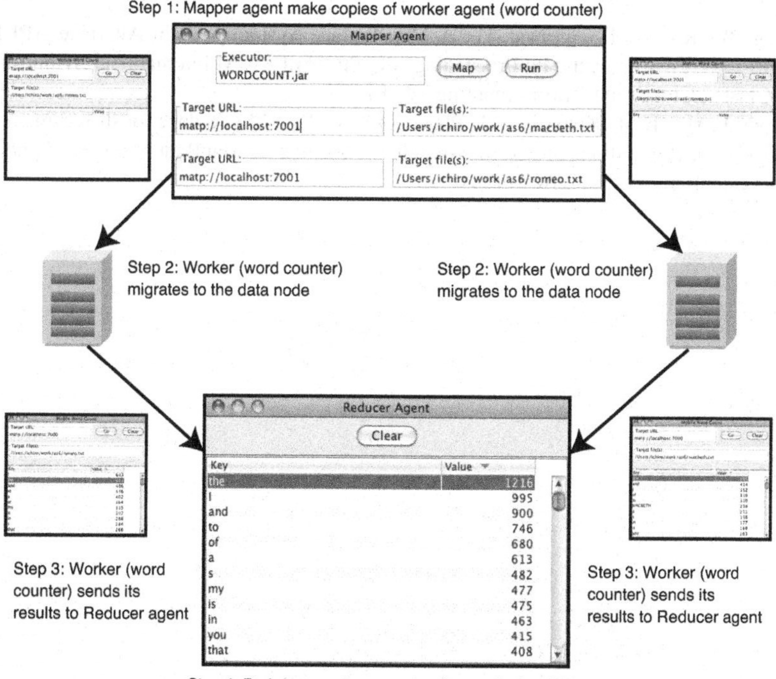

Fig. 3 Mobile agent-based MapReduce processing

6 Conclusion

We presented a novel distributed processing framework inspired from MapReduce processing. It was designed for analyzing data at the edges of networks and constructed based mobile agents. It introduces mobile agent technology so that it distributed data processing tasks to distributed nodes as a *map* process and aggregates their results by returning them to specified servers as *reduce* process.

References

1. Bu, Y., Howe, B., Balazinska, M., Ernst, M.D.: HaLoop: Efficient Iterative Data Processing on Large Clusters. Proceedings of the VLDB Endowment 3(1) (2010)
2. Dean, J., Ghemawat, S.: MapReduce: simplified data processing on large clusters. In: Proceedings of the 6th Conference on Symposium on Opearting Systems Design and Implementation, OSDI 2004 (2004)
3. Ekanayake, J., Li, H., Zhang, B., Gunarathne, T., Bae, S.H., Qiu, J., Fox, G.: Twister: a runtime for iterative MapReduce. In: Proceedings of the 19th ACM International Symposium on High Performance Distributed Computing (HPDC 2010). ACM (2010)
4. Grossman, R., Gu, Y.: Data mining using high performance data clouds: experimental studies using sector and sphere. In: Proceedings of the 14th ACM SIGKDD International Conference on Knowledge Discovery and Data Mining (KDD 2008), pp. 920–927. ACM (2008)
5. Jiang, W., Ravi, V.T., Agrawal, G.: A Map-Reduce System with an Alternate API for Multi-Core Environments. In: Proceedings of 10th IEEE/ACM International Symposium on Cluster, Cloud, and Grid Computing (2010)
6. Talbot, J., Yoo, R.M., Kozyrakis, C.: Phoenix++: modular MapReduce for shared-memory systems. In: Proceedings of 2nd International Workshop on MapReduce and Its Applications (MapReduce 2011). ACM Press (2011)

Migrants Selection and Replacement in Distributed Evolutionary Algorithms for Dynamic Optimization

Yesnier Bravo, Gabriel Luque, and Enrique Alba

Departamento de Lenguajes y Ciencias de la Computación, Universidad de Málaga, España
{yesnier,gabriel,eat}@lcc.uma.es

Abstract. Many distributed systems (task scheduling, moving priorities, mobile environments, ...) can be linked as Dynamic Optimization Problems (DOPs), since they require to pursue an optimal value that changes over time. We have focused on the utilization of Distributed Genetic Algorithms (dGAs), one of the domains still to be investigated for DOPs. A dGA essentially decentralizes the population in islands which cooperate through *migrations* of individuals. In this article, we analyze the effect of the migrants selection and replacement on the performance of dGAs for DOPs. Quality and distance based criteria are tested using a comprehensive set of benchmarks. Results show the benefits and drawbacks of each setting for DOPs.

1 Introduction

Dynamic Optimization Problems (DOPs) are important research challenges appearing in real life applications, many of them in fact linked to distributed systems like task scheduling, moving priorities, mobile environments, etc. A DOP is actually a problem where the definition changes as the solving algorithms is progressing. This forces the constant research in new techniques for tracking the moving optima over time. Several authors have proposed the use of multiple populations for solving DOPs using Genetic Algorithms (GA) [2, 7, 9, 8], with the aim of tracking the optimum changes by specializing and pursuing promising regions of the search space. However, one of the domains still to be investigated is the utilization of Distributed Genetic Algorithms (dGAs) [5, 4], characterized by decentralizing the population in demes, named *islands*, independently evolving and communicating through *migrations* of individuals. Conversely, their use should be valuable for DOPs because of the natural diversity enhancement and speciation-like features [5].

In this article, we study distinct migrants selection and replacement strategies, based on both quality (fitness) and distance based criteria. Specifically, we analyze how they affect the performance of a physically parallel dGA in a comprehensive set of DOP benchmarks. Results show the benefits and drawbacks of each strategy for addressing distinct DOP features.

The remainder of this paper is organized as follows. Sect. 2 provides a brief background on DOP and the dGA model. Sect. 3 summarizes the experimental design,

S. Omatu et al. (Eds.): *Distrib. Computing & Artificial Intelligence*, AISC 217, pp. 155–162.
DOI: 10.1007/978-3-319-00551-5_19

and the obtained results are analyzed in Sect. 4 and Sect. 5. Finally, conclusions and future works are stated in Sect. 6.

2 Background

An optimization problem in which environmental data, constraints and/or objectives change over time, is named Non-stationary, Time-variant or Dynamic Optimization Problem [6]. In contrast to other types of uncertainty in real-world optimization problems, in this domain the fitness function is deterministic at any point in time, but is dependent on time t, i.e.,

$$F(X) = f_t(X). \quad (1)$$

Hence, the algorithm for DOPs should be able to track the optimum changes over time. The following features are commonly used in literature to characterize DOPs:

- **Change frequency:** It determines how often the changes occurs. It is usually measured in number of generations between two consecutive changes.
- **Change severity:** It defines how different the fitness landscape is after a change, the higher this value the more abrupt is the change.
- **Cycle length, cycle accuracy (noise):** They characterize cyclic environments, where a finite set of states recur over time. They denote how often it takes to return to a previous state, and how close this return is to it, respectively.

These features are used later in this article to build the set of DOP benchmarks for analyzing the influence of the migration policies of dGAs for DOPs.

A few authors have proposed distributed (multi-population) models for solving DOPs, but all of these approaches use specific migration policies. For example, Oppacher and Wineberg in [7], send the elite (best) individuals from *colonies* subpopulations to a *core* subpopulation. Other policies used in literature involve a global knowledge of the entire population, like Ursem in [9], by applying the *hill-valley detection* mechanism among the best individuals of each subpopulation, named *nation*. Recently, Park et al. [8] have used two populations with different evolutionary objectives and, given the inconvenience of normal migrations, they applied crossbreeding as a means of information exchange. However, to the best of our knowledge, no coherent and comprehensive study has been accomplished regarding migrants selection and replacement in DOPs. Our work is a contribution in this direction, with the aim at supporting the design of new approaches for DOP based on dGA models.

As aforementioned, the dGA [5], also known as island model, structures the population in *islands*. Each island independently evolves, like a parallel agent, and communicates with the other ones through migration of individuals. The pseudo-code for each island is shown in Alg. 1, which basically consists of a standard GA with an additional communication step for migrations.

As you can notice from Alg. 1, two essential parameters in the specification of a dGA [5] are the criteria for selecting emigrants and for replacing existing individuals in the target subpopulation by incoming migrants, often referred to in the

Algorithm 1. Pseudocode for an island evolution in a dGA model.

```
Initialize & evaluate the subpopulation
while not stop_condition do
   Select parents for reproduction
   Apply crossover, mutation, and evaluation operators
   Select new parents and replace the old subpopulation
   if migration_period is met then
      Select outgoing migrants & send them to the neighbor islands
      Select individuals to replace & replace them by the incoming migrants
   end if
end while
```

literature as the *migration policy*. The importance of understanding their effects on the performance of the dGA has been previously reported for static problems [3], but it gets even more important for DOPs. The reason is that they control the convergence rate of the dGA. An excessively fast convergence removes the ability to react after a change, while a very low one does not allow for adaptation to the new environmental conditions.

In this paper, we will focus on the criteria for selecting outgoing migrants and for selecting existing individuals, at the target subpopulation, for replacing with incoming migrants. We test both quality (fitness) and distance based selection methods, thus influencing the behavior of the dGA from the phenotype and genotype, respectively. Table 1 shows the most representative strategies.

Table 1 Migrants selection and replacement criteria

Migration policy	*Description*
$best-worst$	Good migrants replace poor individuals
$best-random$	Good migrants replace random individuals
$best-distant$	Good migrants replace distant individuals
$rand-worst$	Random migrants replace poor individuals
$rand-distant$	Random migrants replace distant individuals
$worst-worst$	Poor migrants replace poor individuals
$distant-distant$	Distant migrants replace the distant individuals

In particular, the most *distant* individuals are determined in the Hamming space from the centroid of each deme. In the next section we describe the experimental design used to accomplish this study.

3 Experimental Setup

Aiming to analyse the influence of the chosen migrants selection and replacement criteria (see Table 1 from previous section) we use a canonical dGA consisting of eight islands evolving homogenously. In every island, we use a simple generational GA with 64 individuals, binary-tournament selection, one-point crossover (with $pX = 1.0$), without mutation (to remove possible bias in the results). Migrations occur synchronously on a unidirectional ring topology after every generation, and the migration rate is 10%.

The behavior of algorithms is tested using four dynamic functions (Onemax, Royal-Road, P-Peaks, and MMDP) built with the XOR-DOP benchmark generator [10], thus addressing different difficulties. The Onemax Problem requires to maximize the amount of ones in a binary string of length ℓ ($\ell = 100$). The Royal-Road Problem separates each string in 25 equally-sized building blocks, each of which contributes to the total fitness if all the four bits are set to one. The P-Peaks Problem consists of 50 strings, named *peaks*, and the fitness of each candidate solution depends of the distance to the nearest peak in the Hamming space. Lastly, the Massive Multimodal Deceptive Problem (MMDP) is made up of $k = 16$ subproblems of 6 bits each, which contribution depends on the number of ones s_i ($i = 0,\ldots,6$) they have, being extremal cases ($s_i \in \{0,6\}$) better than a more prone intermediate value (deceptiveness). OneMax and Royal-Road are *unimodal* DOPs (only one suboptimal solution, which is also optimal), while P-Peaks and MMDP are *multimodal* DOPs (multiple suboptimal solutions).

For each function, the XOR-DOP generator temporarily applies a binary mask to every candidate solution before evaluating it. Every $\tau = 10$ generations (change frequency), a number $\rho \times \ell$ ($\rho \in \{0.05, 0.1, 0.2, 0.5, 1.0\}$) of bits from the current mask are varied (change severity). The higher the ρ value the severer the change; $\rho = 1.0$ means a random severity in the range $[0.01, 0.99]$. We also test distinct change modes (cyclic, cyclic with noise, and random). The cycle length is 5 changes and the noise sums up a severity of 0.05.

Algorithms and benchmarks were implemented in C++, using the MALLBA library[1]. All experiments were performed in a PC with an Intel Core i7-720QM processor at 1.60GHz, 4GB of RAM, and running GNU/Linux Ubuntu 12.4.

To describe the performance of algorithms we compute the mean *accuracy* or *relative error*, i.e.:

$$Acc = \frac{1}{N}\sum_{i=1}^{N}\frac{f(generation\_best_i) - Min_i}{Max_i - Min_i} \quad (2)$$

where N is the total number of generations, and Max_i and Min_i are the current maximum and the minimum fitness values, respectively. High values of this metric indicate a better adaptation of the algorithm to the changing optimum along the run. In addition, we use the standard deviation (STD) in the Hamming space of the solutions to determine the global (inter-deme) and local (intra-deme) genotypic diversity. In both cases, we average the results of over 100 independent runs and evaluate the statistical significance. First, we use the Kolmogorov-Smirnov test to check whether the data follow a normal distribution or not. If so, then we do an ANOVA test to compare the means; otherwise Kruskal-Wallis test is used to

[1] Online available at `http://neo.lcc.uma.es/mallba/easy-mallba`

compare the medians. In each case, a level of confidence of 95 % is used. In the next section we summarize and discuss the obtained results.

4 Influence of the Migration Policy on the Accuracy

We begin by analyzing the influence of migrants selection and replacement on the performance of the dGA model for dynamic environments. Our goal is to show how different migration policies can help to reach a good balance between convergence (quickly adapt to the optimum movements) and the natural speciation-like behavior (search multiple sub-optima at the same time). This conclusion confirms the importance of tracking multiple suboptimal values on dynamic optimization, since they are candidate optima after a change in the environment, thereby underlining our interest on the dGA model for DOPs.

Table 2 summarizes the results for all DOPs, change modes (**Cy**clic, **Cy**clic with **N**oise, and **Ra**ndom), and the five change severities tested. Since we are interested in the effect of the migration policies, we have grouped results with different change severities, accounting the number of experiments in which the migration policy is statistically better than the rest (values range from 0 to 5). A better value of this metric indicates a better adaptation to a wider range of change severities.

Table 2 Number of experiments in which the migration policy is statistically better than the rest. The best values are boldfaced.

Migration	*Onemax*			*Royal-Road*			*P-Peaks*			*MMDP*			
policy	**Cy**	**CyN**	**Ra**	**Cy**	**CyN**	**Ra**	**Cy**	**CyN**	**Ra**	**Cy**	**CyN**	**Ra**	Total
best-worst	3	1	1	3	0	0	2	0	0	0	0	0	10
best-rand	0	0	0	0	0	0	1	0	0	0	0	0	1
best-distant	0	**4**	**4**	0	**4**	**4**	0	0	0	0	0	0	16
rand-worst	4	3	2	3	2	2	3	1	1	0	0	0	19
rand-distant	1	**4**	**4**	0	**4**	**4**	0	0	0	0	0	0	17
worst-worst	**5**	**4**	2	**5**	**4**	2	**5**	**5**	**5**	**5**	**5**	**5**	**52**
distant-distant	0	3	**4**	0	**4**	**4**	0	0	0	0	0	0	15

You can notice from Table 2 that the *worst-worst* strategy (poor migrants replace poor individuals) reaches the best accuracy level in a wide range of DOP instances (52 of 60 problem instances), most notably in multimodal DOPs (*P-Peaks* and *MMDP*) where the results are significantly better than the rest in all cases. These problems consist of a large number of suboptimal solutions and require not only to search for multiple candidate optima in the search space, but also to pursue their movements. In this scenario, the *worst-worst* strategy promotes an isolated evolution among islands, which becomes basins of attraction for the low-quality migrants and, at the same time, prevents the best solution from dominating the whole population.

However, we also notice that the fitness-based replacement (including the *worst-worst* strategy) quickly degrades the accuracy of the algorithm when larger movements of the optimum are considered (e.g., increasing the severity, including noise,

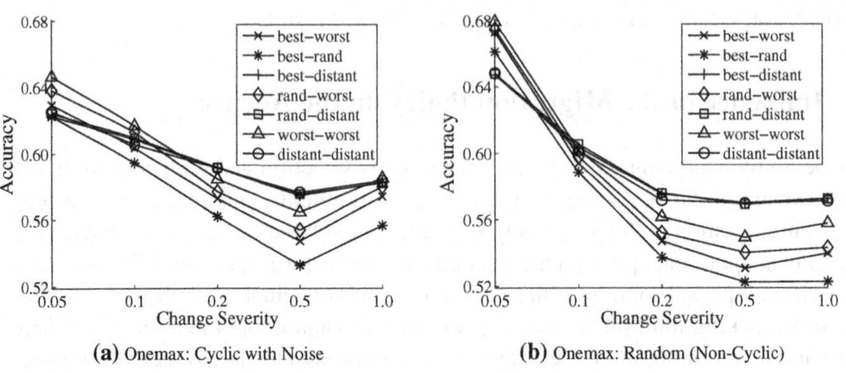

(a) Onemax: Cyclic with Noise **(b)** Onemax: Random (Non-Cyclic)

Fig. 1 Effect of the different migration policies in the performance of the dGA for Onemax with Cyclic with Noise (a) and Random (b) change modes, and severities ($\rho = \{0.05, 0.1, 0.2, 0.5, 1.0\}$)

doing random changes). In these scenarios, replacing the most distant individuals is many times a better migration policy, as you can observe in Fig. 1.

This good performance of distance-based replacement is due the enhancement in the global diversity (low coupling among islands) and the speciation-like behavior of the dGA. The diversity enables the algorithm to react and adapt to abrupt and discontinuous changes, while the speciation helps to store old optima solutions that are useful to bias the search process in the future.

Therefore, in addition to the benefits of the *worst-worst* migration policy, we underline that the replacement criterion is statistically significant and must be taken into account when dealing with DOPs. This is an interesting finding, since such parameter is often referred to as less important in the literature (see, for example, [3] or [1]). In the next section we will deeply examine this issue by analyzing the effect of the migration strategies on the diversity.

5 Influence of the Migration Policy on the Diversity

In this section, we analyze the influence of distinct migration policies on the population diversity. For this purpose, we compute the standard deviation (STD) among the individuals inside every island, and average the obtained values (intra-deme diversity). We also compute the STD among the centroid individuals of the islands (inter-deme diversity). Fig. 2 shows the obtained results for the MMDP with random change mode and severity $\rho = 0.5$ (the same behavior has been observed for the other problem instances).

You can notice from Fig. 2a that the distance-based replacement ensure a high global (inter-deme) diversity throughout all the run. However, the frequent replacement of good individuals, since they are far from the remainder members of the subpopulation, reduces the local (intra-deme) diversity (see Fig. 2b). Consequently,

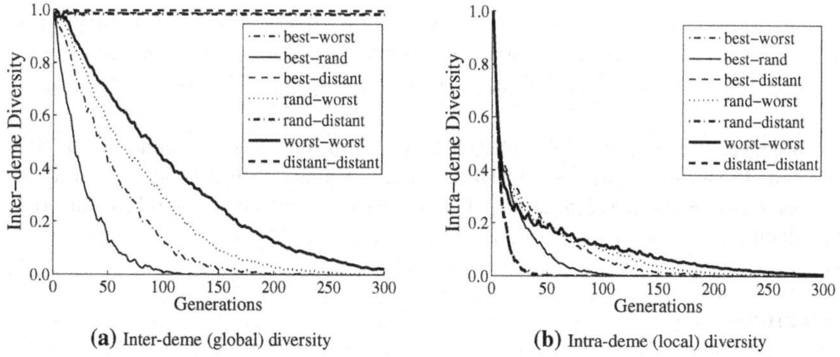

(a) Inter-deme (global) diversity

(b) Intra-deme (local) diversity

Fig. 2 Effect on the population diversity for the MMDP with Random (Non-cyclic) change mode and severity $\rho = 0.5$

each island quickly converges to the local suboptima, and it is unable to track their movements after a change. However, we have previously noticed its benefits for unimodal DOPs with high change severity (see Sect. 4), which is a hard scenario to converge, and a rapid adaptation to the first stationary periods used to succeed.

In addition, you can notice from Fig. 2b that the *worst-worst* criterium also exhibits the loss of diversity over time. Nevertheless, this issue can be solved by considering mutation, or incorporating common dynamic optimization techniques like *random-immigrants*, *fitness sharing*, etc.

6 Conclusions

In this paper, we have analyzed the influence of the migration selection and replacement criteria, two important parameters in the specification of dGA models, for DOPs. We used a canonical version of the algorithm without mutation to remove possible bias in the results and tested combinations of both quality and distance-based migration policies. Finally, we used a comprehensive test environment based on real-world difficulties, with several change modes and severities.

On the one hand, results showed the benefits when both migrant selection and replacement are chosen the least-fit individuals of each subpopulation. The performance was notably better when addressing unimodal DOPs with small changes. However, this migration policy also exhibits the loss of global diversity, denoting unability to track the moving optima for a long time.

On the other hand, distance-based replacement strategies showed to be more robust to track larger number of local optima, improving the speciation-feature of the population. However, it is not good to explore the changing landscape efficiently, since it produces the early loss of local diversity.

In conclusion, we have shown how migration policies allow us to balance between search of optima solutions and track of environmental changes for adapting to distinct dynamic optimization scenarios. In future works, we aim at developing adaptive or self-adaptive dGAs that exploit the main findings of this work with respect to the migration policy, thus enhancing the local behavior of the sub-populations with other techniques (random immigrants, fitness sharing, ...).

Acknowledgement. Authors acknowledge funds from the Spanish Ministry of Sciences and Innovation European FEDER, under contract TIN2011-28194 (roadME *http://roadme.lcc.uma.es*), and from the AUIP as sponsor of the Scholarship Program Academic Mobility.

References

1. Alba, E., Troya, J.M.: Influence of the migration policy in parallel distributed GAs with structured and panmictic populations. Appl. Intelligence 12, 163–181 (2000)
2. Branke, J., Kaussler, T., Schmidt, C., Schmeck, H.: A multi-population approach to dynamic optimization problems. In: 4th International Conference on Adaptive Computing in Design and Manufacture, pp. 299–308. Springer (2000)
3. Cantú-Paz, E.: Migration policies and takeover times in parallel genetic algorithms. In: Proc. of the GECCO. Morgan Kaufman (1999)
4. Homayounfar, H., Areibi, S., Wang, F.: An Island based GA for static/dynamic optimization problems. In: 3rd International DCDIS Conference on Engineering Applications and Computational Algorithms (2003)
5. Luque, G., Alba, E.: Parallel Genetic Algorithms. SCI, vol. 367. Springer, Heidelberg (2011)
6. Nguyen, T.T., Yang, S., Branke, J.: Evolutionary dynamic optimization: A survey of the state of the art. Swarm and Evolutionary Computation 6, 1–24 (2012)
7. Oppacher, F., Wineberg, M.: The shifting balance genetic algorithm: Improving the GA in a dynamic environment. In: Proc. of the Genetic and Evolutionary Computation Conference (GECCO), pp. 504–510. Morgan Kaufman (1999)
8. Park, T., Choe, R., Ryu, K.R.: Dual-population genetic algorithm for nonstationary optimization. In: Proc. of the GECCO, pp. 1025–1032. ACM (2008)
9. Ursem, R.K.: Multinational GAs: Multimodal optimization techniques in dynamic environments. In: Whitley, D., et al. (eds.) Proc. of the GECCO, pp. 19–26. Morgan Kaufmann (2000)
10. Yang, S., Yao, X.: Experimental study on population-based incremental learning algorithms for dynamic optimization problems. Soft. Computing 9(11), 815–834 (2005)

Ubiquitous Computing to Lower Domestic Violence Rate Based on Emotional Support Social Network (Redsiam)

Maria De Lourdes Margain[1], Guadalupe Obdulia Gutiérrez[1], Begoña García[2], Amaia Méndez[2], Alberto Ochoa[3], Alejandro de Luna[1], and Gabriela Hernández[1].

[1] Universidad Politécnica de Aguascalientes, México
{lourdes.margain,guadalupe.gutierrez,alejandro.deluna, gabriela.hernandez}@upa.edu.mx
[2] Universidad de Deusto, País Vasco
{amaia.mendez,mbgarciazapi}@deusto.es
[3] Universidad Autónoma de Ciudad Juárez, México
alberto.ochoa@uacj.mx

Abstract. The Project being presented is related to the use of ubiquitous computing to reduce the domestic violence rate. The general framework presented is for the design and development of an emotional support social network. The technologies used are a combination of artificial intelligence, data mining, speech processing and Android based services. In this same way a comparative is going to be obtaining from the Basque Country in Spain and Mexico so an Iberoamerican emotional support social network is constructed. The assessment is based on satisfaction surveys and longitudinal analysis of stored data.

Keywords: Ubiquitous Computing, Data mining, Smartphones services.

1 Introduction

Through previous support in social networks investigations it has been proven that a social group that suffers due to the same issue tries to get in a group seeking support through their partners. Domestic violence has dramatically increased not only in Europe (Romania and Spain), but also in Latin America (Mexico and Brazil), the consequences tend to be fatal being that the average represents a third of the cause of death in such societies. According to the National Institution of Geography and Statistics (INEGI), in the year 2005 2,159 women deaths registered were caused by family violence in Mexico, figure that exceeded the deaths caused by organized crime, which were 1,776. The French network Voltaire specialized in the analysis of international relations reported worrying figures, given that after Guatemala, Mexico is the country with the highest rate of feminicide. In 2009, 529 women and girls were killed in just 8 states of the republic according to the data of the national femicide citizen observatory. Considering that the technology can be

S. Omatu et al. (Eds.): *Distrib. Computing & Artificial Intelligence*, AISC 217, pp. 163–170.
DOI: 10.1007/978-3-319-00551-5_20

used to support the emotional support of a virtual social network, is intended to implement through the use of mobile devices; the voice recognition during a violent attack from the person's partner will send an alarm signal to family members, the support social network and the police. Since it has a GPS system, the victim could be located immediately and therefore receive the help needed. The project proposal is to carry out a comparative between Spain (Basque country) and Mexico (Juarez City and Aguascalientes City); to determine the causes and motives related to women who suffer from abuse and the reason why they won't leave their partner. An emotional support social network will allow every member to receive help from within regarding to the prevention of attacks and support whether is for shelter where to spend the night, or support with food for her and kids if she has any. The information generated in each society will allow to determine/to ascertain social behavior patterns in each one, determined by the aspects that specify the way in which they are extorted, threaten and outraged by their partners. In Juarez City, located in Chihuahua, the legislation contemplates the oral trials, which allow using as evidence: text messages, calls, and voice mails to demonstrate the guilt of the attacker. In the state of Aguascalientes and its parish of the same name, the legislation in this moment does not allow oral trials, it won't be until 2014 that the state will be cited and start the process so by 2016 it would start following through. This Project is intended to be developed in a time period of five years and it's contemplated that a 20 women group will be conforming the emotional support network, all of them being in a relationship for over a year and being known for the use of violence coming from their partners. The project pretends to develop specialized software and the implementation of the hardware through the purchase of the devices that will allow the voice recognition and could send and receive text messages.

2 Virtual Support Social Networks

The concept of a social network was used for the first time in the 40's by the anthropologists Radcliffe-Brown and Barnes who expressed two slopes. Radcliffe and Brown unveiled the first one defining the social structure as a network between the already existing relationships of people that make up a society, later on to be defined by Barnes as a way to describe the everyday social relationships based on the graph theory . A little shortly after that Girard refers the social networks as a society base that is an important road to integration. Nowadays, a social network often refers to platforms in the internet with their objectives being to make communication easier and share information of common interest, as well as to offer support to the society. In literature, some investigations have documented that the mental health area where the fact that the use of social networks by the victims that suffer any kind of violence, represents a great support in problem solving is highlighted, when its known that there is people close by and ready to provide help in any moment they ask for it . Organizations like the APC (Association for Progressive Communications) and Mindalia are some examples of social networks declared on the internet that establishes ties and connections between users who support instantaneously any person with welfare issues.

3 Family Violence Comparative between Mexico and the Basque Country

Statewide in Spain, the numbers of registered complaints in 2011, was higher than 134,000 for every 10,000 habitants, a figure that has increased in the last years. This data is divided by parishes as it can be seen in figure 1. The number of complaint figures for gender and domestic violence corresponding to Basque country amounts over 4,125, which even though this may seem an alarming number, it is far away from other communities like Cataluña o Andalucía, where figures are above 20.708, 18.475 and 27.727 respectively. The typology of the crimes instructed in Spain can be observed during 2011, highlighting the 95 cases brought due to homicides.

In contrast to Mexico, where violence adopts multiple manifestations, the present work is focused on the one used/ that one used within the family members, which is known as family violence or domestic violence. Leaving aside a simple classification based on the severity of the violence being applied, Johnson for example establishes the following distinctions:" common violence in a couple, intimate terrorism, violent resistance and violent control". In this project we take only the common distinction of violence in which the partner understood that is not connected to a general control pattern but is suggested in the context of a specific fight in which one or both members attack each other. As a consequence of violence in a relationship, the topic of family violence has produced a debate about what Connel (1997) called "the gender balance of violence". There exists evidence exerted by males (husbands, partners) against females with a higher frequency than the opposite. However, there are studies that also show that women have said to have experienced physical and sexual violence as frequently as or more often than males. Finally, in an INEGI study in 2009 it was reported that women at the age of 15 or older had suffered a type of violence against them by their partner in the last 12 months, therefore making emotional violence the mayor type of violence exerted, bringing in second, economic violence, followed by physical and in last place sexual. In the state of Aguascalientes the pattern mentioned by the type of violence persists.

4 Ubiquitous Computing

In the first decades of the last century we would not have been able to imagine such different digital elements linked to the human being in an intelligence way. However, currently, in the beginning of a new century lays a reality of coexistence between devices and human beings. Thus the creation of computer crowded environments and the communications capacity, everything that is integrated in an unappreciated way next to the people are recognized by Mark Weiser in 1991 as Ubiquitous computing. This new paradigm opens up opportunities to the creation of new devices that have been mixed in the everyday life, changing this way the everyday life model by benefiting it, even in assistance to its life necessities in an emergency case or social assistance. In this Project the Ubiquitous systemization is integrated by using mobile applications, to the lives of women needing emotional support. With a series of artifacts and portable equipment connected to a support social network that can give simultaneous assistance.

5 Architecture Based on SOA for RedSiam

For the development of the network, the fundamental work paradigm of the service oriented architecture is taken by allowing to connect services, applications and own information technology. The objective of the construction of this type of architecture is to take advantages of the resources already created and to optimize the new operations. Figure 2 shows the layers network architecture:

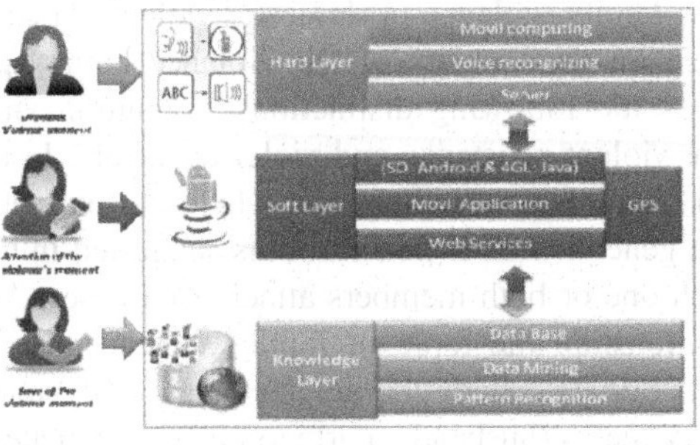

Fig. 1 RedSiam Architecture

Hard Layer: in which mobile computing processes, voice recognition, and the application server can be located. **Soft layer:** stores the platform like the operating system, the programming language, the mobile application and the web services in connection with the GPS. **Knowledge layer:** contains the knowledge of the people which is physically stored in a data base and will be exploited through data mining techniques, by obtaining pattern recognition. In this work the authors only explain the hard layer. This layers the process through mobile devices, voice recognition services and server communications. The soft layer and the knowledge layer will be explained in future works. The assessment is based on satisfaction surveys [10] and longitudinal analysis of stored data. Longitudinal analysis needs a long term pilot which will be published in the future.

Hard Layer: Voice Recognition Process through Mobile Devices.
Nowadays, mobile devices count with a technology that allows voice recognition to happen; such technologies can be useful when being oriented to our application, since it can be modified in such a way that it would be the detonation for the recorded conversations, the voice analysis segments, as well as the sending of the emergency signals. With this, the technology would be taken to modify the

management of domestic violence oriented to the relationship and the access to it with the mobile devices in the population.

Mobile Devices

The mobile technology has brought significant changes in the user's everyday life activities, as it has the Internet, along the way the technology has kept evoling conceiving smart mobile devices with a bigger coverage and connectivity. These devices provide similar advantages to the ones of a personal computer, being the difference that the mobile device allows the user to dispose of any information, to share information and to be in contact with other users in any place or moment, as long as it has an Internet connection. In literature it is documented that the mobile terminals have been considered the new computers of the century and in the past Mobile World Congress, Eric Schmidt, Google CEO has affirmed that in today's time, a thousand millions of people have access to a mobile device making it one of the most bought product in the last social times. Nowadays the mobile market has two great companies leading this area: Apple with its iOS operating system and Google with its Android operating system. Table 2 shows a comparative between this platforms as well as devices of the brand Blackberry and Windows phone.

Table 1 Comparison of mobile

OS	Android	iOS	Windows
Operating System Kernel	Linux	Mac OS X	Windows CE
Programming Language	Java, C++	Objective-C, C++	C++
Software License	free software	proprietary	proprietary
Number of Applications	140,000 60% free	300,000 30% free	4,000

The characteristics for Android makes it accessible for the development of applications, implementing free software for the development of the social network. For the moment the reach of the application developments are focused on the Android operating system and more specifically on such mobile devices as Samsung Galaxy SIII (SG-SIII) since it has the technical capacity to run the application without any problem due to its Exynos 4 Quad processor with 4 nucleus of 1.4 GHz. Also the SG-SIII has a voice recognition system denominated as S-Voice very similar to Apple's with the difference that this one is capable of understanding Spanish and also waking up just by saying the phrase "hello Galaxy", which will be a great help when it comes to programming and analyzing as far as the women voice recognition that in a determined time could suffer an attack.

Voice Recognition Services

The voice recognition service would be of great support in the process, having as its main function to send alarm signals when a violence attack is happening or when there is a situation where certain factors are very specific and are generally considered violent. This is carried out through key words established and certain events described with the voice that would tips of id an act of violence have occurred. As far as voice recognition is concerned some open codes developed will be taken to show this effect on mobile devices, with an adaptation in the phrase recognition that will determine if violence has occurred, as well as in the sending of messages to help the other 14 women, and the DIF monitor. Also the recordings of what happened in that instant that can be evidenced in any given situation. Even when the development of algorithms of voice recognition in mobile devices has demonstrated to be efficient in the word recognition there is still a variety of aspects that has to be resolved, which are concrete for our application:

1. The distance in which the person should be from the device when submitting their voice, since in not all violence attacks the person would carry the phone to pronounce the word. The algorithms that make the voice recognition more robust have to be applied to the system.
2. The running time of the application, the amount of time that it would be on sleep mode and how to go from an inactive mode to an active so it can be able to link a recording application as well as the voice recognition voice application.
3. The way in which a word would be pronounced while in the middle of a violence situation, since all the algorithms developed in mobile applications are based on environments where the user is not involved within a violent situation.
4. The background noise is a crucial factor for the voice recognition system that will be taken place, since in every moment where it is pretended to be available would not be in the majority of the cases of control environments.

All the programs mentioned above should be fought to obtain a good support system for the intervention of lower domestic violence. The voice recognition system would be a great back up in the development of the application since it pretends to use other voice factors such as tone, intensity, speed, and fundamental frequency (pitch) to determine apart from of the word spoken, the precise moment when a person is in a violent situation, this is made with the intention to cover all the possible situations, including the ones that could present themselves when the victim could not pronounce a word. The application will be developed in an open code, for this reason the operating system Android offers the mayor advantage in the sense in front of the proprietary operating systems, as well as the existence of a voice recognition application with the capacity to be adapted to our objective. In the advantages of the usage of this operating system, it is highlighted that the voice recognition application will run constantly and in a transparent way to the user with very little usage of the phone's resources, in comparison to the proprietary operating systems where the application should be initialized before using it, by employing most of the phones mayor resources. Also taking into consideration that the application to be developed should run in continuous time, the Android System offers the advantage of having this tools available at all time for its use. The language in which the application is developed would be at the beginning orientated towards the Android system; even though it will be seeking the implementation in other languages oriented to other proprietary operating systems as a comparison of the usage of the application resources in each one of them.

Server

To carry out the voice recognition process in the instant the victim is being attacked, as well as the behavior pattern analysis that each one of them reflect according to the data saved in the system, it is planned to do a platform where the mobile applications will communicate directly. The design of the general architecture proposed to identify the server communications, the platform and the mobile application.

The server will be in charge of storing and analyzing a part of the recordings of the voice recognition to identify the tone used for help in each one of the victims that go through violence by their partners, also it will detect hidden behavior patterns according to the voice characteristics and/or messages in each one. It will also be in charge of locating to locate the closest help available in case of an alert notification. For the time being, the study will only be applied to Samsung Galaxy III mobiles. Then, after it is planned to be included an iPhone devices, for which that stage will be applied in web services to establish communication between cell phones. According to Mobile applications, it is through mobile application where the victim will be able to link directly the platform , where it can interact with other women that are in the same situation as her by means of the communication tools such as chat and video calling. It is s planned that the alert notifications will be activated by the movements the victim would do in a mobile. With respect to Social networking platform, the victim could Access the social network platform both via web and mobile. In this part there is an option given to interact with other victims and specialized people to help with their treatment. It is also here where the hidden behavior patterns the victims present will be visualized. This way, the server will accommodate different applications and services, mainly the data base, the social network, and web services (mobile device interactions).

6 Conclusions

Several studies about violence in women can acknowledge that this is a social problem not solved. To propose solutions require work from different perspectives and in different countries, in order to contribute to the problem. The proposed solutions consider the emotional, legal and social topics, including the use of information technology (ies). These last two topics lead the way to propose solutions that integrate several disciplines. The first step outlined in this work established the foundation for building an Iberoamerican network of emotional social support: RedSiam. This early work, responds to capture the dramatic moment of violence. For this step is required to establish the design of service-oriented architecture. Additionally, for the project was acquired basic hardware for the network, as mobile equipment (Samsung Galaxy III) for locating mobile computing processes, the operating system selection (Android) can enable speech recognition and application server connections. The preliminary contact with the potential final user has been very successful in terms of their support from the beginning contributing in the specification of the social network, collaboration in the definition of the surveys and the participation during the pilot. As future work is required to develop the mobile application and install the proposed architecture. Subsequently, work on the requirements of the soft layer and layer knowledge layer.

References

1. Radcliffe-Brown, A.R.: Structure and Function in Primitive Society (1965)
2. Barnes, J.: Human Relations. Nueva York. T.VII, pp. 39–58 (1954)
3. Kauchakje, S., Camillo, M., Frey, K., Duarte, F.: Redes socio-técnicas y participación ciudadana: propuestas conceptuales y analíticas para el uso de las TICs. REDES- Revista Hispana Para el Análisis de Redes Sociales 11(3) (2006)
4. Girard, M.: Les jeunes chômeurs et leurs réseaux: une stratégie efficace, une efficacité relative. These de doctorat. Départament de sociologie. Université du Québec a Montréal (2002)
5. Juárez, C., Valdez, R., Hernández, D.: La percepción del apoyo social en mujeres con experiencia de violencia conyugal. Red de Revistas Científicas de América Latina y el Caribe, España y Portugal 28(004), 66–73 (2005)
6. Johnson, M.P., Ferraro, K.J.: Research on domestic violence in the 1990s: Making distinctions. Journal of Marriage and the Family 62, 948–963 (2000)
7. Dobash, R.E., Dobash, R.P.: Women, violence, and social change. Routledge Kegan Paul. Boston (1992)
8. Muehlenhard, C.L., Kimes, L.A.: The social construction of violence: The case of sexual and domestic violence. Personality and Social Psychology Review 3(3), 234–245 (1999)
9. Gironés, J.: El gran libro de Android. marcombo ediciones técnicas, ed. m.e. técnicas. Alfaomega, Barcelona (2012)
10. UPA. Usability Professionals' Association, http://www.upassoc.org/usability_resources/guidelines_and_methods/

Multiobjective Local Search Techniques for Evolutionary Polygonal Approximation

José L. Guerrero, Antonio Berlanga, and José M. Molina

University Carlos III of Madrid, Avda. Universidad Carlos III, 22, Colmenarejo, Spain
jguerrer@inf.uc3m.es, {aberlan,molina}@ia.uc3m.es

Abstract. Polygonal approximation is based on the division of a closed curve into a set of segments. This problem has been traditionally approached as a single-objective optimization issue where the representation error was minimized according to a set of restrictions and parameters. When these approaches try to be subsumed into more recent multi-objective ones, a number of issues arise. Current work successfully adapts two of these traditional approaches and introduces them as initialization procedures for a MOEA approach to polygonal approximation, being the results, both for initial and final fronts, analyzed according to their statistical significance over a set of traditional curves from the domain.

1 Introduction

Segmentation problems are based on the division of a given curve in a set of n segments (being each of these segments represented by a linear model, which points to another common naming convention for this process: piecewise linear representation, PLR) minimizing the representation error. Polygonal approximation techniques [10] are offline segmentation algorithms (since they require the whole curve they will be applied to) and can be divided into three different categories: sequential approaches, split and merge approaches and heuristic search approaches.

Sequential approaches are constructive methods based on a given local search over the current time series, trying to obtain, at each step, a new segment division (where the length of these segments is sequentially increased) which satisfies a certain criterion. Examples of the criteria used may be finding the longest possible segments ([12]). Split and merge approaches perform an initial segmentation over the given time series and afterwards start an iterative process to merge the initial segments until a certain criterion is met. According to their definition, these approaches have to deal with two different issues, the initial segmentation procedure and the merging criterion. An example of these processes is the bottom-up algorithm [7]

S. Omatu et al. (Eds.): *Distrib. Computing & Artificial Intelligence*, AISC 217, pp. 171–178.
DOI: 10.1007/978-3-319-00551-5_21

Heuristic search approaches are based on the development of heuristic methods in order to avoid the exhaustive search of the optimal *dominant points* for the given time series (which is a process with an exponential complexity). Different techniques may be used for this purpose, such as dynamic programming [11] or different metaheuristics, among them different solutions based on evolutionary algorithms [13]. The idea proposed by these works is to codify the time series as a chromosome with n genes, corresponding each of these genes to one of the points in the original data. If the gene value is a "1", it is considered a *dominant point*, and the algorithm tries to find the ideal codification of the chromosome according to a fitness function which evaluates the quality of the given codified segmentation in the chromosome.

Recently, the multiobjective nature of these processes is being explicitly approached from different perspectives [8, 5]. In [5] a multi-objective evolutionary algorithm [1] is proposed for the multi-objective solution of the segmentation issue, while in [4] a comparison between different possible initializations was carried, focusing on the different results between a random initialization aiming at the coverage of the obtained Pareto fronts versus the results from different local search techniques. One of the detailed issues is the single-objective nature of the traditional techniques used, which required different executions with different parameters in order to obtain different individuals from the front, also introducing issues regarding the configuration of these techniques to obtain such different individuals.

Current work will introduce a multi-objective explicit formulation from two traditional techniques for the polygonal approximation domain: bottom-up [7] and top-down algorithms[9]. This implementation will produce a whole Pareto front from a single execution without parametrization required from the user. These initializations will be later tested as initial populations for a MOEA approach, in order to determine whether they have successfully created better initial populations than the approach presented in [4] and how these initial populations translate into the final results of the algorithm.

The structure of this work is divided into three different sections: section two will present the techniques according to their traditional and multi-objective approach. After the new implementation has been presented, section three will present the results when these implementations are used to create the specified initial populations, analyzing the final results of the chosen algorithm. Finally the conclusions which can be extracted from the presented results will be presented in the final section, leading to the future lines of the work.

2 A Multiobjective Perspective to Local Search Polygonal Approximation Techniques

Two different issues can be stated regarding a polygonal approximation problem: $Min - \#$ and $Min - \epsilon$. $Min - \#$ is based on the optimization of the representation error for a previously set number of segments. $Min - \epsilon$, on the other hand, tries to find the minimum number of segments such that the final representation error does not exceed a previously established error ϵ. In [5] it was stated that, according to these two different perspectives, the segmentation issue is in fact a multi-objective problem, and also analyzed, according to different techniques available in the literature, how this nature had been faced. It was also shown that, given that some key dominant points are shared by different solutions with different resolutions, the solutions for $Min - \#$ and $Min - \epsilon$ problems can be closely related and share information between them. This multi-objective nature is faced with a Multi-objective evolutionary algorithm.

Local search algorithms may be introduced to enhance this approach, leading to several issues: the configuration to obtain the different individuals is hard to establish, and each of this individuals requires an independent execution of the local search algorithm, providing disappointing results [4]. This section will present alternative, parameter-free versions of two well known local search algorithms for polygonal approximation which provide a whole Pareto front of solutions: Top-Down and Bottom-up algorithms.

Top Down algorithm [9] is an offline process based on finding the best splitting point (understanding by this that measurement which divides the trajectory into the two segments with the lowest added errors) recursively, until all the resulting segments have an error value bellow a user defined boundary. The Top Down algorithm is applied in a wide variety of domains and fields, being also known by different names[2].

The multi-objective version of the Top Down algorithm suppresses the two issues available in the traditional implementation: the recursive calls (which may prevent the application of the algorithm to figures with a large number of points) and the user configuration (which introduces the issues previously described in the obtaining of a whole Pareto front). At each step, the best splitting point is located (the one which provides with the smallest representation error), a new individual is generated adding that new dominant point and the costs of the possible segments are updated (implying the recomputation of the costs of the segments from the dominant point immediately to the left of the new splitting point and those from the splitting point to the dominant one immediately to its right). Therefore, no recursive calls are included, and each split point choice has a global view of the representation error (as opposed to the partial one available in the traditional implementation). Figure 1 represents the multi-objective version implementation of this algorithm.

Bottom up algorithm[7] is an offline process complementary to Top Down, where the time series is initially divided into every possible segment (composed of two measurements) and finds the best possible segment fusion afterwards (understanding by this the fusion which obtains the segment with the lowest error) until any possible fusion obtains a segment having an error above a user defined boundary. The bottom up algorithm, as well, has

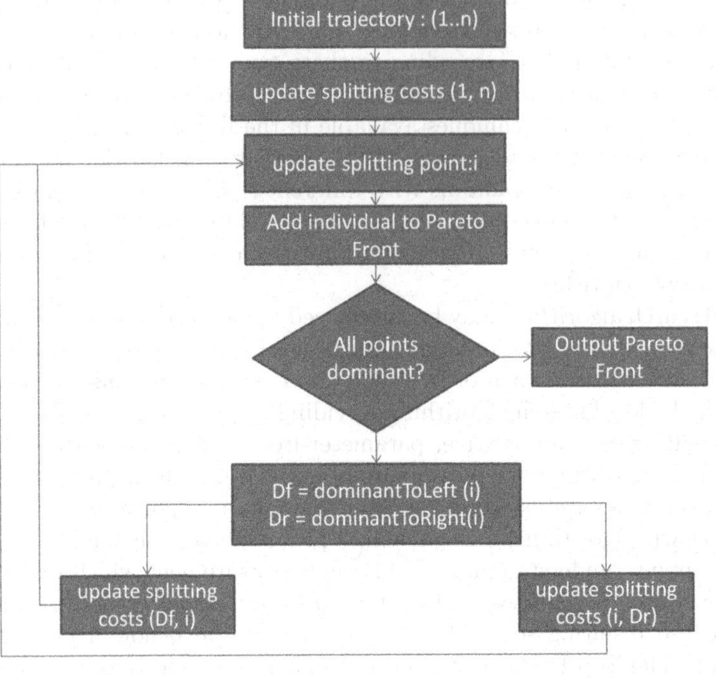

Fig. 1 Top Down algorithm multi-objective implementation

spread to different fields and research areas using different names, such as the computer graphics domain and decimation methods[6].

The multi-objective version of bottom-up algorithm removes the user-defined boundaries for the algorithm termination, being this ending triggered once no further merging can be performed. Figure 2 presents the multi-objective version. It must be noted that each update here triggers only one segment update, while every new splitting point in the top down algorithm triggered the recomputation of all the possible new splitting points for the two new segments created in the representation. Since each of these steps are, in fact, mutations over the chromosome guided by a specific heuristic, the principles for an efficient implementation established in [3] can be applied for the computation of the fitness values of each of the produced individuals.

3 Experimental Validation: Initialization for MOEA Polygonal Approximation

The experimental validation proposed will include the two detailed multi-objective local search procedures to create the initial populations for a Multi-objective evolutionary approach to polygonal approximation. This algorithm

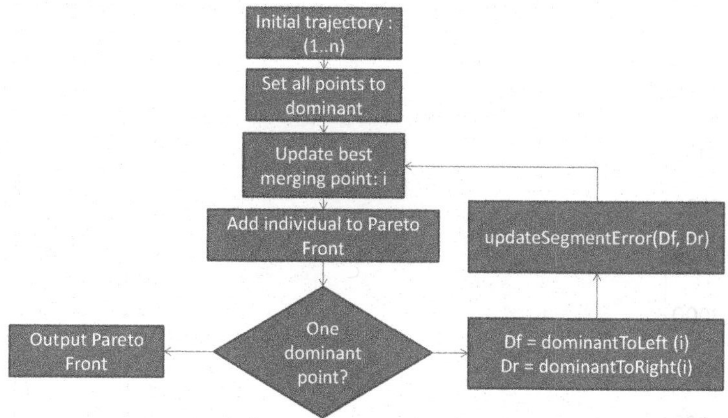

Fig. 2 Bottom up algorithm multi-objective implementation

is based on the SPEA2 [14] MOEA, according to the configuration presented in [5]. The default initialization process creates a uniform Pareto Front in terms of coverage of the objectives, as presented in [4]. This section will cover the comparison between the initial and final populations of the two techniques presented and the suggested initialization process. The dataset used is composed of three traditional curves, usually named chromosome, leaf and semicicle. Their definition, according to ther freeman chain-code representation, can be found in [5]. Figure 3 represents these figures.

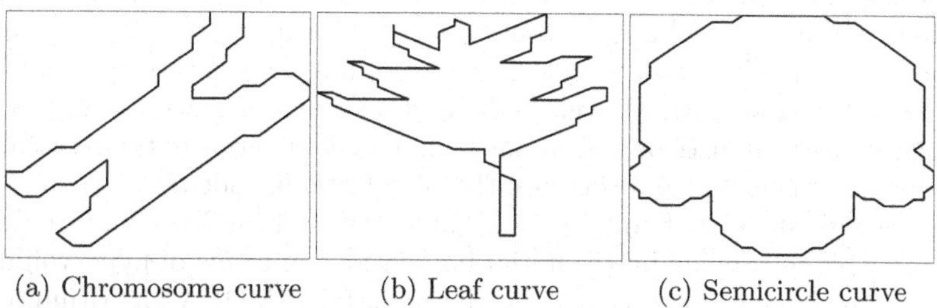

(a) Chromosome curve (b) Leaf curve (c) Semicircle curve

Fig. 3 Curves included in the data set

It is interesting to notice the complementary nature of the two multi-objective techniques presented, since once applied its heuristic with a value of 1 dominant point and applies successive splitting over the figure (Top-Down) and the other begins with a solution with all of its points considered dominant and applies successive merging (Bottom-up). Since the solutions tend to degrade with the successive application of the heuristic, each of them will be more successful at their initial individuals. This can be seen in figure

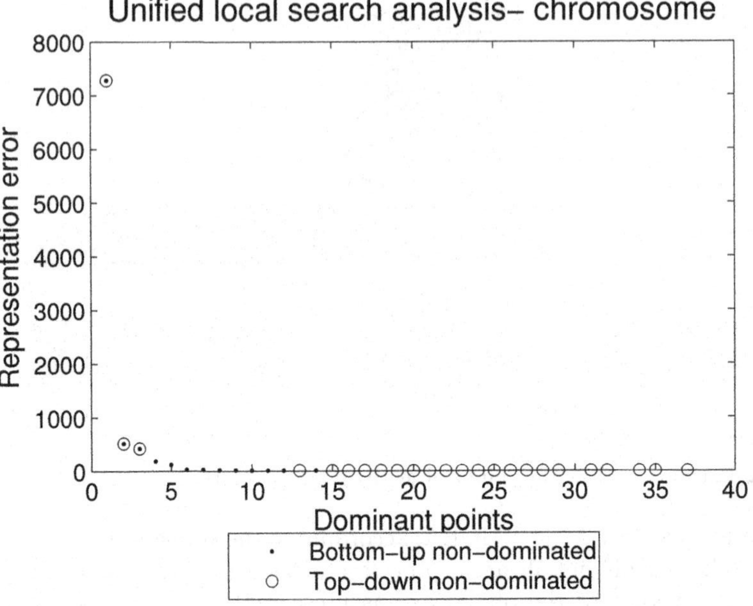

Fig. 4 Analysis of unified local search front non-dominated individuals

4 which shown the non-dominated individuals of a Pareto front composed of bottom-up and top-down initial fronts for the chromosome curve (figure 3a).

The results for the four techniques, including their mean and median values for the hypervolume of the obtained Pareto fronts are included in tables 1 (initial fronts values) and 2(final fronts values). Also, a best technique column has been added. This value is calculated according to a Wilcoxon test with a 95% confidence performed over 30 different executions, since the values do not follow a normal distribution. If one technique is superior to the remaining ones, its name is included, otherwise the '-' value is included.

The results show that, on the one hand, the multi-objective search approach is able to provide better initial populations in terms of hypervolume. When these populations are used by the underlying MOEA algorithm, different cases appear. For easy problems, such as Chromosome, the uniform initialization provides better final results. This happens due to the focus which the local search initialization introduces according to its underlying heuristic. Even though the initial results are clearly improved, the final ones are too guided by the initial heuristic. For harder problems, the initialization provides the algorithm with an important enough advantage such that the final populations are either not statistically significant (leaf) or significantly better (chromosome). These results seem to point to a combination of both techniques to provide initial populations that, while benefiting from the

Table 1 Initial populations comparison

Figure	Bottom-up		Top-down		Local search		Uniform		Best
	Mean	Median	Mean	Median	Mean	Median	Mean	Median	
Chrom.	0,98647	0,98647	0,98646	0,98646	0,98651	0,98651	0,98436	0,98427	L.S.
Leaf	0,99355	0,99355	0,99322	0,99322	0,99365	0,99365	0,99271	0,99281	L.S.
Semi.	0,99157	0,99157	0,99183	0,99183	0,99218	0,99218	0,99101	0,99111	L.S.

Table 2 Final populations comparison

Figure	Bottom-up		Top-down		Local search		Uniform		Best
	Mean	Median	Mean	Median	Mean	Median	Mean	Median	
Chrom.	0.98665	0.98664	0.98667	0.98671	0.98665	0.98667	0.98671	0.98672	Unif.
Leaf	0.99376	0.99376	0.99374	0.99376	0.99376	0.99376	0.99377	0.99378	-
Semi.	0.99206	0.99219	0.99213	0.99217	0.99219	0.99219	0.99213	0.99217	L.S.

enhanced initial populations of local search techniques, can be not hampered by the heuristic focus.

4 Conclusions

Local search techniques have been the focus of polygonal approximation, developing different techniques based on specific heuristics for this issue. However, the new multi-objective approaches require modifications over these techniques in order to efficiently obtained the required Pareto Fronts. This work has modified Bottom-up and Top-down techniques to provide a multi-objective approach with the required characteristics presented. The results have been tested for the initialization of a MOEA algorithm. These results show that the multi-objective techniques are successful in providing statistically better initial populations, however the final results may be too focused on the heuristic used in these techniques, which in some cases hamper their quality. Future lines imply the research of the combination which may be performed over local-search and uniform initialization in order to provide initial populations taking advantage of local search improved initial populations without their excessive focus on their underlying heuristic.

Acknowledgement. This work was supported in part by Projects MEyC TEC2012-37832-C02-01, MEyC TEC2011-28626-C02-02 and CAM CONTEXTS (S2009/TIC-1485).

References

1. Coello, C., Lamont, G., Van Veldhuizen, D.: Evolutionary algorithms for solving multi-objective problems. Springer-Verlag New York Inc. (2007)
2. Duda, R.O., Hart, P.E.: Pattern Classification and Scene Analysis. Wiley (1973)
3. Guerrero, J., Berlanga, A., Molina, J.: Fitness-aware operators for evolutionary polygonal approximation. In: Applied Computing, pp. 283–290. IADIS (2012)
4. Guerrero, J.L., Berlanga, A., Molina, J.M.: Initialization procedures for multiobjective evolutionary approaches to the segmentation issue. In: Corchado, E., Snášel, V., Abraham, A., Woźniak, M., Graña, M., Cho, S.-B. (eds.) HAIS 2012, Part III. LNCS, vol. 7208, pp. 452–463. Springer, Heidelberg (2012)
5. Guerrero, J., Berlanga, A., Molina, J.: A multi-objective approach for the segmentation issue. Engineering Optimization 44(3), 267–287 (2012)
6. Heckbert, P., Garland, M.: Survey of polygonal surface simplification algorithms. In: Multiresolution Surface Modeling Course. ACM Siggraph Course notes (1997)
7. Keogh, E., Chu, S., Hart, D., Pazzani, M.: Segmenting time series: A survey and novel approach. Data Mining in time series databases, 1–21 (2004)
8. Kolesnikov, A., Franti, P., Wu, X.: Multiresolution polygonal approximation of digital curves. In: Proceedings of the 17th International Conference on Pattern Recognition, ICPR 2004, vol. 2, pp. 855–858. IEEE (2004)
9. Ramer, U.: An iterative procedure for the polygonal approximation of plane curves. Computer Graphics and Image Processing 1, 244–256 (1972)
10. Sarfraz, M.: Linear capture of digital curves. In: Interactive Curve Modeling, pp. 241–265. Springer, London (2008)
11. Sato, Y.: Piecewise linear approximation of plane curves by perimeter optimization. Pattern Recognition 25(12), 1535–1543 (1992)
12. Sklansky, J., Gonzalez, V.: Fast polygonal approximation of digitized curves. Pattern Recognition 12(5), 327–331 (1980)
13. Yin, P.: Genetic algorithms for polygonal approximation of digital curves. International Journal of Pattern Recognition and Artificial Intelligence 13(7), 1061–1082 (1999)
14. Zitzler, E., Laumanns, M., Thiele, L.: SPEA2: Improving the Strength Pareto Evolutionary Algorithm. In: EUROGEN 2001, pp. 95–100. International Center for Numerical Methods in Engineering (CIMNE), Athens (2001)

GSched: An Efficient Scheduler for Hybrid CPU-GPU HPC Systems

Mariano Raboso Mateos and Juan Antonio Cotobal Robles

Facultad de Informática, Universidad Pontificia de Salamanca. Compañía 5, 37002 Salamanca, Spain
mrabosoma@upsa.es, jacotobal@hotmail.com

Abstract. Modern and efficient GPUs evolve towards a new integration paradigm for parallel processing systems, where Message-Passing Interfaces (MPI), Open MP and GPU architectures (CUDA) may be joined to perform a powerful high performance computation system (HPC). Nevertheless, this challenge requires much effort to properly integrate both technology and software programming. This paper describes GSched (Grid Scheduler), an optimized scheduler that allows distributing both CPU and GPU processor execution using a previous calculated optimum pattern, obtaining the best elapsed execution time for overall execution. Furthermore, high-level algorithm description is introduced to efficiently distribute processing and network resources.

Keywords: Scheduler, MPI, GPU, Open MP, parallel processing, HPC, CUDA.

1 Introduction

Most of traditional algorithms can be adapted to be executed on uniform and homogeneous parallel clusters based on CPU and GPU [1]. Nevertheless, computer networks installed on university campus and other scientific organizations are often equipped with heterogeneous computers, with different computing capabilities. There are several research advances that are achieving substantial improvements on heterogeneous systems performance. CudaMPI [2] makes available MPI standard functions on GPU systems. Caravela MPI implementation [3] also provides a library to be used on GPU-node based networks. Other technologies optimize synchronization as CUDA Inter-Process Communication (Cuda IPC) [4].

General Purpose GPU computing clusters have dramatically decreased execution time but scheduling has become a serious issue due to the complexity for scheduling complex and time-consuming jobs in a heterogeneous computing network. Scheduler solutions as Merge framework from Linderman et al. [5] achieves performance and energy efficiency across heterogeneous cores via a dynamic library-based dispatch system. Merge also proposes a programming model that abstracts the architecture and requires additional information from the programmer for dispatch decisions. In [6] a dynamic scheduler is proposed using a cost function to evaluate CPU and GPU, designing a resultant optimal chunk-size for

S. Omatu et al. (Eds.): *Distrib. Computing & Artificial Intelligence*, AISC 217, pp. 179–185.
DOI: 10.1007/978-3-319-00551-5_22

various applications and problem sizes. A new task management system is proposed in [7], extracting the workloads from multiple kernels and merging them into a super-kernel. Other solutions as in [8] merge kernels using a queue system that improves CUDA kernel efficiency.

This paper proposes an optimized static scheduler similar as in [6], adding fine coarse grain by testing every computing resource in the grid. This information allows obtaining a weighted decision function to be later used by the allocation policy that is transparent to the end user. It has been considered because modern GPU differ much on compute and bandwidth performance.

Section 2 describes how the scheduler is designed and the general rules to test ad run applications. Measurements, reports and results are shown in section 3 and 4.

2 Scheduler Design

Design objective for the new scheduler is to use a proportional workload depending on the node performance or at least the CUDA compute capability [9]. The compute resource allocation also depends on the task assigned. Some works allow using a fine grain partition policy. For this case a master node is responsible for sending little size fragments to the rest of nodes [10] and assembling the partial solutions generated. This policy allows the fastest nodes to resolve more fragments but also introduces new problems as communication delays, fragment losses or endless waits. The scheduler first analyses the performance of each computing node and then provides and optimal partition of the overall workload offered. This technique allows using a great variety of nodes based on CPU, GPU or multi GPU and mixed CPU-GPU computing resources.

2.1 Process Description

We have developed a network with several heterogeneous computers, interconnected by MPI middleware.

The process (fig. 1) has two stages:

1. - A program "scheduler" that evaluates the nodes of a network and obtains several optimization parameters.

2. - A program "dispatcher" that sends jobs to nodes "worker" for its execution according to the parameters previously obtained.

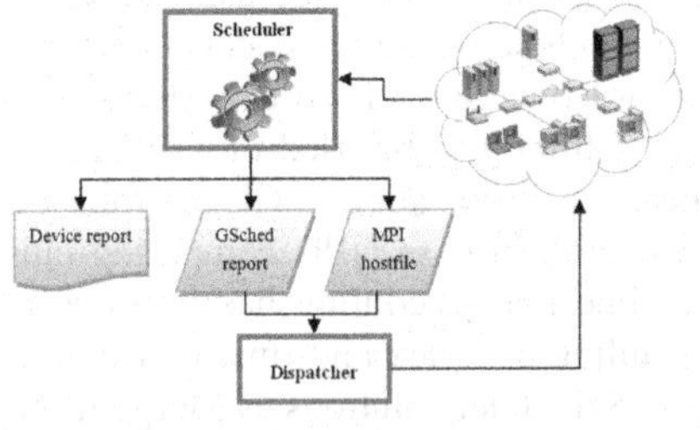

Fig. 1 Scheduler/Dispatcher diagram

The scheduler examines the nodes of a network, obtaining an inventory of its features. We have designed a synthetic benchmark with algebraic matrix operations. This test will run on the CPU on each node and if they have graphics cards

NVIDIA CUDA [11], performs the same operations on each. The benchmark gets standardized scoring for distributing the workload while indicating the use of CPUs or GPUs devices, in the most appropriate way as possible.

The scheduler does not need to calculate an absolute value of the power of each node (as Linpack Benchmark). It is enough to obtain a relative value for the nodes.

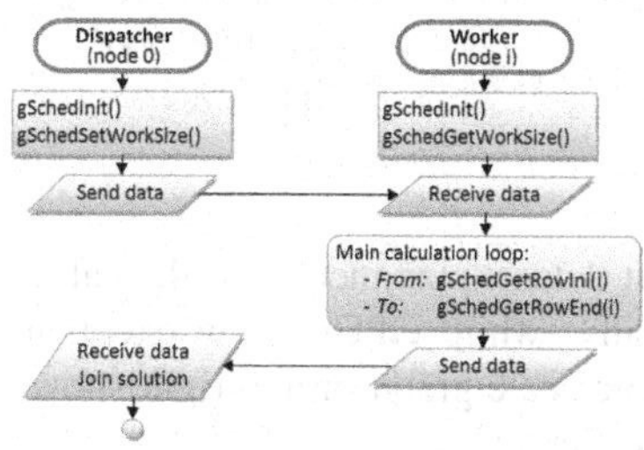

Fig. 2 Structure of an algorithm adapted to GSched scheduler

Issues as delays by MPI communications or data transfer bandwidth between CPU and GPU components have not been considered.

To incorporate the GSched scheduler, the algorithms need to meet a structure based on a dispatcher process and a set of workers. We have defined a C++ library with a simple API to guide its execution using scheduling parameters.

2.2 *Hardware Platform*

The developed system has been implemented in the parallel computing laboratory at the Computer Science Faculty at the Pontifical University of Salamanca, performing a small cluster of computers selected with some different components: a computer with an Intel Core i7 and a powerful Tesla C2070 GPU card, a second Intel Core i7 server computer equipped with two GeForce GTX 560 cards and two VMs under VMWare, with an Intel Core i7 host with no CUDA GPU support.

A first node (node #0) designated as the master or dispatcher, will divide the problem, coordinate communication with other nodes and collect partial solutions. Each computing unit is treated as if it was a separate node. For example, a computer consisting of one CPU and two GPUs will be defined as a three separate nodes: CPU, GPU #1 and GPU #2.

3 Scheduling and Measurement Reports

Once the scheduler has performed the tests and evaluated network nodes, it generates three files with useful information for later job execution:

- **Device report**, with name, MPI id, test time and other node characteristics.
- **Measurement report**, a normalized "power ranking" indicating how much work must address each node and what CPU or GPU device to use, which have different implementations for the same problem.
- **MPI hostfile**, containing the ordered list of nodes according to the scheduler.

Along this work we have made scheduling tests with datasets of different sizes (fig. 3), trying to calibrate their behavior to achieve the best results for the execution of subsequent tasks.

Nodes	Characteristics	Id	100	200	500	750	1000	2000	3000	4000	5000
node1	Intel(R) Core(TM) i7-2600K	0	0,0101	0,0380	0,6152	2,2526	6,2988	61,0332	251,1146	589,4984	1296,3066
node1-GPU1	Tesla C2070	1	0,0643	0,0747	0,0824	0,0900	0,1149	0,3965	1,1898	2,5575	5,2373
node2	Intel(R) Core(TM) i7-2600	2	0,0091	0,0354	0,6138	2,2602	6,2739	60,7619	249,4472	578,9618	1288,7764
node2-GPU1	GeForce GTX 560 1Gb	3	0,0449	0,0620	0,0623	0,0721	0,1152	0,5034	1,5944	3,5004	7,1454
node2-GPU2	GeForce GTX 560 1Gb	4	0,0581	0,0528	0,0696	0,0832	0,1202	0,5162	1,5947	3,4882	7,1445
node3	Intel(R) Core i7 CPU - VMWare	5	0,0082	0,0650	1,1751	4,4098	13,7145	144,0742	552,4320	1366,5967	2838,1218
node4	Intel(R) Core i7 CPU - VMWare	6	0,0082	0,0652	1,1737	4,5750	16,2319	161,1131	612,5109	1511,6269	2845,7425

Fig. 3 Table of measurements for tests 100^2, 200^2,... 5000^2 sizes

The following charts show that, when the test dataset size increases, the calculation times of the CPUs increases geometrically, while GPUs show excellent capabilities to operate in parallel. The logarithmic scale graph on the right shows that only using smaller tests the CPUs are more efficient:

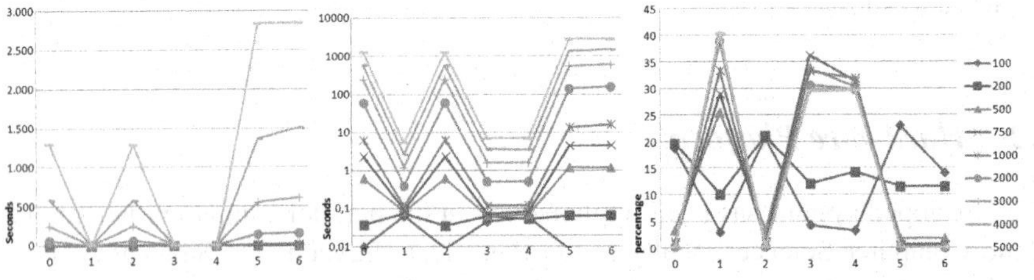

Fig. 4 a) Computation time b) Time in logarithmic scale c) Assignment workspaces

These data lead to a logical conclusion: when the scheduler tests are more costly and complex, the fraction of work assigned to GPU is greater, leaving the CPU with decreasing percentages except in the simplest tests.

4 Execution and Results

The scheduler's efficiency is proved running large jobs. The above approach is best illustrated with an example that will visually show the process time without using the scheduler and the improvement that is achieved thanks to scheduler.

We will apply a filter to a large figure for obtaining a new modified image. Specifically, the process involves applying a convolution kernel to detect and enhance the edges of the original image [12]. Actually, this work results in multiplying the kernel matrix by the matrix of the same size that surrounds each pixel of the original image. The sum of the values of each resulting matrix will be a new image point of the solution.

Fig. 5 Original image

This task, which requires a lot of computing power, can be easily solved by parallel techniques.

This approach is a typical scenario in which "fine grain" partition is not suitable because each image fragment must add additional margins or image boundaries, as large as the kernel to be applied, to compute the outer points of each image fragment. An excessive degree of parallelization implies a redundancy and a useless extra-calculation.

Our scheduler must minimize the number of fragments to treat (one for each element of calculation, CPU or GPU) and setting its most appropriate size for the power of each node. This will result on reducing the final time resolution of work. To better see the effect of different processes, pixels will be clarified or darkened depending on the node id (even or odd, respectively) that realize its treatment.

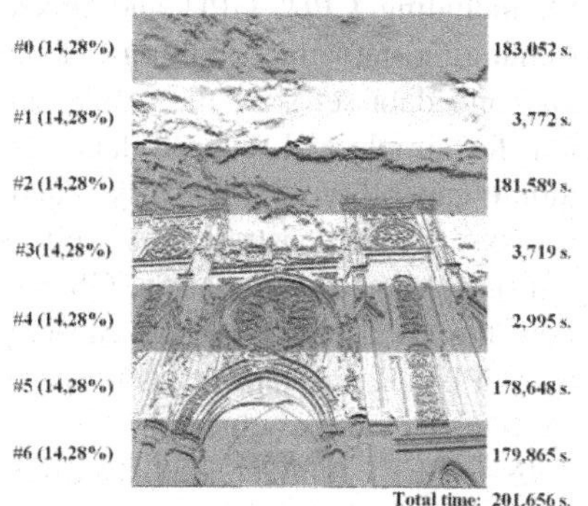

Fig. 6 Image results (no scheduling)

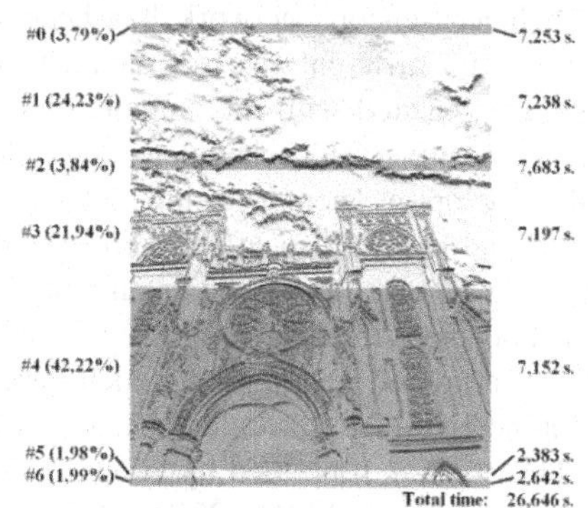

Fig. 7 Image results (with scheduler)

In a first execution without the scheduler, the program divides the original image into fragments of the same size, one for each MPI node available. It is observed that the total processing time is directly related to the elapsed time of slower nodes (in this case, CPUs #0 and #2). As a result, faster nodes remain inactive for too long. Figure 7 shows results running the scheduler. Since the nodes finalize in a similar time, an improvement is obtained in overall execution time.

For the second stage the scheduler program is executed and the Measurement Report obtained with the distribution of work under each node. Figure 8 shows the first node (dispatcher) with a quota of 0%. It is responsible for dispatching the work and collecting results.

```
nodo01; 0;  0.000000  | nodo02; 1; 21.918267
nodo01; 0;  3.799809  | nodo02; 2; 42.220635
nodo01; 1; 24.237837  | nodo03; 0;  1.988695
nodo02; 0;  3.841253  | nodo04; 0;  1.993504
```

Fig. 8 Report for seven computing nodes and a dispatcher

To achieve the optimum time would be necessary that the scheduler had such precision that, in real execution all nodes invest exactly the proportional time calculated. Because of the multitude of factors involved in the process, this goal is very difficult to achieve by automatic means.

5 Conclusions

The scheduler developed optimizes large jobs according to real information measurements obtained, improving the overall execution time. In order to explore the variables involved, a great variety of configurations has been tested including a variable number of processors and cores for the network nodes. Our static scheduler includes detail computing performance information from the whole computing nodes to build the cost function for the allocation policy. Other information, as constraints, will be taken into account for future versions.

A heterogeneous network has been used, including CPU, GPU and mixed nodes, with different sizes of data sets and changing network conditions. If the tests are selected with the suitable complexity and data set size, the results are always positive. The scheduler is less efficient for simple and reduced data sets. Other factors as memory limits, CUDA compute capability, and network (bandwidth and delay) also affects job execution tasks. These constraints as well as dynamic scheduling will be taken into account on future revisions of the GSched scheduler functionality. During the tests made, the GPUs high parallel capacity has been proved running jobs much faster than CPUs. If this capacity is properly combined with the scheduler, speedups greater than 10x can be achieved.

This work is supported by grants: 10MLA-IN-S08EI-1 (Universidad Pontificia de Salamanca), and PON323B11-2 (Junta de Castilla y León).

References

[1] Karunadasa, N.P., Ranasinghe, D.N.: Accelerating high performance applications with CUDA and MPI. In: 2009 International Conference on Industrial and Information Systems (ICIIS), December 28-31, pp. 331–336 (2009)

[2] Lawlor, O.S.: Message Passing for GPGPU Clusters: cudaMPI. In: Proceedings of IEEE Cluster 2009 (2009)

[3] Yamagiwa, S., Sousa, L.: CaravelaMPI: Message Passing Interface for Parallel GPU-Based Applications. In: Eighth International Symposium on Parallel and Distributed Computing, ISPDC 2009, June 30-July 4, pp. 161–168 (2009)

[4] Potluri, S., Wang, H., Bureddy, D., Singh, A.K., Rosales, C., Panda, D.K.: Optimizing MPI Communication on Multi-GPU Systems Using CUDA Inter-Process Communication. In: 2012 IEEE 26th International Parallel and Distributed Processing Symposium Workshops & PhD Forum (IPDPSW), May 21-25, pp. 1848–1857 (2012)

[5] Linderman, M.D., Collins, J.D., Wang, H., Meng, T.H.: Merge: a programming model for heterogeneous multi-core systems. SIGPLAN Not. 43(3), 287–296 (2008)

[6] Ravi, V.T., Agrawal, G.: A dynamic scheduling framework for emerging heterogeneous systems. In: 2011 18th International Conference on High Performance Computing (HiPC), December 18-21, pp. 1–10 (2011)

[7] Chen, L., Villa, O., Krishnamoorthy, S., Gao, G.R.: Dynamic load balancing on single- and multi-GPU systems. In: 2010 IEEE International Symposium on Parallel & Distributed Processing (IPDPS), April 19-23, pp. 1–12 (2010)

[8] Guevara, M., Gregg, C., Hazelwood, K., Skadron, K.: Enabling task parallelism in the cuda scheduler. In: PEMA 2009 (2009)

[9] Jamsek, D., Van Hensbergen, E.: Experiences with hybrid clusters. In: IEEE Int. Conference on Cluster Computing and Workshops, CLUSTER 2009, August 31-September 4, pp. 1–4 (2009)

[10] Kijsipongse, E., U-ruekolan, S., Ngamphiw, C., Tongsima, S.: Efficient large Pearson correlation matrix computing using hybrid MPI/CUDA. In: 8th Int. Joint Conference on Computer Science and Software Engineering (JCSSE 2011), May 11-13, pp. 237–241 (2011)

[11] Paralel Programming and Computing Platform CUDA NVIDIA, `http://www.nvidia.com/object/cuda_home_new.html` (last accessed: January 2013)

[12] Lee, C.-K., Hamdi, M.: Parallel image processing applications on a network of workstations. Paralell Computing 21(1), 137–160 (1996)

Comparison of K-Means and Fuzzy C-Means Data Mining Algorithms for Analysis of Management Information: An Open Source Case

Angélica Urrutia[1], Hector Valdes[1], and José Galindo[2]

[1] Universidad Católica de Maule, TRICAHUE Database Group, Chile
aurrutia@ucm.cl
[2] Dpto. de Lenguajes y Ciencias de la Computación, Universidad de Málaga, Spain

Abstract. This research presents the knowledge discovery using Data Mining from the organization and with a KPI management point of view. The stages presented here are based on techniques and Data Mining models, with emphasis on clustering techniques, such as the C-MEANS algorithm. We both consider the classic and fuzzy perspectives, namely Fuzzy C-MEANS and K-MEANS, and then compare the results based on the level of support which each algorithm provides to information management. The CRISP-DM methodology is used in our implementation, which is then applied to three case studies.

Keywords: Fuzzy C-MEANS algorithm, K-MEANS, Data Mining, management data analysis.

1 Introduction

The main goal of Business Intelligence (BI) is to give support to executive users to make decisions easier, improving the management indicators in an enterprise, and promoting its competitive advantages in the market. In few words, BI allows organizations to make better decisions, with timely, strategic and faster information [6]. Data Mining (DM) is defined as the process of extracting useful and comprehensible knowledge, previously unknown, from big quantities of data stored in different formats [4]. The Knowledge Discovery in Databases (KDD), including DM, is applied to organizations, and it embraces the concept of organizational culture and how it affects the selection of decision-making tools. [5] shows the use of DM and the CRISP-DM methodology on anomalies detection to find possible abuses in the main services given by the Chilean enterprise.

Our research [6] focuses on applying three *K-means* DM techniques and their extension to fuzzy sets, for the analysis of strategic information and the search of knowledge in the enterprise using a BI architecture. Here are presented the following sections: basic concepts, BI architecture proposal for DM, implementation of

S. Omatu et al. (Eds.): *Distrib. Computing & Artificial Intelligence,* AISC 217, pp. 187–195.
DOI: 10.1007/978-3-319-00551-5_23

K-means, Fuzzy K-means and C-means, application of the phases from the CRISP-DM methodology in a case study and finally, conclusions.

2 Basic Concepts

Key Performance Indicators (KPI) is the term used in BI to name the most important measurements, since it is necessary to measure what is thought as relevant when making decisions [6]. KPI is a measure, but a measure is not always a KPI. The difference lies in the fact that a KPI always reflects strategic assessment guides, while metrics can represent a measurement of any economical activity. What is really important about a management indicator is that data on which the indicator relies are consistent and correct, and that those data are available on time [3].

A **Data Warehouse** (DW), in a smaller scale a Data Mart, is a repository of data specially designed to support decision-making. For the creation of DW, data is loaded from the operational database OLTP (On Line Transactions Process) through ETL tools [5]. The **ETL Process (Extraction, Transform and Load)** is fundamental for DW. The well designed process allows meeting quality standards of data [6]. On the other hand, **Clustering** is one of the most common techniques in DM. This technique consists on finding groups between sets of individuals, which have a big similitude with each other, and it sores the group characterization. There are different clustering methods [4]. The **K-Means algorithm** is a partitioning algorithm whose goal is to find a specific number of groups or clusters, which are represented by a centroid [2]. This algorithm can be applied to n-dimensional objects. K-Means has a training phase that can be slow, depending on the number of elements to classify and the dimension of the problem. When training is finished, the classification of new data becomes faster.

Fuzzy Clusters are groups that add an element to a single group (binary), something that is very restrictive. Fuzzy sets can be used, because it assigns elements to groups according to different membership degrees [3]. The clustering algorithm Fuzzy C-Mean (FCM) also known as ISODATA, is a method of partitions of sets based on iterations necessary to optimize the target function, the sum of squared errors [2, 3]. FCM employs a fuzzy partition where each point can be a member of all the clusters with different membership degrees. FCM looks for centroids minimizing the similitude function (distance) [1].

3 BI Architecture Proposal for Data Mining

The problematic, its solution and expected results are presented in this section, showing our approach (BI Architecture). Many organizations need to analyze the behavior of the actors which interact with this information, determining characteristics that allow an effective decision-making process. Problems generated by this analysis occur when we find special situations, like higher sales than averages, returns, decomposition, loss of certain products, and so on. As a solution, the use of DM is proposed in order to analyze big volumes of data stored in databases. This lets us solve countless problems related to data analysis. For the particular case mentioned before, the clustering technique is used, which consists in the generation of groups of elements which share similar characteristics.

For the solution, the Pentaho Open Source tool is used, giving us more flexibility in the implementation, adaptability and costs reduction. Also the clustering algorithms are programmed. As a result of this, the characteristics of the groups are obtained to do analysis and making decisions. This result is obtained using classic and fuzzy methods of classification. Apart from the results obtained from the clusters, all of the used algorithms are compared.

The proposed architecture to solve the clustering problem with DM techniques is shown on Figure 1: Layer 1: It is the lowest layer, where operational data are found (OLTP), which are used as data source. Layer 2: It is the ETL process of data obtained from the KPI. This is one of the phases that require the hardest work, because data must be prepared, including a cleaning of them, for a correct analysis. Layer 3: In this layer the DW is designed through a multidimensional model (OLAP), to store data of strategic information. The DW modeling must be oriented to the problem that is going to be solved and data are loaded by the ETL. Layer 4: In this layer the DM data tasks are done. The techniques of clustering are used in this investigation. For this, three methods from two different tools are chosen: a) Simple K-Means: this method, used by Pentaho DM Community Edition (PDMCE), also known as WEKA, contains some simple variations of the traditional K-Means. b) K-Means: this method is implemented in an application created as part of this investigation, programmed in Java. c) Fuzzy C-Means: is the fuzzy implementation of K-Means algorithm, which is also created and implemented in Java as part of this research. Layer 5: This layer compare between the DM algorithms of Layer 4. Parameters are compared, such as cluster calculation time in the two tools created in this investigation and the results obtained by K-Means (classic and fuzzy).

Applied Methodology: CRISP-DM (*Cross Industry Standard Process for DM*) is a reference guide widely used in DM projects. CRISP-DM is divided in four levels organized in a hierarchical way on tasks that go from a general level to the most specific cases and organizes the development of the DM project, in a series of six phases. The succession of the six phases of the CRISP-DM process is not necessarily rigid. One can go forward or backward between phases [5].

Data Extraction: One of the most important stages is the creation of a Datamart inside of the DW. For the creation of this design, the indicators must be considered (KPI) creating the facts tables of each Datamart. When the Datamart design is obtained, data has to be loaded and the null and white data cleared. To do this, the Open Source ETL tool is used PDMCE, Formally, the ETL process can be defined as:

$$A \xrightarrow{T} B \quad A = \{x_1, x_2, \ldots, x_n\} \quad B = \{y_1, y_2, \ldots, y_m\} \,/ B \neq \emptyset, B \neq null$$

Where A is operational data (OLPT), B are analytic data stored in multidimensional structures (OLAP), and T is the ETL.

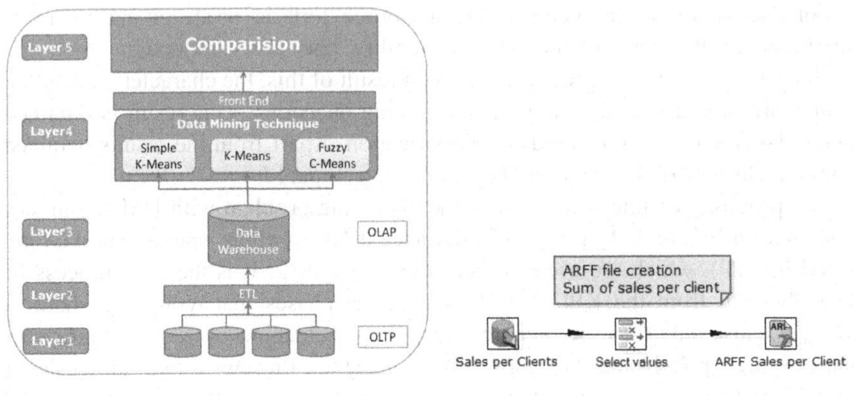

Fig. 1 Proposed BI Architecture for Data Mining

Fig. 2 ARFF file creation ETL, Sales per Client from data filtered

4 Implementation of K-Means, Fuzzy C-Means

Our tools and methods of DM in the three selected and implemented algorithms are chosen and described below.

Simple K-Means: The first tool is PDMCE. The chosen technique is clustering, specifically the Simple K-Means algorithm. This algorithm has these main inputs: the standard deviation of the distances points to the centroid, distance function, number of groups, and number of maximum iterations in case of no convergence.

K-Means: The second tool is called Tricahue DM. We created this tool in Java, to make a fuzzy extension of the K-Means algorithm. The input parameters of this algorithm are: number of clusters, maximum number of iterations, maximum error for convergence, and distance function.

Fuzzy C-Means: In order to program the fuzzy K-Means algorithm the extension to fuzzy sets was used. It assigns a membership degree to each cluster. This algorithm receives all the parameters of K-Means and the influence of the membership degree.

The differences between the pseudo-codes for the algorithms used in this investigation are found in [2, 3]. In the classic K-Means case it is initialized with the random centroids and it has a total membership, on the other hand, for the fuzzy K-Means it begins by initializing the membership matrix and the membership is partial or fuzzy through the membership degree, which makes the grouping more robust.

Evaluation: As a first task in this phase comes the analysis of results, from the business goals and the expected results. If expected results are not completely achieved, a phase can be restructured or even, in the worst case, the best solution may be starting from zero with a new DM project. Another task is the revision of the DM process, with the aim of identifying elements that could be improved, to achieve better results.

5 Application of CRISP-DM Methodology in a Practical Case

CRISP-DM methodology and its phases will be applied on a real case, applying clustering as DM technique, specially the K-Means algorithm and some of its modifications.

Business Understanding: As a practical case, real data are used from operational systems (OLTP) from the sales area of an organization in which they are needed to be grouped to obtain unknown patterns, improving the sales process. The sales system analysis allows evaluating the entities interacting in it, like salesmen, clients and products. It analyzes their performances in certain periods of time.

In order to comprehend the organization in a more effective way, it is necessary to define some KPI's that allow modeling the problem and the Datamart. From the interviews we can define the following KPI's.

KPI 1.- *Quantity* sold by each *Client* and *Salesman* in a certain period of *Time.*
KPI 2.- *Total* of sales for each *Client* and *Salesman* in a certain period of *Time.*
KPI 3.- *Products* sold in a certain period of *Time.*

Data Comprehension: The database consists on the sales made in a period of time, describing prices and quantities, also the client, the salesman, the product and the date of the sale. Another way to analyze the data is by using data distribution graphics of the numerical attributes, relevant for the analysis, one for *Quantity*, and another for *Net* values.

Preparation of data: When data is analyzed and include the operation of the sales area of the organization is understood, data begins to be prepared to be used in DM. The first stage consists in choosing the necessary data for the creation of the Table Dimensions and Facts that compose the Datamart. With the chosen data, the following tables are made: Tables Dimensions Customer, Date, Salesman, Product and Table Fact Sale, each table with its respective attributes.

With the Datamart design made, data starts to be loaded from the OLTP sources. The ETL from Pentaho Community Edition suite [13, 15]. The load of data is done through the creation of transformations inside Kettle, one for dimensions and another for the fact. Also Jobs can be created, to join transformations and assigning data loading periods, e.g., daily at a certain time.

Modeling: Clustering techniques are selected in order to solve the problem of classification, specially the K-Means algorithm. As previous step for the application of DM, it is necessary to create an ARFF file with useful data to be mined with the clustering technique. The tools used for DM support ARFF input files. The same ETL tool (see Figure 2).

Simple K-Means: With the ARFF file we start to mine data. The Simple K-Means algorithm from WEKA is applied as cluster. The parameters are configured and the data going to be mined are selected, only net and quantity since they are numerical. The data are the following: Number of real Iterations = 18; Sum of the mean squared error = 1.144; The centroids of each group depends on de percentage of grouped instances in each cluster.

K-Means: This algorithm is very similar to the one above. The only difference is the application created for this research, called *Tricahue DM*. Loaded de same file ARFF is created before, with the sum of sales per client. With the data loaded, the data going to be mined are chosen, which will be net and quantity, as the configuration used before, to compare both tools. The data are the following: Number of Iterations = 20; Grouped Instances is created depending upon the Centroids.

Fuzzy C-Means: In the Simple K-Means from WEKA and K-Means from Tricahue DM algorithms, partitions are classic, so they do not allow to classify in an effective way the equidistant elements to two or more groups, or atypical elements as a very successful sale or products devolution. This algorithm is applied with the same software as above, Tricahue DM.

In order to begin, the data from the ARFF file are loaded just using the same way as the algorithms used before and selecting the numerical data to be analyzed. The analysis of sales per product or sales per salesman, the same procedure is performed, while only configuring the input parameters of the ARFF file according to the specific problem requirements.

Evaluation: The evaluation is structured in the following way:

1. Comparison of simple K-Means from Weka and K-Means from Tricahue DM, with the respective tools.
2. Evaluation of the programming of the algorithm against the extension of functionalities available on the market.
3. Structure of Fuzzy C-Means and validity of the fuzzy clusters.
4. Evaluation of the results obtained in the optimal case.
5. Validation of the above phases.

On the comparison of the obtained results between Simple K-Means from WEKA and K-Means from Tricahue DM, small differences were found on the centers and grouped instances, but not considerable, due to the random initialization of the initial centroids which obstructs a good comparison. Since they are minimal differences, it allows validating K-Means from Tricahue DM created for this work. Time was used to compare the tools, determining that there is not a considerable difference, having into account that 350 registers were used, corresponding to the number of clients in the organization. The incorporation of fuzzy set, allows the use of partial membership to each one of the groups. This helps to improve the discrimination of best or worst clients, giving more information when making decisions. Therefore a client which is equidistant to two or more groups will belong to more than one, not only to one as in classic logic, as it is with K-Means.

To validate the results presented in the K-Means fuzzy algorithm (Fuzzy C-Means) modeling phase, two cluster validity functions were used: Xie-Beni Measurement and Partition Coefficient.

The Xie-Beni Measurement, measures how compact and separated the clusters are. The smaller the value is, the better is its partition. Optimal groups and the value of influence of the membership degree can be defined here: Number of groups $C = 2$ and Influence of the membership degree $m = 2$.

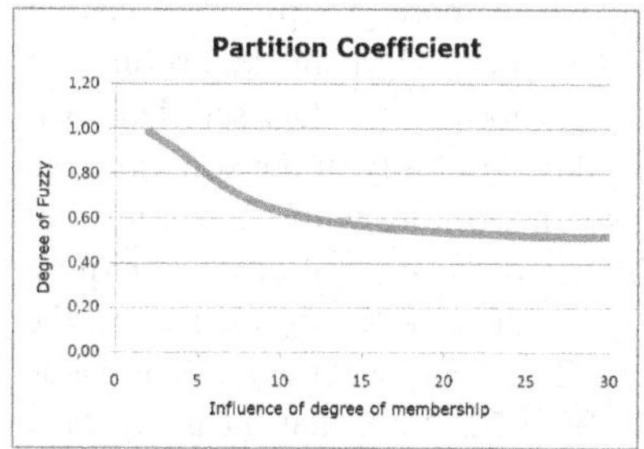

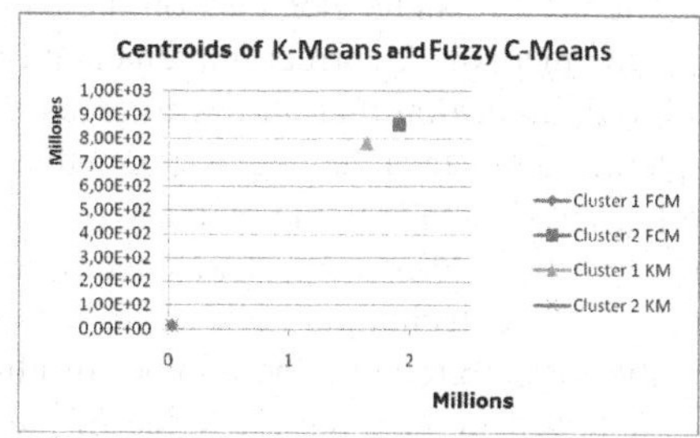

Fig. 3 Partition coefficient, and K-Means and Fuzzy C-Means centroids in two variables

The second function used for validation is the Partition Coefficient that measures the degree of *fuzziness*. That indicates the closer to 1/c, the more fuzzy is the partition. Figure 3 shows that the partition, with the influence of the membership degree m equal to 2 is close to 1, what tells us that it is a good partition, because it is not fuzzy, validating the parameters found with the previous function.

When the optimal case is obtained, according to the cluster validity, we start the analysis of the obtained results with an influence of the membership degree of 2 and 2 clusters; this means that it is necessary to reevaluate the modeling phase. This gives as a result that the 96.8% of the data are close to the mean value, and they are grouped in the same group. Although, there are few clients that show better purchases. These can also receive some offer or benefit for their purchases, to incentive new sales. Also, it can be seen that the difference between the centers of the groups is bigger in Fuzzy C-Means than in K-Means, and this tells us that there is a remarkable separation between classes, which makes it a more optimal partition. It is not necessary to modify any phase or starting a new project, because results were according to what was expected.

Implementation: This phase is not directly applicable in this work. But it does not mean that it can not be implemented in the organization. Since it passed the evaluation tests of the results and used tools. Due to the results obtained are favorable, the implementation of a decision-making support system with DM would be greatly helpful in the organization. For a correct functioning, a monitoring stage and future evaluations are required, aiming to give continuity in time.

6 Conclusions

The advantage of using Fuzzy C-Means, is that data can belong to many classes in different degrees, making it more flexible and useful in practical applications. It defines classes (clusters) with a defined shape. K-Means is unrealistic and non usable in real problems. This extension with fuzzy set gives to clustering, the opportunity to handle elements on the edge of two or more groups, or values far from the median. A correct implementation of the Fuzzy C-Means algorithm was achieved, algorithm that allows to create fuzzy partitions of a data set. This was implemented in Java, having as a big advantage the portability of the software, but the big disadvantage is the memory overflow when working with big volumes of data. For the studied case, it was necessary the use of cluster validity techniques. One of these techniques allows to obtain the value of the influence of the membership degree and the number of clusters, 2 and 3 respectively. This value was obtained minimizing the value of the Xie-Beni function, that measures how compact and separated are the clusters. Also the partition coefficient was used. It measures the degree of *fuzziness*, the obtained result was close to 1, which means that it is only somewhat fuzzy, which is the optimal.

As a conclusion of the studied case, the 96.8% of clients has similar purchases, thus belonging to the same group, and the 3.2% left belongs to other groups. Fuzzy C-Means allowed the partial or fuzzy assignation, which handled the data from the borders of the normal distribution in a more robust way, getting the centers closer making them less influenced to this value.

Future Works: Investigations in the Open Source area, exploring tools that allow us to make BI applications with other algorithms and other extensions.

Acknowledgement. Project Conicyt 79112009 (2012-2014) Strengthening the research in pattern recognition and DM with the insertion of a new Advanced Human Capital to DCI. Project TIN2011-26046 of Ministerio de Ciencia e Innovación, of the Spanish government.

References

1. Albayrak, S., Amasyalı, F.: Fuzzy C-Means Clustering on Medical Diagnostic Systems. In: International XII. Turkish Symposium on Artificial Intelligence and Neural Networks - TAINN 2003 (2003)
2. Bezdek, J.C., Hathawa, R.J., Sabin, M.J., Tucker, W.T.: Convergence Theory for Fuzzy c-Means: Counterexamples and Repairs. IEEE Transactions on Systems, Man, and Cybernetics SMC-17(5) (1987)
3. Chen, J., Mikulcic, A., Kraft, D.H.: An Integrated Approach to Information Retrieval with Fuzzy Clustering and Fuzzy Inferencing. In: Perteneciente al libro Knowledge Management in Fuzzy Databases. Physica-Verlag a Springer-Verlag Company (1999)
4. Feil, B., Abonyi, J.: Introduction to Fuzzy Data Mining Methods. In: Galindo, J. (ed.) Handbook of Research on Fuzzy Information Processing in Databases, vol. I, pp. 55–95. Information Science Reference, Hershey (2008)
5. Reddy, G.S., Srinivasu, R., Chander Rao, M.P., Reddy Rikkula, S.: Data warehousing, Data Mining, OLAP and OLTP technologies are essential elements to support decision-making process in industries. (IJCSE) Int. Journal on Computer Science and Engineering 2(9), 2865–2873 (2010)
6. Héctor, V.: Implementación de los Algoritmos de Minería de Datos K-Means y Fuzzy C-Means para el Análisis de Información de Gestión: Un Caso Open Source. Tesis de licenciatura en Ciencias de la Ingeniería, Facultad de Ingeniería Universidad Católica del Maule, Talca Chile (2012)

Mobile-Based Distributed System for Managing Abandoned or Lost Pets

Daniel Garrote-Hildebrand[2], José-Luis Poza-Luján[1], Juan-Luis Posadas-Yagüe[1], and José-Enrique Simó-Ten[1]

[1] University Institute of Control Systems and Industrial Computing (ai2).
[2] School of Engineering in Computer Science. Universitat Politècnica de València (UPV). Camino de vera, s/n. 46022 Valencia, Spain
{jopolu,jposadas,jsimo}@ai2.upv.es. dagarhil@inf.upv.es

Abstract. This paper presents the work in progress of a mobile-based distributed system which aims to minimize the social impact of abandoned or lost animals. System is based on the use of smart mobile devices to provide message warnings of animals localized. Messages are stored in a database to be processed. In order to enter data such as photography, audio and artificial images, system uses different mobile device interfaces. Data processing consists mainly in matching localized animals with lost animals, assigning abandoned animals at shelters and generating notifications for animal shelters or authorities. Currently, the system is in the development phase. The technical challenges in which we are working are to optimize data and metadata matching, and the management of message warning.

1 Introduction

In most countries, legislation penalizes animal abuse and neglect. However, animal loss or abandonment is a current problem in some communities. Numerous associations work in communities to help animals either welcoming in animal shelters or finding them a new home. Animal Protection Societies (APS) lack common information, databases and channels to receive animal warnings and notify the news. Smart mobile devices, cloud computing and data storage optimization can provide technological support to APS. Currently, to send information about a lost pet can be easy by using a mobile device because of they have photography camera and microphone. Besides, these devices let send geospatial and temporal data. It is possible to optimize the location of the pet's owner, or an adopter, through data-matching by using a distributed data system. In order to do this, it is necessary to use unified data and synchronize correctly different data sources and destinations. The system presented in this paper aims to help APS to manage the animals and optimize the location of owners or adopters.

This paper is organized as follows. Section 2 reviews the current state of the problem and current solutions found by the authors. Section 3 describes the main system data and components. Finally, section 4 exposes the current state and challenges for the future of animal management.

S. Omatu et al. (Eds.): *Distrib. Computing & Artificial Intelligence,* AISC 217, pp. 197–200.
DOI: 10.1007/978-3-319-00551-5_24

2 Related Work

There has been carried out several studies about protocols that must be followed when either a lost animal or abandoned animal is found by someone [1]. Most of these protocols insist specially on both the importance of information management [2] and the huge possibilities of being applied into other technology fields regarding animal researches [3]. In order to follow these protocols, animal protection societies usually use web pages in which they can warn about a lost animal or search one to adopt.

Warnings get to the APS, which must publicize the loss. APS transmit the notice through their own webs as well as in social networks. Currently advises are sent without checking whether lost animals have been located. Besides, users receive messages without picking the geographic area of interest as they can receive advises of lost animals that are located far away from the area to be searched.

It is of special interest the use of the new technologies to optimize the entire process. Some authors stand out the benefits of using mobile applications to locate lost animals [4]. Concerning Android, there are about ten applications that are able to advice of a lost pet or abandoned one. Most of these mobile applications have a data insertion based on forms and they are adapted to specific data bases [5]. Based on the earlier premises, authors are developing a system based on mobile applications expecting to improve current systems that are described straightaway.

3 System Components and Current State

Figure 1 shows the system overview. The system has two types of users: Advertisers and Managers. Advertisers send warnings about abandoned or lost pets by means of a smart mobile device or a personal computer (step 1). Advertiser mobile application sends two types of information: data about the pet characteristics and metadata related by the date, time, geolocation and user contact (step 2). Metadata must be approved by the user. In the first version, the characteristics of animals to transfer are three: animal species (dog, cat, etc.), size and prevailing hair colour.

A Web service receives data and stores it in the database (step 3). When data is stored, system launches the matching process with the new animal data (step 4).

The matching process returns both matching percentage: data and metadata. Data match result is used to determine if an animal exists in database and to discard animal localized (step 5). If a result is above a threshold, the database sends a warning to a Manager User (step 6) that can determine and adds certainty to the result (step 7). Manager User, or system in automated mode, sends the message warning to the APS nearest to the animal localized (step 8).

Metadata is automatically introduced by the application. There are four methods to enter animal data: Form, Virtual Draw, Picture and Recorded Voice. Form is the classical method used by Web Applications and implies an effort by the user. To reduce this effort, the mobile application offers other three methods. Virtual Draws allows the user to model an animal by using only the touch screen. With the Form and the Virtual Draw the application provides directly the animal parameters. The other two methods facilitate data entry but require data preprocessing.

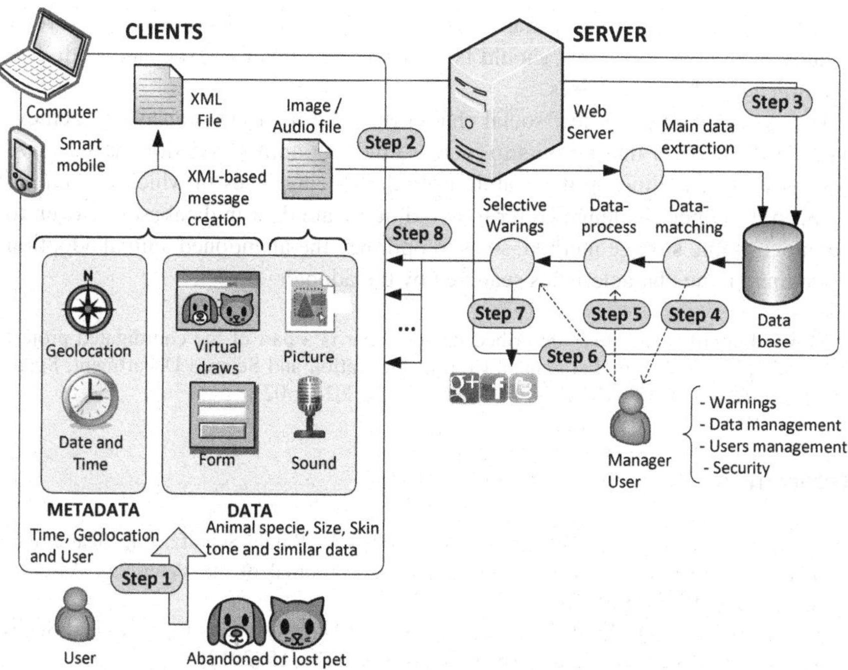

Fig. 1 System overview with Advertiser Users (left) and Manager Users (right)

Picture needs segmented and form detection. Recorded Voice requires keyword detection as "big dog" or "white cat". Both processes are implemented in the server side. Data match is made by means Linear Discriminant Analysis (LDA). Manager User monitors and increases results accuracy.

The first prototype of the mobile application is developed in Java with the Android SDK provided by Google [6]. But, due to the large amount of data to be treated, the trial version is being developed with the framework Sencha Touch [7]. In the server-side, the services applications are being developed in PHP with MySQL to provide the database support [7].

Currently no significant tests have been performed, so that results are not available. However, the first test will measure the percentage of successes in the collation within the pets located and the lost pets. Simultaneously it's necessary to test if the system reduces the time that it takes to locate an animal since the warning occurs compared with current methods used by the APS, and if the system increases animal's warnings.

4 Conclusions and Challenges

Nowadays the system is being developed. In the first phase, we are developing the mobile application and the server-side applications. In second phase, the collation algorithms, with the selective alarms, will be developed. This project has several technological challenges that should be achieved are mainly associated with data collation and selective advises.

The system suggests a few social challenges. The main one consists of achieving a collation level that permits to increase the lost animal location, as selective messages allow delimiting the search up to the recent places in which the animal has been to. Other challenges refer to avoid duplicate data in databases or even to provide a simple storage method, so as to optimize the abandoned animal adoption depending on the characteristics searched by the adopter.

Acknowledgments. The study described in this paper is a part of the coordinated project COBAMI: Mission-based Hierarchical Control. Education and Science Department, Spanish Government. CICYT: MICINN: DPI2011-28507-C02-01/02.

References

1. Lord, L.K., Wittum, T.E., Ferketich, A.K., Funk, J.A., Rajala-Schultz, P.J.: Search methods that people use to find owners of lost pets. Journal of the Veterinary Association 230(12), 1835–1840 (2007)
2. Weiss, E., Slater, M., Lord, L.: Frequency of Lost Dogs and Cats in the United States and the Methods Used to Locate Them. Animals 2, 301–315 (2012)
3. Laplante, P.A.: Exciting Real-Time Location Applications. IT Professional 13(2), 4–5 (2011), doi:10.1109/MITP.2011.22
4. IFPUG (International Function Point Users Group). The IFPUG Guide to IT and Software Measurement. Auerbach Publications (2012)
5. Yun, L., Peiji, S.: Applying RFID to the pet's information management to realize collaboration. In: 7th Int. Conf. on Proc. Serv. Syst. Serv. Manage., Tokyo, Japan, pp. 1–6 (2010)
6. Android SDK, `http://developer.android.com/sdk`
7. Clarck, J.E., Johnson, P.B.: Sencha Touch Mobile Javascript Framework. Packt Publishing (2012)

Robust Optimization of Intradomain Routing Using Evolutionary Algorithms

Vitor Pereira[1], Pedro Sousa[2], Paulo Cortez[2], Miguel Rio[3], and Miguel Rocha[1]

[1] CCTC, School of Engineering, University of Minho, Portugal
mrocha@di.uminho.pt
[2] Algoritmi Center, School of Engineering, University of Minho, Portugal
pns@di.uminho.pt, pcortez@dsi.uminho.pt
[3] Dept. Electrical Engineering, University College London, UK
m.rio@ee.ucl.ac.uk

Abstract. Open Shortest Path First (OSPF) is a widely used routing protocol that depends on weights assigned to each link to make routing decisions. If traffic demands are known, the OSPF weight setting (OSPFWS) problem can be defined to seek a set of weights that optimize network performance, typically by minimizing a congestion measure. The OSPFWS problem is NP-hard and, thus, meta-heuristics such as Evolutionary Algorithms (EAs) have been used in previous work to obtain near optimal solutions. However, the dynamic nature of this problem leads to the necessity of addressing these problems in a more robust manner that can deal with changes in the conditions of the network. Here, we present EAs for two of those tasks, defining objective functions that take into account, on the one hand, changes in the traffic demand matrices and, on the other, single link failures. Those functions use weighting schemes to provide trade-offs between the behaviour of the network in distinct conditions, thus providing robust sets of OSPF weights.The algorithms are implemented in the open-source software NetOpt framework.

1 Introduction

In today's IP intra-domain networks, Open Shortest Path First (OSPF) and Integrated Intermediate System-to-Intermediate System (IS-IS) are the most commonly used Interior Gateway Protocols (IGPs). In both cases, each router uses the Dijkstra's algorithm [2] to compute shortest paths from itself to all other routers and traffic flows are then routed along these shortest paths. A set of weights assigned to each link by a network administrator define those shortest paths and, therefore, how traffic flows in the network. If traffic demands are known, the OSPF Weight Setting (OSPFWS) problem seeks to find a set of OSPF weights (or equivalently IS-IS weights) that optimizes a measure of network congestion [5]. Other Quality of Service (QoS) constraints at the network level can also be taken into account in the objective functions used in the optimization process [10].

The OSPFWS problem is NP-hard and, among others, Evolutionary Algorithms (EAs) have been used to improve routing configurations, with different objective functions [3, 10]. Those strategies, despite providing good solutions, assume that the optimization conditions are static, and do not take into account any changes in

S. Omatu et al. (Eds.): *Distrib. Computing & Artificial Intelligence,* AISC 217, pp. 201–208.
DOI: 10.1007/978-3-319-00551-5_25

the network conditions, resulting from, for example, a link failure or drastic changes in the traffic demands. A possible approach to deal with those scenarios would be to run a new OSPFWS optimization process to adapt the network configuration to the new conditions. However, in many cases, this is not a good practice, as it can lead to a temporary instability in the network [6].

In this work, we propose two approaches that take into account relevant changes in the network. We will present a method to optimize OSPF weights settings that simultaneously work well with two demands matrices, for scenarios where traffic requirements undergo periodic changes, as for example day and night patterns. A second method will be introduced to optimize weight settings that can work well both in the normal conditions and when a single link failure occurs, in this case taking the worst case scenario considering the link with the highest load.

2 The OSPFWS Problem

The mathematical networking model used in this work is a directed graph $G = (N,A)$, which represents routers by a set of nodes (N) and transmission links by a set of arcs (A). A solution to the OSPFWS problem is given by a link weight vector $\mathbf{w} = (w_a)$ with $a \in A$. OSPF requires integer weights from 1 to 65535 $(2^{16} - 1)$[9], but the range of weights can be reduced to smaller intervals $[w_{min}, w_{max}]$, in this work [1,20]. This reduces the search space and increases the probability of equal cost paths [6]. Given a demand matrix D, consisting of several d_{st} entries for each origin and destination pair (s,t), where d_{st} is the amount of data traffic to be sent from a source s to a destination t, the problem consists in routing these demands over paths in the network, minimizing a given measure of network congestion.

For each arc $a \in A$, the capacity is expressed by $c(a)$ and the total load by $\ell(a)$. The total load over a is the sum of flows, $f_a^{(s,t)}$, that represent how much of the traffic demand between s and t travels over arc a. Here, the cost of sending traffic through arc a is given by $\Phi(\ell(a))$. The cost value depends on the utilization of the arc and is given by the linear function proposed by Fortz and Thorup:

$$\Phi_a'(x) = \begin{cases} 1 \ \ for \ \ 0 \leq x/c(a) < 1/3 \\ 3 \ \ for \ \ 1/3 \leq x/c(a) < 2/3 \\ 10 \ \ for \ \ 2/3 \leq x/c(a) < 9/10 \\ 70 \ \ for \ \ 9/10 \leq x/c(a) < 1 \\ 500 \ \ for \ \ 1 \leq x/c(a) < 11/10 \\ 5000 \ \ for \ \ x/c(a) \geq 11/10 \end{cases} \tag{1}$$

The objective of the OSPFWS problem is to distribute the traffic demands in order to minimize the sum of all costs, as expressed in Equation 2.

$$\Phi = \sum_{a \in A} \Phi_a(\ell(a)) \tag{2}$$

To enable results comparison among distinct topologies, a normalized congestion measure Φ^* is used. It is important to note that when Φ^* equals 1, all loads are below $1/3$ of the link capacity, while when all arcs are exactly full the value of Φ^* is 10 2/3. This value will be considered as a threshold that bounds the acceptable working region of the network.

EAs are well suited to handle the OSPFWS problem and, thus, in previous work by the authors they have been used to address several variants of this problem [10]. In these proposed EAs, each individual encodes a solution as a vector of integer values, where each corresponds to the weight of a link. The objective function used to evaluate each individual (solution) in the EAs varied depending on the target of the optimization. The simplest alternative implemented the minimization of the congestion according to the function given above, while other constraints related to propagation delays were also addressed, thus creating a multiconstrained optimization framework that could contain other QoS related measures [10].

3 Methods and Results

The EAs used in this work followed closely the configuration of the ones mentioned above from previous work. The initial population was randomly generated, with weights taken from a uniform distribution in the $[1,20]$ range. Two mutations and one crossover are used as reproduction operators: *Random mutation*, *Incremental/decremental mutation* and *Uniform crossover*. A roulette wheel scheme is used in the selection procedure by converting the fitness value into a linear ranking in the population. In each generation, 50% of the individuals are kept from the previous generation. In the experiments, a population size of 100 was considered. The generic EA used in this work follows the structure given by Algorithm 1.

The presented results were obtained in two networks denoted as 30_2 and 30_4, both with 30 nodes, and with, respectively, 55 and 110 links. Those network topologies were generated using the Brite topology generator [8]. The considered demands matrices (D) have distinct expected mean of congestion (D_p) with values in $\{0.3, 0.4, 0.5\}$, where larger values imply more difficult problems (details in [10]).

Algorithm 1. Overall structure of used EAs

```
t = 0
INITIALIZE P(0)
EVALUATE P(0)
while t is less than Maximum number of generations do
    t = t + 1
    SELECT parents for reproduction
    APPLY REPRODUCTION operators to create offspring
    EVALUATE offspring
    SELECT the survivors from P(t) to be kept in P(t+1)
    INSERT offspring into P(t+1)
end
```

3.1 OSPFWS Optimization for Two Traffic Demands

For a given network, the traditional routing problem deals with the selection of paths to route given amounts of demands between origin and destination routers. On this definition, it is assumed that the volume of traffic between each pair source, destination is known and fixed. However, the increase and variety of services in the contemporary networks, translates into traffic variations that hinder the planning and management of trusted networks based on static traffic demands.

On a daily basis, for example, there are major differences between daytime and night-time traffic. However, on different days, these variations are relatively similar, that is, traffic undergoes periodic changes in a predictable daily basis [4, 1]. Thus, it becomes necessary to find a weight configuration that promotes an acceptable level of congestion for all periods. An alternative solution could be to overestimate traffic requirements to accommodate day and night demands. However, this solution would lead to overprovisioning network resources.

To deal with this type of periodic changes, we suggest optimizing the configuration of weights for two representative traffic matrices (e.g. that model day and night traffic). In this context, it is necessary to redefine the optimization problem. For a given network topology, defined as above, and two demand traffic matrices D_1 and D_2, we aim to find a set of weights w that simultaneously minimize the functions $\Phi_1^*(w)$ and $\Phi_2^*(w)$, where $\Phi_i^*(w)$ represents the function $\Phi^*(w)$ considering the traffic demands in D_i. This multi-objective optimization is achieved with this aggregated objective function (instantiating the evaluation step of Algorithm 1):

$$f(w) = \alpha\Phi_1^*(w) + (1-\alpha)\Phi_2^*(w), \ \alpha \in [0;\ 1] \tag{3}$$

Several simulations were made to evaluate the EA performance. Some of the obtained results are presented in Table 1 for a daytime traffic demands matrix $D0.5_1$, and two night-time traffic demands matrices $D0.4$ and $D0.5_2$. All experiments are run 10 times and the results shown are the means of the values.

Table 1 $f(w)$ optimization results for two traffic demands matrices

	Day:$D0.5_1$ & Night:$D0.4$		Day:$D0.5_1$ & Night:$D0.5_2$	
	Φ_1^*	Φ_2^*	Φ_1^*	Φ_2^*
$\alpha = 1$ (only daytime opt.)	2.894	7.560	2.894	52.859
$\alpha = 0$ (only night-time opt.)	43.575	1.696	28.440	3.894
$\alpha = 0.5$ (daytime and night-time opt.)	3.590	2.068	4.722	4.886

An optimization process is usually executed for a specific demand matrix, and therefore, solutions that are near optimal for one matrix may not be good enough for other distinct matrices. In these results, a solution for a demand matrix $D0.5_1$, with congestion measure 2.894, is inadequate for a demands matrix $D0.5_2$, as the congestion measure reaches over 50. It is possible, however, to obtain a suitable configuration for both matrices, only slightly compromising the congestion level in each individual scenario (last row). The parameter α is used to tune the trade-off between both components. A network administrator may wish to sacrifice night-time traffic congestion level favouring daytime traffic. In this context, the value of α should translate this trade-off with a value closer to 1. In the experiments, α was set to 0.5, giving equal importance to both conditions. This approach could easily be generalized to k matrices by defining $k-1$ trade-off parameters.

3.2 OSPFWS Optimization for Single Link Failure

Besides changes in traffic demands, other conditions can change within a network environment, such as a link failures, which can have a severe impact on the congestion levels and other network performance parameters. After a link failure, affected routers start to advertise the new state, forcing all routers to recalculate shortest paths (SPs), based on the weight configurations defined for the topology without failure. However, an OSPF weight setting configuration, optimized for the original topology, may not be suitable for the new topology with link failure. Traffic that previously flowed through the failed link is shifted to other recalculated routes, but these new paths may be inefficient, causing congestion in some parts of the network, especially in case of higher demands.

3.2.1 Cost Function Based on Congestion Measure

Single link failure scenarios require dealing with two states of a network: a first state where all links are functional and another where a link has failed. For each solution w, the proposed algorithm assesses the congestion level of the network without failure (Φ_n^*) and after the failure (Φ_{n-1}^*). The evaluation of a solution that supports both states is performed by the aggregation function given in Equation 4.

$$f(w) = \alpha \Phi_n^*(w) + (1-\alpha)\Phi_{n-1}^*(w), \; \alpha \in [0;\, 1] \tag{4}$$

The failure of the link with the highest load on a network configures, in principle, the worst single link failure scenario for congestion. So, this was the selected scenario for the experiments performed in this work. In [7, 11] alternatives approaches that resort to other optimization methods are proposed, differing in the configuration of the weighting factor. While our proposal considers a weighting factor α that can assume any value in the range [0, 1], in [7, 11] the weighting factor is constant and equal to 0.5. The weighting scheme induced by α offers a more flexible alternative to network administrators. When $\alpha = 1$, the optimization is only performed for the normal state topology, without any link failures. In optimization processes configured with $\alpha = 0.5$, the same level of importance is given to the two topology states. However, as the link failure optimization can compromise the network congestion level in a normal state, a network administrator may wish to focus on the performance of the normal state, at the expense of the failed state, that may not occur (setting the variable α with a value between 0.5 and 1).

A set of experiments was devised to evaluate this approach, and the results are presented in Table 2. The algorithm was applied to the network 30_2, considering

Table 2 Link failure optimization with $\alpha = 1$, 0.5, 0.75 and 0.85

Demands	Without link failure optimization $\alpha = 1$		With link failure optimization $\alpha = 0.5$	
	Before Failure	After Failure	Before Failure	After Failure
D0.3	1.395	30.429	1.479	1.504
D0.4	1.726	55.589	1.803	1.824
D0.5	3.915	199.636	6.059	5.414
Demands	$\alpha = 0.75$		$\alpha = 0.85$	
	Before Failure	After Failure	Before Failure	After Failure
D0.3	1.455	1.479	1.450	1.536
D0.4	1.754	1.810	1.807	1.850
D0.5	4.926	6.453	4.933	7.559

traffic demands with D_p levels of 0.3, 0.4 and 0.5, and weighting factors 1 (without link failure optimization to provide a reference value), 0.5, 0.75 and 0.85.

Comparing the results, one can observe that a slightly worse behaviour on the congestion level in its normal state is compensated by a large gain in all scenarios with link failure. For $\alpha = 0.85$, with a slight penalty on the congestion levels, from 1.395 to 1.450 (D0.3), 1.726 to 1.807 (D0.4) and 3.915 to 4.933 (D0.5), the gains on the congestion levels of the link failure network are very significant, reducing from 30,429 to 1.536 (D0.3), from 55.589 to 1.850(D0.4), and from 199.636 to 7.559 (D0.5). These results indicate that this may be a good solution to optimize OSPF weights that accommodate a failure of the link with the highest load.

3.2.2 Cost Function with Delay Constraints

A new cost function γ can be introduced [10], to evaluate the fulfilment of delay constraints between each pair of nodes in the network, which are also defined by a matrix (DR). This function applies the same type of penalties as Φ, through the substitution of the link utilization by an end-to-end delay measure Del_{st}/DR_{st}, where Del_{st} is the mean traffic delay and DR_{st} is the maximum allowed delay between nodes s and t. This new cost function is also normalized to γ^*.

In this case, the evaluation function uses another weighting factor β to quantify the trade-off between congestion (Φ^*) and delay (γ^*), while as before α maintains the trade-off between the original and the failure scenarios, as given by:

$$f(w) = \alpha \quad (\beta \Phi_n^*(w) + (1-\beta)\gamma_n^*(w)) + \\ (1-\alpha)\left(\beta \Phi_{n-1}^*(w) + (1-\beta)\gamma_{n-1}^*(w)\right), \; \alpha, \beta \in [0,1] \qquad (5)$$

Some results contemplating congestion and delay optimization for a specific problem are presented in Table 3, for scenarios with and without failure of the link with the highest load. While these results are only a showcase of this approach, it is visible that there are gains in both measures for the failure scenario with only limited losses in the normal one. These gains are far more impressive when looking at

Table 3 Optimization of congestion and delay for a link failure scenario on network 30_4 with demands D0.3 and delay restrictions DR4.0

α	β	Before failure		After failure	
		Φ^*_n	γ^*_n	Φ^*_{n-1}	γ^*_{n-1}
1		2.182	2.411	73.938	8.583
0.5	0.5	3.273	3.470	3.272	4.921
0.75		2.754	2.689	2.839	6.224

congestion, a result that would be expected given the link that was selected for the failure (the one with the highest load) that has a greater influence on this parameter. Other configurations should be tested in the future regarding the selection of this link to fully evaluate the potential of the approach.

4 Conclusions

Changes in network conditions, such as link failures or changes in the traffic demands, can deteriorate its performance. This work presented two evolutionary approaches for preventive robust optimization of intradomain routing. The algorithm proposed for two traffic demands presents a solution for scenarios where periodic changes on traffic demands occur. The link failure optimization introduces greater tolerance to scenarios of single link failure, allowing more acceptable congestion values even in extreme events. It is also possible to take into account other constraints such as end-to-end delay requirements in both cases (here only results for the second are shown). The presented methods were included in NetOpt, a user friendly and open source framework, available at `http://darwin.di.uminho.pt/netopt` together with appropriate documentation. Here, a number of configuration options for these tasks are made available for network administrators.

Although the results obtained so far seem promising, a more thorough evaluation of these approaches will be done for a wider range of test scenarios. Also, since the optimization problems stated here are of a multi-objective nature, the implementation of specialized algorithms, such as multi-objective EAs seems natural. Indeed, two of those (SPEA2 and NSGA II) are already implemented in NetOpt, but the analyses of their results is left as future work.

Acknowledgement. This work is funded by National Funds through the FCT - Fundao para a Cilncia e a Tecnologia (Portuguese Foundation for Science and Technology) within project PEst-OE/EEI/UI0752/2011.

References

1. Cortez, P., Rio, M., Rocha, M., Sousa, P.: Multiscale internet traffic forecasting using neural networks and time series methods. Expert Systems 29(2), 143–155 (2012)
2. Dijkstra, E.: A note on two problems in connexion with graphs. Numerische Mathematik 1(1), 269–271 (1959)

3. Ericsson, M., Resende, M., Pardalos, P.: A Genetic Algorithm for the Weight Setting Problem in OSPF Routing. Journal of Combinatorial Optimization 6, 299–333 (2002)
4. Feldmann, A., Greenberg, A., Lund, C., Reingold, N., Rexford, J., True, F.: Deriving traffic demands for operational ip networks: methodology and experience. IEEE/ACM Transactions on Networking 9(3), 265–280 (2001)
5. Fortz, B.: Internet traffic engineering by optimizing ospf weights. In: Proceedings of IEEE INFOCOM, pp. 519–528 (2000)
6. Fortz, B., Thorup, M.: Optimizing ospf/is-is weights in a changing world. IEEE Journal on Selected Areas in Communications 20(4), 756–767 (2002)
7. Fortz, B., Thorup, M.: Robust optimization of OSPF/IS-IS weights. In: Proceedings of the International Network Optimization Conference, pp. 225–230 (2003)
8. Medina, A., Lakhina, A., Matta, I., Byers, J.: BRITE: universal topology generation from a user's perspective. Technical report 2001-003 (January 2001), `http://citeseer.ist.psu.edu/article/medina01brite.html`
9. Moy, J.: OSPF Version 2. RFC 2328 (Standard), Updated by RFC 5709 (April 1998)
10. Rocha, M., Sousa, P., Cortez, P., Rio, M.: Quality of Service Constrained Routing Optimization Using Evolutionary Computation. Applied Soft Computing 11(1), 356–364 (2011)
11. Sqalli, M., Sait, S., Asadullah, S.: Ospf weight setting optimization for single link failures. International Journal of Computer Networks & Communications 3(1), 168–183 (2011)

A MAS for Teaching Computational Logic

Jose Alberto Maestro-Prieto[1], Mª Aránzazu Simón-Hurtado[1], Juan F. de-Paz-Santana[2], and Gabriel Villarrubia-González[2]

Dept. de Informática, Universidad de Valladolid, Spain
{jose,arancha}@infor.uva.es
Dept. de Informática y Automática, Universidad de Salamanca, Spain
{fcofds,gvg}@usal.es

Abstract. In this paper, an Intelligent Tutoring System (ITS) for teaching computational logic called SIAL is described. Several basic topics in computational logic are covered. The more complex part in SIAL is the module in charge of the diagnosis, which performs model-based diagnosis although sometimes, a knowledge-based (expertise) model is necessary in order to yield a more accurate diagnosis. The inherent complexity of the ITS is approached using a Multi-Agent System (MAS). The classical approach in ITS, which divides them into four independent modules, is adapted to a MAS creating an agent for each module and other agent for any other subsystem needed. The results obtained from an experiment of usage of SIAL are presented.

Keywords: Multi-Agent Systems, Intelligent Tutoring Systems, Computational Logic, Model Based Diagnosis, Knowledge Based Diagnosis.

1 Introduction

As software and hardware systems increase its complexity, processes that are more complex are involved in the specification and design of this kind of systems. Using logic-based descriptions is a suitable solution in this case [1]. Some kinds of logic are powerful enough to make complete descriptions and other complex tasks such as model checking, theorem proving and feature verification. Moreover, some of them can be automated, such as First Order Logic (FOL). However, dealing with practical problems of Automated Theorem Proving (ATP) such as memory limits, speeds or other performance constraints is not easy, as it can be seen in the CADE[1] ATP System Competition (CASC) reports [2].

The more complex theorem provers, such as General Proof [3] or OpenProof [4], provides a serious environment to use formal specifications. However, as it was stated by [5, 1], such kind of complex theorem provers will usually need to be guided, as they could not always be able to reach the

[1] CADE: International Conference on Automated Deduction.

S. Omatu et al. (Eds.): *Distrib. Computing & Artificial Intelligence*, AISC 217, pp. 209–217.
DOI: 10.1007/978-3-319-00551-5_26

solution without external (expert) assistance. Thus, using logic makes raise other problems: firstly, it cannot be automated completely, and hence, it is necessary people with a background good enough to be able to use it in order to get the correct results and interpret the correct conclusions.

The Intelligent Tutoring Systems (ITS) approach to self-paced learning can be valid to enhance the knowledge acquisition about Computational Logic and ATP. An ITS is a computer program based on cognitivist theories [6] that usually behaves proposing exercises to the student, and adapting its behavior (the exercises proposed, tips provided, changing the objectives if a lack of knowledge is detected, etc.) as the student acquire more knowledge and skills. An ITS usually exhibits an intelligent behavior and it is able of detecting of the student mistakes, diagnosing the detected errors, assessing the student knowledge, planning the set of exercises having into account the actual knowledge of the student, and maybe others.

An ITS fulfils most of the requirements to become a Multi-Agent System (MAS). The typical design of an ITS splits the functionality in separated, independent modules working with each other for achieving the ITS goals. In [7] agents are described as flexible problem solvers, operating in an environment over which they have only partial control and observability. One of the goals of a MAS is to construct systems capable of autonomous and flexible decision-making, and of cooperating with other systems [8]. Using the MAS approach can help to organize the complex behavior of the ITS. In [9, 10, 11, 12, 13] different proposals for implementing an ITS as a MAS can be found.

The next sections are organized as follows: computational logic is introduced, then SIAL is presented and the MAS architecture is sketched, a simple example is included and finally, the conclusions.

2 Teaching Computational Logic

FOL allows representing knowledge in a domain by stating the general relationships in that domain (called axioms) and some concrete facts for representing the specific problem. Thus, a problem can be declaratively described just stating facts and relationships, a process called formalization. It is also possible to obtain new knowledge using logic rules. Logic rules are transformation operations that get logic expressions (such as axioms and maybe facts), and yield new expressions, which it is said to be inferred by the rule. Sound, valid rules yield sound, valid new knowledge from the existing one. Sound logic rules, axioms and facts together with a proper strategy (or an algorithm) can let you prove a theorem. Automated theorem proving was a landmark in the development of the Artificial Intelligence (AI) which still appears in the standard curricula for an introductory course to AI -for example, the proposed in [14]-, as an example of an automated reasoning procedure. Classical theorem proving is based on the use of the Resolution Rule, and may be others. In order to use the Resolution rule, logical expressions should have a specific format, termed clause form, which can be achieved by applying a well-known sequence of logical transformations to a Well-Formed Formula (WFF). At the final of the nine steps, a set of clauses are obtained. A clause is just a disjunctive set of literals. A literal is a predicate, maybe negated.

The general idea of adding a graphical representation to text descriptions is still valid [15]. Graphics can help to ease the learning of concepts and procedures and to improve the knowledge transfer, and it can be more helpful for novices [16].

This is probably the first time the students come face to face with a declarative programming language [17]. There exist several tools which goal is to ease the learning of logic and to get practice solving logical problems, such as HyperProof [18], Deep Thought [19] or WinKE [20]. More recently, a new graphical representation based on puzzle pieces for the SWI-Prolog interpreter output has been proposed in [17]. All these tools share a feature: it is available some kind of graphical output for representing the developed internal process.

Formalization has also been approached using ITSs, for example in [21]. However, to the best of our knowledge, transforming a WFF into the clause form has only been approached twice. In [21] it is described a computer program able to process a WFF up to one specific step (of the nine steps process for obtaining the clause form from a WFF), on demand of the student. Then, the student can check its own, hand written, response and visually comparing it against the result yielded by the program. The other alternative is our own solution, called SLI, which is also able to expose the individual steps of the process for transforming a set of WFF into a set of clauses.

3 SIAL: An ITS for Computational Logic

SIAL is an ITS designed to perform as a practical tool, that automatically can check the student's solution to a proposed exercise. SIAL provides computational support in order to help the student to fix some concepts and procedures in a practical way; the student is proposed exercises that should be solved using usual techniques and methods of computational logic.

SIAL is designed to be able to propose exercises ranging from obtaining the clause form from FOL expressions up to being able to use the hyperresolution rule. SIAL is organized in 15 thematic levels (Table 1). The levels are arranged from the simplest skills to the most complex ones. Levels 1 to 6 include converting a set of WFF to clause form, predicate unification, binary resolution, resolution refutation and the factoring rule. These levels constitute the basic skills to be acquired, thus forming the main basis for the following levels.

Levels 7 to 12 deal with methods for selecting clauses to yield a new resolvent and reducing the number of generated clauses. Pure literals, tautologies and subsumed clauses should be detected and removed. In addition, the set of support strategy is included as a way to control the progressive increase for the generated clauses. Level 12 is devoted to the hyperresolution rule.

SIAL levels can be grouped depending on the interaction the student is allowed to do. Levels 1 to 6 are designed for novice students. These levels are more guided and the student is encouraged towards a stepwise interaction. Despite this, student interaction is still wide enough, and this makes it impossible obtaining an accurate diagnosis in every case. Levels 7 to 12 are designated for middle-level students and tje guidance is weakened. In these levels, some restrictions are removed and students can provide a solution without some intermediate steps. There exists also an automatic mode in which SIAL solves a problem proposed by the student. SIAL behaves as a usual ATP program.

SIAL is designed to increase the complexity of the developed process in two ways: firstly, exercises should be arranged into each level from the easier to the most difficult. Second, the first three levels only deal with one process (although a complex process), whereas levels from 4 to 12 are defined to combine several processes. This approach makes the learning process progressive as upper levels lie on previous ones. Each level above the level 4 can assume the functionality of each level below it. In this way, the user must integrate the knowledge already practiced jointly with a new technique, in the current problem solving process.

4 The ITS as a Multi-agents System

SIAL is composed of several agents which are capable of interact with the student while collect, process and evaluate new data together with historical data. The agents of SIAL know the student history and they use these data to adapt the training plan. SIAL architecture is based on the classical four modules ITS architecture [22]. The same as in [13], classical modules

Table 1 Description of each level defined in SIAL, guiding and topic classification

Level	Guiding	Topic	Description
1	Strong	Single	Getting the clause form (without Skolemization process).
2	Strong	Single	Getting the clause form (with Skolemization process).
3	Strong	Single	Unification procedure.
4	Strong	Compound	Resolution rule.
5	Strong	Compound	Strongly guided refutation resolution.
6	Strong	Compound	Factoring rule.
7	Weak	Compound	Weakly guided refutation resolution.
8	Weak	Compound	Pure literal removing.
9	Weak	Compound	Tautology removing.
10	Weak	Compound	Support set strategy.
11	Weak	Compound	Subsumption.
12	Weak	Compound	Hyperresolution (positive/negative hyperresolution).

have become in agents. Together with these main agents, any other support subsystem used in SIAL has also been wrapped as an agent of the system.

4.1 The Main Agents

The **interface** agent is responsible for showing the environment for problem solving and controlling the user interaction. Besides the program interface, the interface agent also contains the *Help manager*. In order to develop its functionality, the interface agent must communicate with the others main agents, those acting as the expertise module, the student model and the pedagogical module.

The agent performing the **expertise model** is responsible for interpreting the actions the student takes. It is composed of two automated theorem provers that represent the domain model which detect mistakes and identify errors. It also includes an *expert system engine* (CLIPS) for helping in error identification. This agent is in charge of inferring the student's knowledge by analyzing the student's actions. SIAL applies up to two different processes in order to obtain an accurate diagnosis [23]: a model-based diagnosis, based on ATP, to ensure the student's answer validity and maybe obtain a diagnosis and, whenever it is not possible to get an accurate diagnosis, an expertise-based diagnosis (using expert systems) is tried.

The agent in charge of the **student model** is responsible for maintaining a representation of the student's knowledge. It stores the actions taken by the student, the mistakes and errors identified, the exercises solved and so on, in a database.

The agent acting as the **pedagogical module** is responsible for the instructional support. The pedagogical module agent uses the information contained in the student model in order to plan a sequence of exercises, to decide if a reinforcement exercise is necessary or whether any advice message should be sent to the student. It also decides when a student is promoted to the next level.

A knowledge-based implementation has been chosen for the pedagogical behavior. The knowledge for selecting the next exercise to be shown is represented as an expert system.

4.2 The Support Agents

Together with the main agents, SIAL also has some other agents in order to wrap other software programs and obtain some typical features in MAS architecture such as, an autonomous behavior, high modularity and interoperability. Wrapping the external programs as agents can also help to replace some of these programs, by newer versions or by other programs without changing the whole application.

The expertise model agent rely on three other agents, both of them are automated theorem provers (SLI and OTTER) and the other is the *expert*

Table 2 Number of errors identified by SIAL, grouped by exercise level and diagnosis method

SIAL Level	N. errors	% Diagnosed by Formal Model	% Diagnosed by E. S.
1	1626	84′38	15′62
2	1957	87′79	12′21
3	976	100′00	
4	261	100′00	
5	402	100′00	
7	482	100′00	
Total	5702	91′35	8′65

system engine CLIPS. The wrapping agents for SLI and OTTER make the communication uniform between them and the other agents despite the fact that both provers do not share a common interface. In addition, some of these agents communicate with several agents, for instance, the CLIPS agent should communicate with the expertise model agent and with pedagogical module agent. Providing them with a high-level communication protocol (a subset of the KQML standard language [24]) eases the communication among them.

The student model agent also interacts with a database. The database has also been wrapped by its own agent. In one hand, this allows to change the physical database reducing the changes in the whole program. SIAL has been used with several Databases: MySQL, MS SQLServer and PostgreSQL whereas MS Access is used when developing. This also allows running SIAL maintaining a central database. SIAL agents can communicate via standard HTTP requests and hence, the agent database can be placed in the computer containing the physical database whereas the others agents and specially, the program interface, can be placed in other computer.

5 A Practical Experiment

SIAL has been used in the practical sessions of an introductory course of Artificial Intelligence. 33 students regularly assisted to practical sessions and use SIAL together with SWI-Prolog to solve exercises. The experiment consisted of 120 practical exercises classified in levels (from 1 to 6) some of them mandatory (61) and others (59) for reinforcement. Each student made an average of 70.16 exercises ($\sigma = 18.56$). Data collected about the errors identified during the experiment is shown in Table 2.

Only 270 (4,74%) out of the 5702 detected errors were syntactic errors. All the other errors were errors due to a wrong proposed solution: wrong expressions, wrong use of operators, lack of literals, wrong use of rules, and so on. Hence, the implemented interface in SIAL seem to be able to minimize the number of errors due to expression manipulation. The huge difference in the number of errors detected between the first two levels and the others can be explained because of the complexity of transforming WFF into the clause form with respect to the unification and resolution processes. The expert systems are designed to deal with errors at levels 1 and 2 (clause form), and they are invoked only after the model based diagnosis. The model based diagnosis diagnoses the 91,35% of the errors and provides SIAL with a great detection power and accuracy.

The students received a survey once they have solved some exercises using SIAL and SWI-Prolog. The survey inquired students about their opinion about usefulness, usability and a comparison of SIAL and SWI-Prolog. Students were also asked to mark SIAL. Questions have four possible answers.

Most of the students said that SIAL is very useful (31.82%) or useful (54,54%) for learning computational logic. Only 1 student (4,54%) said that SIAL is not useful at all.

All the students said that they have solved exercises using both SIAL and SWI-Prolog. A 31.82% of the students said that SIAL cannot be replaced by SWI-Prolog, and a 50.0% said that SIAL only can be replaced very partially by SWI-Prolog. None said that SIAL and SWI-Prolog were equal.

The 77.27% of the students said that SIAL is very easy to use and the 13.64% of the students said that is easy to use. None of the students said that SIAL is difficult to use. SIAL was marked as B (Good) by the 90.91% of the students. The rest of the students (9.09%) mark SIAL as C (Pass).

6 Conclusions

SIAL, an ITS for learning computational logic, has been described. SIAL has been implemented as a MAS. The different modules of the ITS have been converted in agents together with the auxiliary subsystems. One of the most important modules in SIAL is the expertise module, which is made up of two automatic theorem provers and expert systems. The chosen approach allows us to provide students with an accurate tutor. The chosen model-based approach lets SIAL detect the most of user's errors accurately, inform the user about its existence, and provide some hint to localize and fix the mistake.

The results obtained in the practical experiment are promising. The student's answers to the survey show a clear difference between SIAL and SWI-Prolog. Student's answers also say that SIAL is useful for learning computational logic and it is easy to use.

References

1. Bundy, A., Moore, J.D., Zinn, C.: An Intelligent Tutoring System for Induction Proofs. In: Melis, E., Scott, D., et al. (eds.) CADE-17 Workshop on Automated Deduction in Education, pp. 4–13 (2000)
2. Sutcliffe, G.: The CADE-23 Automated Theorem Proving System Competition - CASC-23. AI Communications 25, 49–63 (2012)

3. Aspinall, D., Lüth, C., Winterstein, D.: A Framework for Interactive Proof. In: Kauers, M., Kerber, M., Miner, R., Windsteiger, W. (eds.) MKM/CALCULEMUS 2007. LNCS (LNAI), vol. 4573, pp. 161–175. Springer, Heidelberg (2007)
4. Barker-Plummer, D., Etchemendy, J., Liu, A., Murray, M., Swoboda, N.: Openproof - A Flexible Framework for Heterogeneous Reasoning. In: Stapleton, G., Howse, J., Lee, J. (eds.) Diagrams 2008. LNCS (LNAI), vol. 5223, pp. 347–349. Springer, Heidelberg (2008)
5. Aitken, J.: Problem Solving in Interactive Proof: A Knowledge-Modelling Approach. In: Wahlster, W. (ed.) Proceedings of the 12th European Conference on Artificial Intelligence, pp. 335–339. John Wiley & Sons, Budapest (1996)
6. Driscoll, M.P.: Psychological foundations of instructional design. In: Trends and Issues in Instructional Design and Technology, 2nd edn., pp. 36–44. Pearson Education, Inc., Upper Saddle River (2007)
7. Jennings, N.R., Wooldridge, M.: Agent-Oriented Software Engineering. In: Bradshaw, J. (ed.). AAAI/MIT Press (2000)
8. Rodríguez, S., Pérez-Lancho, B., de Paz, J., Bajo, J., Corchado, J.: Ovamah: Multiagent-based Adaptive Virtual Organizations. In: Proceedings of the 12th International Conference on Information Fusion, Seattle, USA, pp. 990–997 (2009)
9. Cheikes, B.: GIA: An agent-based architecture for Intelligent Tutoring Systems. In: Proceedings of the CIKM 1995 Workshop on Intelligent Information Agents (1995)
10. Capuano, N., Marsella, M., Salerno, S.: ABITS: An Agent Based Intelligent Tutoring System for Distance Learning. In: Proceedings of the International Workshop in Adaptative and Intelligent Web-based Educational Systems, pp. 17–28 (2000)
11. Hospers, M., Kroezen, E., Nijholt, A., op den Akker, R.: Developing a generic agent-based intelligent tutoring system. In: Devedzic, V., Spector, M., Sampson, D., Kinshuk, M. (eds.) The 3rd IEEE International Conference on Advanced Learning Technologies. IEEE Computer Society, Los Alamitos (2003)
12. Fernández-Caballero, A., Gascueña, J., Botella, F., Lazcorreta, E.: Distance learning by intelligent tutoring system. Part I: Agent-based architecture for user-centred adaptivity. In: Proceedings of the 7th International Conference on Enterprise Information Systems, pp. 75–82 (2005)
13. González, C., Burguillo, J., Vidal, J.C., Llamas, M., Rodríguez, D.: ITS-TB: An Intelligent Tutoring System to provide e-Learning in Public Health. In: Proceedings of the 16th EAEEIE Annual Conference on Innovation in Education for Electrical and Information Engineering, EIE (2005)
14. Association for Computing Machinery (ACM): Computer Science Curriculum 2008: An Interim Revision of the CS 2001 (2008)
15. Mayer, R.E., Moreno, R.: Nine Ways to Reduce Cognitive Load in Multimedia Learning. Educational Psychologist 38, 43–52 (2003)
16. Clark, R., Mayer, R.: e-Learning and the Science of Instruction: Proven Guidelines for Consumers and Designers of Multimedia Learning, 3rd edn. Pfeiffer (2011)
17. Mondshein, L., Sattar, A., Lorenzen, T.: Visualizing prolog: a "jigsaw puzzle" approach. ACM Inroads 1, 43–48 (2010)

18. Stenning, K., Cox, R., Oberlander, J.: Contrasting the Cognitive Effects of Graphical and Sentential Logic Teaching: Reasoning, Representation and Individual Differences. Language and Cognitive Processes 3, 333–354 (1995)
19. Croy, M.J.: Graphic Interface Design and Deductive Proof Construction. Journal of Computers in Mathematics and Science Teaching 18, 371–385 (1999)
20. Endriss, U.: The interactive learning environment winke for teaching deductive reasoning. In: Manzano, M. (ed.) Proceedings of the 1st International Congress on Tools for Teaching Logic. University of Salamanca (2000) (invited talk abstract)
21. Hatzilygeroudis, I., Giannoulis, C., Koutsojannis, C.: A Web-Based Education System for Predicate Logic. In: Proceedings of the IEEE International Conference on Advanced Learning Technologies, pp. 106–110. IEEE Computer Society, Washington, DC (2004)
22. Mitrovic, A., Martin, B., Suraweera, P.: Intelligent tutors for all: The constraint-based approach. IEEE Intelligent Systems 22, 38–45 (2007)
23. Ferrero, B., Fernández-Castro, I., Urretavizcaya, M.: Multiple Paradigms for a Generic Diagnostic Proposal. In: Gauthier, G., VanLehn, K., Frasson, C. (eds.) ITS 2000. LNCS, vol. 1839, p. 653. Springer, Heidelberg (2000)
24. Finin, T., Weber, J., Wiederhold, G., Genesereth, M., Fritzson, R., McGuire, J., Pelavin, R., Shapiro, S., Beck, C.: Specification of the KQML Agent-Communication Language. Technical Report, DARPA, DRAFT (1993)

Fast Post-Disaster Emergency Vehicle Scheduling

Roberto Amadini[1], Imane Sefrioui[2], Jacopo Mauro[1], and Maurizio Gabbrielli[1]

[1] Department of Computer Science and Engineering/Lab. Focus INRIA, University of Bologna, Italy
{amadini,jmauro,gabbri}@cs.unibo.it

[2] Information and Telecommunication Systems Laboratory, Faculty of Sciences, Abdelmalek Essaadi University, Tetuan, Morocco
sefrioui.imane@gmail.com

Abstract. Disasters like terrorist attacks, earthquakes, hurricanes, and volcano eruptions are usually unpredictable events that affect a high number of people. We propose an approach that can be used as a decision support tool for a post-disaster response that allows the assignment of victims to hospitals and organizes their transportation via emergency vehicles. Exploiting Operational Research and Constraint Programming techniques we are able to compute assignments and schedules of vehicles that save more victims than heuristic based approaches.

1 Introduction

Disasters are unpredictable events that demand dynamic, real-time, effective and cost efficient solutions in order to protect populations and infrastructures, mitigate the human and property loss, prevent or anticipate hazards and rapidly recover after a catastrophe. Terrorist attacks, earthquakes, hurricanes, volcano eruptions etc. usually affect a high number of people and involve a large part of the infrastructures thus causing problems for the rescue operations which are often computationally intractable. Indeed, these problems have been tackled by using a plethora of different approaches and techniques, ranging from operational research to artificial intelligence and system management (for a survey please see [3]).

Emergency response efforts [15] consist of two stages: pre-event responses that include predicting and analyzing potential dangers and developing necessary action plans for mitigation; post-event response that starts while the disaster is still in progress. At this stage the challenge is locating, allocating, coordinating, and managing available resources.

In this paper we are concerned with post-event response. We propose an algorithm and a software tool that can be used as a decision support system for assigning the victims of a disaster to hospitals and for scheduling emergency vehicles for their transportation. Even though our algorithm could be used to handle daily ambulance responses and routine emergency calls, we target specifically a disaster scenario where the number of victims and the scarcity of the means of transportation are usually overwhelming. Indeed, while for normal daily operations the ambulances can be sent following the order of the arrival of emergency calls, when a disaster

S. Omatu et al. (Eds.): *Distrib. Computing & Artificial Intelligence,* AISC 217, pp. 219–226.
DOI: 10.1007/978-3-319-00551-5_27

happens this First In First Out policy is not anymore acceptable, since the number of victims involved and the quantity of damages require a plan and a schedule of rescue operations, where usually priority is given to more critical cases, trying in any case to maximize the number of saved people. In this context there are clearly also essential ethical issues which we do not address in this paper (for example, is ethically acceptable not to save a person immediately if this behavior allow us to save two people later on?).

Our tool assumes a simplified scenario where the number, the position and the criticality of victims is known. The tool compute solutions that try to maximize the global number of saved victims. In many practical cases, finding the optimal solution in not computationally feasible, hence we use a relaxation of the pure optimization problem. Our approach uses a *divide-et-impera* technique that exploits both Mixed Integer Programming (MIP) and Constraint Programming (CP) in order to solve the underlining assignment and scheduling problem.

To evaluate the effectiveness of our approach we have compared it against a baseline greedy approach based on the heuristic that sends the ambulances first to the most critical victims and later on to the others. Empirical results based on random generated disaster scenarios show that our approach is promising: it is able to compute the schedule usually in less than a minute and almost always save more victims than the greedy approach.

The remaining of this paper is organized as follows. In Section 2 we define the model we are considering while in Sections 3 and 4 we present the algorithm and the test we have conducted. In Section 5 we present some related work. We conclude giving some directions for future work in Section 6.

2 Model

In the literature a lot of models have been proposed to abstract from a concrete disaster scenario. Some of them are extremely complex and involve a lot of variables or probability distributions [6, 7]. For the purposes of this paper we adapt one of the simplest models, following [5], which considers only three entities: victims, hospitals, and ambulances. Thus we assume to have the following data:

- the spatial coordinates (a two dimensional value) of every entity;
- the capacity of ambulances and hospitals;
- an estimate of the time to death of every victim;
- the dig-up time of every victim, i.e the time needed by the rescue team to be able to rescue the victim as soon as the ambulance arrives on the spot;
- the time needed by an ambulance to reach the hospitals or the victims;
- the initial time an ambulance becomes available (an ambulance may be dismissed or already busy when the disaster strikes).

We are well aware that, especially in a disaster scenario, these data may be difficult to retrieve, imprecise and unreliable. Nevertheless our model can use these data to compute a first solution and then later, when the information become more precise, it can be rerun to improve the computed solution. Moreover, in order to get these informations one can use the results of such works like [11, 13, 14] that allow to esteem the time to death of a civilian or to find the best routes to reach the victims.

Assuming that all the above information are known, our goal is then to find an optimal scheduling of the ambulances in order to bring the maximal number of (alive) victims to the hospitals.

3 Procedure

Solving optimally the scheduling of ambulances may be computationally unfeasible, especially in case of a large number of victims. Moreover, in our scenario, a fast response of the scheduling algorithm is important for different reasons. First of all, the quicker the response is, the faster we can move the ambulances and therefore more victims may be saved. Moreover, waiting for a long time may be useless because usually information rapidly changes (i.e. more victims come, the criticality of the patients vary, the hospitals may have damages or emergencies, ambulances can be broken). Hence, spending a lot of time for computing an optimal solution that in few seconds could become non optimal may result in a waste of resources and then lead to the impossibility of saving some victims. On the other hand, a purely greedy approach that at each stage makes the locally optimal choice (according to heuristics such as the seriousness or the location of the victims) would be definitely faster, but could result in a smaller global number of victims saved.

For these reasons we developed an approach that at the same time allows to compute a solution within a reasonable time limit and still allows us to save more victims than greedy strategies. Our approach basically lies in the interaction of two phases: the *allocation phase*, in which we try to allocate as many victims as possible to ambulances and hospitals, and the *scheduling phase*, in which we compute the path that each ambulance must follow in order to bring the victims to the hospitals.

In the allocation phase, we relaxed some constraints of the problem assuming that every ambulance can save in parallel all the victims it contains (in other terms, each ambulance with capacity c can be seen as the union of c distinct ambulances with capacity 1). The allocation of every victim to an ambulance and a hospital is performed by solving a Mixed Integer Programming problem. More precisely, the constraints that we enforced are the following:

- a victim can not be assigned to more than one ambulance and hospital;
- the number of patients on a given ambulance and in a hospital must not exceed its capacity;
- a victim assigned to an ambulance is also assigned to a hospital and vice versa;
- the time an ambulance needs to reach a victim, dig up and bring her to a hospital, is enough to save the victim.

Since the objective of the MIP problem is to be able to maximize the number of rescued victims, we defined an objective function which takes into account both the seriousness and the location of the victims. Solving this problem gives an esteem of the victims that could be saved and a preliminary allocation of every victim to an ambulance and a hospital. It is worth noticing that since this is a relaxation of the original problem, it may be possible that not all the victims allocated to an ambulance may be saved. Anyway, the allocation guarantees that at least one victim for ambulance can be rescued. Also, there are no restrictions on the number of hospitals that an ambulance can visit.

Once the victims have been allocated by the first phase, the scheduling phase allows to define the path that each ambulance must follow in order to maximize the number of victims saved. In other words, for each ambulance we have to solve a scheduling problem that can be seen as the problem of finding a minimal cost Hamiltonian path in a direct and weighted graph where:

- each node represents the spatial coordinates of one entity: either an ambulance, or a victim, or a hospital. In particular, the start node of the path has to be the one representing the start position of the ambulance, while the end node must represent the coordinates of an hospital.
- each arc represents the possibility of moving from one node to another, while its weight represents the estimated time to do so.

The resulting path must be constrained with respect to the time to death of each victim. The scheduling phase can therefore be mapped into a Constraint Optimization Problem (COP) and solved by using constraint programming techniques.

As already stated, it may be the case that not all the victims allocated to an ambulance may be saved, since differently to what happen in the relaxed problem now an ambulance has to save the victims sequentially. When this happens we have to compute a schedule that saves a maximal subset of such victims. However, instead of considering as maximal subset the one which contains the greater number of elements, we chose the one which has the maximum *priority value* which is calculated as follows. We first compute the remaining time (RT) of each victim by subtracting her dig-up time from the expected time to death. Then, given a subset of victims, we set its priority to the sum of the reciprocals of their RT (bigger values means higher priority). We decided to use this sum to evaluate the priority because, analogously to what happens for the harmonic average, the sum of the reciprocal gives priority to the victim having least RT and it mitigates at the same time the influence of large outliers (i.e. victims with big RT that can be easily saved later).

When an ambulance is scheduled, the model is updated and the allocation phase is possibly restarted in order to try to allocate the victims which have not yet been allocated. The procedure ends when no more victims can be saved.

From the computational point of view this procedure cyclically solves MIP and COP problems. We are aware that these are well known NP-hard problems. However, by exploiting the relaxation of the MIP problem, on one hand, and the limited size of the COP problems, on the other (the capacity of an ambulance is usually a small number) it is possible to get quickly optimal solutions by using current MIP and COP solvers.

4 Tests Results

We did not find in the literature extensive benchmarks of disaster scenarios that we could use to evaluate and compare the performances of our approach. For this reason, in order to evaluate our approach, we built some random generated scenarios obtained by varying the number of hospitals in the set $\{1,2,4\}$, the number of ambulances in $\{4,8,16,32,64\}$, and the number of victims in $\{32,64,128,256,512\}$. The position of each entity was randomly chosen in a grid of 100×100 by using the

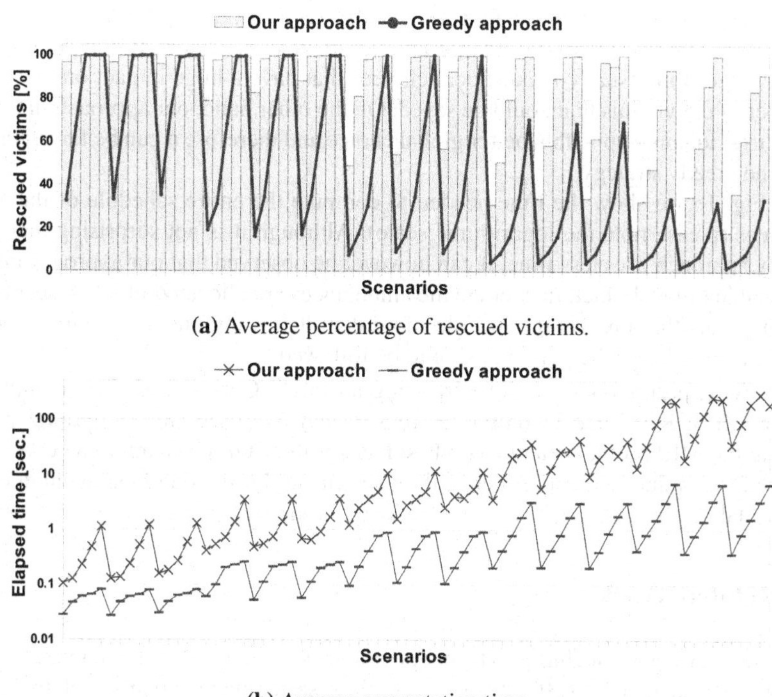

(a) Average percentage of rescued victims.

(b) Average computation time.

Fig. 1 Test results

euclidean distance to estimate the time needed for moving from one point to another. The capacities of the ambulances and the hospitals were selected randomly in the intervals [1..4] and [300..1000], respectively, while the dig-up time and the time to death of every victim were randomly chosen in [5..30] and [100..1000], respectively.

To increase the accuracy and the significance of the results we tested our approach by running the experiments 20 times for each different scenario and measuring the average number of rescued victims as well as the time required to solve the problem. For every scenario we compared the results obtained with those obtained by running a greedy algorithm that at each time assigns the most critical victims to the closest available ambulance and then the ambulance to the closest available hospital. In total we tested 75 different scenarios.

Fig. 1a shows, for each scenario, the average percentage of rescued victims obtained by using our approach and the greedy one. The x-axis represents the scenarios sorted lexicographically by increasing number of ambulances, victims and hospitals (labels are omitted for the sake of readability, since each x-value is actually a triple of values). In only 4 cases (5.3% of the scenarios) the greedy algorithm is better than our approach. However in this few cases the difference of saved victims is less than 1% while the gap between the greedy approach and our approach can reach peaks of about 80%. In average our approach is able to solve 37.32% of more victims than the greedy approach. From the plot we can also see that our approach is especially better for scenarios involving a large number of victims. Indeed in these case the greedy algorithm usually makes local choices that have a huge impact on the total number of the victims that could be saved. On the other hand, our approach in these cases tries to come up with a better global choice and therefore it can be far superior than the greedy strategy.

In Fig. 1b we show the time needed to compute the entire schedule of the ambulances (please note the logarithmic scale). Although it is not surprising that the greedy approach uses less time, it can however be observed that our approach takes reasonable times. In fact, in average the ambulances are allocated in 32.34 seconds, which means that on average in little more than half a minute all the ambulances will be able to know the path that should be followed.

All the experiments were done by using an Intel®Core™ 2.93 GHz computer with 6 GB of RAM and Ubuntu operating system. We used Gurobi optimizer for solving the MIP problems and Gecode solver for the COP problems. The code developed to conduct the experiments, the experimental results and technical details are available at `http://www.cs.unibo.it/~amadini/dcai_2013.zip`

5 Related Work

In the literature many techniques from operational research and artificial intelligence have been used to tackle different aspects of the disaster management problem. Most approaches are trying to develop and study pre-event solution to decrease the severity of the disaster outcome. As an example, in [4] the authors study the best allocation of deposits that allows to handle in the most efficient way the rescue operations in case natural disaster happens. In [7] the authors use MIP in order to schedule the operation rooms and the hospital facilities in case of a disaster. These paper however have a different goal from ours: we are not concerned with considering preventing measures that could allow to mitigate the consequences of future disasters. We are instead concerned with saving more victims, after the disaster happened.

There is a large literature also related to the problem of deciding the initial location of ambulances in order to decrease the average response time for ambulance calls. However, very few papers deal with the computation of the schedule for an ambulance. Some authors focus just on computing the best path for an ambulance toward the victim. For instance in [11] the authors use graph optimization algorithms in order to find a path for an ambulance while in [9] a multi-agents system is used to retrieve the best route for organ transportation. In our work we assume to have such a path and we are concerned with the problem of defining the order of the victims that an ambulance should pick up. In [8, 16] the authors propose a routing algorithm for ambulances but, differently from our case, their model is probabilistic.

In [5] the authors proposed the use of an interactive learning approach which allows rescue agents to adapt their preferences following strategies suggested by experts. The decision of the ambulance is based on a utility function incrementally improved through expert intervention. Differently from our approach the authors here use an heuristic to dispatch the ambulances which rely on expert decision makers, while we rely only on optimization techniques.

In [12] is solved a task scheduling problem in which rescuing a civilian is considered as a task and the ambulances are considered as resources that should accomplish the task. The goal is to perform as many tasks as possible by using the Hogdson's scheduling algorithm to compute the solutions. Differently from our case, the authors considered here only the execution cost of the task and its deadline, ignoring important constraints such as the capacity of the hospitals and ambulances.

The authors in [10] proposed a model based on a Multi-Objective Optimization Problem. They adjust controllable parameters in the interaction between different classes of agents (hospitals, persons, ambulances) and resources, in order to minimize the number of casualties, the number of fatalities, the average ill-health of the population, and the average waiting time at the hospitals. Then, they use Multi-Objective Evolutionary algorithms for producing good emergency response plans. Their underline model is completely different from ours and we argue that is not very adaptable to deal with continuous changes and unexpected situations.

Finally we are aware of the existence of commercial application for Emergency Dispatching (e.g. [1, 2]). The technical details explaining how these software are working unfortunately are always missing.

6 Conclusions

In this work we have described a procedure that can be used as a decision support tool for a post-disaster event when a big number of victims need to be transported to the hospitals. The proposed algorithm takes into account the position of the victims and their criticality, and schedule the ambulance to maximize the number of saved victims. Even though there is no guarantee that the solution obtained is the optimal one, experimental tests confirm that the number of saved victims is greater than the one that could be obtained by using a greedy, priority based heuristic. Moreover the proposed solution is usually fast enough to assign all the available ambulances in less than a minute.

As a future work we are planning to evaluate our approach in a more dynamic scenario. In particular we are planning to develop a discrete event simulator that will allow the comparison of our decision support tool with stochastic approaches such as the one described in [6, 10] or other disaster scenario models such as [12]. Moreover we would like to integrate the model with heuristics developed by domain experts. Adopting these heuristics will allow the system to be able to react to a change very quickly, by using a default behavior that later can be changed if a better solution is found solving the optimization problem. Another direction worth investigating is to study the performance and the scalability of the algorithm proposed taking into account also the robustness of the solutions.

References

1. GeoFES website,
   http://www.dhigroup.com/MIKECUSTOMISEDbyDHI/GeoFES.aspx
2. Odyssey website,
   http://www.plain.co.uk/index.php?option=com_content&task=view&id=67&Itemid=98

3. Altay, N., Green, W.G.: OR/MS research in disaster operations management. EJOR 175(1), 475–493 (2006)
4. Andersson, T., Petersson, S., Värbrand, P.: Decision Support for Efficient Ambulance Logistics. Department of Science and Technology (ITN), Linköping University (2005)
5. Chu, T.-Q., Drogoul, A., Boucher, A., Zucker, J.-D.: Interactive Learning of Independent Experts' Criteria for Rescue Simulations. J. UCS 15(13), 2701–2725 (2009)
6. Erdogan, G., Erkut, E., Ingolfsson, A., Laporte, G.: Scheduling ambulance crews for maximum coverage. JORS 61(4), 543–550 (2010)
7. Issam, N., Jean-Christophe, N., Jolly, D.: Reactive Operating Schedule in Case of a Disaster: Arrival of Unexpected Victims. In: WCE, pp. 2123–2128 (2010)
8. Lopez, B., Innocenti, B., Busquets, D.: A Multiagent System for Coordinating Ambulances for Emergency Medical Services. Intelligent Systems 23(5), 50–57 (2008)
9. Moreno, A., Valls, A., Bocio, J.: Management of Hospital Teams for Organ Transplants Using Multi-agent Systems. In: Quaglini, S., Barahona, P., Andreassen, S. (eds.) AIME 2001. LNCS (LNAI), vol. 2101, pp. 374–383. Springer, Heidelberg (2001)
10. Narzisi, G., Mysore, V., Mishra, B.: Multi-Objective Evolutionary Optimization of Agent Based Models: an application to emergency response planning. In: Computational Intelligence, pp. 228–232 (2006)
11. Nordin, N., Kadir, N., Zaharudin, Z., Nordin, N.: An application of the A* algorithm on the ambulance routing. In: CHUSER, pp. 855–859 (2011)
12. Paquet, S., Bernier, N., Chaib-Draa, B.: Multiagent Systems Viewed as Distributed Scheduling Systems: Methodology and Experiments. In: AI, pp. 43–47 (2005)
13. Suárez, S.A., Quintero, C.G., de la Rosa, J.L.: A Real Time Approach for Task Allocation in a Disaster Scenario. In: PAAMS, pp. 157–162 (2010)
14. Teruhiro, M., Weihua, S., Keiichi, Y., Minoru, I.: Transportation Scheduling Method for Patients in MCI using Electronic Triage Tag. In: eTELEMED, pp. 1–7 (2011)
15. Tufekci, S., Wallace, W.: The Emerging Area of Emergency Management And Engineering. Engineering Management 45(2), 103–105 (1998)
16. Ufuk, K., Ozden, T., Saniye, T.: Emergency Vehicle Routing in Disaster Response Operations. In: POMS (2012)

Creating GPU-Enabled Agent-Based Simulations Using a PDES Tool

Worawan Marurngsith and Yanyong Mongkolsin

Department of Computer Science, Faculty of Science and Technology, Thammasat University
99 Phaholyothin Road, Pathum Thani, 12120, Thailand
wdc@cs.tu.ac.th, 5109035054@student.cs.tu.ac.th

Abstract. By offloading some computation to graphical processing units (GPUs), agent-based simulation (ABS) can be accelerated up to thousands of times faster. To exploit the power of GPUs, modellers can use available simulation frameworks to auto-generated GPU codes without requiring any knowledge of GPU programming languages. However, such frameworks only support computation on the GPUs of a particular vendor. This paper proposes techniques, implemented in a synchronous parallel discrete event simulation (PDES) tool, to allow modellers to create ABS models, and to specify computation regions in the models for multiple vendor's GPUs or CPUs. The technique comprises a set of meta-language tags and a compilation framework to convert user-defined GPU execution regions to OpenCL. A well-known cellular ABS models, the Conway's Game of Life, have been implemented and evaluated on two platforms *i.e.*, the NVIDIA GeForce 240M LE and AMD Radeon HD6650M. The preliminary results demonstrate two findings: (a) the proposed technique allows the example ABS model to be executed on a PDES engine successfully; (b) the generated GPU-enabled ABS model can achieve fourteen times faster than its multicore version.

Keywords. OpenCL, GPU, Agent-based simulation, PDES, Acceleration.

1 Introduction

Exploiting the computational power of graphic processing units (GPUs), high performance agent-based simulations (ABS) permit modellers to study large-scale complex phenomena faster than before. Development of languages and tools has made the integration of GPU computation to ABS models become more user friendly. Among several GPU languages[1] (*e.g.*, Brook, CUDA, and OpenCL), CUDA is probably the most popular language used in ABS as they are several parallelisation techniques and optimisation available[2, 3]. Despite its popularity,

CUDA-based ABS models suffer from lack of portability as they are limited to executing only on NVIDIA GPUs. To overcome this limitation, recent GPU languages including, OpenCL[4] and C++AMP[5], can be a good candidate. Both languages allow programmers to express computation on heterogeneous processing platforms including CPUs and GPUs (and any compliant devices). The

S. Omatu et al. (Eds.): *Distrib. Computing & Artificial Intelligence,* AISC 217, pp. 227–234.
DOI: 10.1007/978-3-319-00551-5_28

OpenCL language has been used in Multi2Sim [6] framework to accelerate parallel discrete event simulations (PDES). However to the best of the authors' knowledge, there are still no ABS frameworks based on OpenCL or C++ AMP.

As significant performance improvements are necessary, the implementation of many legacy ABS simulation frameworks *e.g.*, EcoLab, NetLogo, Mason, Repast (HPC version), ABM++ and FLAME have shifted towards parallel performance and scalability using multi-threading and multiprocessing techniques [7]. Only two ABS frameworks *i.e.*, the work of Lysenko and D'Souza [8] and FLAME[9], provide performance acceleration of ABS models using GPUs.

The first framework [8], developed at the Michigan Technological University, is a completely GPU-based ABS framework using C++, OpenGL and GLSL. The framework composed of three parts *i.e.*, a system for handling environments, mobility mechanism in agents, and visualisation. Mobile agents are off-loading to GPUs and encoding into textures. Some algorithms have been employed to perform three basic tasks of mobile agents on GPUs, including: storing and updating the mobile agent state and connecting the mobile agents to their environment. The experimental results show that using the framework, an ABS model can achieve a speed up factor of over 9,000.

FLAME [9] is the flexible large-scale agent modelling environment developed at the University of Sheffield. The framework is template-based, ideally suited for creating ABS models of cellular systems running on multicore and GPUs. Simulations are advanced by using discrete time steps. Modellers use the FLAME's template, based on XML syntax, to describe all GPU computations without having to have any specialist knowledge of the underlining GPU platform. The template is then automatically translated into the corresponding CUDA codes. Using the auto-generated GPU codes in FLAME, simulation of cellular models could achieve hundreds of times faster than they multicore version. Model data can also be saved and analysed post simulation; or viewing on the fly. Published experimental results have shown that, using FLAME, cellular models can be massively accelerated using the parallel GPU architecture beyond that of CPU-based frameworks.

Existing simulation frameworks have confirmed that GPU computing can be a novel cost-effective approach towards high-performance ABS models, with less programming effort from modellers. Some research has shown the successful parallelism techniques for achieving high-performance ABS models on multicores and GPUs *i.e.*, cellular automata models using CUDA with OpenMP[3]. adaptive Swarm model using Java-binding OpenCL [10]. However, developing a simulation framework for ABS models to exploit computational power of multicores and GPUs at the same time is still a challenge.

This paper presents the use of a synchronous parallel discrete-event simulation (PDES) tool, called P-HASE[1], as a framework for creating GPU-based ABS models using OpenCL C++ binding. Our main contributions are the following:

---

[1] Available at `http://parlab.cs.tu.ac.th/P-Hase`

- The mapping of ABS components to the discrete event simulation (DES) components is presented,
- A technique to convert user-defined GPU computation regions to OpenCL kernels is proposed,
- An example GPU-based ABS model is presented; and the performance analysis experiment of the model on two GPUs models has shown the speedup up to fourteen-fold, in comparison to its multicore counterpart.

The rest of the paper is organised as follow. The next section presents the techniques used to create ABS models and to integrate them with GPU computation. Section 3 presents the detail of compilation framework for generating OpenCL codes. An example ABS model and the experimental results are presented in Section 4. Section 5 gives the conclusion of the paper.

2 GPU-Enabled ABS on P-HASE Simulation Tool

P-HASE [11] is a parallel extension of HASE [12] (the Hierarchical computer Architecture design and Simulation Environment), which is a framework for discrete-event simulation (DES) and visualisation, developed at the University of Edinburgh. The key idea of HASE is two folds. First it aims to provide modellers facilities to allow a simulation model to be hierarchically structured. Second it provides modellers the visual verification of a model's behaviour via an animation. The P-HASE simulation tool derives all functionalities from HASE. Thus, it supports the creation, execution of simulation models, and post-mortem debugging via an animation. Extra to HASE is that P-HASE provides a scalable PDES engine using conservative synchronous technique [11]. The tool generates a PDES from the HASE DES definition. Thus, sequential models can be accelerating by the parallel engine provided in P-HASE.

Table 1 Mapping between components of P-HASE model to the ABS model

DES Components in P-HASE	ABS Components
Entity	Environment, static agent
Ports	Interaction between environments/static agents
Task object inside an Entity	Agents
Compound entity	Hierarchical environments
Design template	Multiple environments connected via a topology

There are several ways to represent ABS models in a specification which conform with the discrete event system specification (DEVS) formalism **[13]**. As diagram shown in Fig.1, to create a simulation model in P-HASE, modellers first have to provide a structural definition of model's components using a file written in the entity description language (EDL). Components of a model in P-HASE are atomic units called *Entities* that communicate by passing events via *Ports*. Each

Entity can be viewed as an independent thread. Since it is better to group agents which frequently communicate with each other on one thread, the Entity is used to represent an Environment (or virtual space) in ABS. So that Ports can be used to represent the communication between Environments.

Modellers can group together sets of Entities that are logically related in two ways: (a) aggregating them into a *compound entity*, or (b) using a *design template*. Compound entities describe the vertical composition of a more complex entity from its basic subunits up to its higher abstract-level units. A design template describes functional relationships among entities and also provides the means of connection between them. Thus, a group of Environments in ABS can be represented in three ways: (a) as multiple Entities connected via ports, (b) as a compound entity or (c) as a design template. Table 1 shows the possible mapping between DES components used in P-HASE to the ABS components.

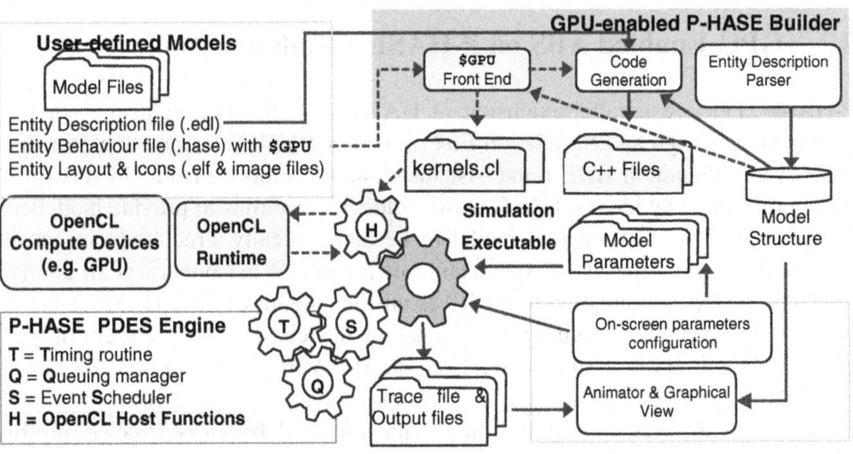

Fig. 1 Modules of the P-HASE tool with GPU-enabled extension

The top left corner of Fig.1 depicts all types of input file needed for creating a model. The key file type for adding the GPU computation to a model is the behaviour file (*.hase*). Once the structure of model components (*.edl file*) has been defined, modellers insert C++ codes to mimic the behaviour of each entity in *.hase* file. The behaviour file is a text file organised in sections separated by tags (e.g., **`$startup`**, **`$pre`**, **`$phase_0`**, **`$phase_1`**, **`$report`**). Modellers write C++ code into each section to implement the behaviour for each particular time step. For example, in synchronous models using two-phase clocking, the behaviour of each clock phase must be added under the **`$phase_0`** and **`$phase_1`** sections respectively. Codes specified under these sections will be generated by the HASE Builder into a clock-phase routine. In a clock phase routine, a special tag (**`$GPU`**) is used to specified the beginning of a GPU computation. The tag must be immediately followed by a **`for`** loop which implements the behaviour of agents. As show in Fig.1, when modellers compile the model, the behaviour file with **`$GPU`** tag is passed to the **`$GPU`** Front End module. The tag instructs the module to generate loop iterations as agents. The module performs some semantic analysis on the following `for` loop and generates the corresponding OpenCL codes. The generated codes have two parts. The first part is generated as a file with *.cl* extension, called *kernels*. It is a file containing C-style functions each of which describes a unit of task to be offloaded to GPUs (so called *devices*). The body of the `for` loop is generated as a kernel function. All array references are flattening into one dimension to fit with the OpenCL device architecture. The second part, called *host*, is source codes used to initialise the OpenCL execution environment (*Context*), and to build and compile the kernel. The host codes are generated and inserted into the behaviour file to replace the `for` loop.

The modified behaviour files are passed to the original code generation process in HASE to build the simulation executable. During a simulation run, the simulation executable acts as the host of the OpenCL environment. Thus, it will create a context and launch kernels to GPUs.

3 Compilation Framework and the GPU-Enable PDES

Fig. 2 depicts the structure of the compilation framework implemented in P-HASE for translating each `$GPU` tags to an OpenCL kernel, and inserting the calls from host. Similar techniques used in our previous work have been adapted to perform static analysis and code generation [14]. Fig.3 shows the mechanism of GPU-enabled PDES. During a simulation run, each Entity represents an Environment where agents reside. Each Environment is created as a thread executing on CPUs, deriving from a clock abstract class for timing control. Events are used for time advancing. At the start of a simulation, the Environment initialises the OpenCL host functions. These includes querying for platforms (CPUs or GPUs) or *compute devices*, and creating *Context* on the devices found; creating command queue in the context; reading kernels.cl into a stream and creating a program object from it; specifying a kernel function from the program object and creating it as a kernel object; creating memory buffers for passing inputs and receiving results. At each clock phase routine, Environment starts agents by submitting the kernel object to the command queue for executing on GPUs. Each kernel object will be created as an execution unit, called *work-item*, and assigned to each ALU of the compute device. When agents finished the tasks of each time step, results are written back to Environment via buffers.

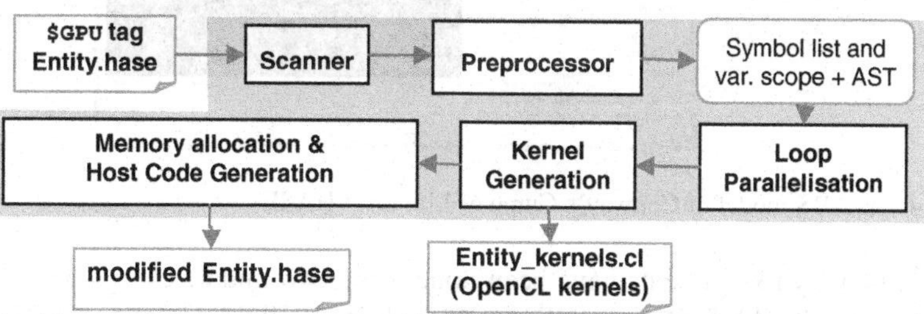

Fig. 2 Overview of the compiler framework implemented in the P-HASE Builder

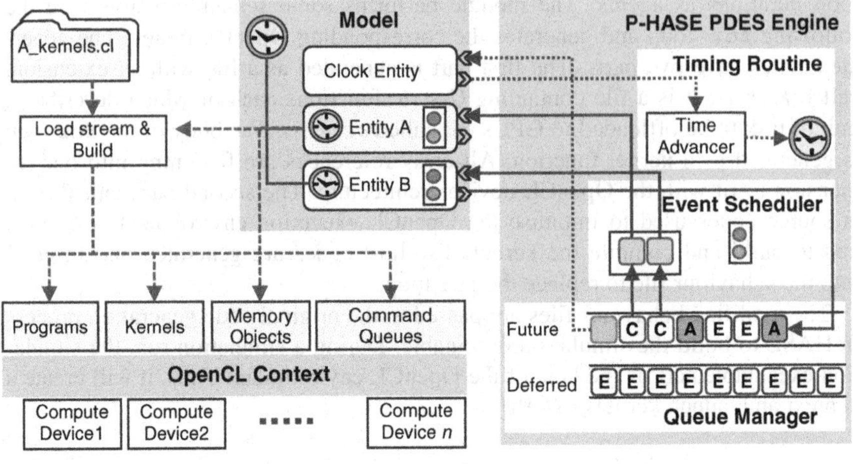

Fig. 3 Structure and mechanism of the P-HASE Tool (GPU related modules are highlighted)

4 Preliminary Experiments and Results

Our proposed platform is still limited to cellular models which tasks of agents are written in a loop. The Conway's Game of Life (GOL) with two-dimensional grid has been developed in the P-HASE tool (Fig.4) in two versions, the model with 1- and 2- environments (GOL-1, and GOL-2). The environment represents the $n \times m$ grid of cells. In the environment, a cell lives in each grid entry and cannot move. Each version receives the number of agents as a parameter; and has been validated by matching the model results against the CPU-only models.

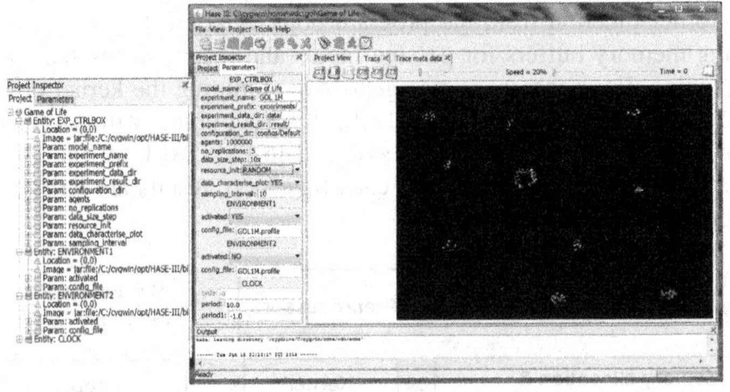

Fig. 4 The ABS model of Conway's Game of Life on P-HASE

Preliminary experiments have been done using two middle class graphics card for laptops, the AMD Radeon HD6650M and the NVIDIA GeForce 240LE. The setup of the experiments aimed for finding the speedup achieved on using

GPU-enabled models generated by P-HASE if modellers are mobile. That is if the simulation is run on-site where any high performance computation platform is not applicable. The experiments used the GOL-1 model (Fig.5) and GOL-2 model (Fig.6) with four different numbers agents. Each model was repeated five times, and executed on a core-2 duo CPU, a Core i7-2620M CPU, and on the two graphics cards. Execution time was recorded for each run and the speedup was calculated by using the dual-core execution time as a base line.

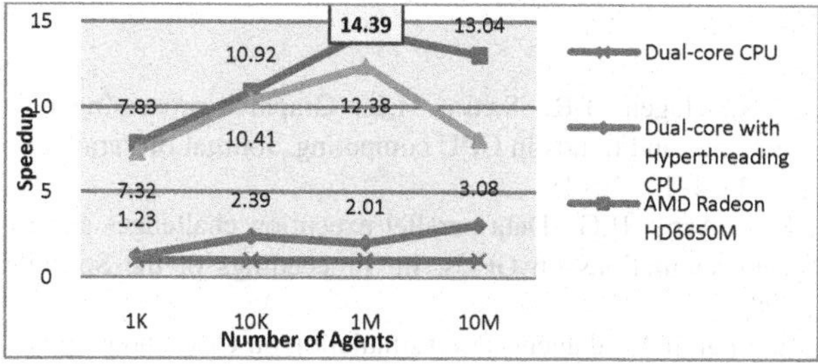

Fig. 5 Speedup of three platforms compared to dual-core model for single environment model

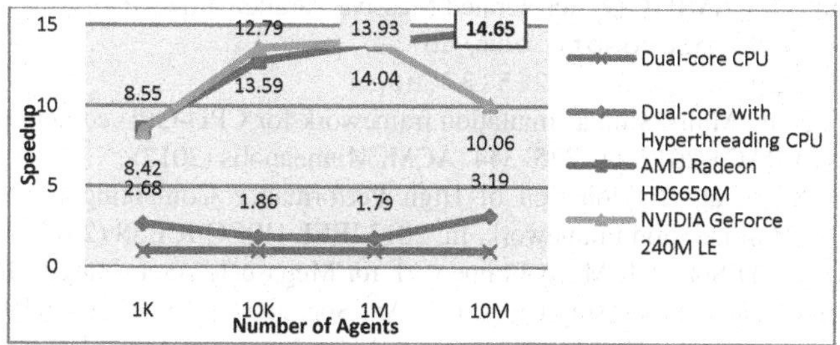

Fig. 6 Speedup of three platforms compared to dual-core model for two environments model

Overall results show that GPU-enabled models outperform the dual-core versions for 7 – 14 times, and outperform the Core i7 model for 5 – 12 times. The speedup plots of GPU-enabled models drop when the agent size is 10,000,000, which may cause by the synchronisation overhead. Detail analysis has to be done to quantify the overhead and to improve the code generation process.

5 Conclusion

The use of a synchronous PDES tool, called P-HASE, as a framework for creating a GPU-based ABS models using OpenCL C++ has been presented. A cellular ABS model was evaluated and confirmed the feasibility to use the PDES framework to create efficient GPU-based models. The work presented in this paper is a preliminary wor. The framework is still limited to represent cellular models with simple behaviours. Thus our immediate future work is to explore the ways to cover more realistic models to understand the performance limitation.

Acknowledgement. We wish to thank Professor Roland Ibbett and David Dolman for allowing us to extend the original HASE tool. We thank the referees for valuable comments.

References

1. Brodtkorb, A.R., Hagen, T.R., Sætra, M.L.: Graphics processing unit (GPU) programming strategies and trends in GPU computing. Journal of Parallel and Distributed Computing 73(1), 4–13 (2013)
2. Perumalla, K.S., Aaby, B.G.: Data parallel execution challenges and runtime performance of agent simulations on GPUs. In: Proceedings of the SpringSim. SCS, pp. 116–123 (2008)
3. Falk, M., et al.: Parallelized agent-based simulation on CPU and graphics hardware for spatial and stochastic models in biology (2011)
4. Group, K.: OpenCL - The open standard for parallel programming of heterogeneous systems (2013), http://www.khronos.org
5. Microsoft. C++ AMP (C++ Accelerated Massive Parallelism) (2013), http://msdn.microsoft.com/en-us/library/vstudio/hh265137.aspx
6. Ubal, R., et al.: Multi2Sim: a simulation framework for CPU-GPU computing. In: Proceedings of PACT 2012, pp. 335–344. ACM, Minneapolis (2012)
7. Coakley, S., et al.: Exploitation of High Performance Computing in the FLAME Agent-Based Simulation Framework. In: 2012 IEEE HPCC-ICESS (2012)
8. Lysenko, M., D'Souza, R.M.: A Framework for Megascale Agent Based Model Simulations on Graphics Processing Units. J. of Art. Soc. and Soc. Sim. 11(4), 10 (2008)
9. Richmond, P., et al.: High performance cellular level agent-based simulation with FLAME for the GPU. Briefings in Bioinformatics 11(3), 334–347 (2010)
10. Laville, G., et al.: Using GPU for multi-agent multi-scale simulations, pp. 197–204 (2012)
11. Mongkolsin, Y., Marurngsith, W.: P-HASE: An Efficient Synchronous PDES Tool for Creating Scalable Simulations. In: Xiao, T., Zhang, L., Fei, M. (eds.) AsiaSim 2012, Part III. CCIS, vol. 325, pp. 231–245. Springer, Heidelberg (2012)
12. Coe, P.S., et al.: Technical note: a hierarchical computer architecture design and simulation environment. ACM Trans. Model. Comput. Simul. 8(4), 431–446 (1998)
13. Tauböck, S., et al.: The <morespace> Project: Modelling and Simulation of Room Management and Schedule Planning at Universityby Combining DEVS and Agent-based Approaches. J. on Developments and Trends in Mod. and Simulation 22(2), 11–20 (2012)
14. Makpaisit, P., Marurngsith, W.: Griffon - GPU programming APIs for scientific and general purpose computing (Extended version). International Journal of Artificial Intelligence 8(12 S), 223–238 (2012)

Agent and Knowledge Models for a Distributed Imaging System

Naoufel Khayati and Wided Lejouad-Chaari

SOIE Laboratory, Optimization Strategies and Smart Computing,
University of Tunis, Tunisia
naoufel.khayati@soie.rnu.tn, wided.chaari@ensi.rnu.tn

Abstract. Program Supervision aims at automating the use of complex programs, independently of any particular application domain. Program Supervision Systems offer original approaches to plan and control program processing activities. Since real applications imply more and more participants on various sites, we worked on the distribution of such systems. Therefore, given distributed data, programs and knowledge, our aim is to propose convenient and efficient models for distributing program supervision systems dealing with image processing and image analysis. In this paper, we present the Agent and Knowledge Models we used for the distributed, collaborative and intelligent assistant for image processing and its analysis. It is collaborative because the participants (knowledge-bases designers, program developers and other users) can work collaboratively to enhance the quality of programs and then the quality of the results. It is intelligent since it is a knowledge-based system including, but not only, a knowledge base, an inference engine said "supervision engine" and ontologies.

Keywords: Program Supervision, Distributed Program Supervision Systems, Mobile Agents, Knowledge Model, Ontologies, Image Analysis.

1 Introduction

Image analysis is widely needed in many application fields, like medicine and astronomy. This analysis helps specialists in image processing in developing more effective programs and more efficient approaches. In our work, we determined an "image protocol" for various image resolutions [4][6]. Setting such an image protocol consists in planning a sequence of programs and tuning their input values. This constitutes a tedious, time consuming task which requires both users (clinicians and radiologists, for example) and image processing experts to collaborate. Our solution was to provide an interactive tool which relies on artificial intelligence approaches to build image protocols in different situations. Moreover, we wish to make this system accessible to radiologists, a geographically scattered community. In this paper, we will present our technological and architectural choices in terms of mobile agents, knowledge model and ontologies.

S. Omatu et al. (Eds.): *Distrib. Computing & Artificial Intelligence*, AISC 217, pp. 235–242.
DOI: 10.1007/978-3-319-00551-5_29

2 Program Supervision

Several program libraries have been developed by specialists in various domains with an aim of automating the image processing. But, the user of these libraries of image processing and medical imaging does not have competences in data and image processing allowing him to use them in an effective way. Moreover, users (physicians) must focus themselves on the interpretation of the results and not on the way in which these programs are carried out and scheduled. Thus, approaches of Artificial Intelligence were proposed in order to assist a non-specialist in data processing for correct use of these programs in its field. These approaches are known as "Program Supervision" [9] which consists of the automation of management and the use of pre-existent programs. These programs are considered as "black boxes" and their application domain or their programming language is not relevant. The goal is not to optimize the programs themselves, but to assist program *usage* [9].

To carry out a supervision task, a subset of programs is chosen, scheduled, and applied to a specific problem. This selection and this scheduling in various configurations are ensured by a supervision system, which, thanks to the reasoning of its engine and the knowledge contained in its base can free the user (the physician) to make this management manually. This enables a physician to run programs, to check the consistency of some image analysis methods, to compare algorithms, to evaluate results, to reconsider some parameters and to readjust them.

Program supervision may be applied to different domains related to image, signal processing, or scientific computing like Astronomical Imaging (e.g. automatic galaxy classification [10]), Vehicle Driving Assistance (e.g. road obstacle detection [7]), and Medical Imaging (e.g. chemotherapy follow-up based on Factorial Analysis of Medical Image Sequences [1][8], and the segmentation of 3D MRI images of the brain [1][8], osteoporosis detection in bone radiographies [4][5]).

3 Architecture of the Distributed and Collaborative System

In our previous applications we used a centralized supervision system in which data, programs and the knowledge-based system (KBS) have the same location. However, applying program supervision to real applications as osteoporosis detection, require distributing it. Indeed, both data (images from different hospitals) and programs (developed by different teams) come from various places. In most cases it would be fairly inefficient to move data and executable code to the same place as the KBS, and in some case it is even not possible (e.g., programs may execute only on a specific hardware). Therefore, we have developed a distributed version of the assistant, based on mobile agents, where clinicians from different countries may safely use the system and work collaboratively (run programs of other teams, check consistency of image analysis methods and evaluate results); where medical imaging experts may manage different versions of their programs; and where knowledge engineers may capitalize knowledge and adapt it to program changes. The distributed system, whose architecture is proposed in [3][4] and given by figure 1, is a triple (S, A, KM) defined by:

- S, a set of three types of servers :
  - A session server playing the role of an interface allowing end-users to access the supervision services and to communicate with the other components of the distributed system.
  - A set of resource servers hosting programs, knowledge and supervision engine of the centralized system: a supervision server hosting the supervision engine, some program servers, some knowledge servers and some data servers.
  - And possibly a set of execution servers, on which programs are executed. For example, when dealing with MatLab programs, their execution requires at least, the presence of the MatLab tool on one of the servers.
- A, a set of agents who are responsible for updating the previous components and for performing requests. This multi-agent system combines stationary and mobile agents.

  KM, a set of knowledge models in the form of metadata and ontologies used to locate resources in order to define the mobile agents itinerary, to define access permissions and to analyze the user request so that it is properly treated.

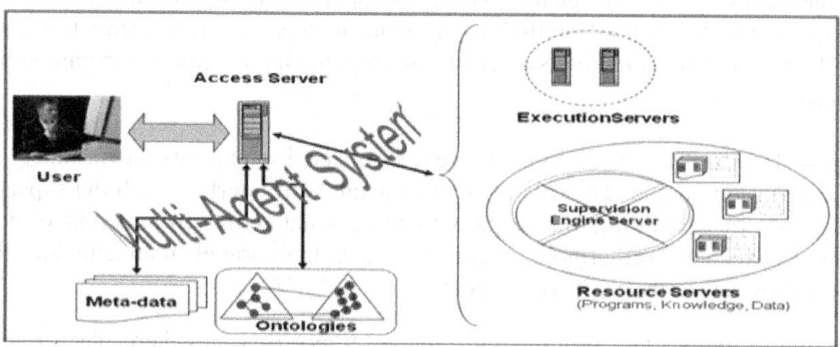

Fig. 1 Distributed Assistant Architecture

4 Agent Model

When distributing the supervision system, mobile agents are used to implement the communications between its components. Metadata help localize the various resources (programs, data, execution servers, etc.).

The architecture of the multi-agent system as presented on figure 2 consists of several types of servers. First, one *Program Supervision Server* runs the supervision engine named PEGASE; then possibly several *Execution Servers* enable to execute the planned programs; *Resource Servers* contain remote resources needed by executions (e.g., data files, scripts); finally, the end-user interacts directly with a *Session Server.*

The agents are classified according to their roles into three categories: *Interface agents, Processing agents* and *Communication agents.*

4.1 Interface Agents

Supervisor Agents. They are called also *Session Managers*; they are stationary and are coupled with the access web server. Each one manages the whole user session and the whole process of solving the supervision query in its charge. This means that there is one Supervisor agent by query. Such an agent:

- Reads the metadata to identify and localize the involved resources in the resolution of the query;
- Creates the other stationary agents for interfacing with the supervision engine (Engine agent) and the different execution servers (Execution agents);
- Determines the number of needed Solver agents and creating them. This computation is done according to some information given by the Supervision Engine: the $N\delta\epsilon\pi$ dependency sets and the $N\pi\alpha\rho$ parallel operators by set.

For the last point, in a YAKL code (*YAKL for Yet Another Knowledge Language, used by PEGASE*), we may note parallel tasks in the *Body* section of a composite operator. We may also note dependency links between the sub-operators of a composite one, in its *Flow* section. These concepts are necessary when determining the number of solver agents. The dependencies will allow building some dependency sets that will allow finding the right number of solver agents to create. We define a *dependency set* as a set of operators having at least one common parameter.

Engine Agents. There is an Engine Agent by query; it represents an "instance" of the supervision engine. The role of such an agent is to interface with the supervision engine by submitting to it, a query to process and getting from, a plan of programs to execute. Furthermore, it communicates the parallel treatments and the dependency sets to the Supervisor agent.

Execution Agents. Once sent to their running servers, they stay there. Their role is to interface with these servers by passing the instructions to execute (received from a Solver agent) and retrieving the obtained results.

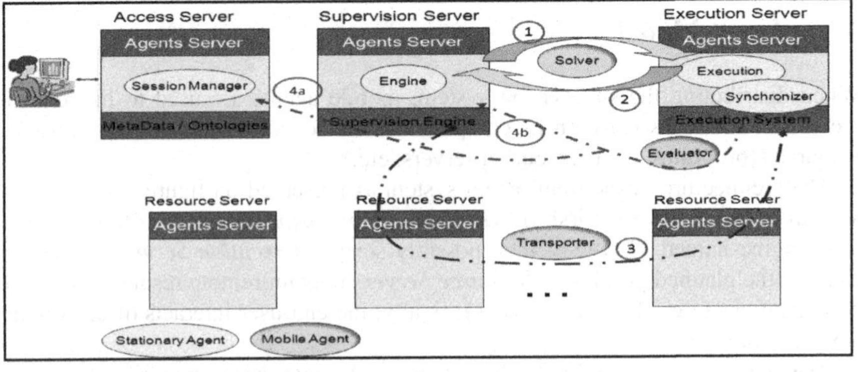

Fig. 2 Architecture of the Multi-Agent System and Scenario of Use

4.2 Processing Agents

Solver Agents. They are mobile and have to launch the execution of the remote programs already planned by the supervision engine.

For example, while being under the control of the Supervisor agent, the Solver agent migrates to the supervision server (path 1 on figure 2) looking for the next instruction (the program and its call syntax). Then it seeks the necessary resources for this instruction from the Transporter agents, migrates to the corresponding execution server (path 2 on figure 2), and then waits until all the resources are available (information received from a synchronizer agent). At this stage, it starts executing the instruction, and finally, it sends the results to the Supervisor.

In general, the number of solver agents is determined according to these rules:

- For a dependency set of N elements, if it contains *Npar* parallel elements ($Npar \leq N$), then the maximum number of solver agents will be equal to *Npar* (one agent per parallel task, then one of them will continue with the other tasks of the same set). Otherwise it will be equal to 1.
- For *Ndep* dependency sets the maximum number of solver agents will be equal to $\sum_{i=1}^{Ndep} Npar\_i$. If no set has parallel tasks, such number shall be equal to *Ndep*.

Evaluator Agents. They are created by Solver agents on the program sites or on the execution sites. They are created only if they are needed, i.e. if some programs require the evaluation of their results.

An agent of this class stores the result to evaluate in its context and then must go to another server to perform its task. For its migration, the destination depends on the assessment type. Thus, if the evaluation is automatic, i.e. made by the supervision engine based on the knowledge base rules, it must migrate to the supervision server (path 4b on figure 2). Otherwise, if the assessment is interactive, i.e. it requires the user intervention; it migrates to the access web server (path 4a on figure 2). In the case of an interactive assessment, the user response will be sent to the engine agent so it can decide the next step (continue with a new program or re-run the same program with repaired values for its parameters).

4.3 Communication Agents

Transporter Agents. They are created by the solver and are responsible for searching the necessary knowledge for the supervision engine in order to select the programs and their sequencing. In addition, they search the necessary resources to execute the current instruction, and transport them to an execution server (path 3 on figure 2). Since an instruction may need resources located on different nodes, there may be, simultaneously, several active Transporter agents.

Synchronizer Agents. They are created by the Solver agent to synchronize the operations of the resources transport performed by the Transporter agents. Indeed, since many Transporters may be active simultaneously, they must register with the associated Synchronizer. When they are all registered, this agent starts them and waits until they do their jobs. Finally, when all the needed resources are available, it indicates to the Solver that it can continue its work.

5 Knowledge Model

In order to share a common understanding of our application domains and to solve problems related to the semantic of the supervision queries, we have equipped our distributed supervision system with ontologies.

5.1 Ontological Architecture

The global ontological architecture (figure 3) of our distributed system consists of three interoperable ontologies: an ontology for the supervision domain and its knowledge, a second for the elements involved in its distribution (distribution ontology) and a third for the application domain (medical imaging and osteoporosis detection). For their integration into the system, our ontologies undergo the application of a reasoner which offers, in case of inconsistency, possible repair operations. Thus, we obtain semantic files reflecting them (Step 1 on figure 4).

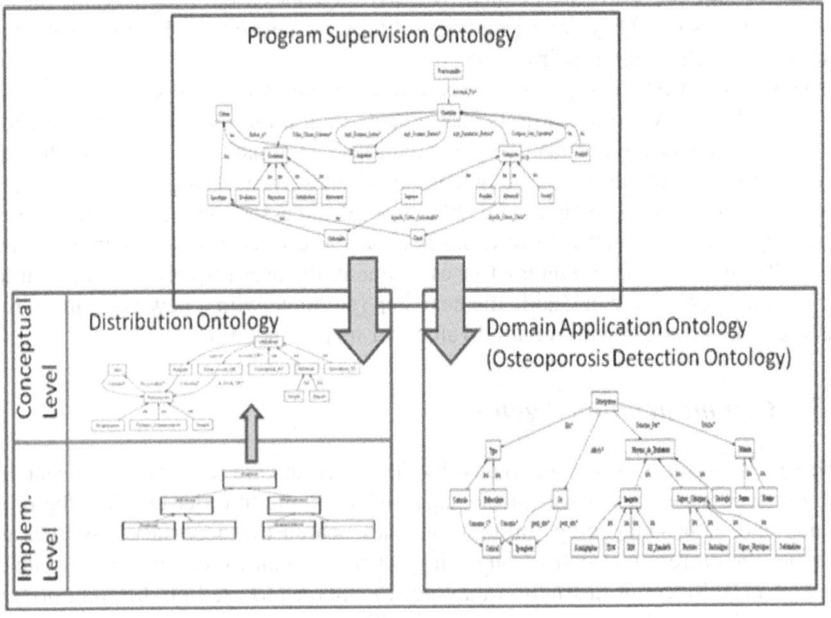

Fig. 3 Ontological Architecture

5.2 *Asking the Ontologies*

The query language defines the syntax and the semantics required to express queries and the possible forms of the results. We expressed interrogative SELECT queries type, which extract a sub-graph corresponding to a set of resources satisfying the conditions specified in the WHERE clause. Indeed, the user expresses his query in textual form, for example, "check the status of the bone". This sentence will be split into words; neutral words will be eliminated (the, of, etc.). Then, the application domain ontology will be queried. This gives that the word "check status bone" is subsumed by the "Osteoporosis" concept. For testing, we added white concepts such as "diabetes", "Hepatitis", etc. to ensure that the query asked to the ontology receives the good answer.

Then, the supervision ontology has to fetch the functionality which is responsible for the osteoporosis detection and subsequently the appropriate composite operator (Step 2 on figure 4). Once determined, this functionality will be communicated to the supervision engine (Step 3 on figure 4) so it can decide the "good" knowledge files (Step 4 on figure 4) and thereafter, the right resources needed for the resolution of the current query. At this stage, agents will be launched (Step 5 on figure 4) to start a supervision process.

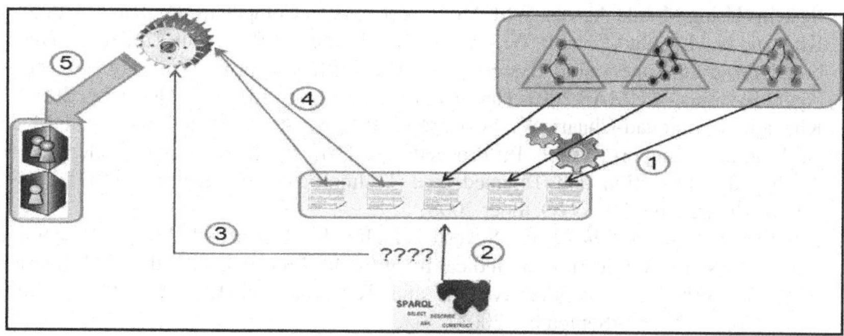

Fig. 4 Ontologies Integration and Preparation Phase

6 Conclusion

Distributing program supervision systems for medical image processing is fundamental because physicians, image processing programs, images, inference engine, knowledge bases, etc. are generally located at different sites. Our distributed environment is based on a Web server, mobile agents for the communication inter-components and Semantic Web ontologies to facilitate physician access and knowledge exploration.

Hence, the strength of the proposed system comes from the following points:

- Taking advantages from the technological advances in medical image processing.
- Facilitating the experiments by radiologist and physicians and allowing them to adjust and repair the values of the programs parameters.
- Making different experts work together.
- Consolidating the diagnosis of the disease.
- Taking into account the heterogeneity of data (images, knowledge, ontologies, programs, etc.).

As prospects, we plan to improve the performance of our distributed and collaborative intelligent system when scaling with multiple queries, several images, etc. and this, in order to test its ability to maintain its functionalities and performance in important demand.

References

1. Crubézy, M., Aubry, F., Moisan, S., Chameroy, V., Thonnat, M., Di Paola, R.: Managing complex processing of medical image sequences by program supervision techniques. In: Proc. of SPIE Medical Imaging 1997, Newport Beach, CA, vol. 3035-85, pp. 614–625 (February 1997)
2. Khayati, N., Lejouad-Chaari, W., Moisan, S., Rigault, J.P.: Distributing Knowledge-Based Systems Using Mobile Agents. WSEAS Transactions on Computers 5(1), 22–29 (2006)
3. Khayati, N., Lejouad-Chaari, W., Moisan, S., Rigault, J.P.: Agent Model for Distributed Program Supervision Systems. In: The Fifth European Workshop on Multi-Agent Systems - EUMAS 2007, Hammamet, Tunisia, pp. 155–164 (December 2007)
4. Khayati, N., Lejouad-Chaari, W., Sevestre-Ghalila, S.: A Distributed Interactive Medical Diagnosis Support System. In: Proceedings of the 2nd International Conference on Advanced Information and Telemedecine Technologies for Health (AITTH 2008), Minsk, Belarus, pp. 59–63 (October 2008)
5. Khayati, N., Lejouad-Chaari, W., Sevestre-Ghalila, S.: A Distributed Image Processing Support System: Application to Medical Imaging. In: Proceedings of the IEEE International Workshop on Imaging Systems and Techniques (IEEE-IST 2008), Chania, Greece, pp. 261–264 (September 2008)
6. Lejouad-Chaari, W., Moisan, S., Sevestre-Ghalila, S., Rigaut, J.P.: Distributed Intelligent Medical Assistant for Osteoporosis Detection. In: Proc. of the International Conference of IEEE Engineering in Medicine and Biology Society, Lyon, France (August 2007)
7. Shekhar, C., Moisan, S., Thonnat, M.: Real-Time Perception Program supervision for Vehicle Driving Assistance. In: Kaynak, O., Ozkan, M., Bekiroglu, N., Tunay, I. (eds.) ICRAM 1995 Intl. Conference on Recent Advances in Mechatronics, Istanbul, pp. 173–179 (1995)
8. Thonnat, M., Moisan, S., Crubézy, M.: Experience in Integrating Image Processing Programs. In: Christensen, H.I. (ed.) ICVS 1999. LNCS, vol. 1542, pp. 200–215. Springer, Heidelberg (1998)
9. Thonnat, M., Moisan, S.: What can Program Supervision do for Software Reuse? IEEE Proc. Special Issue on Knowledge Modelling for Software Components Reuse 147(5), 179–185 (2000)
10. Vincent, R., Thonnat, M., Ossola, J.C.: Program supervision for automatic galaxy classification. In: Proc. of the Intl. Conference on Imaging Science, Systems, and Technology, CISST 1997 (June 1997)

DPS: Overview of Design Pattern Selection Based on MAS Technology

Eiman M. Salah, Maha T. Zabata, and Omar M. Sallabi

Dept. of Computer Science, University of Benghazi, Libya
{eiman.sahly,maha.zabata,omar.sallabi}@benghazi.edu.ly

Abstract. The design patterns have attracted increasing attention in the field of software engineering, since effectively selecting the fits pattern for a given problem can seriously improve the quality of the software, on the contrary of the expert developers selecting the suitable pattern process consider to be critical phase especially for novice developers which have to be provided with mechanism to help them find a suitable pattern to a particular solution. This paper introduces a design pattern selection architecture (DPS) based on a Multi-Agent System (MAS) that aim to obtain the appropriate recommendation to reduce development efforts, facilitate and assist the developers in selecting the suitable patterns for their problems.

1 Introduction

According to [1] design patterns are a generic proven solution to recurring common design problems. It is consider as a powerful tool for a developer and important concept in software engineering, due to its ability to represent a highly effective way to improve the quality of software systems. Based on [2] [3] design patterns benefits can be summarized as following:

- Increase the flexibility of maintenance and reusability of software design.
- Contribution in extensibility of software through reducing the problem of architectural decay.
- Capture the design knowledge based on real experience of software design.
- Document the best practices in solving many different kinds of problems.
- A communication tool that improves the communication among software developers and provide a mechanism to share workable solutions between developers and organizations.

This work continues the proposed research in [4]; which is used three possible scenarios aiming to decide which pattern to select, based on four steps. A developer starts with the problem, which is formulated as a problem description; the problem description is used as a query. After retrieving patterns, the next step is to choose among these patterns, this process can be repeated several times, until the desired state (i.e. find the accepted pattern that solve the problem). Then, generate

S. Omatu et al. (Eds.): *Distrib. Computing & Artificial Intelligence*, AISC 217, pp. 243–250.
DOI: 10.1007/978-3-319-00551-5_30

recommendations that help to applied and use the patterns. The final step is the system evaluation according to the user feedback in hand to enhance and refinement the system.

In the next section, the selection pattern problem is described. Section 3 mentions related work. In section 4, introduces our approach. Section 5, explain conclusions and future work.

2 Problem Statements

One of the main challenges in the researches is how to choose suitable patterns that can solve a particular problem, and how to apply the selected pattern during system design.

Actually expert developers who have a deep knowledge of patterns can select a fitting pattern to specific design problems [5]. Conversely, for the novice developers it is quite difficult to select the suitable pattern for a given design problem, as a result this has led to significantly reduce the use of patterns due to the following reasons:

- Novice developers don't have tangible definition or a clear understanding of the problem domain [6].
- Novice developers not common to the design pattern and do not have enough knowledgeable about it for decide whether reuse patterns or develop a special-purpose solution [5].

Obviously, the lack of knowledge about patterns and knowing only very few of them, can confronted us and affect our ability choosing which pattern to apply or choosing patterns that is not exactly fitting to the design problem within a program, the situation can be worse when applying a pattern wrongly or applying the wrong pattern for a problem.

In fact, recent active researches in field of software engineering stated that determining the suitable design patterns to a specific problem is a considerable challenge facing the software designers. However, the GoF book [1]; authors explains that it not easy selecting the pattern which addresses a particular design problem among 23 different design patterns especially with progressively increasing number of patterns due to the several researches at conferences and workshops, and many published books of the new patterns and online repositories, selecting the suitable design pattern become a critical issue.

Table 1 Shows sample of published patterns

Publication	Year	Number of Pattern
Almanac book [7]	2000	1200 Software Patterns
Henninger & Corrêa Survey [8]	1994 – 2007	2241 Software Patterns
SOA book[2]	2008	60 Patterns
Bunke el. Survey [9]	1997 - mid 2012	415 Security Patterns
Neil book [10]	2012	70 Mobile Design Pattern

Consequently, the growing number of discovered patterns that exceed our ability to understand and analysis the exciting design patterns call for our need of necessary tools to assist in this extremely important process. Recently, the number of documented patterns has steadily increased through the new publications, consequently the need for automated and supporting processing tool of design patterns become more important as well as supporting the learning process for the understanding of concrete patterns. However, to gain benefits from design patterns, it is necessary to solve the previously mentioned problems by supporting the developers working with them. The main problem that motivated this paper is the need for effectively supporting approach-based software development by using multi-agent system (MAS) technology .MAS architectures could be seen as a social organization of independent software (agent) that flexible, autonomous, learning, reasoning and corporate with the environments. In addition, characteristics of agent structure can assist to decide which action is appropriate and enable to recommend the suitable suggestion patterns for a particular design problem.

3 Related Work

This section describes existing approaches that try to support resolving the problems discussed previously. There are many different approaches used to addressing the pattern selection problem. Table 2, shows some approaches in the literature [11- 20].

Table 2 Illustrates the different approaches of exist for design pattern selections

Approach	Summary of Purposed	Technique
ESSDP [11]	Represent a methodology for constructing expert system which can support design patterns to solve a designer's design problem through Used dialog with the designer to narrow down the choices(question-answer session), that implements the All GoF pattern.	- Expert system
ESSSDP [12]	Develop of a prototype expert system for the selection of design patterns that are used in object-oriented software for providing an automated solution regarding the design pattern application problem, and implements only five GoF patterns.	- Expert system. -Object-oriented environment
RMDP[13]	Developed a information retrieving model of two major parts, the analysis of all GoF pattern document to create search index and the calculation of index weight, this model used to retrieving correct design patterns.	- Information Retrieval (IR)
SRSDP[14]	Introduce a simple recommender system to help user in choosing among the 23 design patterns from the GoF. This approach is based on extract important words by analyzing the textual GoF patterns.	- Recommender System
IC-Pattern[15]	Developed a multi-agent system for recommending patterns, which takes into account the social part of the problem, providing users with suggestions from other community members about patterns that were supports conventional IR and Case-Base Reason methods for finding design pattern suitable for the given problem.	- Recommender System - MAS - IR - CBR

Table 2 (*continued*)

RDP[16]	Proposed an initial model to solve the problem by using Case-Based Reasoning (CBR) and Formal Concept Analysis (FCA). They use FCA as the way of inductive for extracting embedded knowledge in case base and allows the flexibility to maintain index.	- CBR - FCA
DPRA [17]	Presented an interactive tool to assists the designers choosing their suitable design patterns; through gives a bitty to draw a design fragment, present the problem, re-phrases the problem in order to obtain the intention of a certain pattern via use of WordNet which lexical dictionaries. Then, it explores the candidate solutions by filtering patterns that meet the intentions through the use of recommendation rules.	- Recommender System - Draw a design fragment
RSVEISIG[18]	Described a recommendation tool embedded in a visual environment for pattern-based design which aims at suggesting patterns to help novice designers to produce better designs and understand the language.	- Recommender System - collaborative filtering
DPR[19]	Proposes a DPR prototype for suggesting design patterns, based on a simple Goal-Question-Metric (GQM) approach, and knowledge-base (KB) for pattern details and relative information. They presented a sample interactive session with the designer. DPR prototype was inspired from a previous work [11].	- Recommender System - Expert system

In the previous approach, we noticed that each has its properties and features. The goal of our idea is integration of these features and combines them together in new approach which it contribute the improvement of the selection process.

4 The Proposed DPS Approach

4.1 The Architecture Overview

In this section, we propose the architecture of our DPS based on MAS technology. As shown in Fig.1, it has Three-Layered services can be divided into the following distinct logical layers:

- **Interface Layer**. This layer contains the user functionalities responsible for an interaction between the user and the system. It consist two components which are Login User and Description Problem.
- **Application Services Layer**. This layer it consists of eight components used to implement the core functionality of the system.
- **Infrastructure Layer.** This layer provides access to data. It consists of different of resources such as databases, repository, and etc.

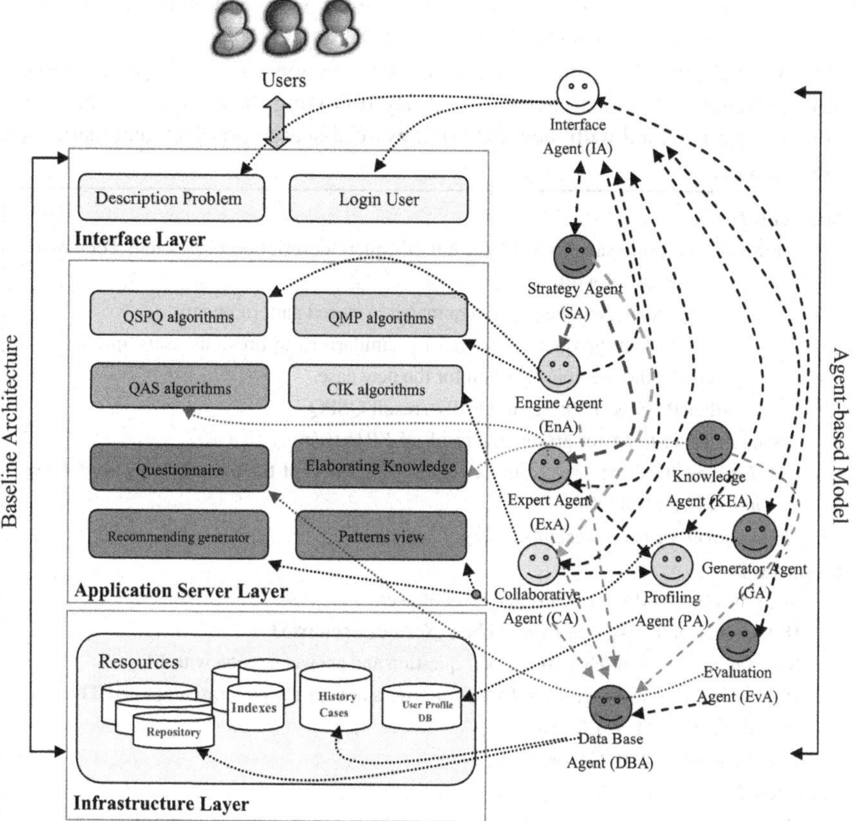

Fig. 1 DPS Conceptual Overview based on MAS.

From the previous Figure, ten agents have been used to achieve functionalities efficiently. Each agent has responsibility of performing function as follows:

- **Interface Agent (IA):** as an intermediary between the user and other Agents.
- **Profiling Agent (PA):** is responsible for the determine level of user (See in [4, table 1]) and updating an active user profile.
- **Engine Agent (EnA):** It Applies two algorithms parallel *QMP*, *QSPQ* (See [4, Figure 3, 5]), filter process by intersect result two algorithms, send this result to *IA*. If *IA* informs *EnA* Accept then inform *SA* for accept and inform *DBA* for storage. Else inform SA for failure.
- **Expert Agent (ExA):** It Applies *QAS* algorithms which performing the question and answer session with *IA*. If result *IA* is Accept then inform *SA* for accept and inform *DBA* for storage. Else invoke *CA* to begin.

- **Collaborative Agent (CA):** It uses *CIK* algorithm in [4, Figure 8], (i.e. This work similar as; Yahoo! Answers[1], stack overflow[2]).
- **Strategy Agent (SA):** Incorporates situation action rules to switch between three Scenarios as illustrated in Fig. 2. By following these rules will be avoid problems associated with: new pattern, keyword-search problem, and learn uses the patterns.

Scenarios 1:

IF ***SA*** received message from ***IA*** for a user's query description of problem **Then** *Send* message to ***EnA*** for work.

- *QMP*: Analyzed query to find a match in stored pattern intents,
- *QSPQ*: used the query to search for similarities to previous users' queries and reusing the previous solution for the new case.
- Filter Process (*FP*): result QMP $\cap$ result QSPQ

Send message to ***IA*** for present the result of *FP* to user.

IF ***EnA*** received message from ***IA*** *inform* user's **accept** the patterns suggested **Then** *inform* ***GA***, ***DBA***, ***PA***

Else *inform* ***SA*** is not **accept** the Apply Scenarios2

Scenarios 2:

Send message to **PA** ***for*** determine level of user.

IF the received message $\in$ *Expert* **Then** executes **Scenario3**

Else *Send* message to ***ExA*** for work // question and answer session with ***IA***.

IF ***ExA*** received message from ***IA*** *inform* user's **accept** the patterns suggested **Then** *inform* ***GA***, ***DBA***, ***PA*** for work.

Else *Send* message to ***CA*** for work.

Scenarios 3:

CA Send message to **PA** for find $\bigcap_{k=1..n} E(D, P_i)$ // for more details see in [4]

Send message to ***IA*** to ***inform*** user *Suggest* $\bigcap_{k=1..n} E(D, P_i)$

IF ***CA*** received message from ***IA*** *inform* user's **accept** the patterns suggested **Then** *inform* ***GA***, ***DBA***, ***PA*** for work.

Fig. 2 Strategy Agent action rules

- **Generator Agent (GA):** After the user fined the appropriate pattern, SA sends massage to *GA* for generate three recommendations to the user.
  - Recommending pattern sequences: Showing the recommendations consist of sequences of patterns usually apply together that depending on relationship between patterns (i.e. patterns language).
  - Recommending apply patterns: provide recommendations how to implement the pattern (s) (i.e. by giving guidance how to implement pattern).

---

[1] http://answers.yahoo.com/

[2] http://stackoverflow.com/

  - Recommending practical: provide a list of users who have used this pattern(s) before (i.e. Suggests users who have previous experience in the application of the same pattern which proposed).

- **Data Base Agent (DBA):** It's responsible for the access to databases, data documentation, and informs the *KEA* about update the database.
- **Knowledge Agent (KEA):** Is an autonomous agent which works to update the knowledge base (for more detail see in [4]). It's depends at two equation as following:

$$R(D,P_i) \notin KB \rightarrow \text{Build}[\bigcap_{k=1..n} \text{E}(D,P_i)]: i \in D \tag{1}$$

$$R(D,P_i) \in KB \rightarrow \text{Evaluation}[\bigcap_{k=1..n} \text{E}(D,P)]: i \in D \tag{2}$$

- **Evaluation Agent (EvA):** Evaluate the effectiveness of the system, from the user feedback every time you using the system (See in [4, table 2].

5 Conclusion and Future Work

The selection a suitable design pattern is a worthwhile topic, important to the design community and it is an area that needs addressing design pattern selection problem and improving this process. In this paper, we presents DPS approach based MAS technology for generate advice for using design patterns and assists developers to learning a design patterns.

As future work we would like to find a solution for searching patterns problems [20] via the following points:

- Documentation techniques to:
  - Summaries patterns descriptions in a standard form to allow easy comparing of the patterns and allow the search using patterns forms.
  - According to researchers conducted in [21]; we thing of applying a patterns classification and clustering techniques to enhance the indexing, searching and selection of the patterns.
- Aggregation algorithms to collect patterns from different sources (books, proceedings, journals, etc), to single patterns repositories.

References

1. Gamma, E., Helm, R., Johnson, R., Vlissides, J.: Design Patterns: Elements of Reusable Software. Addison-Wesley, Reading (1994)
2. Erl, T.: SOA Design Patterns. Prentice Hall, New York (2008)
3. Knox, J.: Adopting Software Design Patterns in an IT Organization: An Enterprise Approach to Add Operational Efficiencies and Strategic Benefits. M.S. thesis, AIM program, Dept. of Computer and Information Science, University of Oregon (Spring 2011)

4. Saleh, E.M., Sallabi, O.: Design Pattern Selection: A Solution Strategy Method. In: 2012 IEEE International Conference on Computer Systems and Industrial Informatics (ICCSII 2012), Sharjah, UAE (2012)
5. Sommerville, I.: Software Engineering. Addison-Wesley, Boston (2004)
6. Kim, D.-K., El Khawand, C.: An approach to precisely specifying the problem domain of design patterns. Journal of Visual Languages and Computing 18(6), 560–591 (2007)
7. Rising, L.: The Pattern Almanac. Addison-Wesley, Boston (2000)
8. Henninger, S., Corrêa, V.: Software pattern communities: current practices and challenges. In: PLOP 2007: Proceedings of the 14th Conference on Pattern Languages of Programs. ACM, NY (2007)
9. Bunke, M., Koschke, R., Sohr, K.: Organizing Security Patterns Related to Security and Pattern Recognition Requirements. International Journal on Advances in Security 5, 46–67 (2012)
10. Neil, T.: Mobile Design Pattern Gallery UI Patterns for Mobile Applications. O'Reilly Media (March 2012)
11. Kung, D.C., Bhambhani, H., Shah, R., Pancholi, G.: An expert system for suggesting design patterns: a methodology and a prototype. In: Software Engineering with Computational Intelligence (2003)
12. Moynihan, G.P., Suki, A., Fonseca, D.J.: An expert system for the selection of software design patterns. Expert System Journal 23(1) (2006)
13. Sarun, I., Weenawadee, M.: Retrieving Model for Design Patterns. ECTI Transactions on Computer and Information Technology, 51–55 (2007)
14. Guéhéneuc, Y., Mustapha, R.: A simple Recommender System for Design patterns. In: The 1st EuroPLoP Focus Group on Pattern Repositories (2007)
15. Birukou, A., Weiss, M.: Patterns 2.0: a Service for Searching Patterns. In: EuroPLoP 2009, Irsee Monastery, Germany (2009)
16. Sarun, I., Weenawadee, M.: Retrieving Design Patterns by Case-Based Reasoning and Formal Concept Analysis. In: ICCSIT International Conference on Computer Science and Information Technology, pp. 424–428 (2009)
17. Nadia, B., Kouas, A., Ben-Abdallah, H.: A design pattern recommendation approach. In: CORD Conference Proceedings, pp. 590–593 (2011)
18. Díaz, P., Malizia, A., Navarro, I., Aedo, I.: Using Recommendations to Help Novices to Reuse Design Knowledge. In: IS-EUD, pp. 331–336 (2011), doi:10.1007/978-3-642-21530-8_35
19. Palma, F., Farzin, H., Guéhéneuc, Y.G., Moha, N.: Recommendation System for Design Patterns in Software Development: An DPR Overview. In: 2012 Third International Workshop on Recommendation Systems for Software Engineering (RSSE) (2012)
20. Birukou, A.: A Survey of Existing Approaches for Pattern Search and Selection. In: Proceedings of the 15th European Conference on Pattern Languages of Programs (EuroPLoP 2010), Irsee Monastery, Germany (2010)
21. Hasheminejad, S.M.H., Jalili, S.: Design patterns selection: An automatic two-phase method. Journal of Systems and Software 85(2), 408–424 (2012)

Representing Motion Patterns with the Qualitative Rectilinear Projection Calculus

Francisco Jose Gonzalez Cabrera, Jose Vicente Álvarez-Bravo, and Fernando Díaz

Dpto. de Informática, Universidad de Valladolid, Escuela Universitaria de Informática, Campus María Zambrano, Pza. Alto de los Leones, 1, 40005, Segovia, Spain
{fjgonzalez,jvalvarez,fdiaz}@infor.uva.es

Abstract. The Qualitative Rectilinear Projection Calculus (QRPC) is a novel representation model for describing qualitatively motion patterns of two objects through the possible relationships among the rectilinear projection of their trajectories. The paper introduces the key issues of the model (i) the set of geometric relations defined in terms of the front-back and left-right dichotomies, (ii) how it can be possible enumerate an exhaustive set of qualitative states by the composition of these relations and (iii) the possible transitions among states based on the notion of conceptual neighborhood. The representational ability of the model is illustrated by an example extracted from the traffic engineering field where the relative motion of two objects is analyzed and described in terms of the QRPC-states.

Keywords: Qualitative representation model, Motion pattern, Qualitative Spatial Reasoning, Cognitive Modeling.

1 Introduction

If anyone would have to explain how objects are moving around, sentences such as: "the object is currently to my left moving toward the right" or "at this instant of time it is crossing just in front of me, moving toward the right" may be used. All these kind of descriptions contain spatiotemporal information in terms of spatial and temporal statements and can be used for distinguishing more complex behaviors and making reasoning processes.

The qualitative formalism has been proved to be suitable since it can be used in any context, even when only partial or imprecise information or knowledge is available [1, 2]. Thus, this modeling approach has been applied in robotics, in spatial reasoning, in spatial scene modeling for ambient intelligence environments, in Robocop or in a traffic engineering context.

Several authors have worked previously on different models of qualitative representation. An exhaustive revision of these works is available in [3]. Only a few of these models deals with a complete representation model for overcoming the

S. Omatu et al. (Eds.): *Distrib. Computing & Artificial Intelligence*, AISC 217, pp. 251–258.
DOI: 10.1007/978-3-319-00551-5_31

difficulties of describing and reasoning properly about motion in any context and, furthermore, these ones do not consider some kinematical behaviors such as rotational motions. Specifically, the QTC (*Qualitative Trajectory Calculus*) [4] and Gottfried's model [5] are based on the relationship between the trajectories of two moving point objects and these qualitative relationships are described in terms of two dichotomies: the front-back and left-right dichotomies. OPRA (*Oriented Point Relation Algebra*) is based on the notion of geometry-based orientation [6] and the geometric alternatives derived from the relative position of two dipoles in the plane [7] which allow to the authors to introduce a powerful algebra and the notion of "conceptual neighborhood" [8–10].

A new intuitive approach in this regard is the QRPC (*Qualitative Rectilinear Projection Calculus*) has been recently defined in [3]. This novel qualitative model is based on the relationships between the trajectories of two objects in two dimensions and it provides a richer description of motion characterizing the trajectories of the objects in terms of an oriented rectilinear projection than the aforementioned models. Thanks to this feature, the QRPC model is able to describe motions that have not been considered previously such as, for instance, the rotation of an object in motion with respect to itself.

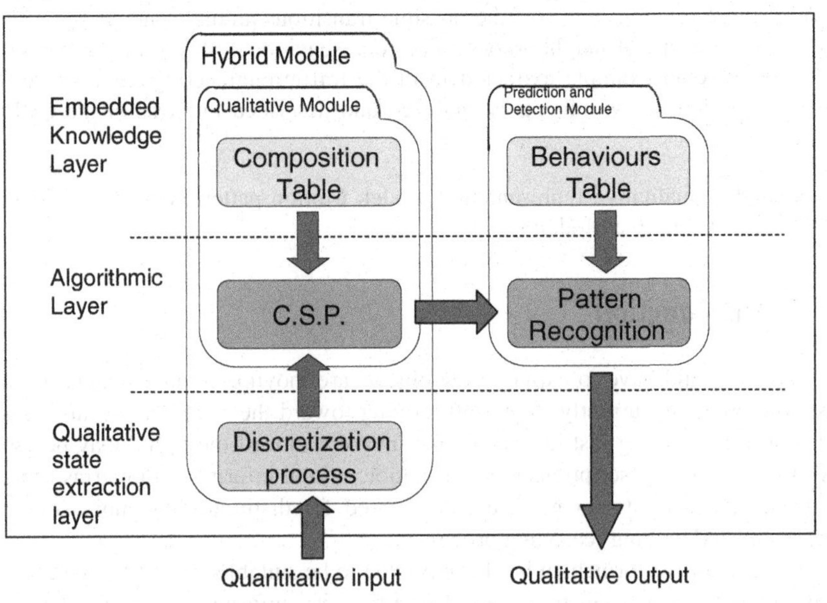

Fig. 1 An overview of the Qualitative-based Detection and Prediction System

In this context, it is assumed that the proposed qualitative model of representation would be embedded in the Detection and Prediction System which governs the autonomous behavior (motion) of a hypothetical intelligent agent (e.g. a robot) which is assumed to be the reference object. As it is sketched in Figure 1, the Qualitative-based Detection and Prediction System comprises several functional modules whereas the proposed abstract model of representation is the formal basis of the *embedded knowledge layer* across the whole system. Briefly, the *qualitative module* consists of the constraint satisfaction algorithm and a composition table. The *qualitative module* receives as input a collection of measurements from different types of sensors of the intelligent agent to determine directly the QRPC states with regard to other objects which are present in his environment. Once the overall state of the agent with regard to the other agents is known, this information will be fed to the *prediction and detection module* to compute the qualitative output of the whole Detection and Prediction System. At this point we need to introduce the notion of inference. In this context, an object in motion A can know (through its sensors) its relative position with respect to other object in motion B (the QRPC state of A with respect to B). Although object A can not determine its state with respect to a third object C (since C is outside of the range of its sensors), object B can communicate to A, its relative position with respect to C (the QRPC state of B with respect to C). Then, assuming that the state of A with respect to B and the state of B with respect to C are known by object A, it can be possible to infer (by object A) the qualitative state of A with respect to C. This inference task will be the main functionality of the *qualitative module* and it will be supported by the *composition table*.

Then, Section 2 includes a concise description of the key issues of the QRPC model, and Section 3 presents a real example (described in terms of the qualitative states defined by the QRPC model) in order to illustrate its representational ability. The paper concludes with a summary and the further work of current research.

2 The Qualitative Representation Model

Basically, the QRPC model defines four spatiotemporal features within a system (initially considering only two objects in motion) in terms of the possible relationships among the rectilinear projections of their trajectories. In this context:

- The objects are depicted as oriented points presenting two qualitative spatial dichotomies with respect to its object-face direction: the front-back and left-right dichotomies [6] (see Figure 2a). These two distinctions allow us to consider a richer description for a single object in motion.
- The trajectory of an object is a spatiotemporal entity characterized, at a given instant of time, through the oriented rectilinear projection containing the tangential component of the velocity vector.
- Under this approach, the object-face direction always coincides with the tangential component of the velocity vector. If this tangential component is zero then the object trajectory is characterized through the object-face direction.
- Then, the motion between two objects is described qualitatively as a sequence of spatiotemporal classes. Each one depicts a primitive kinematical behavior

expressed in terms of a set of features derived from the relationships between the two trajectories (see Figure 2b).

- A generalization of the proposed model can be achieved when more than two objects are considered and they are related to each others by mean of the inference task above introduced.

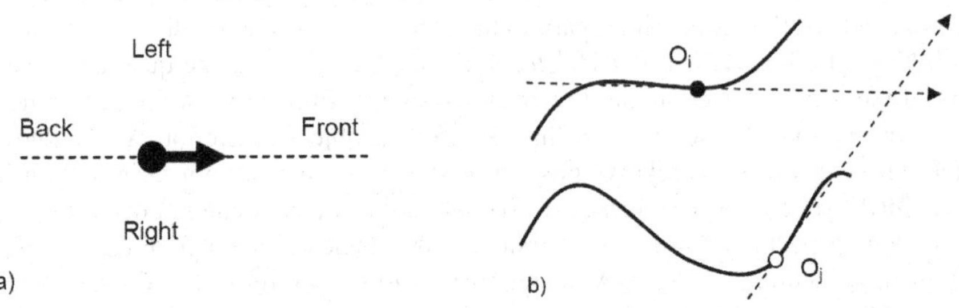

Fig. 2 a) An oriented object and its two qualitative spatial dichotomies. **b)** Relationship between the trajectories at a given instant of time.

2.1 *The Qualitative Features*

These set of features represent the relative positions among objects in terms of some relevant spatial concepts: the oriented rectilinear projection (*P*) defined previously, the intersection point of the two oriented rectilinear projections (*C*) and the objects (*O*). These last ones are also described, as it was mentioned, through its object-face direction (*D*), so any relationship involving two objects is expressed in terms of its front/back (FB) and left/right (LR) dichotomies. Then the following notation is introduced, whose formal definition can be found in [3]:

- $(\boldsymbol{P_iP_j}) \in \Sigma_P = \{X, \uparrow\uparrow, \uparrow, \uparrow\downarrow, \updownarrow\}$. The relative disposition between the oriented rectilinear projection of the O_i and the oriented rectilinear projection of O_j and the semantic of each symbol corresponds to their intuitive graphical meaning.
- $(\boldsymbol{O_iO_j^{LR}}) \in \Sigma_D = \{+, 0, -\}$. The relative position of O_i with respect to the left-right dichotomy of O_j., which are represented by the symbols '+' (on the right), '–' (on the left), and '0' (on the same trajectory).
- $(\boldsymbol{CO_i^{FB}})\ (\boldsymbol{CO_j^{FB}}) \in \Sigma_D \times \Sigma_D$. The relative position of the *C* with respect to the front-back dichotomy of both objects, where the symbol '+' represents "*C* is in front of the object", the symbol '–' represents "*C* is behind the object" and the symbol '0' represents "the object is on the intersection of trajectories".
- $(\boldsymbol{O_iO_j^{FB}}) \in \Sigma_D$. The relative position of O_i with respect to the front-back dichotomy of O_j when the trajectories are superimposed.

Under this notation, the subscripts *i* and *j* are used to denote two different objects and the object O_i will be represented graphically by a hollow dot whereas object O_j will be represented by a solid dot.

2.2 The Qualitative Relationships (Primitives)

Under the proposed approach, a primitive relationship, also known as an atomic pattern, can be represented by an n-tuple composed by the ordered set of symbols associated to each one of the qualitative features defined in Section 2.1. An atomic pattern describes qualitatively one instant (or state) of the motion of the (target) object O_i with respect to the (reference) object O_j. Initially, the lexicographical representation of an atomic pattern α, $R_{ij}(\alpha)$,can be defined as:

$$R_{ij}(\alpha)=\begin{cases}\left[(P_iP_j)\,(O_iO_j^{LR})(D_iO_j^{LR})\,(CO_i^{FB})\,(CO_j^{FB})\,(O_iO_j^{FB})\right] & \text{if } (P_iP_j)=X \wedge (CO_i^{FB}=0 \vee CO_j^{FB}=0)\\ \left[(P_iP_j)\,(O_iO_j^{LR})\,(CO_i^{FB})\,(CO_j^{FB})\,(O_iO_j^{FB})\right] & \text{otherwise}\end{cases} \tag{1}$$

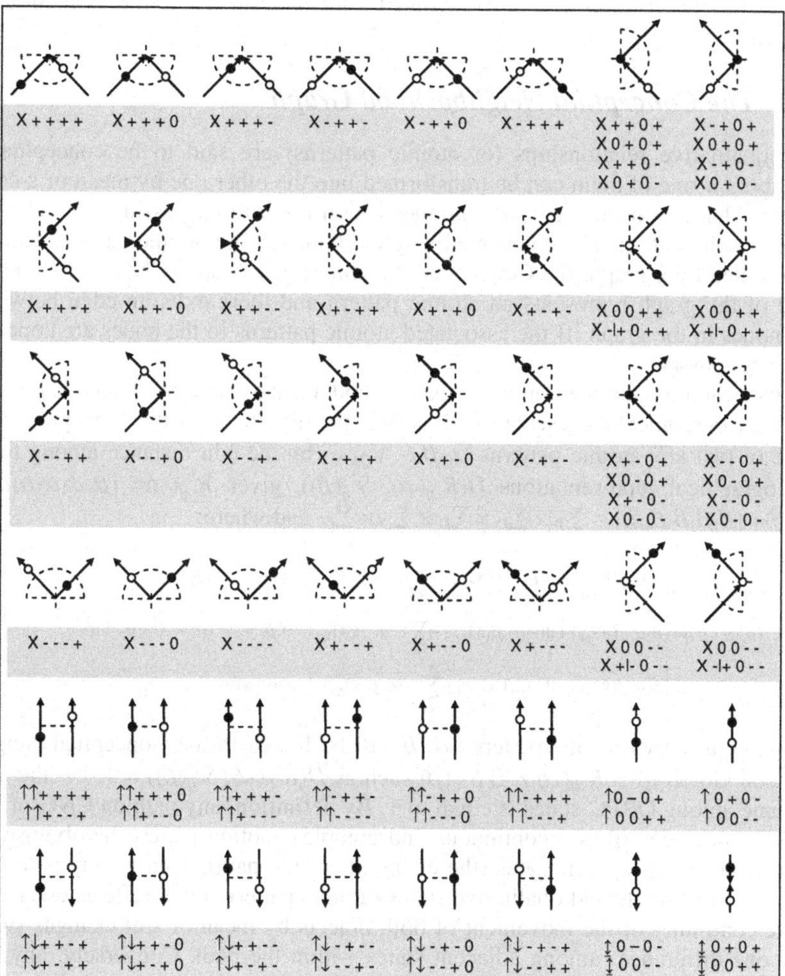

Fig. 3 The 48 atomic patterns of the QRPC model and their corresponding iconic and compact lexicographical representations

The original representations considers an additional qualitative feature referred to as the directionality of the target object, $(D_iO_j^{LR})$, but after considering and introducing intended ambivalent representations (see [3] for details) the compact notation of an atomic pattern α, $R^*_{ij}(\alpha) \in \Sigma_P \times \Sigma_D \times \Sigma_D \times \Sigma_D \times \Sigma_D$ can be defined as:

$$R^*_{ij}(\alpha)=\left[(P_iP_j)\,(O_iO_j^{LR})\,(CO_i^{FB})\,(CO_j^{FB})\,(O_iO_j^{FB})\right] \quad (2)$$

Since $R^*_{ij}(\alpha)$ is a string of symbols, it will be referred to as the compact lexicographical representation of the atomic pattern, where the semantics of each symbol is defined by the corresponding qualitative feature. Figure 3 presents the complete set of atomic patterns of the QRPC model, hereafter V_{QRPC}, by enumerating the lexicographical representation of each pattern. In order to improve the comprehension of this description, it is also included a meaningful iconic representation of each pattern.

2.3 The Conceptual Neighborhood Graph

Two qualitative relationships (or atomic patterns) are said to be conceptually neighbors if one of them can be transformed into the other one by mean of a continuous change (in our case, a continuous motion) without passing through any other atomic pattern. The conceptual neighborhood among atomic patterns can be represented by a graph, the Conceptual Neighborhood Graph (CNG), where each node of the graph represents an atomic pattern and there exits an edge between two nodes in the graph iff the associated atomic patterns to the nodes are conceptually neighbors.

Since each atomic pattern of the QRPC model has associated at least one compact lexicographical representation then, it is possible to asses the conceptual distance of two any atomic patterns $\alpha, \beta \in V_{QRPC}$, by the edit distance among their lexicographical representations $D(R^*_{ij}(\alpha), S^*_{ij}(\beta))$, given $R^*_{ij}(\alpha)= [\alpha_1\alpha_2\alpha_3\alpha_4\alpha_5]$, $S^*_{ij}(\beta)= [\beta_1\beta_2\beta_3\beta_4\beta_5] \in \Sigma_P \times \Sigma_D \times \Sigma_D \times \Sigma_D \times \Sigma_D$, and where:

$$D(R^*_{ij}(\alpha),S^*_{ij}(\beta))=\begin{cases} d(\alpha_1,\beta_1|G_{\Sigma P})+d(\alpha_2,\neg\beta_2|G_{\Sigma D})+\sum_{k=3}^{5} d(\alpha_k,\beta_k|G_{\Sigma D}) & \text{if } \alpha_3='\text{-}'\wedge(\beta_1='X'\wedge\beta_3='0') \\ d(\alpha_1,\beta_1|G_{\Sigma P})+d(\neg\alpha_2,\beta_2|G_{\Sigma D})+\sum_{k=3}^{5} d(\alpha_k,\beta_k|G_{\Sigma D}) & \text{if } \beta_3='\text{-}'\wedge(\alpha_1='X'\wedge\alpha_3='0') \\ d(\alpha_1,\beta_1|G_{\Sigma P})+d(\alpha_2,\beta_2|G_{\Sigma D})+\sum_{k=3}^{5} d(\alpha_k,\beta_k|G_{\Sigma D}) & \text{otherwise} \end{cases} \quad (3)$$

Formally, any two atomic patterns α, β will be linked in the Conceptual Neighborhood Graph iff $\exists\, R^*_{ij}(\alpha) \wedge \exists\, S^*_{ij}(\beta)$ such as $D(R^*_{ij}(\alpha), S^*_{ij}(\beta)) = 1$, i.e. the edit distance among QRPC-states is equal to 1. By definition, any path on CNG of the QRPC model describes a continuous and complex motion pattern involving two objects among two given states (the endpoints of the path). That is, any path is a sequence of discrete and qualitative states (atomic patterns) that reflects essentially the continuity of the movement of both objects by mean of soft changes (with only one difference) among adjacent states within the path. Conversely, any sequence of QRPC-states where the edit distance among any two consecutive states is one, then necessarily is a valid path in the CNG. Since each atomic pattern in the CNG has a lexicographical representation, the CNG can be also interpreted as a collection of syntactic rules for discussing and determining the correctness of any proposed motion pattern.

3 An Example of the Representational Ability of the QRPC Model

Once the representation model has been concisely described, it is just the moment to show the practical relevance of this approach by analyzing and describing the motion between objects in motion in a specific context. The aim is to show how the model is able to distinguish between similar behaviors and how it is possible to assign a semantic meaning for each described situation.

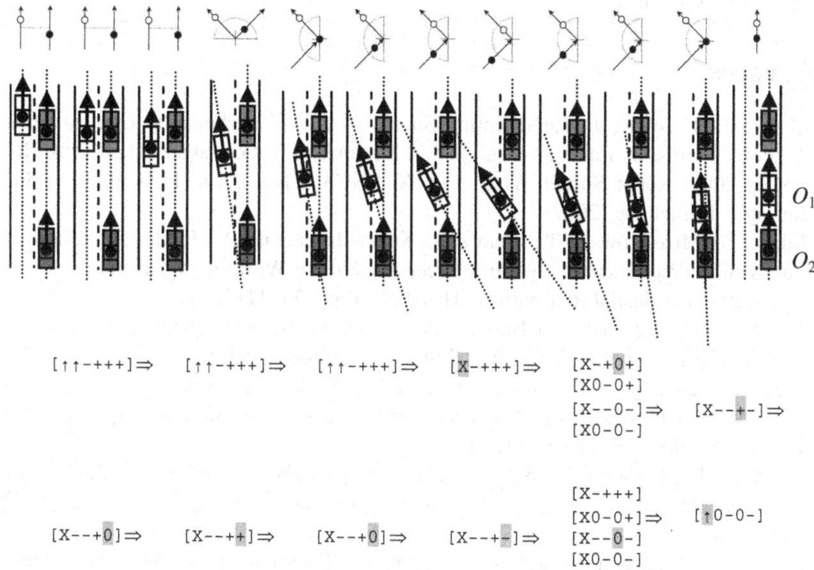

Fig. 4 The parking maneuver

From the analysis of this well-known maneuver and according to the CNG associated to the QRPC model, a sequence of twelve atomic patterns is derived from the relative motion of object O_1 wrt O_2. Figure 4 shows the sequence of motion pattern in terms of their graphical representations (and of their lexicographical representation) which is associated to the parking maneuver of the vehicle. As it is can be seem from the sequence, any two states differs at most in only one symbol in their lexicographical representation (highlighted in gray in the sequence). Therefore, by the definition of the CNG, the associated sequence of QRPC states to the parking maneuver is also a valid path in the CNG from the first state to the last one. Moreover, the states with ambivalent representations allow us to alter the reasoning chain by choosing a different representation from the original one which was matched with the previous state in the sequence. The ambivalent states (when a different representation is selected) can be viewed as landmarks of a subsequence with a proper meaning: from the fifth to the eleventh state is properly the subsequence which corresponds with the parking maneuver itself.

4 Conclusions

In this work a new qualitative representation model for describing motion patterns is introduced. The QRPC model, which is based on the relationship between the two trajectories associated to the objects in motion, is able to catch the main features of any physical system involving motion. Furthermore, it provides a rich description that can be verbalized in terms of a natural language as a mean of analyzing and reasoning about motion patterns, as it is illustrated with the given example. Further work will be centered on the definition of the inference process introduced in Section 1.

References

1. Gottfried, B., Witte, J.: Representing Spatial Activities by Spatially Contextualised Motion Patterns. In: Lakemeyer, G., Sklar, E., Sorrenti, D.G., Takahashi, T. (eds.) RoboCup 2006: Robot Soccer World Cup X. LNCS (LNAI), vol. 4434, pp. 330–337. Springer, Heidelberg (2007)
2. Lücke, D., Mossakowski, T., Moratz, R.: Streets to the OPRA - Finding your destination with imprecise knowledge. In: Proceedings of the Workshop on Benchmarks and Applications of Spatial Reasoning at IJCAI 2011, pp. 25–32 (2011)
3. Glez-Cabrera, F.J., Álvarez-Bravo, J.V., Díaz, F.: A new qualitative model for representing motion patterns. Expert Syst. Appl. (in press, 2013)
4. Van de Weghe, N., Cohn, A.G., Maeyer, P.D., Witlox, F.: Representing moving objects in computer-based expert systems: the overtake event example. Expert Systems with Applications 29, 977–983 (2005)
5. Gottfried, B.: Reasoning about intervals in two dimensions. In: Proceedings of the IEEE International Conference on Systems, Man and Cybernetics, pp. 5324–5332. IEEE (2004)
6. Moratz, R.: Representing Relative Direction as a Binary Relation of Oriented Points. In: Brewka, G., Coradeschi, S., Perini, A., Traverso, P. (eds.) Proceedings of the 2006 Conference on ECAI 2006: 17th European Conference on Artificial Intelligence, pp. 407–411. IOS Press (2006)
7. Dylla, F., Moratz, R.: Exploiting qualitative spatial neighborhoods in the situation calculus. In: Freksa, C., Knauff, M., Krieg-Brückner, B., Nebel, B., Barkowsky, T. (eds.) Spatial Cognition IV. LNCS (LNAI), vol. 3343, pp. 304–322. Springer, Heidelberg (2005)
8. Dylla, F., Wallgrün, J.O.: Qualitative Spatial Reasoning with Conceptual Neighborhoods for Agent Control. Journal of Intelligent and Robotic Systems 48, 55–78 (2007)
9. Dylla, F., Wallgrün, J.O.: On Generalizing Orientation Information in OPRAm. In: Freksa, C., Kohlhase, M., Schill, K. (eds.) KI 2006. LNCS (LNAI), vol. 4314, pp. 274–288. Springer, Heidelberg (2007)
10. Frommberger, L., Lee, J.H., Wallgrün, J.O., Dylla, F.: Composition in OPRAm - Technical Report No. 013-02/2007 (2007)

Event Management Proposal for Distribution Data Service Standard

José-Luis Poza-Luján, Juan-Luis Posadas-Yagüe, and José-Enrique Simó-Ten

University Institute of Control Systems and Industrial Computing (ai2). Universitat Politècnica de València (UPV). Camino de vera, s/n. 46022 Valencia, Spain
{jopolu,jposadas,jsimo}@ai2.upv.es

Abstract. This paper presents a proposal to extend the event management subsystem of the Distribution Data Service standard (DDS). The proposal allows user to optimize the use of DDS in networked control systems (NCS). DDS offers a simple event management system based on message filtering. The aim of the proposal is to improve the event management with three main elements: Events, Conditions and Actions. Actions are the new element proposed. Actions perform basic operations in the middleware, discharging the process load of control elements. The proposal is fully compatible with the standard and can be easily added to an existing system. Proposal has been tested in a distributed mobile robot navigation system with interesting results.

1 Introduction

Currently, the Event-Based Control (EBC) paradigm technology (also called event-driven control) is adopted to implement systems where the periodic sampling is not possible (i.e. when no discrete-time model is available) or recommended (i.e.: in distributed control systems to improve communications performance decreasing the load of the network) [1].

In networked control systems (NCS), distributed control elements are connected by a network. A middleware enhances and offers to control elements a set of services in order to facilitate the access to the network. Therefore, if EBC is used in NCS, the middleware will be an essential component [2].

Middleware architecture is based on communications paradigms: message passing, client-server, Publish/Subscribe (P/S) or blackboard. The Data Distribution Service for Real-Time Systems (DDS) is an Object Management Group (OMG) standard based on the P/S paradigm. DDS offers time-controlled communications between components using Quality of Service (QoS) policies [3].

NCS needs certain event management, i.e.: to select only messages with a meaningful value to trigger the control action. DDS can be used to send any kind of Data, including event data. DDS allows the application (control component in NCS) to perform flexible filtering of events but DDS does not define a built-in event type and advanced event management. To provide support to NCS using

S. Omatu et al. (Eds.): *Distrib. Computing & Artificial Intelligence*, AISC 217, pp. 259–266.
DOI: 10.1007/978-3-319-00551-5_32

DDS we are developed an extension of the DDS event components. This article presents the specifications of this extension. The proposal is fully compatible with the DDS model and maintains the OMG philosophy of simplicity and ease of use.

The paper is organized as following. Next section outlines the DDS elements involved on event management. Section 3 presents the proposal elements to add to the current DDS model. Section 4 presents an example of the use of the proposed elements, Section 5 the results of a simple experiment. Finally, evaluation and future work are discussed in conclusions.

2 Data Distribution Service Event Management

Data Distribution Service (DDS) provides a platform independent model that is aimed to real-time distributed systems. DDS is based on publish-subscribe communications paradigm. Publish-subscribe components connect information producers (publishers) and consumers (subscribers) and isolate publishers and subscribers in time, space and message flow [4].

When an application (producer) wants to publish some information, it should write it in a "Topic" by means of a component called "Data Writer" which is managed by another component called "Publisher". Both components, Data Writer and Publisher, are included in another component called "Domain Participant". On the other side of the communication, a Topic can be received by two kinds of components: "Data Readers" and "Listeners" by means a "Subscriber". A Data Reader provides the messages to application when the application requests them, A "Listened" sends the messages without waiting for the application request (Figure 1).

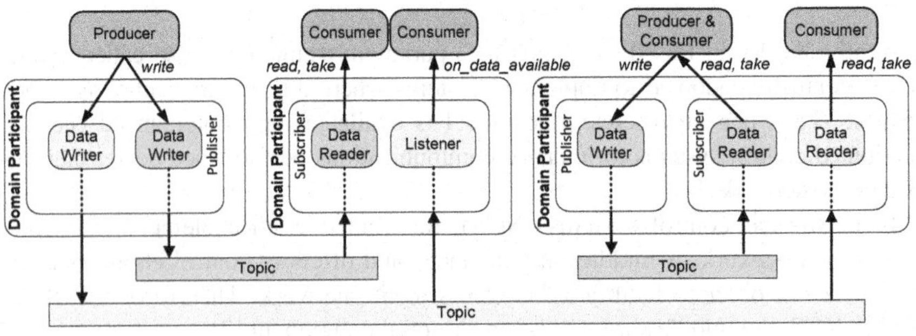

Fig. 1 DDS elements and functions

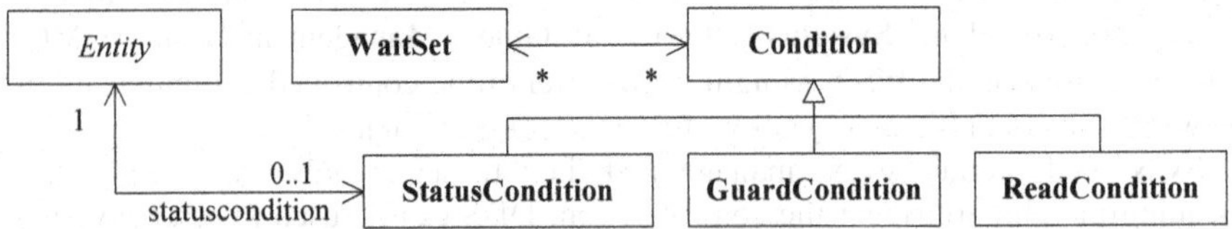

Fig. 2 UML class diagram of the DDS elements involved in event management: conditions and WaitSet

To configure communications, DDS uses QoS policies. A QoS policy describes the service behaviour according to a set of parameters defined by the system characteristics or by the administrator user. An Entity is the base class for communication components: Publisher, Subscriber, Data Writer, Data Reader and Listener. Details about the protocol can be obtained from [5]. Figure 2 shows the main elements of DDS involved on the event management.

Entities apply for relevant information by creating one of the types of Condition objects (StatusCondition, GuardCondition or ReadCondition) and attaching it to a WaitSet object. So that, a WaitSet object allows an Entity to wait until one of the attached Condition objects has a “triggervalue” of TRUE or else until the timeout expires.

3 Event Management Proposal

The event management used in DDS standard allows to perform a wide set of operations, such as filtering messages to be transmitted from the middleware to the application. The DDS QoS policies allow the application to change the characteristics of the communications between nodes, such as the message frequency or the deadline. When EBC is applied in a NCS, each control component sends messages based on the internal events (i.e.: when a new control action is calculated) but in distributed systems it is necessary that control nodes coordinates messages between them. In NCS, coordination requirements are based on the internal characteristics of control algorithms. DDS offers adequate support to coordinate communications between nodes, but it does not provide a mechanism to change the internal characteristics. To extend the power of DDS, a new component, called Action, is added.

Figure 3 shows the event management proposed. Main components are “Events”, “Conditions” and “Actions”. Events are situations that are necessary to be monitored. The Event component is similar to the Condition DDS element. Conditions are group of events using logical operations; the Condition component increases the WaitSet ability. Finally, Actions are the component that implements the effects on the system that are associated with Conditions. Events are categorized on three types: component operations, quality alarms and message filters.

A component operation happens when a middleware or control component method is called; for example, when a control component starts the control action or when a middleware element is disconnected from a communications channel. Alarms are events associated with the compliance of the quality parameters. In our proposal we include the QoS and Quality of Control (QoC) parameters. QoC parameters [6] are associated to the efficiency of the control action and are directly related with the QoS parameters [7]. Thus, both parameters are considered. An example of QoS alarm is when a message arrives after the deadline. QoC alarm is related to the content of the message, for example when the reference value exceeds the reference value. Finally, message filters events are triggered when the content of the message is (or not) identical compared to a content pattern; for example, when the message field “source” (user defined) corresponds to a particular control node.

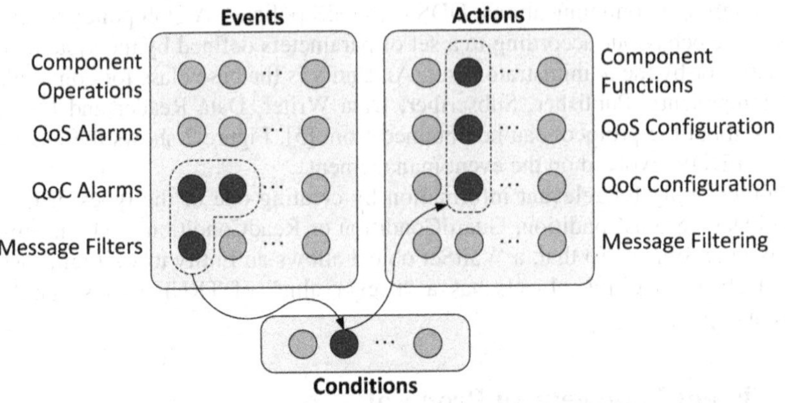

Fig. 3 Conceptual model of event management with the source of events, conditions and actions

Conditions are associations of Events by using basic logic operations (AND, OR). For example, if a message from the node one (message filter) arrives in time (QoS alarm) and the internal value doesn't exceed the reference value (QoC alarm). The Condition is associated with a DDS element; so that, when the condition occurs, the element knows the existence. The method to trigger the alarm and to filter the message is the same used on the Remote Network Monitoring protocol (RMON) [8] due to its simplicity and efficiency.

When an element is triggered by a Condition, the element can run some function to process the trigger. In our proposal we add a new element called Action. Actions are associated to a Condition and are processed internally in the middleware. Initially four types of actions have been considered: component functions, QoS and QoC configuration and message queue actions. Component functions are the same functions that can produce events. For example, an action can be disconnecting a middleware element.

The QoS and QoC configurations are the variation of the parameters. For example, in order to increase the temporal limit of messages (QoS configuration) or the control error (QoC configuration).

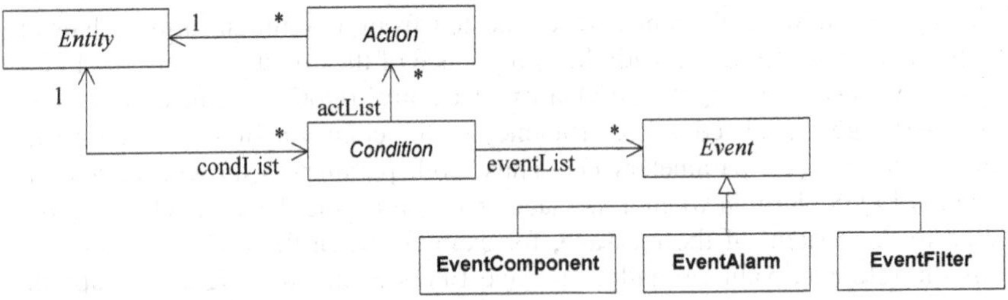

Fig. 4 UML class diagram of the proposal elements

Finally, message queue actions are functions that change the behaviour of the queue, such as priority message, or message removal. So that, Actions discharge elements by processing of simple operations. Figure 4 shows the UML class diagram of the proposed elements. In our proposal Condition is similar to WaitSet DDS element, Event is similar to Condition and Action is the new element.

4 Implementation in a Distributed Control Architecture

A distributed mobile robot navigation environment has been used to test the validity and usefulness of the proposed model. The environment has been used in previous studies [9]. The mobile robot is controlled by a set of control algorithms. Some algorithms, generally algorithms that implement the reactive behaviours, are embedded within the robot and other algorithms, commonly the deliberative algorithms, are implemented in distributed nodes. However, in the case of study, all control algorithms are placed in distributed nodes in order to use a reduced dataset. The robot used in the study consisted of a simple two wheels and one distance eight sensors ring. The sensor distribution (figure 5) corresponds to a Khepera robot [10], thus developed algorithms can be compared with existing algorithms. The algorithm used is the "obstacle avoidance" based on Braitenberg vehicles behaviours [11].

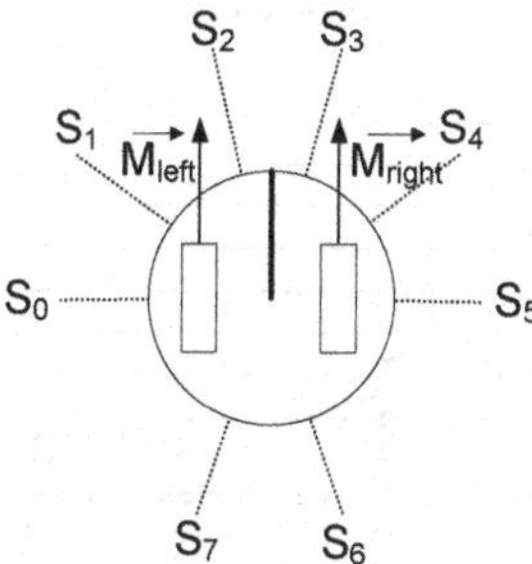

Fig. 5 The sensor distribution model used

The obstacle avoidance algorithm is based on the speed variation of the motor depending on the distance detected by all sensors. For each motor, each sensor has a weighting value depending on its position on the robot. The motor speed is obtained combining the sensor weights by means the equation 1 where *m* is the concrete motor (left or right), *K* is the concrete weight factor applied to sensor *i* and motor *m*.

$$MotorSpeed_m = \sum_{i=0}^{7} K_i \cdot S_i \quad (1)$$

The maximum linear speed of the robot depends on the distance to the nearest obstacle and the sampling period to update motor velocities. The linear speed can be obtained from the equation 2, where *Sd* is the obstacle distance detected on the current robot path and T is the sampling period.

$$SpeedLimit = Sd/T \tag{2}$$

Figure 6, shows the results of the equation 2 to obtain the speed limit based on the sampling period and different distances. The communication channel defines the maximum sampling period, for example: a sampling period of 10 milliseconds needs a bandwidth to transfer at least 100 messages by second. The frequency of messages sent can be changed through the DDS QoS policies. Therefore Actions objects can increase or decrease this value automatically without the intervention of control components only with the distance value obtained from the sensors messages. The error in the control of obstacle avoidance behaviour is based on maximizing the distances to obstacles. Therefore, when the robot navigates far an obstacle, the robot does not need a high sampling rate. So, it can decrease the communications load without losing QoC.

This experiment is based on previous experiments performed in mobile robot navigation; these experiments are detailed in [12]. In previous works [9], QoC has been used to optimize the path regardless of the type of obstacle. The experiment presented below does have in mind the type of obstacle.

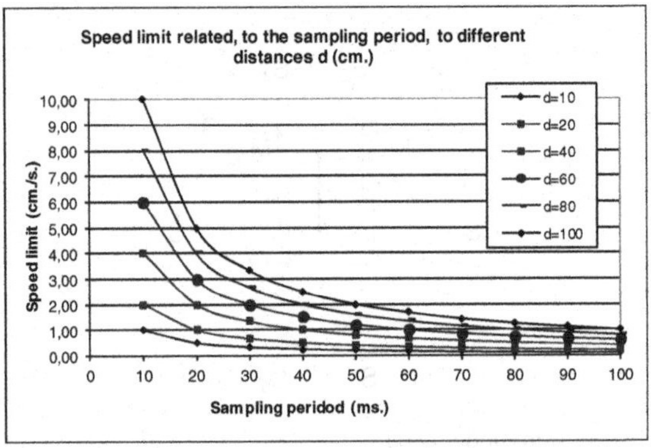

Fig. 6 Speed Limit obtained from the distance d and the sample period T

5 Experiment and Results

Two scenarios have been tested: robot navigation without event detection in the middleware and the same navigation with event detection and one Action linked by means a Condition. The Event detected in the middleware is the distance obtained from the robot by means a filter. The Condition set involves comparing the distance with the robot speed to obtain the optimal sampling period based on the equation 2. When middleware doesn't detect events, the sampling period is set to 10 milliseconds, and motor speed is obtained only with the equation 1. The experiment measures the sampling period and distance to an obstacle along time.

In the case studied the sampling period changes internally in the middleware without the control component intervention. Figure 7 shows the graphs obtained for two obstacles and table 1 shows the summarised results.

Table 1 Experimental results for the two obstacles tested (average values)

Scenarios	Sampling period (1)	Distance average (1)	Sampling period (2)	Distance average (2)
a. Robot in corridor	10,0	2,1	62,6	2,1
b. Wall in front of the robot	10,0	1,1	9,4	0,6

When the robot navigates through a corridor (scenario a) the average distance is exactly the same for both middleware. This is because the corridor navigation doesn't involve avoiding an obstacle and the robot can run at maximum speed. However, the absence of obstacles allows navigating with the same efficiency but with fewer messages. Besides, robot reaches the target in less time. In this scenario the proposal does not improve the robot navigation, but reduces the network load. Scenario b is more complex, robot needs avoid an obstacle. As a result, the sampling period is similar. Nevertheless, the robot has more accuracy performing the manoeuvre.

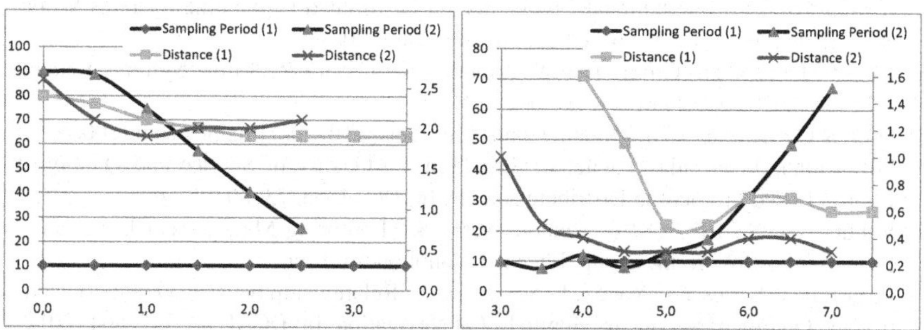

Fig. 7 Sampling period and distance variations along time robot navigation, without the proposal (1) and using the proposal (2) for two different obstacles

6 Conclusions and Future Work

This article has presented a proposal to increase the event management system proposed in the DDS standard. The most significant contribution is the inclusion of a new object called Action. Actions automatically make changes on the middleware based on a combination of events.

The Action object has been tested with a simple mobile robot system. The test is based on the automatic variation of the QoS settings in function of the distance measured without control component intervention. The number of messages sent is reduced, and, as a result, the communications load is also reduced.

Future work is to test the middleware with complex combinations of events that generate different actions. The problem of generating different actions is the possibility to obtain contradictory actions, i.e.: increase and decrease the deadline QoS Policy. This problem can be solved by using priorities in Conditions or in Actions, but probably the inclusion of Actions in the middleware might be limited.

Acknowledgments. The study described in this paper is a part of the coordinated project COBAMI: Mission-based Hierarchical Control. Education and Science Department, Spanish Government. CICYT: MICINN: DPI2011-28507-C02-01/02.

References

1. Sánchez, J., Guarnes, M.Á., Dormido, S.: On the Application of Different Event-Based Sampling Strategies to the Control of a Simple Industrial Process. Sensors 9, 6795–6818 (2009)
2. Sandee, J.H., Heemels, W.P.M.H., van den Bosch, P.P.J.: Case Studies in Event-Driven Control. In: Bemporad, A., Bicchi, A., Buttazzo, G. (eds.) HSCC 2007. LNCS, vol. 4416, pp. 762–765. Springer, Heidelberg (2007)
3. Hadim, S., Nader, M.: Middleware Challenges and Approaches for Wireless Sensor Networks. IEEE Distributed Systems Online 7(3) (2006)
4. Pardo-Castellote, G.: OMG Data-Distribution Service: architectural overview. In: Proceedings of 23rd International Conference on Distributed Computing Systems Workshops, Providence, USA, vol. 19-22, pp. 200–206 (2003)
5. Object Management Group. Data Distribution Service for Real-time Systems Version 1.2 (2007), http://www.omg.org/
6. Dorf, R.C., Bishop, R.H.: Modern Control Systems, 11th edn. Prentice Hall (2008)
7. Poza-Luján, J., Posadas-Yagüe, J., Simó-Ten, J.: Quality of Service and Quality of Control Based Protocol to Distribute Agents. In: DCAI, pp. 73–80 (2010)
8. Waldbusser, S.: RFC 2819 - Remote Network Monitoring Management Information Base. Network Working Group. Lucent Technologies (2000)
9. Poza-Luján, J., Posadas-Yagüe, J., Simó-Ten, J.: Relationship between Quality of Control and Quality of Service in Mobile Robot Navigation. In: DCAI, pp. 557–564 (2012)
10. K-Team Corporation. Khepera III robot, http://www.k-team.com
11. Braitenberg, V.: Vehicles: Experiments on Synthetic Psychology. MIT Press, Cambridge (1984)
12. Poza-Luján, J.: Propuesta de arquitectura distribuida de control inteligente basada en políticas de calidad de servicio. Universitat Politècnica de València Press (2012)

Trajectory Optimization under Changing Conditions through Evolutionary Approach and Black-Box Models with Refining

Karel Macek[1,2], Jiří Rojíček[1], and Vladimír Bičík[1]

[1] Honeywell Laboratories, V Parku 2326/18, Praha 4, 148 00, Czech Republic
{karel.macek,jiri.rojicek,vladimir.bicik}@honeywell.com
[2] Institute of Information Theory and Automatization, Academy of Scienced of the Czech Republic, Pod Vodárenskou věží 4, Praha 8, 182 08, Czech Republic

Abstract. This article provides an algorithm that is dedicated to repeated trajectory optimization with a fixed horizon and addresses processes that are difficult to describe by the established laws of physics. Typically, soft-computing methods are used in such cases, i.e. black-box modeling and evolutionary optimization. Both suffer from high dimensions that make the problems complex or even computationally infeasible. We propose a way how to start from very simple problems and - after the simple problems are covered sufficiently - proceed to more complex ones. We provide also a case study related to the dynamic optimization of the HVAC (heating, ventilation, and air conditioning) systems.

Keywords: Empirical function minimization, black-box modeling, simplification, refining, dynamic building control.

1 Introduction

Real world application scheduling and planning suffers from the lack of knowledge of the system structure as well as from the well known curses of dimensionality [6, p. 3-6]. There are already some approaches addressing these issues. For the modeling part, black-box models can be mentioned: e.g. neural networks [7], local regression [1], and others. All those black-box approaches to modeling reach their limits whenever the number of parameters is high in contrast to the quantity and quality of available data.

Another curse of dimensionality is related to complexity of the optimization problems. In case of black-box modeling, it is not possible to assure that the derived optimization problem will be linear, convex, or without local optima. For such cases, wide class of evolutionary techniques has arisen, including differential evolution [8], covariance matrix adaptation [2], and others. There are also numerous attempts to optimize those optimization techniques, e.g. [10] or [5]. In case of dynamic problems, techniques of approximate dynamic programming are being improved continuously [6]. Even though those methods and approaches lead to improved results both in testing suites and practical applications, the essential issue remains: the higher dimensions, the more difficult problem.

S. Omatu et al. (Eds.): *Distrib. Computing & Artificial Intelligence,* AISC 217, pp. 267–274.
DOI: 10.1007/978-3-319-00551-5_33

This paper provides no new method of black-box modeling or evolutionary optimization. We offer a way how to make complex problems computationally feasible while the user might decide about the simplification. After the system is able to solve simpler problems, the input space might be step-by-step refined, possibly to the original problem.

The paper is organized as follows: We start in Sect. 2 with a very simple case study that motivates the methodology. In Sect. 3, we introduce formal notation that is used in Sect. 4 for the problem formulation. Consequently, the basic approach is provided in Sect. 5 and extended in Sect. 6. Then, the case study in Sect. 7 is used for numerical illustration of the approach. The work is concluded in Sect. 8 where also further work is sketched out.

2 Case Study

We introduce the case study before the formal notation in order to improve the readability of the abstract definitions. Let us consider a building control system with indoor zone temperature T_{za} which stands for the temperature in the room. We will assume a hot season and the upper zone temperature T_{zau}. The control is considered for the whole day, i.e. 24 time instants. Thus we write $T_{za,t}$ and $T_{zau,t}$, where $t = 1,2\dots,24$. The comfort that has to be assured is defined as $T_{za,t} \leq T_{zau,t}$. Whenever $T_{za,t} > T_{zau,t}$, chillers are started in order to cool the zone down so $T_{za,t} = T_{zau,t}$. There is related power consumption and cost to this operation. Typically, the $T_{zau,t}$ is defined as high as allowed, i.e. $T_{zau,t} = T_{\max,t}$. However the pre-cooling, i.e. choosing $T_{zau,t} < T_{\max,t}$ might be beneficial due to more appropriate ambient profiles of dynamic power prices and this fact motivates the dynamic optimization of $T_{zau,t}$.

The goal of this case study is to determine $(T_{zau,t})_{t=1}^{24}$ that will minimize the overall costs of the building operation while keeping $T_{zau,t}$ in bounds, i.e.:

$$T_{\min,t} \leq T_{zau,t} \leq T_{\max,t} \tag{1}$$

Note that the case study omits important facts which are considered in other publications, such as internal heat gains, or indoor thermal capacity [3], since we are striving to provide an approach that is able to optimize the system with limited knowledge only.

3 Notation

In this section, formal notation is introduced and demonstrated in the relationship to the above discussed case study.

Definition 1 (Trajectory). Let $n_x \in \mathbb{N}$ and $n_t \in \mathbb{N}$. The matrices $X \subset \mathbb{R}^{n_t,n_x}$ are called trajectories, their row indices are called times and column indices inputs. The elements of the trajectories will be denoted as $x_{t,i}$.

In our case study, we have $n_x = 1$, $n_t = 24$, $x_t \equiv T_{zau,t}$, and X is given by (1).

Definition 2 (Conditions). Let (Ω, Σ, P) be a probability space called conditions where Ω is a sample space containing all possible outcomes, Σ is set of events where event is subset of Ω, and $P : \Sigma \to [0,1]$ assigns probabilities to all events.

The conditions in the considered case study can be both internal (occupancy) and external (weather) factors that influence the operation of the system during next 24 hours.

Definition 3 (Cost function). Let $c : X \times \Omega \to \mathbb{R}$ be a mapping called cost function.

The costs in the case study are the costs that the building owner will pay to the utility company. These costs are given by the chiller input power and actual power price.

Definition 4 (Evidence). Let e be a random vector of size n_e over (Ω, Σ, P), such that exists, be called evidence. Let set $E \subset \mathbb{R}^{n_e}$ satisfying $P(E) = 1$ and $\forall \bar{E} \subset \mathbb{R}^{n_e}$ hold $E \subset \bar{E}$.

Obviously, E is set of all considerable evidences, other have zero probability. The evidence provides information about the future conditions.

The evidence in the case study can be embodied as a sequence of weather forecast $\hat{T}_{oa,t}$. The more precise weather forecast, the more valuable the evidence will be.

Definition 5 (Optimal Trajectory). If a trajectory $x^* \in X$ satisfies for a given evidence e the following condition:

$$x^* = \arg\min_{x \in X} \mathscr{E}\left[c|e,x\right] \tag{2}$$

we call it optimal for the evidence e.

Definition 6 (Data). Let us define data as a sequence of triples $D = (\tilde{x}^{(j)}, \tilde{e}^{(j)}, \tilde{c}^{(j)})_{j=1}^{n_j}$ where matrix $\tilde{x}_j$ is a trajectory used for x_j j-th experiment, vector $\tilde{e}_j$ is the evidence available for the j-th experiment, and scalar $\tilde{c}_j$ is the realized cost for the given trajectory and evidence.

The examples of the optimal trajectory and data for given case study are straightforward: the goal is to set-up the indoor temperature profile minimizing the expected costs. The data records involve the profile used, corresponding weather forecast, and related costs.

4 Problem Formulation

The challenge addressed in this work has two aspects. The first one is that the model of $c|e,x$ is not known exactly (as well as its structure) and we have only limited data set D available. Next issue is related to the fact that n_t and n_x are relatively large[1]. The first issue makes it difficult to formulate the mathematical problem (2), while the second complicates the search of the optimal solution.

To resolve these issues, we introduce two more concepts:

[1] The case study can be extended to optimization of 10 zones with 15 minutes sampling, i.e. $n_t = 24 \cdot 4 = 96$ and $n_x = 10$. Empirical optimization of trajectories with 960 parameters is numerically infeasible.

Definition 7 (Data-Centric Model). Let $c|e,x,D$ be for each e,x,D a random variable. We denote it as data-centric model.

Note that this definition involves wide class of models, including those that have been mentioned in the introduction.

Definition 8 (Simplification). Let $s = (s_x, s_e)$ be a pair of mappings where $s_x : V_x \to X$ where $V_x \subset \mathbb{R}^{n_{sx}}$. Next, $s_e : E \times \mathcal{D} \to V_e$, where $V_e \subset \mathbb{R}^{n_{se}}$ and $\mathcal{D}$ set of all possible data sets. We call s simplification and the set of all considerable simplifications S. A trajectory $x \in X$ satisfies given simplification $s \in S$ iff $\exists v \in V_x : s_x(v) = x$.

The concept of simplification is intended to reduce the dimension of both black-box modeling and consequent reconstruction of relevant trajectory. Thus, typically $n_{sx} + n_{se} \ll n_e + n_t \cdot n_x$.

In the case study, the simplification is considered as

$$s_x(v_x) = \begin{cases} v_x \text{ for } & t = 1,2,\dots 8 \\ 24 \text{ for } & t = 9,10,\dots 22 \\ 30 \text{ otherwise} \end{cases} \tag{3}$$

where $15 \leq v_x \leq 30$. The s_e could return e.g. the average temperature of the past day (depends on data D) and average value of the forecast for the next day (depends on evidence e). For the visualization purposes we will adopt the average temperature of the past day only.

The quality of the simplification impacts the quality of suboptimal data-centric optimal simplified trajectory, defined as follows, in terms of realized cost:

Definition 9 (Data-Centric Optimal Simplified Trajectory). Let $s \in S$ be a simplification. Let D be a given data set where all $\tilde{x}_j$ satisfy the simplification s. Let $c|e,x,D$ be a data-centric model for given simplification and given data set. The trajectory $x^* \in X$ satisfying

$$x^* = s_x(\arg\min_{v \in V_x} \mathcal{E}\left[c|s_e(e,D), v, D_s\right]) \tag{4}$$

where we assume all records in D contain a trajectory satisfying the simplification s and D_s stands for data D transformed by the simplification s_e, s_x^{-1}, is called data-centric optimal simplified trajectory or DCOS trajectory.

DCOS trajectories are the objective of the proposed method. In the case study, we will focus on the optimization of the $T_{zau,t}$ for $t < 9$, i.e. the upper bound between midnight and time when first occupants arrive.

5 Basic Algorithm

The basic approach can be summarized in the following steps:

1. Define the simplification $s \in S$.
2. Define the data set as empty, i.e. $D = \emptyset$.

3. Set $j = 1$
4. Generate a random trajectory $\tilde{x}^{(1)}$ satisfying s, store the related evidence $\tilde{e}^{(1)}$.
5. Apply the trajectory $\tilde{x}^{(j)}$ to the system and observe the costs $\tilde{c}^{(j)}$.
6. Extend the data set D by new $(\tilde{x}^{(j)}, \tilde{e}^{(j)}, \tilde{c}^{(j)})$.
7. Obtain and store new evidence $\tilde{e}^{(j+1)}$
8. Optimize the new DCOS trajectory $\tilde{x}^{(j+1)}$ for given s, D, and $\tilde{e}^{(j+1)}$. This involves:
   a) Simplification of the data using s_x and s_e to D_s.
   b) Identification of the model $c|s_e(\tilde{e}^{(j+1)}, D), v, D_s$
   c) Solution of $v^* = \arg\min_{v \in V_x} \mathscr{E}\left[c|s_e(\tilde{e}^{(j+1)}, D), v, D_s\right]$.
   d) Putting back from the simplified world, i.e. $\tilde{x}^{(j+1)} = s_x(v^*)$
9. Set $j = j + 1$ and go to 5.

Note that during the whole optimization the data are limited to trajectories $\tilde{x}^{(j)}$ that satisfy the given simplification s.

6 Refining and More Complex Trajectories

We propose to extend the basic algorithm by application of refining to the used simplification. Let us define the concept first:

Definition 10 (Refining). Let $s, \tilde{s} \in S$ be two simplifications. Iff $\forall x \in X$ where s is satisfied also $\tilde{s}$ is satisfied, and there is a surjective mapping $\gamma : \tilde{V}_e \rightarrow V_e$, then $\tilde{s}$ is a refining of s.

The refining might lead to broader set $\tilde{V}_x$ and might lead to better results approaching closer to the minimizer of (2). In the algorithm, the refining can be applied between step 8 and 9. First, it has to be tested whether the refining may be applied. Consequently, the refining has to be chosen.

An example of the simplification in the case study would be $\tilde{s} = (\tilde{s}_x, \tilde{s}_e)$ where $\tilde{s}_e \equiv s_e$ and $\tilde{V}_x \equiv [15, 30]^2$ and

$$\tilde{s}_x(\tilde{v}_x) = \begin{cases} \tilde{v}_{x,1} \text{ for } & t = 1, 2, \ldots 4 \\ \tilde{v}_{x,2} \text{ for } & t = 5, 6, \ldots 8 \\ 24 \text{ for } & t = 9, 10, \ldots 22 \\ 30 \text{ otherwise} \end{cases} \tag{5}$$

In an analogical way, this refining can be refined again to dimension 4 and again to dimension 8.

Before refining the actual simplification, it has to be tested whether the procedure achieves for the given simplification the best possible results. For this purpose we propose the application of the following test. However, other tests can be considered, too. Let $\rho \in \mathbb{R}_+$ and $n_b \in \mathbb{N}$ be two parameters. Let b_j be defined as follows:

$$b_j = \begin{cases} 0 & \text{if} \quad j = 1 \\ b_{j-1} + 1 & \text{if} \quad \|\tilde{x}_j - \hat{x}_j\| < \rho \\ 0 & \text{otherwise} \end{cases} \tag{6}$$

where $\hat{x}_j$ is the DCOS found for $\tilde{e}^{(j)}$, s, and data $D\backslash(\tilde{x}^{(j-1)}, \tilde{e}^{(j-1)}, \tilde{c}^{(j-1)})$. The refining is carried out after $b_j > n_b$. The value of n_b will be discussed in numerical examples in Sect. 7.

The refining itself is related either to the trajectories X or evidences E. In both cases the refining consists in adding more information into the problem definition. In case of trajectories, the simplification restricts the trajectories to X_1 and the refining uses $X_2 \supset X_1$. In case of evidences, the refining consists in involving additional information. In both cases the refining leads to an increase of the dimension of the black-box model for cost c.

Of course, some refining might lead to no improvements, therefore some testing can be carried out and the simplification might be rejected and another can be tried. However, deeper discussion on this topic is out of the scope of this paper.

7 Numerical Example

The case study has been introduced step-by-step in the previous sections. In order to demonstrate the results numerically, we provide basic information about the set-up since detailed description is beyond the scope of this work. At the end of this section, achieved results are discussed.

First, a first-principle simulation model has been adopted for a single zone building with one chiller. We adopted a version of [9] where the thermal capacity of the zone is influenced by ambient air temperature, heat gains from the occupants (people in the zone), internal thermal inertia given by gains and the chiller itself. The parameters have been determined based on experience of modeling of large-scale single zone buildings. The weather profile have been used from a real building, slightly shifted so it had values between 20 and 35°C.

The adopted surrogate black-box model was based on local polynomial regression with Gaussian kernel with covariance matrix $C_{i,i} = 1$ and $C_{i,j} = 0, i \neq j$ and degree 2 (quadratic regression). More details can be found in [11]. The considered model had single output, namely the costs c and several regressors: first was related to the data D while the others represented v_x. The first regressor was the first PCA [4] component from data containing: (i) weather profile from the last day and (ii) internal temperatures (zone air, building construction). As a procedure for optimization of v_x, the covariance matrix adaptation [2] was selected and the limit of iterations was 200 for each optimization.

We worked with 30 days of data and carried out 2 experiments with different settings as illustrated on Fig.1. The first experiment with the setting $n_b = 0$ and $\rho = 1$ leads to fast refining and the problem has comparable complexity as if the 8-dimensional problem would be addressed directly. The other experiment with $n_b = 5$ leads to slower refining. It can be observed that better results, i.e. lower costs, are obtained when the dimensionality grows slowly with increasing information in the data.

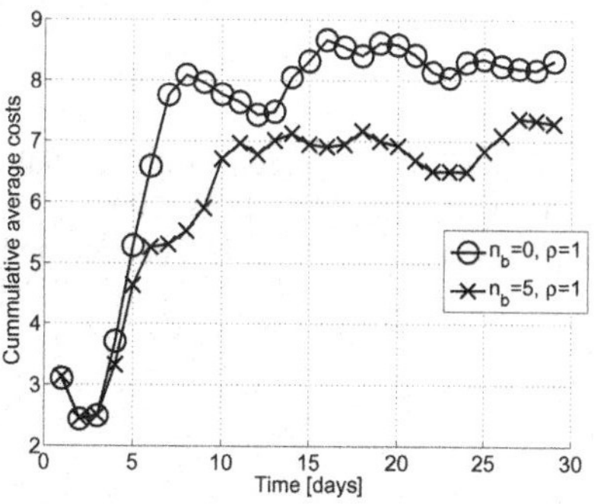

Fig. 1. Comparison of fast and slow refining

8 Conclusions and Further Work

In this paper, we have addressed the trajectory optimization problem for cases when the prior knowledge about the problem is limited. We offered an approach based on simplification in both actions (using a set of simplifications) and states (using first PCA factors). The theoretical approach has been illustrated in a case study related to optimal cooling in a building. It has been demonstrated that slow refining leads to better results than addressing the high-dimensional problem directly.

The promising results challenge further research. From the theoretical point of view, more precise statistical tests for next refining shall be established. Next, the applicability of alternative black-box models and optimization algorithms shall be evaluated. Then, more extensive tests shall provide more evidence about the benefits of the proposed methodology, especially on comparison to model predictive control. Finally, the approach might be applied also in another domains, such as inventory management.

References

1. Cleveland, W.: Robust locally weighted regression and smoothing scatterplots. J. Am. Stat. Assoc. 74(368), 829–836 (1979)
2. Hansen, N., Ostermeier, A.: Adapting arbitrary normal mutation distributions in evolution strategies: The covariance matrix adaptation. In: Proceedings of IEEE International Conference on Evolutionary Computation, pp. 312–317. IEEE (1996)
3. Henze, G., Kalz, D., Liu, S., Felsman, C.: Experimental analysis of model based predictive optimal control for active and passive building thermal storage inventory. HVAC&R Res. 11(1), 183–213 (2004)
4. Jolliffe, I.: Principal Component Analysis. Springer, New York (1986)
5. Macek, K., Boštík, J., Kukal, J.: Reinforcement learning in global optimization heuristics. In: Mendel, 16th International Conference on Soft Computing, pp. 22–28 (2010)
6. Powell, W.B.: Approximate Dynamic Programming: Solving the Curses of Dimensionality. John Wiley & Sons, Hoboken (2007)
7. Rojas, R.: Neural Networks - A Systematic Introduction. Springer, Berlin (1996)
8. Storn, R., Price, K.: Differential evolution - a simple and efficient heuristic for global optimization. J. Global Optim. 11, 341–359 (1997)

9. Střelec, M., Macek, K., Abate, A.: Modeling and simulation of a microgrid as a stochastic hybrid system. In: Proceedings of the IEEE PES Innovative Smart Grid Technologies, ISGT 2012 (2012)
10. Tvrdík, J.: Adaptation in differential evolution: A numerical comparison. Appl. Soft Comput. 9(3), 1149–1155 (2009)
11. Wasserman, L.: All of Nonparametric Statistics. Springer, New York (2006)

Modelling Agents' Risk Perception

Nuno Trindade Magessi and Luis Antunes

GUESS/ LabMag/ Universidade de Lisboa, Portugal
nmaggessi@hotmail.com, xarax@di.fc.ul.pt

Abstract. One of the open issues in risk literature is the difference between risk perception and effective risk, especially when the risk is clearly defined and measured. Until now, the main focus has been given on the behaviour of individuals and the evidences of their biases according to some stimulus. Consequently, it is important to analyse what are the main reasons for those biases and identify the dimensions and mechanisms involved. To that purpose, we tackle the classic problem of tax fraud as a case study. In this paper, we will look into how agent based modelling methodology can help unfold the reasons why individuals commit errors of judgment when risk is involved.

1 Introduction

Risk and its perception have been on the centre of much research in recent decades, because of the impact it has on the decision-making process. Individuals make decisions based on their own perception on the subject to be decided. One of its issues is the gap between risk perception and effective risk. From psychology, we can identify four reasons why some risks are perceived to be more or less serious than they actually are [1, 2]. First, individuals overreact to intentional actions, and underreact to accidents, abstract events, and natural phenomena. Second, individuals overreact to things that offend their moral. Third, individuals also overreact to immediate threats and underreact to long-term threats. Fourth, they also underreact to changes that occur slowly and over time [1]. Instead, individuals exaggerate largely on rare risks and downplay common risks. They exhibit trouble on estimating risks for subjects distinct from their normal situation. Personified risks are perceived to be greater than anonymous risks. People also underestimate risks they willingly take and overestimate risks in situations they cannot control. And finally, they overestimate risks that are being talked about and remain an object of public scrutiny [2]. Those aspects launch the important questions: (1) what are the dimensions and factors involved on these biases? (2) How can artificial intelligence contribute to the understanding of this issue?

The article is organised as follows. Next section describes the evident differences between perceived risk and effective risk. On section 3, we discuss the non-existence of integrative theory of risk perception and the urgency for one. Section 4 introduces the economic classic problem of tax fraud as case study where the

S. Omatu et al. (Eds.): *Distrib. Computing & Artificial Intelligence*, AISC 217, pp. 275–282.
DOI: 10.1007/978-3-319-00551-5_34

risks perceived by taxpayers are distinct from the effective risks mentioned on literature. Section 5 explains how artificial intelligence can contribute to the understanding of these biases. Finally, on section 6, we present our conclusions.

2 The Gap between Perceived and Effective Risk

A comparison between perception and effective risk is emerging, as the result of the new developments in the field of cognitive neuroscience.

In the literature, this gap is often attributed to the observation of differences in perception of risk between experts and common people [3, 4]. The authors suggest that biases come from poor quality of information, from the inability of individuals or from insufficient structuring of the environment, leading to an over or underestimated perception. Another explanation suggests that experts go beyond the limits of available data and assess risk based on intuitions [5]. Moreover, the difference can have origin on unjustifiable trust [6]. Another interesting fact is that the disagreement about risk does not disappear even when the risk is obvious. The initial prospects are resistant to change because they influence how future information is interpreted. These studies have referenced only heuristics and psychological components for biases, hence neglecting social interaction, culture impacts and the neurobiological dimension.

Risk perception has become an obstacle to rational decision-making, since people see risks where they do not exist, which is consistent with master estimation [4]. Subsequently, there has been a clash between different notions of risk perception among people, leading to the dilemmas on social risk [4]. Clearly, the subjective probabilities on a given risk vary significantly from the calculated probabilities [7]. Also, uncovered within political context, risk perception is probably not only a cognitive bias [4]. People also tolerate more risk when they volunteer to an event with associated risk [8]. This is due to their feeling of control. People who feel as having control over a certain situation, perceive a lower risk. The gap is time and again outlined as failure [8].

People have a robust yet unreasonable feeling about subjective invulnerability [9]. Psychometric approaches support that risk is plainly subjective. "Risk does not exist out there, independent of our minds and cultures, waiting to be measured" [3]. Given this subjectivity, it is really unreasonable to argue that people should be more rational about the risks they undertake. It is the argument that is itself wrong since it is supported by beliefs. This argument goes against overwhelming evidence, which is obvious on daily events from the real world. The way we perceive and react to a risk is a mixture of facts and feelings, reason and instinctive reaction, cognition and intuition.

Despite the overall success of our subjective system of risk perception, the fact is that we commit errors when assessing risks. There is unyielding confirmation of emotional aspects in risk perception [10]. On his work, Damasio [10] depicts the case of Elliott, who required brain surgery to relieve attacks. The fact was that Elliott could not satisfy his conduct to social standards, and could not settle a decision about anything because no different alternative was preferable. In this sense, subjective elucidations and emotions are inalienable necessities for rational behaviour. As Damasio puts it: "While emotions and feelings can cause havoc in reasoning processes… the absence of emotion and feeling is no less damaging, no less capable of compromising rationality".

3 The Need for an Integrative Framework

As was portrayed above, psychological, social and cultural dimensions influence risk perception. However, the aforementioned dimensions are all interconnected, strengthening up or attenuating each other [11]. Considering these interactions, [12] improved a structured system that provides an integrative and systematic perspective of risk perception. They additionally proposed four connected levels, initially displayed by [13] and adapted from the generic model of Breakwell [14].

All four levels of influence are relevant in order to gain a better and more accurate understanding of risk perception. In spite of many questions and ambiguities in risk perception research, one conclusion is beyond any doubt: abstracting the risk concept to a rigid formula and reducing it to two components "probability and consequences" does not match people's intuitive thinking of what is important when making judgments about the acceptability of risks [11]. Paul Slovic stated this point quite clearly: "to understand risk perception, one needs to study the psychological, social and cultural components and, in particular, their mutual interactions. The framework of social amplification may assist researchers and risk managers to forge such an integrative perspective on risk perception. Yet a theory of risk perception that offers an integrative, as well as empirically valid, approach to understanding and explaining risk perception is still missing" [3]. So it is clear that at this moment an integrative theory of risk perception does not exist, but only a framework sustained on multiple studies from different areas. Notwithstanding, the described framework clearly misses physical aspects of risk perception. Risk perception, like other perception types, use physical mechanisms of brain, like the memory or the limbic system. In place, different brain sub-systems help individuals to improve their judgements about assessing risk and making better decisions about it. Perception has some features and working patterns similar to awareness [15]. Of course, awareness in individuals has a critical role on risk perception. Only conscious individuals can formulate judgements about risk, and this is not referenced on the risk perception literature, even on the integrative framework suggested on [12]. Perception manifests according to the neuronal assemblies theory [15]. Neuronal assemblies are defined as large scale and highly transient neuronal coalitions, establishing connections between the neurobiological molecules and cellular components. The process occurs when several neurons from various neuronal columns within different sub-systems of brain unify and act harmoniously together. Therefore, the more complex the stimuli are, the greater the complexity of neuronal sets and number of recruited neurons from each subsystem of brain. Neurons are able to join quickly, forming functional groups to accomplish a certain task. Once this task is completed, the group dissolves and neurons are again able to fit on other sets, to fulfil a new task. More specifically, most of the literature agrees that our sensory perceptions, as well as our abstract thoughts, correspond to the activity of neuron assemblies, an activity that is subject to a complex dynamics.

Risk aversion is also important factor, unfortunately neglected by [12]. According [16], the activity of right inferior frontal gyrus correlates with risk aversion and proves its existence. Individuals who are more risk averse also have higher responses to safer options [16].

4 Application to Tax Fraud

Tax fraud is a classic economic problem on literature, which reveals good conditions to be used as case study, taking into account its integrative character, its scope and its intrinsic characteristics. In this problem and according with literature, the risk perceived by taxpayers, is the expectation of being caught by tax authorities and punished to pay effective tax amount plus a fine. Subjective versus objective probabilities of being caught should therefore be of great interest in tax research. [17] Why taxpayers perceive the risk differently from it is? What are the factors that induce this situation?

Reviewing the literature, we can find evidence of this reality. For example, Sandmo contends that the likelihood of being discovered is subjective and the perception of being discovered on cheating tax code is higher than the target probabilities of detection. This contrast may happen because of a simple misconception of risk, or because the individual's ability to evade taxes varies among subgroups of the population [18]. Tax fraud can also create a social stigma on taxpayers, when they are caught as tax cheaters [19]. This happens if the society works as a system and each citizen has a responsible role or contributes to its functionality. It has implicit an emotional context of shame that induces tax compliance.

The perception of neighbourhood behaviours is another example where individuals may commit fraud if someone of his/her relations had also committed fraud without being caught or punished. In this case, the perceived probability of being caught is derives from the individual's own evaded income and the perceived honesty of its social pairs. Individuals pay their taxes if they perceive that other taxpayers pay as well [20]. Tax compliance can assume contours of a psychological contract between taxpayers and tax authorities [20]. If so, evasion depends on the perception of the benefits and taxes paid to government. Instead, taxpayers can perceive that taxes are a contribution to guarantee the government functions on the community. And, in this case, people report the proper income, even if they do not receive back the public goods they were expecting. So, wealth redistribution is dependent on how individuals perceive the fairness and legitimacy of government policies. Another point is the perception by taxpayers of an existing bargaining power of tax authorities rather a relation of partnership [20]. These authors defend that evasion can emerge if taxpayers perceive that tax authorities treat them as inferiors, but they miss that the same can happen if they perceive they are superiors.

Subjective probability that an audit will occur may partly depend on prior experience [21]. The experience of an audit induces a learning process for evaluating audit probabilities [22]. Learning and understanding objective probabilities of uncertain events by experiencing or observing their occurrence can be understood as applying the availability heuristic [23]. On the other hand, it was frequently observed in tax experiments that compliance sharply decreases after an audit [17]. Results suggest that bad perception of chance is the major cause for the strong decrease in compliance immediately after an audit [17]. Propensity for tax evasion was not related to the perceived severity of fines.

5 A Model of Agents Risk Perception

Taking into account the risk perception problem, should artificial intelligence (AI) pull this into their field of study? In our opinion, AI has a word to say about biases of risk perception through agent-based model simulation (ABMS). ABMS applied to tax evasion should become the mainstream of described researches and can now be seen on incipient studies explaining tax fraud at the aggregate level. These works contain models that, while assuming that tax evasion depends on the degree of deterrence, have the key advantage of allowing the formalisation of other effects, such as social interaction [24, 25, 26, 27]. The EC* series shows increasingly complex models, developed by introducing progressive modifications over the standard economic model [28].

There is also the NACSM model, which analyses the relationship between tax compliance and the existence of social networks using the Moore neighbourhood structure [29]. Another approach is the proposal to adapt the ISING physical model to tax research [30]. Yet another is the TAXSIM model, which presents an especially rich and detailed design that includes four types of agents and some innovative factors as the degree of satisfaction with public services, depending on previous experiences of individuals, as well as those who are in their social network [31]. Given that satisfaction depends on the individual's perception, this proves that perception has a core importance in multiple aspects related to human being. However, perception was not taken into account by the authors. In short, the social simulation using ABM is a promising approach to a field in which, despite the abundant literature, results have been isolated and often poorly coordinated with each other, sustaining the claim that investigation on tax evasion is "still in its infancy" [17].

Accordingly, none of cited studies integrates the risk perception of committing fraud when it is critical to understand tax fraud. "Perception is reality" is commonly said in Wall Street. Clearly, the decision of committing fraud is a function of the perceived risk expressed by agents' subjective probability of being caught. This probability derives from the agents' risk perception process, which in turn is influenced by factors grouped into five dimensions.

The first dimension is related to the physical component of perception where agents have an episodic memory, risk aversion and emotions, reflecting different subsystems of the brain. Emotions have been associated to an important role in perception and in decision-making [10]. Emotions influence the balance between potential benefits and harms, to overlapping beliefs, given the inherent causes and consequences of risk [11]. Stigmas, like being recognised as a cheater by society also stimulate emotional responses [11].

The second dimension of agents is heuristic. Agents apply heuristics throughout their judgement process. Heuristics are reasoning strategies based on common sense, and suffering metamorphoses from the biological and cultural evolution of individuals [23]. But heuristics also derived from the acquisition of knowledge, by learning new strategies. Learning is an outcome of social interaction and new information updates, which are processed and integrated within episodic memory. The greater the knowledge and experience in logical reasoning, the greater will be the use of heuristics following this type of methodology, instead of intuition. Regardless of nominal value that these heuristics can offer, they are in fact mechanisms of how agents select, storage and process exogenous stimuli coming from environment that interferes on the magnitude of risk and respective mental representations. These may work temporarily regardless of the type of risk, beliefs or other patterns of conscious perception [11]. Taxpayers with certain beliefs have several restrictions in terms of heuristic options, since they don not formulate heuristics against their own beliefs, although empirical research demonstrates that people possess common heuristics [11].

The third dimension refers to cognition. Cognition rules the allocation of a specific risk-qualitative characteristic. It determines the effectiveness of these features in terms of the magnitude about perceived risk and consequent acceptability [3]. Individuals can be more fearful on their reaction when they have less knowledge and individual experience, for the respective risk [11]. They create expectations about what risk they are going to face.

The fourth dimension is related to social stimulus and interactions among taxpayers. It reports that the level of trust in the government and the tax system affects the way people perceive risk. For example, consider the perception of fairness and justice in the allocation of benefits and risks for different individuals and social groups. Fairness and justice have become more relevant to the perception of risk [11].

The cultural environment is the fifth dimension, and refers to the cultural factors governing or co-determining the remaining levels of influence described above. The "cultural theory of risk" supports cultural differences in risk perception and argues that there are four or five prototypical responses to risk [9]. Specific prejudice-based preferences are a driving factor in risk perception. Institutions assess risks according to their own vested interests and manipulate society in order to force it to accept them.

6 Conclusion

To build an ABM is a particularly adequate method for building a risk perception model for tax behaviour which includes factors like social networks, social influence, heuristics, cognition, emotions, episodic memory, biases in the perception of the tax system, heterogeneity of tax motivations and tax morale among the agents, and other features that may generate complex social dynamics. Those factors have been traditionally neglected in the classical econometric models that aimed to explain the observed levels of tax fraud. Instead of effective probabilities used on the majority of studies, we should use the subjective probability. Subjective probability is the output of the risk perception process, subject to the interaction of interdependent dimensions. Only with this model can we face effective risk and understand the mechanisms for biases which influence its perception

References

1. Gilbert, D.: Stumbling on Happiness, Knopf (2006) ISBN 1400042666
2. Schneier, B., Fear, B.: Thinking Sensibly about Security in an Uncertain World, pp. 26–27. Copernicus, New York (2003)
3. Slovic, P.: Perception of Risk Reflections on the Psychometric Paradigm. In: Krimsky, S., Golding, D. (eds.) Social Theories of Risk, pp. 117–152. Praeger, Westport (1992)
4. Sjöberg, L.: Consequences of perceived risk: Demand for mitigation. Journal of Risk Research 2 (1999)
5. Slovic, P., Fischhoff, B., Lichtenstein, S.: Why study risk perception. Risk Analysis Review 2(2) (1982)
6. Slovic, P., Fischhoff, B., Lichtenstein, S.: Facts and fears, Understanding the perceived risk in Scoeital risk assessment: How safe is safe enough? In: Schwing, R., Albers Jr., W.A. (eds.). Plenum, New York (1980)
7. Kahneman, D., Tversky, A.: A The Framing of Decisions and the Psychology of Choice Science. New Series 211(4481), 453–458 (1981)
8. Brun, W.: Risk perception: Main issues, approached and findings. In: Wright, G., Ayton, P. (eds.) Subjective Probability, pp. 395–420. John Wiley and Sons, Chichester (1994)
9. Douglas, M.: Risk and Blame: Essays in Cultural Theory. Routledge, London (1992)
10. Damasio, A.: Descartes' Error Emotion, Reason, and the Human Brain. Penguin Group (USA), Inc. (1994)
11. Wachinger, G., Renn, O.: Risk Perception and Natural Hazards. Cap-Haz-Net WP3 Report, DIALOGIK Non-Profit Institute for Communication and Cooperative Research, Stuttgart (2010)
12. Renn, O., Rohrmann, B.: Cross-Cultural Risk Perception Research: State and Challenges. In: Renn, O., Rohrmann, B. (eds.) Cross-Cultural Risk Perception: A Survey of Empirical Studies, pp. 211–233. Kluwer, Dordrecht (2000)
13. Renn, O.: Risk governance. Coping with uncertainty in a complex world. Earthscan, London (2008)
14. Breakwell, G.M.: The psychology of risk. Cambridge University Press, Cambridge (2007)

15. Freeman, W.: The brain transforms sensory messages into conscious perceptions almost instantly Chaotic, collective activity involving millions of neurons seems essential for such rapid recognition. Scientific American 264(2), 78–85 (1991)
16. Christopoulos, G., Schultz, W.: Neural Correlates of Value, Risk, and Risk Aversion Contributing to Decision Making under Risk. Journal of Neuro-science 26(24), 6469–6472 (2009)
17. Kirchler, E., Muehlbacher, S., Kastlunger, B., Wahl, I.: Why Pay Taxes? A Review of Tax Compliance Decisions Andrew Young School of Policy Studies. Georgia State University (2007)
18. Sandmo, A.: The Theory of Tax Evasion: A Retrospective View. National Tax Journal LVIII(4) (December 2005)
19. Allingham, M.G., Sandmo, A.: Income tax evasion: A theoretical analysis. Journal of Public Economics 1(3-4), 323–338 (1972)
20. Feld, L.P., Frey, B.: Trust Breeds Trust: How Taxpayers Are Treated. Economics of Governance 3, 87–99 (2002)
21. Spicer, M.W., Hero, R.E.: Tax evasion and heuristics: a research note. Journal of Public Economics 26(2), 263–267 (1985)
22. Mittone, L.: Dynamic behaviour in tax evasion: An experimental approach. The Journal of Socio-Economics 35(5), 813–835 (2006)
23. Kahneman, D., Tversky, A.: Judgment under Uncertainty: Heuristics and Biases. Science, New Series 185(4157), 1124–1131 (1974)
24. Spicer, M.W., Lundstedt, S.B.: Understanding tax evasion. Public Finance 21(2), 295–305 (1976)
25. Mittone, L., Patelli, P.: Imitative behaviour in tax evasion. In: Stefansson, B., Luna, F. (eds.) Economic Simulations in Swarm: Agent-based Modelling and Object Oriented Programming, pp. 133–158. Kluwer, Amsterdam (2000)
26. Davis, J.S., Hecht, G., Perkins, J.D.: Social behaviors, enforcement and tax compliance dynamics. Accounting Review 78, 39–69 (2003)
27. Bloomquist, K. M.: Modeling taxpayers response to compliance improvement alternatives. Paper presented at the Annual Conference of the North American Association for Computational Social and Organizational Science (NAACSOS), Pittsburgh, PA (2004)
28. Antunes, L., Balsa, J., Urbano, P., Moniz, L., Roseta-Palma, C.: Tax compliance in a simulated heterogeneous multi-agent society. In: Sichman, J.S., Antunes, L. (eds.) MABS 2005. LNCS (LNAI), vol. 3891, pp. 147–161. Springer, Heidelberg (2006a)
29. Korobow, A., Johnson, C., Axtell, R.: An Agent Based Model of Tax Compliance with Social Networks. National Tax Journal LX(3), 589–610 (2007)
30. Zaklan, G., Lima, F.W.S., Westerhoff, F.: Controlling tax evasion fluctuations. Physica A: Statistical Mechanics and its Applications 387, 5857–5861 (2008)
31. Zaklan, G., Westerhoff, F., Stauffer, D.: Analysing tax evasion dynamics via the Ising model. Journal of Economic of Coordination and Interaction 4, 1–14 (2009a)
32. Szabó, A., Gulyás, L., Tóth, I.J.: TAXSIM Agent Based Tax Evasion Simulator. In: 5th European Social Simulation Association Conference, ESSA 2008 (2008)

On the Use of PSO with Weights Adaptation in Concurrent Multi-issue Negotiations

Kakia Panagidi, Kostas Kolomvatsos, and Stathes Hadjiefthymiades

Pervasive Computing Research Group (p-comp), Department of Informatics and Telecommunications, National and Kapodistrian University of Athens
{mop10297,kostasks,shadj}@di.uoa.gr

Abstract. In this paper, we deal with automated multi-issue concurrent negotiations. A buyer utilizes a number of threads for negotiating with a number of sellers. We propose a method based on the known PSO algorithm for threads coordination. The PSO algorithm is used to lead the buyer to the optimal solution (best deal) through threads team work. Moreover, we propose a weights adaptation scheme for optimizing buyer behavior and promoting efficiency. This way, we are able to provide an efficient mechanism for decision making in the buyer's side. This is proved by our results through a wide range of experiments.

Keywords: Multi-Issue Negotiations, Concurrent Negotiations, PSO, Optimization.

1 Introduction

In E-Commerce applications one can find virtual places, called Electronic Marketplaces (EMs), where users can exchange products for specific returns. Users usually are involved to interactions for exchanging products. An interaction between entities is usually called negotiation. Negotiations are defined as a decentralized decision making process that seeks to find an agreement satisfactory to the requirements of two or more parties. Intelligent Agents (IAs) can undertake the responsibility of representing users in EMs. In EMs, we can identify three types of users: the buyers, the sellers, and the middle entities. Such intelligent autonomous components can take specific user preferences and return the desired result which is the product (for buyers) or the return (for sellers). In literature, one can identify bilateral (one-to-one) or one-to-many negotiations. In the first case, one buyer negotiates with one seller. In the second case, a buyer can negotiate, in parallel, with a number of sellers. Additionally, the negotiation could involve a single product issue (e.g., the price) or multiple issues (e.g., price, delivery time, quality, etc).

In this paper, we focus on multi-issue concurrent negotiations between a buyer and a number of sellers. We propose specific methodologies for enhancing the intelligence of the discussed IAs. Our aim is to develop a decision making mechanism which could be applied to real life negotiations. We assume that there is

S. Omatu et al. (Eds.): *Distrib. Computing & Artificial Intelligence*, AISC 217, pp. 283–290.
DOI: 10.1007/978-3-319-00551-5_35

no knowledge over the entities characteristics. Such characteristics could be their deadline, their strategies, and so on. In this scenario, buyers have direct interactions with sellers through a number of threads. We propose specific algorithms that can dynamically change the buyer's strategy without the need of a coordinator.

The rest of the paper is organized as follows: Section 2 discusses the related work to the aforementioned problem while Section 3 presents the examined scenario. We present the entities behavior and analyze our methodologies for changing utility function weights. We also describe the PSO approach for reaching to the optimal deal. Section 4 elaborates on our results through an analysis of our experiments. These results reveal the efficiency of the proposed model. Finally, in Section 5, we conclude this paper by giving future work directions.

2 Related Work

Many researchers have proposed models for handling bilateral or one-to-many negotiations. In (Faratin et al. 1998; Fatima et al. 2005; An et al. 2006; Chen et al. 2009; Da Jun & Xian 2002; Sun et al. 2007) models for bilateral multi-issue negotiation are presented. The presented models are mainly based on Game Theory, Fuzzy Logic or Machine Learning techniques. The authors define the strategies of the entities as well as the interaction protocol. Usually, a specific deadline is set for each entity. Multi-issue negotiations are studied in (Fatima et al. 2005; Robu et al. 2005; Lau 2005). The authors in (Robu et al 2005) propose a method for complex bilateral negotiations over many issues between a buyer and a seller. The negotiation can be further improved by incorporating a heuristic proposed (Jonker et al. 2004). Based on the proposed model, an IA uses the history of the opponent's bids to predict her preferences. Lau (2005) proposes a multi-agent multi-issue mechanism. A decision making model is described based on a genetic algorithm. In (Wu et al. 2009), the authors present an automated multi-agent multi-issue negotiation. In their model three agents bid sequentially in consecutive rounds. Additionally, in (Türkay & Koray 2012), the authors present a multi-issue negotiation mechanism which adapts a modified Even-Swaps method. A fuzzy logic inference system is responsible for bargaining on several issues simultaneously.

The authors in (Rahwan et al. 2002; Ngugen & Jennings 2004) discuss one-to-many negotiations by using a coordinator for changing the strategies of the threads. In (Rahwan et al. 2002), similar to our work, there is a number of threads that negotiate on the buyer's behalf with a number of sellers. However, prerequisite for the Intelligent Trading Agency (ITA) is the existence of a coordinator. After each negotiation circle, every IA reports back to the coordinator and receives specific instructions for concluding the negotiation. A similar scenario (use of coordinator for defining threads strategies) also stands in (Ngugen & Jennings 2004).

In our model, the buyer decision process is based on a number of issues (multi-issue negotiation). The buyer utilizes a number of threads each of them exchanging offers with a seller. We assume absolutely no knowledge of the players' characteristics. In contrast to other research efforts, we do not need any coordinator to specify the strategy for each thread. Threads by following the Particle Swarm Optimization (PSO) algorithm try through a team effort to find the optimal solution (best deal). The impact of our work is that the proposed scheme is based on a self organization technique adopted by threads in order to reach to the best result. Thus, the buyer saves resources. Moreover, there is not any need for exchanging a large number of messages as in the coordinator case. The decision making of each thread is made independently and is automatically adapted to each seller strategy.

3 Multi-Issue Concurrent Negotiations

In the buyer side, the best agreement is defined as the agreement that maximizes the final utility. In Fig. 1, we can see the architecture of our model. At every round, each thread sends / receives an offer (a bundle of values for the examined issues). If an agreement is true in a specific thread then the agreement message is sent to the rest of them (the message is received by every thread that currently participates in an active negotiation). Based on the PSO algorithm, the remaining threads changes their strategy in order to pursue a better agreement than the previous. Moreover, the remaining threads change the weights for the utility calculation in order to pay more attention on specific issues and, thus, to achieve a better utility in a possible future agreement.

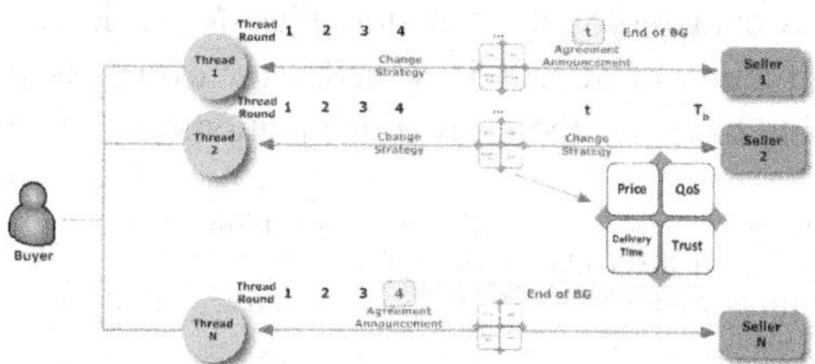

Fig. 1 Concurrent Negotiations Scenario

The IAs, representing each party, have no knowledge about the preferences of their opponents (limited knowledge). There is no need for a central decision maker which collects the information of IAs and transmits directions to them. Every seller has the same product in her property retrieved by a specific cost and tries to sell it in the highest possible profit. Similarly, the buyer is interested in purchasing the product that is close to her preferences. The product has a number of characteristics (issues) that affect the final utility. These issues are categorized as proportional (P) or inversely proportional (IP) to the utility. When examining proportional issues the greater the issue value is, the greater the utility becomes. The opposite stands in the inversely proportional products.

The buyer has a specific deadline defined by her owner. The same stands for the seller. Let us denote the buyer deadline with T_b while the seller deadline is depicted by T_s. In each negotiation, the seller starts first and the buyer follows if the proposed offer is rejected. The seller proposes an offer at odd rounds and the thread makes a counter offer at even rounds. If a player is not satisfied by the proposed offer, she has the right to reject it and issue a counter-proposal. Every offer involves specific values for the examined issues. This approach is defined as the package deal (Rahwan et al., 2002; Torroni & Toni, 2001). If a deadline expires and no agreement is present then the negotiation ends with zero profit for both. Both entities utilize a specific utility function (U) defined as follows:

$$U = \sum_{i=1}^{m} w_i \cdot v_i \tag{1}$$

where m is the number of issues, w_i and v_i are the weights and values respectively. Additionally, both players have their own strategy for offers calculation. We adopt the approach described in (Fatima et al. 2002; Oprea 2002). Each entity has her own reservation values for every issue. We consider an interval $[min_i, max_i]$ where every issue i takes its values. These values differ on the buyer as well as on the seller side. Both entities generate their offers based on the following equations:

$$O_i = min_i + \varphi(t) \cdot (max_i - min_i)\,, O_i = min_i + (1 - \varphi(t)) \cdot (max_i - min_i) \tag{2}$$

for the buyer and the seller side respectively. In the above defined equations, O_i depicts the next offer for issue i. As we can see, our model involves a time dependent strategy that is depicted by the function $\phi(t)$. More details for the ϕ function can be found in (Fatima et al. 2005). Finally, for every issue, we calculate the corresponding utility based on the following equations:

$$U(v_i) = \frac{v_i - min_i}{max_i - min_i}, \text{if issue i is P} \quad \text{or} \quad U(v_i) = \frac{max_i - v_i}{max_i - min_i}, \text{if issue i is IP} \tag{3}$$

We propose a model where buyer threads are automatically organized to change their strategy in order to reach to the best deal. Our proposal is based on the known PSO algorithm (Kennedy & Eberhart 1995). PSO is a computational method that optimizes a problem by iteratively trying to improve a candidate solution with respect to a given measure of quality. In our case, every thread is a particle that can move in the N-dimensional space. The N dimensions of space are assigned to the N issues. Moreover, every threads modifies her position according to the current velocity, current position, her distance between current position and global best position and her distance between current position and local best position. Combining PSO algorithm with the VFA algorithm (Howard et al., 2002), we assume that every issue can create a force F_i on particle i. The combination of FV_i, particle's global best position and particle's local best position will move the particle to her next position x_i^{t+1} at time t+1.

We extend the proposed model by defining a weights adaptation scheme to deal with utility maximization. The aim is to have a trade off between issue values in order to reach to a better agreement. When a thread closes a deal then an agreement message is sent to the rest of the threads. The remaining threads, if necessary (if the utility of the agreement is larger of the utility defined by the current offer), reassess the weights of the utility function aiming at higher utility values. If the weight of an issue, which has greater value than the corresponding agreement issue, decreases, then its proportion in the utility will decrease too. Thus, we emphasize on those issues that have value worse than the agreement value. The reason is that the utility will increase depending on such issues (the utility will be increased if we increase the weight of the specific issue and achieve better value in the negotiation). This way, the thread tries to force the seller to give better offers. The utility value will be increased if and only if the seller will improve those values.

We propose two methodologies for calculating the weights: a methodology based on the Simplex optimization (Kennedy & Eberhart 1995) and a methodology based on the Analytic Hierarchy Process (AHP) (Dagdeviren & Yuksel, 2008). Actually, in our scenario, we implement the revised Simplex to reassess weights. The optimal equation is the following:

$$\text{Maximize } U, \text{ subject to } \sum_{j=1}^{n} w_j = 1 \quad , w_j \geq 0 \quad \text{and } U \geq U_A$$

where U_A is the utility retrieved by the agreement. The discussed calculation is held every time an agreement is announced and the result is the appropriate values of weights that solve the above problem. We also implement the AHP process. At the beginning, every weight is initialized to a very small value close to zero. Then the distance between thread's value and the agreement value is calculated and normalized in [0,1]. As a next step, we utilize the approach of estimating weights presented by (Dagdeviren & Yuksel 2008). A fuzzy comparison matrix differs from Saaty's scale in that we use membership scales, instead of the classic Saaty's scales. Thus, the classic Saaty's table is transformed in order to contain the discussed fuzzy values that show the connection between issues. Thus, the weight for each issue is calculated as the normalized final value.

4 Experimental Evaluation

The performance metrics for our model are: **a)** ***The agreement ratio (AG)***: The AG indicates the number of negotiations that end with an agreement out of a number of negotiations, **b)** ***Average Buyer Utility (ABU) and Average Seller Utility (ASU)***: The ABU is the maximum utility that the buyer gains from all the successful threads (threads with an agreement). The ASU is defined as the utility gained in the seller side, and **c)** ***Average Rounds (AR)***: AR is defined as the number of rounds needed to reach an agreement out of the full horizon $T = \min(T_b, T_s)$. Actually, we examine the percentage of the required rounds on T.

We run experiments for different values of the buyer valuation (V) about the product. This value affects the product price. We run 300 negotiations for $N_T = 50$

(threads), $I = 4$ (issues) and $V \in \{50, 300\}$. It should be noted that at the beginning of each experiment, we randomly choose intervals $[min_i, max_i]$ for each side (buyer and seller). These values are chosen in [0,100]. Moreover, we randomly choose T_b and T_s in the interval [50,100]. The seller's parameters are also randomly selected in every experiment (e.g., the cost is randomly selected in the interval [10, 50]). For offers calculation, we are based on Eq(2).

Table 1 presents our comparison results. Concerning the AG values, all the approaches achieve a large number of agreements (the majority is above 90%). AHP and PSO start from a small value for ABU, which increases gradually. Simplex method is more stable. PSO reaches the Simplex value when $V = 300$ while AHP remains at lower levels. Particles have more space to converge to an optimal deal. Moreover, AHP and Simplex have constant values ranging near to ASU = 0.3. The PSO achieves the greatest ASU value especially for large V. In general, PSO seems to be the most efficient for the buyer as it leads to higher ABU values even for small V while it achieves the higher ASU values (more 'fair' approach).

In Fig. 2, we see our results for different I values. In general, the AG value is not mainly affected by I (is close or over 90%), however, the PSO achieves larger ABU when $I \rightarrow 32$. Thus, it is natural as the PSO is not profitable for the seller (see the ASU plot). The PSO requires more time than the rest when I is small ($I < 8$) while the opposite stands when I is large ($I \rightarrow 32$). Finally, in Table 2, we see our model performance for different N_T values. AHP achieves the best performance concerning AG and is also stable for ABU and ASU values. PSO requires the most time for all N_T. The reason is that in these cases the number of particles increases as well and, thus, reaching the optimal solution is more complicated.

Fig. 3 presents our results related to the optimality of our model. We run 10 experiments for different N_T values. When $N_T = 32$ our model has an average distance 0.0325 from the Pareto optimal with the minimum equal to 0.0017. The average distance is 0.1957, 0.1295 and 0.0736 for $N_T \in \{4,8,16\}$. For large N_T the buyer gains higher utility and our model approaches very close to the Pareto optimal. The greater the N_T is, the closer to the optimal the provided solution becomes as the buyer faces larger number of sellers and, thus, has more opportunities to reach to the optimal solution.

The proposed model is very efficient for the buyer as the ABU is larger than the ASU. Additionally, the model needs very little time to result an agreement especially when AHP is used. Our ABU results outperform those presented in (Lau 2005) where the maximum payoff is equal to 0.64. Additionally, in (Lau 2005) deadlines were equal to 200 rounds and agreements were concluded in the 84% of the deadlines (in our case the maximum AR value is 48%). Additionally, in (Nguyen & Jennings 2004), the maximum ABU is equal to 0.75 while in our case is equal to 0.91. Finally, in (Robu et al. 2005), the presented model reaches the 97% of the optimal value while our model reaches the 99% of the optimal when $N_T = 32$ (see Fig. 3).

Table 1 Results for different V values

V	Simplex Optimization				AHP				PSO			
	AG	**ABU**	**ASU**	**AR**	**AG**	**ABU**	**ASU**	**AR**	**AG**	**ABU**	**ASU**	**AR**
50	0.98	0.83	0.29	0.48	0.88	0.65	0.39	0.22	0.90	0.73	0.36	0.33
300	1.00	0.91	0.31	0.10	1.00	0.85	0.32	0.21	0.99	0.91	0.49	0.28

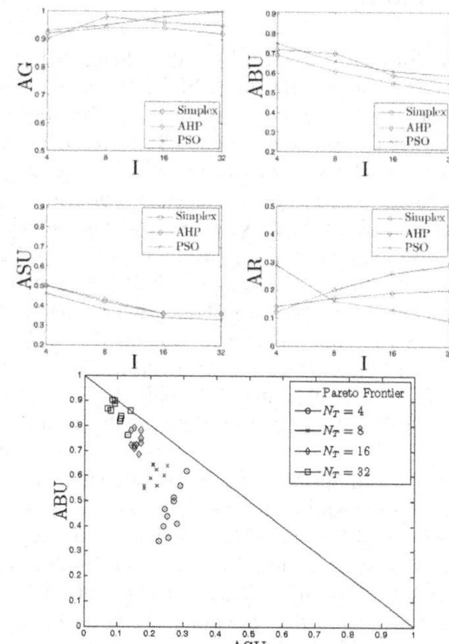

Fig. 2 Results for different I values

Fig. 3 Optimality of the proposed model

Table 2 Results for different N_T values

N_T	Simplex Optimization				AHP				PSO			
	AG	**ABU**	**ASU**	**AR**	**AG**	**ABU**	**ASU**	**AR**	**AG**	**ABU**	**ASU**	**AR**
5	0.51	0.33	0.24	0.19	0.79	0.53	0.36	0.13	0.44	0.25	0.16	0.22
50	0.96	0.88	0.40	0.16	0.96	0.72	0.43	0.15	0.93	0.76	0.46	0.28

5 Conclusions

In this paper, we focus on one-to-many, concurrent negotiations. The proposed solution is an automated process of a dynamic and independent change of the adopted strategy of the IAs involved. We present our method for optimizing negotiations focusing on the buyer's side. A model for dynamically change utility function weights is also analyzed. We propose a PSO approach for leading the buyer threads to the optimal solution. A large number of experiments (negotiations) show that the Simplex method leads to an increased buyer utility. The PSO approach is mainly affected by the number of threads, however, it is an efficient technique when threads number is small. Moreover, by using PSO, when issues number increases, the buyer utility decreases. Future work involves the definition of relevant function for weights adaptation in the seller side. Moreover, concerning PSO, a combination of the proposed technique with other intelligent models, like fuzzy logic, could increase the efficiency of the proposed framework.

References

An, B., Sim, K., Tang, L., Li, S.Q., Cheng, D.J.: Continuous-time Negotiation Mechanism for Software Agents. In: IEEE TSMC-B, vol. 36, pp. 1261–1272 (2006)

Chen, Y.M., Huang, P.N.: Agent-Based Bilateral Multi-Issue Negotiation Scheme for E-Market Transactions. Applied Soft Computing 9, 1057–1067 (2009)

Dagdeviren, M., Yüksel: Developing a fuzzy analytic hierarchy process (AHP) model for behavior-based safety management. Information Sciences 178(6), 1717–1733 (2008)

Da-Jun, C., Liang-Xian, X.: A Negotiation Model of Incomplete Information Under Time Constraints. In: AAMAS, Bologna, Italy, pp. 128–134 (2002)

Faratin, P., Sierra, C., Jennings, N.R.: Negotiation Decision Function for Autonomous Agents. Int. Journal of Robotics and Autonomous Systems 24, 159–182 (1998)

Fatima, S.S., Wooldridge, M., Jennings, N.: Bargaining with Incomplete Information. Annals of Mathematics and Artificial Intelligence 44(3), 207–232 (2005)

Howard, M.M., Sukhatme, S.: Mobile Sensor Network Deployment Using Potential Field: a distributed scalable solution to the area coverage problem. In: Proc. of ICDARS (2002)

Jonker, C., van der Meij, L., Robu, V., Treur, J.: Demonstration of a Software System for Automated Multi-Attribute Negotiation. In: AAMAS, New York City, USA (2004)

Kennedy, J., Eberhart, R.C.: Particle swarm optimization. In: Proc. IEEE International Conference on Neural Networks, pp. 1942–1948 (1995)

Lau, R.Y.K.: Towards Genetically Optimised Multi-Agent Multi-Issue Negotiations. In: Proceedings of the 38th Annual Hawaii International Conference on System Sciences, HICSS 2005 (2005)

Nguyen, T.D., Jennings, N.: Coordinating multiple concurrent negotiations. In: Proc. AAMAS, pp. 1064–1071 (2004)

Oprea, M.: An Adaptive Negotiation Model for Agent Based Electronic Commerce. Studies in Informatics and Control 11(3), 271–279 (2002)

Rahwan, I., Kowalczyk, R., Pham, H.: Intelligent agents for automated one-to-many e-commerce negotiation. In: IEEE Intern. Conf. on Privacy, Security and Data Mining, pp. 197–204 (2002)

Robu, V., Somefun, D.J.A., La Poutré, J.A.: Modeling Complex Multi-Issue Negotiations using Utility Graphs. In: AAMAS, New York, USA, pp. 280–287 (2005)

Sun, T., Zhu, Q., Li, S., Zhou, M.: Open, Dynamic and Continuous One-to-Many Negotiation System. In: 2nd International Conf. on Bio-Inspired Computing, pp. 87–93 (2007)

Torroni, P., Toni, F.: Extending a Logic Based One-to-One Negotiation Framework to One-to-Many Negotiation. In: Proc. of the WESAW, London, UK, pp. 105–118 (2001)

Türkay, D., Koray, A.: Modified Even-Swaps: A novel, clear, rational and an easy-to-use mechanism for multi-issue negotiation. Computers & Industrial Engineering 63(4), 1013–1029 (2012)

Wu, M., Weerdt, M., Poutre, H.: Efficient Methods for Multi Agent Multi-Issue Negotiation: Allocating Resources. In: 12th Intern. Conference on Principles of Practice in MAS, pp. 97–112 (2009)

Monitoring Weight and Physical Activity Using an AmI Setting

João Ferreira[1], Rafaela Rosário[2], Ângelo Costa[1], and Paulo Novais[1]

[1] CCTC - Department of Informatics, University of Minho, Portugal
pg18904@alunos.uminho.pt, {acosta,pjon}@di.uminho.pt

[2] Higher Education Nursing School, University of Minho, Portugal
rrosario@ese.uminho.pt

Abstract. We have an increasingly sedentary population without the care to make a healthy diet. Therefore, it becomes necessary to give the population the opportunity, despite living a very busy and tiring life, to have control over important aspects to their health. This work aims to present a model of an ambient intelligence system for monitoring the weight and physical activity in active individuals. To accomplish this objective we have developed a mobile application that allows users to monitor their weight over a period of time, identify the amount of food they consume and the amount of exercise they practice. This mobile application will give information to users about dietary and physical activity guidelines in order to improve their lifestyles. It is expected that students improve their lifestyles.

Keywords: Monitoring Weight, Physical Activity, AmI, Mobile, Lifestyle, BMI.

1 Introduction

Currently, obesity is increasing across all the population. Besides adults, more than 30% of children are considered as overweight or obese [1]. This increase can be associated with an imbalance between energy intake and energy expenditure. According to the American College of Sports Medicine and the American Heart Association [2], it is recommended that adults, between the ages of 18 and 65, perform exercises of moderated intensity during, at least, 30 minutes per day over 5 days a week, or in alternative exercise of elevated intensity during, at least, 20 minutes per day over 3 days a week. Now, more than ever, it is important to develop and implement interventions that aim to improve physical activity in all the population [3].

Besides physical activity, interventions should improve dietary intake [3]. There are no forbidden food, however following a healthy diet rich in fruit and vegetables and poor in low nutrition, energy-dense foods is associated to a reduced risk of obesity [4], cancer [5], asthma [6] and cardiovascular disease [7].

It is known that there is a need to encourage the population to improve physical activity and eating habits, nevertheless best practice is far from complete. In our view, current projects in this area are very generic and do not take in consideration

S. Omatu et al. (Eds.): *Distrib. Computing & Artificial Intelligence,* AISC 217, pp. 291–298.
DOI: 10.1007/978-3-319-00551-5_36

the personal needs of the users and their daily routines. These applications simply suggest a diet for the users to follow and in some cases dont even keep a record of their evolution while using that application.

1.1 Lifestyles Impact

Lifestyles are important determinants of chronic diseases. Dietary habits may influence cardiovascular disease (CVD), which continues to be the main cause of death in Europe [8], through an effect on risk factors such as serum cholesterol, blood pressure and body weight. In addition regular physical activity is associated to a reduced risk of CVD [6, 9] and a sedentary behavior to the prevalence of obesity [10]. Therefore, these chronic diseases are preventable.

Although age, gender and genetic susceptibility are non-modifiable risk factors, others such as diet and physical activity play an important role on the prevention of these health problems [3]. There is a need to develop and implement effective programs in order to improve eating habits and physical activity behavior. Evidence suggests that the childhood period is an important opportunity to achieve healthier eating habits [11]. In addition multi-component and adapted to the local context interventions are likely to be successful, as well as those which use social structures of a community such as schools [12, 13, 14]. It is also known that interventions which provide feedback tailored to an individuals needs are more likely to be used [14]. The novelty of this study is the development of an application to be used over time and which gives the opportunity to reinforce information about healthier lifestyles to users and analyze their progress.

1.2 Ambient Intelligence

Ambient Intelligence (AmI) has been used most recently due to an increase in the number of devices that are available to users, like for example, computers, tablets, smartphones, etc.

One of the main purposes of AmI is to reduce the need to be constantly providing information to the system and that he can be able to make decisions without needing great interaction with the users. The decisions, being based in data provided by the sensors from the behavior of the users, may be incorrect, which will lead to the users having to perform slight alterations to the system [15]. In order to design an AmI system, we needed to determine what elements to consider (see Fig. 1).

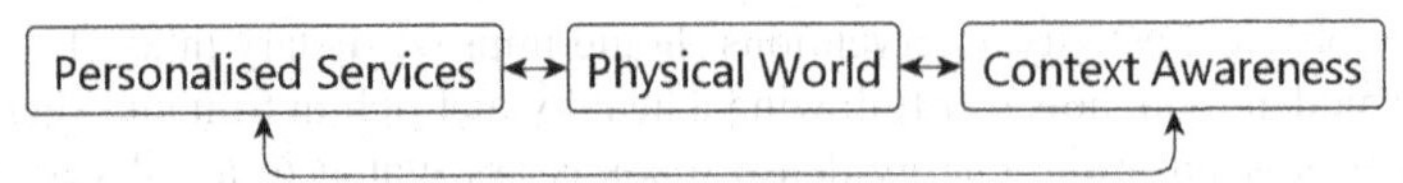

Fig. 1 Required elements in an AmI system

The Figure 1 presents to the following elements:

- Physical World: in this element we consider the environment that surrounds the user, every object and how they communicate between each other. Every chair, desk or plant that surrounds the system is taken in consideration. The users are the most important element of the systems because they are the ones that use the system. Many aspects of the users life are very important for the AmI system such as his daily routine or what type of job he develops. With this information the AmI system can make decisions that will benefit and improve the way that user lives.
- Context awareness: Context Aware Computing is defined to be the ability of computing devices to detect and sense, interpret and respond to aspects of a users local environment and the computing devices [16]. This means that a context-aware system should be attentive to all that surrounds him, with the purpose of adapting to it.
- Personalized services: one very important element of an AmI system is that these systems should adapt themselves to the users habits. A system of this type should learn from all the inputs that the user or the surrounding environment gives it, determine which ones are more relevant. After that, the AmI system should be able to adapt its process and execute all the actions that the user normally would perform, releasing the user from that task.

In the next section is presented part of the model developed, being the mobile application. It is explained its importance and the preliminary development results

2 Platform Architecture and Technologies

Currently, from the complete system, it is being developed the mobile platform. In Fig. 2 can be seen a simple system architecture.

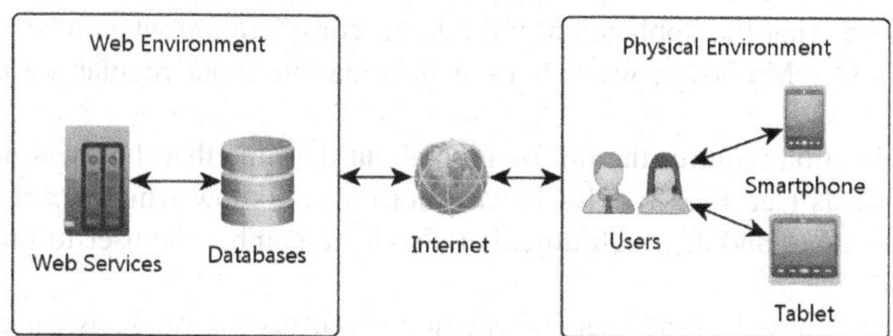

Fig. 2 System architecture

The proposed mobile platform is implemented in Android OS. The Android and mobile device provide access to the accelerometer and GPS, which are vital to the movement mesures.

The accelerometer is a sensor that measures the acceleration force that is applied to a device on all three physical axes (x, y and z), including the force of gravity. Or in other words, detects the motions of the devices (shake, tilt, etc.). The GPS keeps track of the user outdoor position. Both these sensors may provide information that determines if the if the user is performing exercise or not. The platform is built to interact with any type of user, being technological savvy or not [17, 18, 19].

All the information that is retrieved from the use of the application is saved in Json files. This format allows us to serialize and transmit structured data over a network connection. Therefore, the information will be sent to a JAVA web service through any Wi-Fi network to which the device is connected. The web service will intermediate between the user and the database as way to keep the data of the users secure. This information will be important to recognize data patterns, so that it will be easier in the future when the application encounters similar situations.

The platform created for this work is divided into three modules: personal data gathering, monitoring system and decision making. The platform easily adapts to different scenarios, by only changing the configuration and database files. This operation translates in a simple exchange of secure files, which retain all the information saved over the platform operation.

2.1 Personal Data Gathering

The application can correctly adapt to the users personal and work life, by gathering some information essential to its future work. First of all, the application needs to know the users name, so that the contact with the user can be more personal. This way, when in the regular use of the application, the contact between the application and the users is more personal and, therefore, more easily taken in consideration by the user. In order to preserve the identity of the user, when the information is saved in the database, the user is identified by a number and not by his name.

Besides the name, the application needs the information about the height and weight of the user so that we can compute their Body Mass Index (BMI), kg/m2, which will be what the application will take in consideration when analysing the users progress . Moreover, we will have information about regular weight (see Fig. 3).

The application requires the information about the time that the users normally take their meals (see Fig. 3). This is used not only to know which are the meals that the user takes and at which time, but also to remember the user to give those information.

Finally, it is asked to the user information about his/her work, as for example the work schedule. This information is important in order to understand great discrepancies in the values that the accelerometer returns. It is also asked how many hours the user is standing up. The reason behind this question is that on this position he/she is expending more energy at the same time.

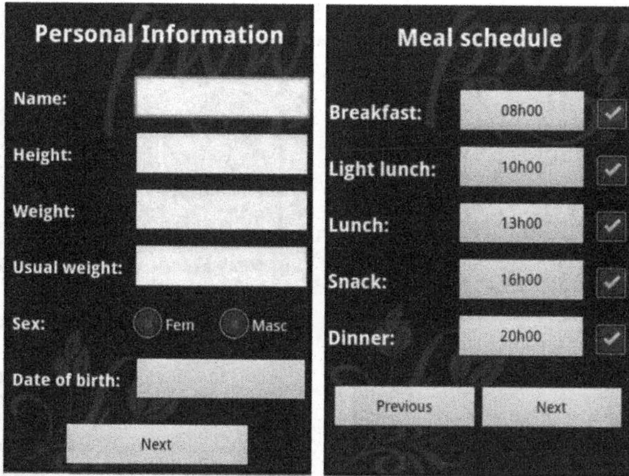

Fig. 3 Personal Info Application Screen

2.2 *Monitoring System*

The monitoring system is constituted of two types: online or offline. The online system uses the information that comes from the smartphone sensors (accelerometer and GPS) and using data mining algorithms, finds patterns in the information, thus calculating the exercise done. This way the user will not have to constantly be inserting information in the application.

The offline system makes the user insert all the information about the exercise that he performs. The user can insert the information daily or, if the user performs exercise in a weekly basis, the user can insert one time and the values will be saved for the correspondent week, as shown in Fig. 4a.

Shared by both is the need to introduce information about the weight and eating habits of the user. Although the information related with users meals needs to be inserted daily, the information about the user weight should only be inserted at the end of the week because the user weight variations are only noteworthy in long periods of time.

When inserting the information about the meal, the user will have to choose the two most representative elements of his meal between the following: fruit, lettuce, meat/fish/vegetables, rice/pasta or cake. Then the user will be asked to insert the quantity of the eaten food. Moreover, the user also has to insert the type of beverage that he/she had and the quantity. Finally, the user will have to insert whether or not if he/she had a coffee (seen in Fig. 4b).

The user can see his BMI evolution in the last 10 weeks as he can also see energy intake by meal in the last 10 days. This gives the user the opportunity to keep track of his/her performance in energy intake since he/she has the application.

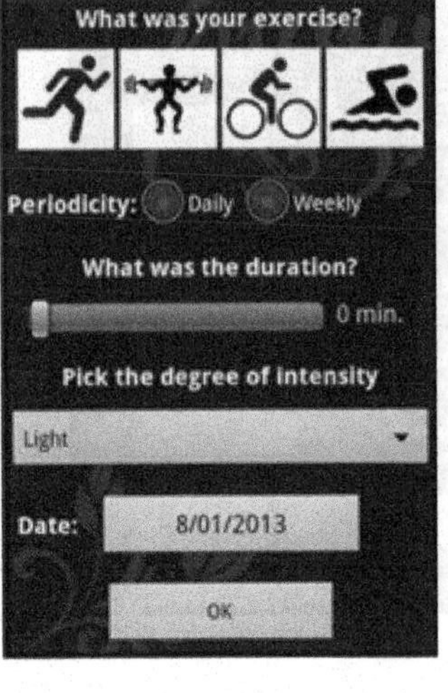

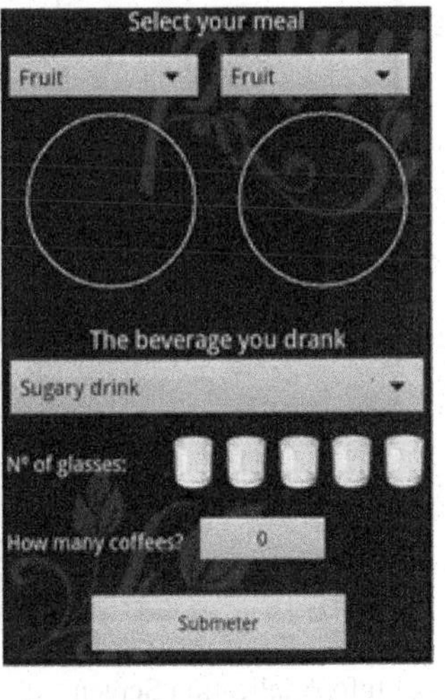

Fig. 4 a) Exercise information b) Meal information

2.3 *Decision Making*

The decision making module will be responsible for giving advice to the user. All the decisions will be based on the information saved in the knowledge base that is stored in the users smartphone memory card. This is a very important module because this is expected to affect the diet of the user when following the tips that the application shows. Therefore, it is expected that the user will start losing weight and having a healthier lifestyles than what he/she used to have.

The application will advise the user about his eating patterns. For example, if the user usually skips breakfast he/she will be advised by the application to change this routine, and will be encouraged to eat the breakfast, as it is considered the most important meal of the day. Another possible advice the application gives refers to the need of the user eat fruit at any meal in the day, especially when the application notices that he/she does not eat this food. Then, the application will recommend that he/she starts eating fruit at least two meals per day, due to the number of nutrients that is available in fruits.

In order to determine the amount of energy intake at every meal, the application sums the values of each element of the meal. These values are calculated differently depending on the element the user chooses. For example, if the user inserts that he/she ate rice/pasta, the application takes the percentage of the plate that the rice occupied. We consider that a portion of rice is 110 gr which represents 140kcal. The user will identify the portion that he/she ate, according to the figure of the plate. At the end we will sum all the eaten foods and compute the final energy of the meal.

$kcalingredient = (\%ofplate/12.5) * kcalperportion$

The formula varies between elements because of the size of the portion. When we are adding all the values so we can determine the total of calories ingested in a meal, all the meal elements have the same importance.

3 Conclusions and Future Work

Currently it is being developed and fine-tuned the mobile application. The model architecture sustains the base structure to the further developments. That is, we are developing the webservices to manage the data coming from the mobile application.

While the mobile application is currently in an alpha stage, it is able to collect and do basic operations, demonstrated in the section 2. We expect acceptance from the users by because of two main motives: personal interest and attractive interfaces. Therefore, work is being done towards optimizing calculations and using profiles. To achieve the main goal, that is being used by a user with intention but without being annoying.

Tests are in the development roadmap. University students will be invited and selected, in a first stage, to participate in this study. All of the students must be considered "active students", according to our selection criteria. Students with physical disabilities will be excluded from the study.

This first trial will give us information to determine if the system has any faults and to know the opinion of each person about the system and the improvements needed. This is will result in a test of a greater magnitude that will outcome in more information. Therefore, allowing the system perfecting and improving any bugs.

Acknowledgement. This work is partially funded by National Funds through the FCT-Fundação para a Ciencia e a Tecnologia (Portuguese Foundation for Science and Technology) within projects PEst-OE/EEI/UI0752/ 2011. Project AAL4ALL, co-financed by the European Community Fund FEDER through COMPETE - Programa Operacional Factores de Competitividade (POFC).

References

1. Padez, C., Fernandes, T., Mourão, I., Moreira, P., Rosado, V.: Prevalence of overweight and obesity in 7-9 year-old Portuguese children: trends in body mass index from 1970-2002. American Journal of Human Biology: the Official Journal of the Human Biology Council 16(6), 670–678 (2004)
2. Haskell, W.L., Lee, I.M., Pate, R.R., Powell, K.E., Blair, S.N., Franklin, B.A., Macera, C.A., Heath, G.W., Thompson, P.D., Bauman, A.: Physical activity and public health: updated recommendation for adults from the American College of Sports Medicine and the American Heart Association. Medicine and Science in Sports and Exercise 39(8), 1423–1434 (2007)
3. World Health Organization: Diet, nutrition and the prevention of chronic diseases. World Health Organization Technical Report Series 916, 149 (2003)
4. Gentile, D.A., Welk, G., Eisenmann, J.C., Reimer, R.A., Walsh, D.A., Russell, D.W., Callahan, R., Walsh, M., Strickland, S., Fritz, K.: Evaluation of a multiple ecological level child obesity prevention program: Switch what you Do, View, and Chew. BMC Medicine 7, 49 (2009)
5. Tantamango, Y.M., Knutsen, S.F., Beeson, W.L., Fraser, G., Sabate, J.: Foods and food groups associated with the incidence of colorectal polyps: the Adventist Health Study. Nutrition and Cancer 63(4), 565–572 (2011)

6. Hamer, M., Chida, Y.: Active commuting and cardiovascular risk: a meta-analytic review. Preventive Medicine 46(1), 9–13 (2008)
7. Oliveira, A., Lopes, C., Rodríguez-Artalejo, F.: Adherence to the Southern European Atlantic Diet and occurrence of nonfatal acute myocardial infarction. The American Journal of Clinical Nutrition 92(1), 211–217 (2010)
8. European Heart Network: Diet, physical activity and cardiovascular disease prevention in Europe. European Heart Network (2011)
9. Warren, T.Y., Barry, V., Hooker, S.P., Sui, X., Church, T.S., Blair, S.N.: Sedentary behaviors increase risk of cardiovascular disease mortality in men. Medicine and Science in Sports and Exercise 42(5), 879–885 (2010)
10. Robinson, T.N.: Reducing children's television viewing to prevent obesity: a randomized controlled trial. JAMA: the Journal of the American Medical Association 282(16), 1561–1567 (1999)
11. Birch, L.L., Fisher, J.O.: Development of eating behaviors among children and adolescents. Pediatrics 101(3 Pt. 2), 539–549 (1998)
12. Anderson, J., Parker, W., Steyn, N., Grimsrud, A., et al.: Interventions on diet and physical activity: what works: summary report. World Health Organization (2009)
13. Rosário, R., Araújo, A., Oliveira, B., Padrão, P., Lopes, O., Teixeira, V., Moreira, A., Barros, R., Pereira, B., Moreira, P.: The impact of an intervention taught by trained teachers on childhood fruit and vegetable intake: a randomized trial. Journal of Obesity, 342138 (2012)
14. Skinner, C.S., Campbell, M.K., Rimer, B.K., Curry, S., Prochaska, J.O.: How effective is tailored print communication? Annals of Behavioral Medicine: a Publication of the Society of Behavioral Medicine 21(4), 290–298 (1999)
15. Cook, D.J., Augusto, J.C., Jakkula, V.R.: Ambient intelligence: Technologies, applications, and opportunities. Pervasive and Mobile Computing 5(4), 277–298 (2009)
16. Yan, Z., Subbaraju, V., Chakraborty, D., Misra, A., Aberer, K.: Energy-Efficient Continuous Activity Recognition on Mobile Phones: An Activity-Adaptive Approach. In: 2012 16th International Symposium on Wearable Computers, pp. 17–24. IEEE (2012)
17. Costa, A., Novais, P.: Mobile Sensor Systems on Outpatients. International Journal of Artificial Intelligence 8(12), 252–268 (2012)
18. Novais, P., Costa, A., Costa, R., Lima, L.: Collaborative Group Support in E-Health. In: 2010 IEEE/ACIS 9th International Conference on Computer and Information Science, pp. 177–182. IEEE (2010)
19. Rosário, R., Oliveira, B., Araújo, A., Lopes, O., Padrão, P., Moreira, A., Teixeira, V., Barros, R., Pereira, B., Moreira, P.: The impact of an intervention taught by trained teachers on childhood overweight. International Journal of Environmental Research and Public Health 9(4), 1355–1367 (2012)

An Emotional Aware Architecture to Support Facilitator in Group Idea Generation Process

João Laranjeira[2], Goreti Marreiros[1,3], João Carneiro[1], and Paulo Novais[4]

[1] GECAD – Knowledge Engineering and Decision Support Group
[2] CCTC – Computer Science and Technology Center
[3] Institute of Engineering – Polytechnic of Porto
[4] University of Minho
{jopcl,mgt,jomrc}@isep.ipp.pt, pjon@di.uminho.pt

Abstract. In an idea generation meeting, the facilitator role is essential to obtain good results. The emotional context of the meeting partially determines the (un)success of the meeting, so the facilitator needs to obtain and process this information. Thus, the facilitator role is to assist the participants to reach their goals, i.e., to generate ideas with quality. In this paper is proposed an emotional aware architecture whose aim is to assist the facilitator in the process of maximizing the results of the meeting.

Keywords: emotional context-aware, idea generation meeting, facilitation.

1 Introduction

An idea generation meeting is composed of two types of elements: participants and facilitators. The main goal of the participants is to generate new ideas, while facilitators must manage the meeting and maximize the results.

According to Nunamaker et al. [1], the facilitator perform four tasks: to provide technical support for the applications used to support idea generation process; to mediate the meeting, maintaining and updating the meeting agenda; to assist in planning the agenda, and finally to provide organizational continuity, defining rules and maintaining an organizational repository. The electronic facilitation intends to expand the human facilitation horizons. Its mission is more extensive than human facilitation and its goal is to influence the group idea generation process, i.e., to make actions to increase the group's performance and the quality of the idea generated [2].

In the past years, several Group Support Systems (GSS) were developed and some of them focus on the support of idea generation processes [3][4][5]; however, just a few of them address the issues related to the affective context of participants and facilitators. Nevertheless, individuals' affective context has been assuming in the past years an important role in several cognitive activities (e.g., decision making, learning, planning and creative problem solving). More

S. Omatu et al. (Eds.): *Distrib. Computing & Artificial Intelligence*, AISC 217, pp. 299–306.
DOI: 10.1007/978-3-319-00551-5_37

specifically, in idea generation meetings it is possible to conclude that the emotional context of an idea generation meeting influences the performance of the participants. Several studies found in the literature prove that when the participants are in a positive mood, they generate more ideas and more creative ideas [6][7][8][9].

In our previous work, we develop a social idea generation system which incorporates social and emotion aspects of the participants [10][11]. We also analyze the positive impact that considering affective context has in the number of ideas generated [12]. In this paper the focus is on the facilitation processes. The facilitator has an important role in keeping participants in an adequate mood. Therefore, what we are proposing in this work is an emotional aware architecture to support facilitator work in group idea generation meetings.

The rest of the paper is organized as follows: firstly, we present a brief overview of the emotional context of group idea generation meeting and the facilitation modeling to this process. Then, we propose an emotion-aware architecture to the group idea generation process, and in the last section, we conclude our work and present the future work.

2 Emotional Context Modeling

In order to give an adequate and efficient support to participants, the facilitator should be able to understand the emotional profile of each participant in particular. Knowing participants' emotional context, it will allow the facilitator to take actions aiming to maintain the participants in a positive mood. These actions contribute directly to the maximization of the participants' performance and, consequently, to the maximization of the idea generation meeting results.

In idea generation meetings several events may occur and affect participants' emotional profile (e.g., the introduction of new ideas, the evaluation of the ideas, the visualization and analysis of the performance). The participants may be positively or negatively affected by those events, according to the desirability they have for that event.

To model participants' emotions, we use the OCC (Ortony, Clore, Collins) model [13] which is a model to infer the users emotions. However emotions are very volatile and they have a short duration. And in literature, what is more referred as having impact in the creative process is the mood, therefore we use emotions to infer participants' mood. Participant's mood represents the participants' emotional context over the time: if the participant is in a negative mood, then recommendations should be generated regarding the events that led the participant to that mood.

In Figure 1 one can see the emotional model proposed by Laranjeira and his colleagues to apply to an idea generation meeting context [10]. This model is based on events, i.e., the input of the model are the events triggered during the meeting and the output of this model comprises the events that generated negative emotions in the participants. This information will be important to help the facilitator recommending the participants to maximize their performance. It will be approached in the next section.

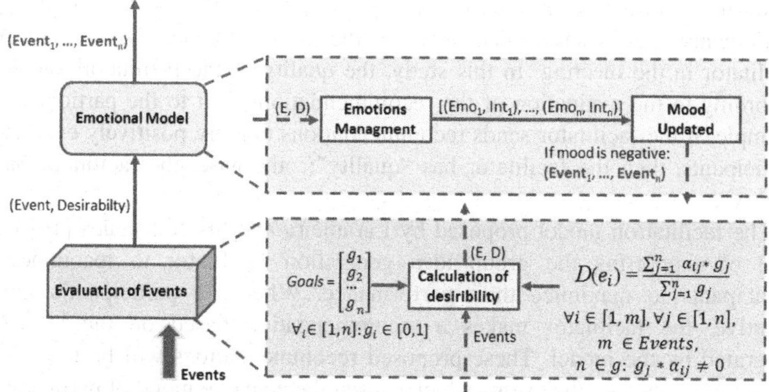

Fig. 1 Emotional Context Model [10]

After the events' evaluation, the model knows if an event generates a positive or negative emotion in the participant. The intensity of the emotion will depend on the desirability of the event. For instance, if a participant has the desirability of 0.7 on the occurrence of event X and this event does not occur, then the emotion will be a potential intensity of 0.7. More information on the model mechanisms can be consulted in [10].

Finally, the model stores the events that generate negative emotions in the participants. With this information, if a participant will be in negative mood, the facilitator can make a recommendation based on negative events.

3 Facilitation Process Modeling

The literature proves us that the use of a facilitator (electronic or human) is essential for the decision-making process, and it is also shown that when the facilitator is good, the results obtained are better than when a facilitator is not used [14]. However, the participants' behaviors are not considered, i.e., the result of interactions of the participants is not considered. This is the reason why we think it is essential to consider the emotional context of the meeting, more specifically, the idea generation meeting. Considering this context, the facilitator will have useful information to develop actions in order to maximize the performance of each participant, and consequently of the whole group.

Dickson et al. [14] proved that the inclusion of a facilitator in a GDSS improve the results obtained at the level of the participants' performance; however when these facilitators have no quality, the results become worse. Despite not knowing what "no quality" means exactly, in the past we did a simulation of an idea generation meeting, in which we test the process when the emotional context is considered and a facilitator is included [12]. Through the results of this study it was concluded that when there is a facilitator with quality, then the participants' performance increase. On the other hand, when the facilitator has no quality, the participants' performance decrease for the level obtained when there is no facilitator in the meeting. In this study, the quality of the facilitator was defined according to the evaluation of the recommendations sent to the participants. For example, if the facilitator sends recommendations that are positively evaluated by participants, then the facilitator has "quality"; otherwise, the facilitator has "no quality".

The facilitation model proposed by Laranjeira and his colleagues [10] has the goal of supporting the group idea generation facilitator to recommend the participants to maximize their performance. When the participant's mood is negative, the facilitator makes a recommendation based on the information generated by the model. These proposed recommendations will be based on the negative events, i.e., the events which caused the negative mood of participant.

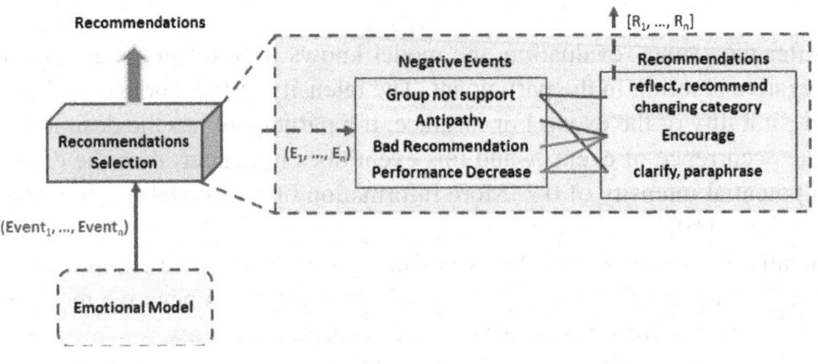

Fig. 2 Facilitation Model [10]

This model pretends to maximize the participants' performance, generating actions to maintain them in a positive mood. Thus, the participants will generate more ideas and more creative ideas [6][7][8][9].

4 Proposed Architecture

The architecture proposed in this section intends to apply the context emotional model and the facilitation model (Figure **2**) to a multi-agent system. However, the architecture is patterned in order to be independent of the system in which it will be inserted.

The **Meeting Context** and **Interface** modules are independent of the system in which they will be inserted. On the contrary, the **Emotional Context** and **Facilitator Advisor** modules are dependent to the system. However, in this paper we present a solution to a multi-agent system. Thus, the Emotional Context and Facilitator Advisor modules will be presented in a multi-agent system approach.

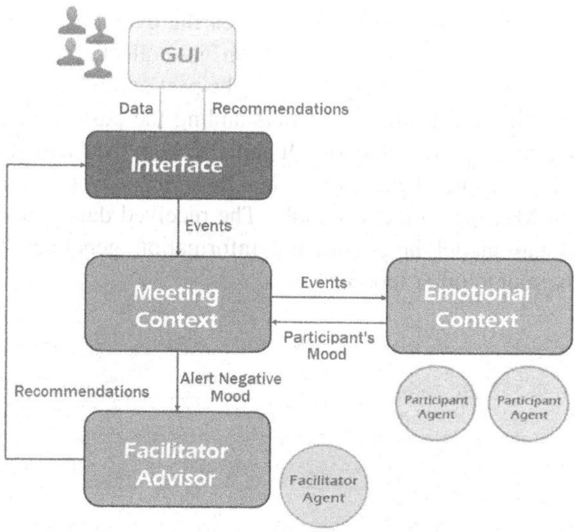

Fig. 3 Proposed Architecture

The **Interface** module is the bridge between the logical layer and the graphic interface. This module receives the information inserted by the meetings' users and shows to the facilitator the recommendations generated by the system. The **Meeting Context** module is a middle module as all information collected by the system and generated by the system passes through it. For example, when an idea is evaluated the information is analyzed by this module.

The **Emotional Context** module represents the emotional context model presented in Figure 3. This module analyzes all emotional events of the system. For example, when an idea or participant is evaluated, this module will analyze if any emotion is generated.

The **Facilitator Advisor** module implements the facilitation model presented in Figure 2. This module generates recommendations for supporting the meeting facilitator, in order to maximize the results of the group idea generation meeting. When a participant is in a negative mood state, analyzing the past events, this module will generate a recommendation. For example, if an idea was strongly rejected, then a recommendation can be sent to the participant to change the category of his ideas.

The architecture proposed in this paper intends to be integrated in a multi-agent system. Thus, we will present two types of agents, essential in this architecture dynamic: participant agent and facilitator agent.

The **Participant Agent** is a type of agent which represents and assists the participant of an idea generation meeting. It is modeled regarding the emotional context of the meeting. Every action performed by the participant in the system will be transmitted to this agent. In this way, it is possible to infer the participant's mood. The *Emotional Model* layer represents the implementation of the emotional model represented above. This layer receives the events triggered in the system and generates emotions. This information is collected to transform it into knowledge, by the *Knowledge* layer. Thus, the past emotional information may be used to predict future actions, i.e., understanding the past participants' behaviors may help predicting future behaviors. It can also help the facilitator understanding the reason for certain behaviors. The *Interface* layer receives and sends information to Meeting Context module. The received data is prepared to be sent to the emotional model layer and the information generated by this layer is prepared to be sent to other module.

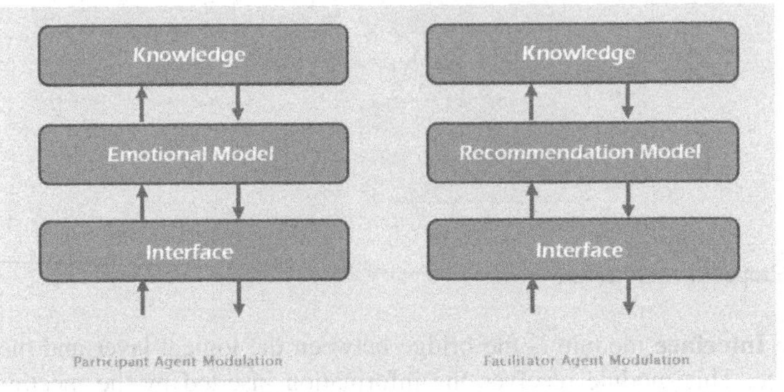

Fig. 4 Participant and Facilitator Agent Modulation

The **Facilitator Agent** represents the human facilitator of the electronic idea generation meeting. His goal is not to replace the human facilitator, but to support him/her, so that he/she can do recommendations to maximize the performance of each participant of the meeting. If a participant agent is in a negative mood, then the facilitator agent is informed about it. Considering the events which contributed to the negative mood, the facilitator agent will generate recommendations. These recommendations will send to the human meeting facilitator, who in turn will detail the recommendation and send it to the meeting participant.

The *Recommendation Model* layer applies the recommendation model presented above. It generates recommendations to the facilitator who, in turn, develops recommendations to maximize the idea generation meeting results. The information generated in this layer is used to create knowledge in *Knowledge* layer. The past recommendations will influence the recommendations generated, and here the information contained in Knowledge layer is essential. It can help the facilitator recommending participants with more quality and assertiveness. The *Interface* layer of the facilitator agent has the same function as the participant agent, i.e., it transforms the information received and prepares the information to send.

With this architecture, we believe the human facilitator will have more information to maximize the results of an idea generation meeting. He will know the meeting context and the emotional context of the participants. This information can help him making recommendations with more quality and assertiveness. The purpose of including this architecture in a multi-agent system comprises the optimization of the information processing in group, such as the emotional context and the meeting context. The information sharing between agents and their own characteristics, such as cognitive and emotional characteristics, may increase the facility and quality of the information processing.

5 Conclusions and Future Work

This paper proposes an emotional aware architecture, which objective is to support the facilitator in the group idea generation process. The goal of this architecture is to generate recommendations to maximize the results of the process by the constant control of the participants' actions. The suggested recommendations will be based on the past and present participants' behaviors.

Our objective is to use this architecture to improve the group idea generation systems. We believe that improving the facilitator support is the key to improve the group idea generation results.

In what concerns the future work, we intend to implement this architecture to a multi-agent system.

Acknowledgments. This work is funded by National Funds through the FCT - Fundação para a Ciência e a Tecnologia (Portuguese Foundation for Science and Technology) within projects PEst-OE/EEI/UI0752/2011 and PTDC/EEI-SII/1386/2012. The work of João Laranjeira is also supported by a doctoral grant by CCTC – Computer Science and Technology Center (UMINHO/BI_1_2013_ENG_Agenda_2020).

References

[1] Nunamaker, J., Briggs, R., Mittleman, D., Vogel, D., Balthazard, P.: Lessons from a dozen years of group support systems research: A discussion of lab and field findings. Journal of Management Information Systems 13(3), 163–207 (1997)

[2] Dennis, A., George, J., Jessup, L., Nunamaker, J., Vogel, D.: Information technology to support electronic meetings. MIS Quarterly 2(4), 591–624 (1988)

[3] Freitas, C., Marreiros, G., Ramos, C.: IGTAI- An Idea Generation Tool for Ambient Intelligence. In: 3rd IET International Conference on Intelligent Environments, Ulm, Germany (2007)

[4] Chen, M., Liou, Y., Wang, C.W., Fan, Y., Chi, Y.: Team Spirit: Design, implementation and evaluation of a Web-based group decision support system. Decision Support Systems 43, 1186–1202 (2007)

[5] Shih, P.C., Nguyen, D.H., Hirano, S.H., Redmiles, D.F., Hayes, G.R.: GroupMind: supporting idea generation through a collaborative mind-mapping tool. In: ACM 2009 - International Conference on Supporting Group Work, Sanibel Island, Florida, USA, pp. 139–148 (2009)

[6] Hirt, E., Levine, G., McDonald, H., Melton, R., Martin, L.: The role of mood in quantitative and qualitative aspects of performance. Single or Multiple Mechanisms? Journal of Experimental Social Psychology 33, 602–629 (1997)

[7] Isen, A., Baron, R.: Positive affect as a factor in organizational behavior. In: Staw, B.M., Cummings, L.L. (eds.) Research in Organizational Behaviour, vol. 13, pp. 1–54. JAI Press, Greenwich (1991)

[8] Frederickson, B.: The role of positive emotions in positive psychology. American Psychologist 56, 18–226 (2001)

[9] Vosburg, S.: Mood and the quantity and quality of ideas. Creativity Research Journal 11, 315–324 (1998)

[10] Laranjeira, J., Marreiros, G., Freitas, C., Santos, R., Carneiro, J., Ramos, C.: A proposed model to include social and emotional context in a Group Idea Generation Support System. In: 3rd IEEE International Conference on Social Computing, Boston, USA (2011)

[11] Laranjeira, J.: Group Idea Generation Support System considering Emotional and Social Aspects. Master Thesis, Polytechnic Institute of Porto, Portugal (2011)

[12] Carneiro, J., Laranjeira, J., Marreiros, G., Novais, P.: Analysing Participants' Performance in Idea Generation Meeting Considering Emotional Context-Aware. In: Novais, P., Hallenborg, K., Tapia, D.I., Rodríguez, J.M.C. (eds.) Ambient Intelligence - Software and Applications. AISC, vol. 153, pp. 101–108. Springer, Heidelberg (2012)

[13] Ortony, A.: On making believable emotional agents believable. In: Trapple, R.P. (ed.) Emotions in Humans and Artefacts. MIT Press, Cambridge (2003)

[14] Dickson, G., Partridge, J., Robinson, L.: Exploring modes of facilitative support for GDSS technology. MIS Quarterly 17(2), 173–194 (1993)

Social Networks Gamification for Sustainability Recommendation Systems

Fábio Silva, Cesar Analide, Luís Rosa, Gilberto Felgueiras, and Cedric Pimenta

{fabiosilva,analide}@di.uminho.pt,
{luisrosalerta,gil.m.fell,cedricpim}@gmail.com

Abstract. Intelligent environments and ambient intelligence provide means to monitor physical environments and to learn from users, generating data that can be used to promote sustainability. With communities of intelligent environments, it is possible to obtain information about environment and user behaviors which can be computed and ranked. Such rankings are bound to be dynamic as users and environments exchange interactions on a daily basis. This work aims to use knowledge from communities of intelligent environments to their own benefit. The approach presented in this work uses information from each environment, ranking them according to their sustainability assessment. Recommendations are then computed using similarity and clustering functions ranking users and environments, updating their previous records and launching new recommendations in the process.

Keywords: Sustainability, Ambient Intelligence, Reasoning, Gamification, Social Networks.

1 Introduction

Sustainability is a multi-disciplinary area based in fields such as economy, environment and sociology. These fields of research are interconnected, but humans have different psychological approaches to them. Thus, is necessary to perceive the behaviors behind each multi-disciplinary area. A computational platform to support and promote a sustainable environment, together with an approach to the energetic and economic problems, must take the decisions as smoothly as possible so as not to cause discomfort to the user. This topic triggered several psychological researches [1], [2] and a common conclusion indicates that humans are not always conscious about their behavior [3]. This field, called psychology of sustainable behavior, despite focusing on measurement and understanding the causes of unsustainable behavior it also tries to guide and supply clues to behavior change. Manning, shows some aspects that are necessary to consider promoting and instilling in people sustainable behaviors [4]:

- All behavior is situational, i.e., when the situation or event changes, the behavior changes, even if exists intention to perform a certain behavior, circumstances can make it change;

S. Omatu et al. (Eds.): *Distrib. Computing & Artificial Intelligence*, AISC 217, pp. 307–315.
DOI: 10.1007/978-3-319-00551-5_38

- There is no unique solution, i.e, people are all different because they have different personalities, living in a specific culture, with distinct individual history;
- Fewer barriers leads to a great effect, i.e., when a person is facing social, physical and psychological obstacles, his attitude tends to flinch, for instance, the lack of knowledge about a procedure leads to a retreat;
- There is no single approach to make an action attempting achievement of sustainability, there are many sustainable possible options that a person can choose.

To overcome these barriers to sustainability, it is suggested the engagement of multiple users in a competitive environment of positive behaviors so that participants have the need to strengthen their knowledge of sustainable actions. In this work, elements of gamification and information diffusion are explored in communities of intelligent environments with the objective of achieving behavioral and physical change. Automatic selection of recommendations based on the known examples, assessed via sustainable indicators, inside such communities can be achieved by constantly comparing and ranking users and physical environment inside intelligent environments. The dissemination of this information helps user inside these communities of intelligent environment take advantage of it and increase the community overall performance.

1.1 Gamification

There are already many studies in regard to gamification, where people use IT to change the behavior of the systems in order to make them more efficient. Still, there is a common trait among them, they are oriented to efficient actions of a system and not to the efficient actions of the user [5]. Changing the former is determining what should be its behavior, while changing the latter means changing their habits, the behaviors that they acquired. In order to tackle this problem, two main concepts will be put in practice: Gamification and Information Diffusion.

In [6], gamification is applied in education where the authors try to take the elements from the games that lead to the engagement and apply them inside the school to the students to keep them motivated. Another example uses a framework that allows users to share their daily actions and tips, review and explore others people actions, and compete with them for the top rank by playing games and puzzles [7]. On another example authors developed a service-oriented and event-oriented architecture framework where all participants communicate via events over a message broker. This system is composed by a set of game rules that define game elements like immediate feedback, rank/levels, time pressure, team building, virtual goods and points (karma points, experience points). Completing game rule generates a reward event for the user over the message broker. There is also an analytical component that may be used to analyze user behavior in order to improve game rules and optimize long-term engagement [8].

As for the second concept, Information Diffusion, this will be applied specifically to social networks. What various studies have proven [9] [10] [11] is that social networks have the potential to diffuse information at a high rate. Besides this point, they can also influence other peers to participate by sharing content. The use of social networks, also mentioned above, has the goal of enhancing the engagement of the users to higher levels by bringing the results to public (respecting user's authorizations) and making each user responsible for his actions at the eyes of the respective network.

As we can see through the examples presented, the application of gamification can raise the levels of loyalty of the users and keep them engaged in our objective by making it more enjoyable.

1.2 Sustainability Indicators

Sustainability is a multidisciplinary concept related with the ability to maintain support and endure something at a certain rate or level. The United Nations have defined this concept as meeting the needs of the present without compromising future generation to meet their own needs.

Due to the importance of sustainability different author have defined measures to assess and characterize sustainability. A popular consensus is based on 3 different indicators used to measure the sustainability of a given environment [12]. This approach is based on three different types of indicators, social, economic and environmental with the specific restriction that until all those values are met a system cannot be deemed sustainable. From this perspective sustainability concerns a delicate equilibrium between different indicators which action to optimize one indicator might severely affect one of the other two.

The presence of indicators to assess sustainability is an established practice [13] and [14], however it does not give any information on how to guarantee or plan sustainability. In reality indicator only inform about the current status of a system.

2 Studies on Sustainability Assessment

2.1 People Help Energy Savings and Sustainability (PHESS)

PHESS concerns a multi-agent platform (figure 1) developed to perform sustainability assessment on both users and environments. The platform establishes an ambient sensorization routine upon the environment, constantly updating sustainability indicators. The use of sustainability indicators represents the current, real time assessment of the environment taking into account historic data. The aim of the platform is not only to assess and identify unsustainable practices but also act with the objective of improving sustainability indicators. For such to happen, user behavior and environment might need to be changed. However, how the change is conducted cannot be determined by sustainability indicators alone.

The data gathering level in the PHESS platform includes sensing agents responsible for controlling the access and delivery of ambient sensor data model and reason agents in the reason context level. Model agents are responsible to monitor changes in the environment creating models with patterns common pattern and predictors for sensor value. Moreover, model agents may also be responsible for maintaining user or environment sustainable indicators updated. Reason agents use context information to formulate hypothesis in order to create recommendation, optimize environments and behaviors. This knowledge inferred from agents is then used in acting agents in the Acting level in this platform.

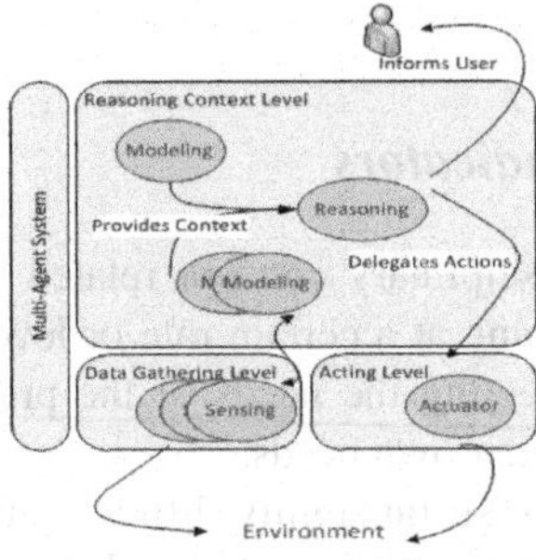

Fig. 1 Multi-Agent System for Deliberation and Sustainable Assurance

In this paper, the process of using indicators from different environments to create and promote recommendation that can be explained is detailed in next sections. However the definition of physical environments and user behavior is extracted from agents in this multi-agent system and the application of the recommendation engine detailed in section 3 is done in the Reasoning and Context Level, as well as, the profile management. Before detailing the recommendation system, an initial explanation about the sustainable indicators and sustainable assessment is necessary to understand the process of creating recommendations.

2.2 *Sustainability Assessment*

The sustainable assessment used in PHESS, uses different indicators within each dimension of the sustainability definition. This approach was also used by some authors, which used these indicators to guide strategic options and perform decisions based on the foreseeable impact of such measures [13], [14]. These indicators represent a ratio between a positive and negative contribution to sustainability and their values are computed in the -1 to 1 range, equation 1. As a consequence, all indicators use the same units of calculation and can be aggregated within each dimension through the use of weighted averages. The use of these indicators is made within each division in the environment and aggregated through average in the environment.

$$\text{Indicator(positive,negative)}=\begin{cases}\dfrac{\text{positive}}{\text{negative}}-1\rightarrow\text{positive}<=\text{negative}\\ 1-\dfrac{\text{negative}}{\text{positive}}\rightarrow\text{positive}>\text{negative}\end{cases} \quad (1)$$

In order to deliberate about sustainability performance it is needed to rank solutions rewarding each solution with a sustainable score, equation 2. Indicators within each dimension of sustainability are averaged according to weights defines in each dimension. The use of ranking formulas enables the use of fitness functions and distance functions to help calculate distances from one sustainable solution to another. Such approach in explored in section 3, integrated in a case based reasoning algorithm and custom sustainable indicators used to perform a proof-of-concept analysis on the proposed algorithm.

$$\begin{gathered}S_{index}=\alpha * I_{economic}+\beta * I_{enverinmental}+\gamma * I_{social}\\ \alpha+\beta+\gamma=1 \wedge 0<\alpha<1 \wedge 0<\beta<1 \wedge 0<\gamma<1\end{gathered} \quad (2)$$

3 Case Based Reasoning to Promote Sustainability

The work here detailed is intended to help communities of intelligent systems let users promote practices from different physical environments with high sustainability indexes to others with a recommendation engine. In order to summarize each environment, it was designed a sustainability profile, stating environment and individual room sustainability. Environment indicators are calculated from the use of aggregated individual room indicators, taking advantage of the indicator structure detailed in section 2 and three sample indicators for each dimension of sustainability. For the social indicator a positive value is represented by the amount of time spend inside the room whereas the negative value is represented by the time outside. Likewise for economic indicator a positive value is represented by the current budget available and the negative the total amount spent. Regarding the environmental indicator, emissions are derived from the CO_2 emission derived from electricity report for the negative value and emissions avoided as the positive value. Each case is maintained in a profile database and it is updated using the PHESS multi-agent platform.

The case based reasoning used in this situation uses a two-step process to evaluate and calculate new solutions for the user. As an initial step, the type of environment is contextualized, for instance, sustainable index, number of divisions and room indicators. A second step concerns the recommendation phase, and uses room indicators to obtain the best solution for the planning of energy use and appliance substitution.

The action flow is detailed in figure 2, where from an initial set of grouped environments a target environment can be compared to environments in higher ranked groups. The initial grouping of environments is made using K-means algorithm on the sustainable index of each environment with a fixed size for number of groups. The retrieval of comparative cases is extracted with the help of similarity functions. In this case, similarity is computed using environments from higher ranked groups and an average Euclidean distance from the distance value, computed for the three sustainability indicators, in every room. This procedure is used taking in consideration the room type, as distances are only calculated for rooms of the same type. The selection of environments favors the longest similarity distance for the value of the indicators in order to help the impact of possible recommendations in the environment. Finally, the list of alternative recommendations is obtained, comparing the room types of the target environment to rooms of the same type in the selecting environment. Any differences found are matched as possible change scenarios, favoring the options taken in the selected environment.

It is useful to remember that sustainable indicators are calculated from data acquired from each environment on a timely basis. The natural consequence is that as time progresses the values of these indicators which might result in environments exchanging the group they were previously.

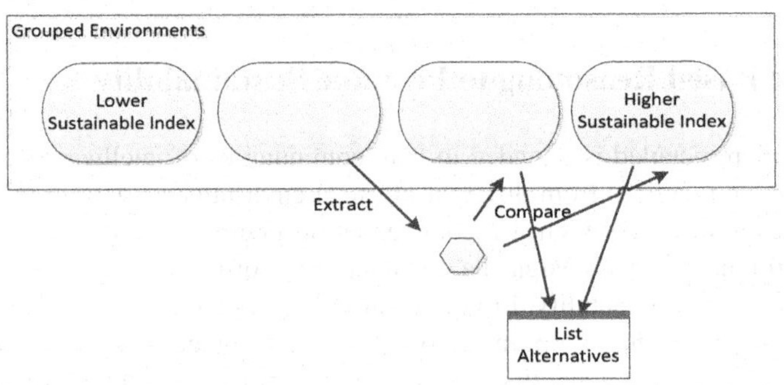

Fig. 2 User suggestion from social database

This dynamic works for the benefit of the system as the selected cases for comparison within each group are changed each time these variations occur enticing environments users to adopt behaviors that do not lead their environment to move to lower ranked groups.

3.1 Results

The results provided in this paper consider an implementation of different intelligent systems inside a community of users. For this purpose and due to current lacking infrastructures and users the environments were simulated defining different environments with different configurations generating user behavior inside them and creating sustainable index using the PHESS platform on such environments.

In order to test the recommendation system within communities, a set of environments was simulated. The setup recreated typical environments commonly found, such as apartments with a bedroom, living-room, kitchen, bathroom and a hall connecting all the other rooms. Inside each room, a set of appliances was also defined ranging from lights and computers to ovens and refrigerators with different consumption patterns. The consumption of appliances was defined from their active use and explicit turn on/off actions from user action simulated in the environment.

In this test 3 environments were defined and divided across 3 groups using the algorithm detailed in section 2.3. The initial step requires information about each environment, namely sustainable indexes for each environment and sustainable indicators for each room inside each environment. This was accomplished running each environment with sample users with sample routines inside each environment in the PHESS system. With information about sustainability on each environment groups was generated resulting with the first group concentrating two of four environments, and one for each of the remaining two groups. Focusing on one of the environment on group with poorer sustainable index, a comparison was made using the environment on the middle group in terms of sustainable index value. For each room possible changes were computed generating a report as defined in table 1 for the living room.

A total of six recommendations were proposed on the target environment in the living room, as seen on table 1, in the kitchen and in the bedroom areas.

Table 1 Example of Recommendations for Living Room

Appliance	Target Room (Average Consumption)	Best Case (Average Consumption)	Decision
Lights	120W	65W	Change
Computer	49W	55W	Remain
Television	60W	30W	Change
TV Box	55W	-	-

Using the PHESS system it was possible to assess that using recommendations on the living room alone was sufficient to improve the target environment sustainability index. In fact, iterating the recommendation algorithm one more time it can be found that if recommendations are followed and user behavior remains equal, the environment would be selected for the middle group, thus showing improvement.

3.2 *Discussion*

Recommendation calculated can be interpreted as using knowledge created within a community to its benefit. The best cases are used as examples to lower ranked cases which provide sense of sympathy from one to another. Also, with this approach, it is not necessary to maintain a database of efficient objects like appliances or lightning. As soon as they appear in the community they may tend to be selected for recommendation as part of someone's environment definition.

In order to further promote the adoption of recommendations and foster better behaviors, a social game could be devised using a points system where an environment has a default number of points due to the group it is fitted complemented with more points as recommendations purposed by the system were followed. It is believed that the devised algorithm for sustainability recommendation should work on gamification platform providing dynamic objectives and goals which are partial dependent on the acceptance of recommendations updated for every environment on a timely basis.

4 Conclusion

With the proliferation of social networks, users share significant amounts of information. Taking advantage of the number of users inside a community to develop a recommendation engine that promotes sustainability as global objective is the objective of the work here presented. The algorithm results and theoretical background support the idea that it possible to use such strategies to drive a social community of user to optimize itself if recommendations are followed.

Nevertheless, practical validation under real environments and a real user base is still needed to validate simulation results. This should be accomplished using field tests in a community focused on increasing their sustainability.

Acknowledgements. This work is funded by National Funds through the FCT - Fundação para a Ciência e a Tecnologia (Portuguese Foundation for Science and Technology) within projects PEst-OE/EEI/UI0752/2011 and PTDC/EEI-SII/1386/2012. It is also supported by a doctoral grant, SFRH/BD/78713/2011, issued by the Fundação da Ciência e Tecnologia (FCT) in Portugal.

References

1. Bartlett, D., Kane, A.: Going Green: The Psychology of Sustainability in the Workplace Green buildings: Understanding the role of end user behaviour (February 2011)
2. Gifford, R.: Environmental psychology and sustainable development: Expansion, maturation, and challenges. Journal of Social Issues 63(1), 199–213 (2007)
3. Sloman, S.A.: The empirical case for two systems of reasoning. Psychological Bulletin 119(1), 3–22 (1996)
4. Manning, C.: The Psychology of Sustainable Behavior (January 2009)
5. Gupta, P.K., Singh, G.: Energy-Sustainable Framework and Performance Analysis of Power Scheme for Operating Systems: A Tool. International Journal of Intelligent Systems 5 (2013)

6. Simões, J., Redondo, R.D., Vilas, A.F.: A social gamification framework for a K-6 learning platform. Computers in Human Behavior (2012)
7. Vara, D., Macias, E., Gracia, S., Torrents, A., Lee, S.: Meeco: Gamifying ecology through a social networking platform. In: 2011 IEEE International Conference on Multimedia and Expo (ICME), pp. 1–6 (2011)
8. Herzig, P., Ameling, M., Schill, A.: A Generic Platform for Enterprise Gamification. In: 2012 Joint Working IEEE/IFIP Conference on Software Architecture (WICSA) and European Conference on Software Architecture (ECSA), pp. 219–223 (2012)
9. Bakshy, E., Rosenn, I., Marlow, C., Adamic, L.: The role of social networks in information diffusion. In: Proceedings of the 21st International Conference on World Wide Web, pp. 519–528 (2012)
10. Goyal, A., Bonchi, F., Lakshmanan, L.V.S.: Learning influence probabilities in social networks. In: Proceedings of the Third ACM International Conference on Web Search and Data Mining, pp. 241–250 (2010)
11. Myers, S.A., Zhu, C., Leskovec, J.: Information diffusion and external influence in networks. In: Proceedings of the 18th ACM SIGKDD International Conference on Knowledge Discovery and Data Mining, pp. 33–41 (2012)
12. Todorov, V., Marinova, D.: Modelling sustainability. Mathematics and Computers in Simulation 81(7), 1397–1408 (2011)
13. Silva, F., Cuevas, D., Analide, C., Neves, J., Marques, J.: Sensorization and Intelligent Systems in Energetic Sustainable Environments. In: Fortino, G., Badica, C., Malgeri, M., Unland, R. (eds.) Intelligent Distributed Computing VI. SCI, vol. 446, pp. 199–204. Springer, Heidelberg (2012)
14. Afgan, N.H., Carvalho, M.G., Hovanov, N.V.: Energy system assessment with sustainability indicators. Energy Policy 28(9), 603–612 (2000)

Web-Based Solution for Acquisition, Processing, Archiving and Diffusion of Endoscopy Studies

Isabel Laranjo[1,2], Joel Braga[1], Domingos Assunção[1], Andreia Silva[1], Carla Rolanda[2,3], Luís Lopes[4], Jorge Correia-Pinto[2,5], and Victor Alves[1]

[1] Department of Informatics, University of Minho, Braga, Portugal
{isabel,valves}@di.uminho.pt,
{joeltelesbraga,dassuncao1990,andreia56837}@gmail.com
[2] Life and Health Sciences Research Institute (ICVS), School of Health Sciences, University of Minho, Braga, Portugal; ICVS/3B's - PT Government Associate Laboratory, Braga/Guimarães, Portugal
{crolanda,jcp}@ecsaude.uminho.pt
[3] Departments of Gastroenterology, Hospital de Braga, Portugal
[4] Departments of Gastroenterology, Santa Luzia Hospital, Viana do Castelo, Portugal
luis.m.lopes@mac.com
[5] Department of Pediatric Surgery, Hospital de Braga, Portugal

Abstract. In this paper we present a distributed solution for the acquisition, processing, archiving and diffusion of endoscopic procedures. The goal is to provide a system capable of managing all administrative and clinical information (including audiovisual content) since the acquisition process to the searching process of previous exams, for comparison with new cases. In this context, a device for the acquisition of the endoscopic video was designed (*MIVbox*), regardless of the endoscopic camera that is used. All the information is stored in a structured and standardized way, allowing its reuse and sharing. To facilitate this sharing process, the video undergoes several processing steps in order to obtain a summarized video and the respective content characteristics. The proposed solution uses an annotation system that enables content querying, thus becoming a versatile tool for research in this area. A streaming module in which the endoscopic video is transmitted in real time is also provided.

Keywords: Distributed Computing; Endoscopy; e-Health; Video Streaming; Video Processing; Endoscopic Video Acquisition.

1 Introduction

In general, technological progress has enabled the improvement of old methodologies, by allowing the development of new systems capable of organizing information in a more efficient and compact way as well as distributing it more easily. Users are spending less time with their tasks and the results tend to improve. This process optimization can be applied in health care, improving the working conditions of health care professionals [1]. The use of computer science solutions to

S. Omatu et al. (Eds.): *Distrib. Computing & Artificial Intelligence,* AISC 217, pp. 317–324.
DOI: 10.1007/978-3-319-00551-5_39

improve the performance of hospital services and their respective tasks is already common practice in many institutions that provide health care services. The information systems available in hospitals are acquiring new features that directly influence the quality of care provided to patients. This increase in quality is very important for institutions and governmental services, and has motivated investments in the development of software solutions that address increasingly specific tasks [2].

Nowadays it is hard to imagine making a diagnosis without resorting to therapeutic and diagnostic test [3]. According to "*Statistical elements - General Information: Health 2008*" [4], clinical analysis represented about 85% of all exams performed in Portugal in 2008. The remaining 15% were distributed among the remaining therapeutic and diagnostic tests (e.g. X-Rays, CT scans, Ultrasound, endoscopy).

Among the therapeutic and diagnostic tests, endoscopy has an increasingly important role because of its low cost and results obtained. It is a technique that allows the physician to observe, study and record images of organs (e.g. stomach, lungs, bladder), using a thin, flexible tube, called endoscope, which has its own lens and light, allowing the visualization of the mucosa through the other end of the device or in a video monitor [5, 6].

While performing a gastroenterological endoscopy, the physician visualizes the real time video on a monitor and, at the same time, he captures the most relevant images for his report. At the end of the examination, the physician uses the selected images and the video is discarded. This process presents some problems for the patient as well as for the physician, because, if the physician wants to review the performed diagnosis later on, or compare video segments with other procedures, he cannot do so. Thus, the patient is forced to repeat an endoscopic procedure. Moreover, there are other problems related with sharing information between different professionals and entities.

These issues served as motivation to design and create *MyEndoscopy*, a web-based distributed system for acquisition, processing, archiving and diffusion of endoscopy studies.

2 MyEndoscopy

MyEndoscopy can be used by any health professional of an entity if properly registered and authorized by the system. It was designed with the intention of linking entities, and standardizes the patient's clinical process management, especially gastroenterology clinical data. Therefore the sharing of information between the different entities is promoted (e.g. hospital, clinic, physician office, research center). For the system implementation, an architecture based on cloud computing concepts was idealized (Fig. 1). This architecture is based on the integration of all *MIVboxes* (developed endoscopic acquisition device) of the system, facilitating the availability of the overall information to users via an application set.

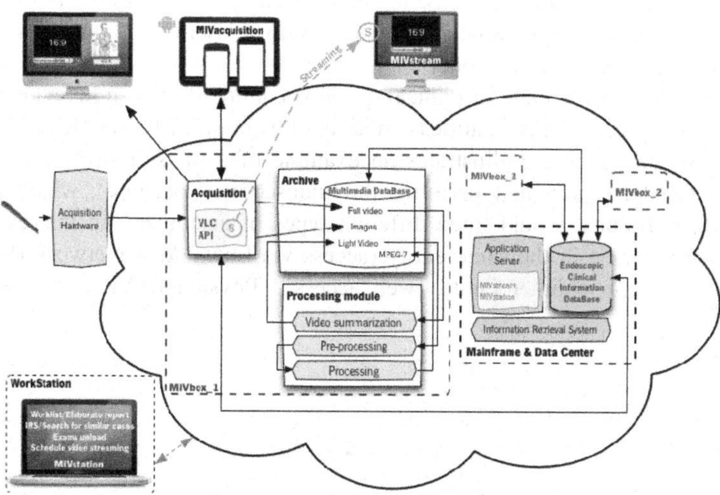

Fig. 1 *MyEndoscopy* overall architecture

During the endoscopy procedure, video is captured in real time via acquisition hardware (Fig. 2). This hardware comprises an endoscope that is connected to the processor and a light source. During the examination, the mucosal surface is illuminated using the existing light source in the endoscope and regardless of the system used to recognize the color (sequential or non-sequential [5]), the electrical signal captured by the Charge Coupled Device (CCD) is transmitted to the processor, which in turn will process it and turn it into an image.

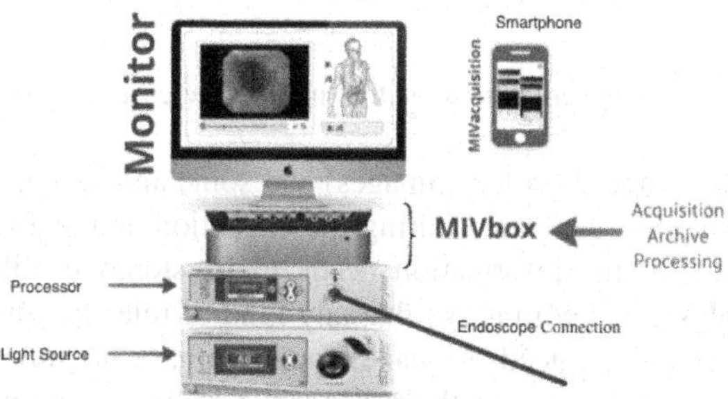

Fig. 2 Set of hardware for video acquisition

The video is received by the *MIVbox* (which is connected to the equipment set, referred above) and then, is transmitted from there to an existing monitor in the room.

The examiner physician may consult information related to the patient through the application *MIVacquisition* (which can be accessed through a smartphone or tablet), as well as manage some of the operations performed on the video during the exam. With this application, the practitioner can also capture the relevant images. These images will appear on the monitor, inserted by location in the gastrointestinal tract in a 3D model.

Sometimes, for academic or clinical questions, a broadcast can be made in real time to other entities. These entities can access this broadcast through the application *MIVstream.* Figure 3 illustrates an example of a transmission between the service provider (entity that performs the endoscopic procedure), which in this case is the MIVbox_1, and three different types of customers. In this case, the client can either be in the same entity (access via Local Area Network (LAN) - MIVbox_2) or in another entity (web access - Physician Office or Research Center).

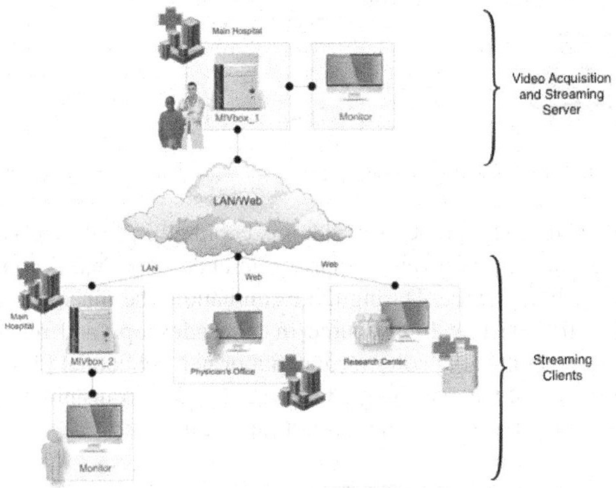

Fig. 3 Potential cases of connection between the supplier and customers streaming

The multimedia material (videos, images) and some annotations are considered as the basic information generated during the execution and analysis of an endoscopic procedure. All this information needs to be stored to allow subsequent visualization, analysis and editing, so that any kind of relevant information is not lost and, in case the patient needs to make a new exam, or monitor the progress of a diagnosis, professionals have at their disposal a basis for comparison between the new procedures and exams already carried out.

In *MyEndoscopy*, there are two kinds of information stored. The first is the storage of clinical and administrative information in a relational database called *Endoscopic Clinical Information Database*. The second concerns the storage of media files, which are linked to clinical information that exists in the *Multimedia Database* located in the *MIVbox*. Figure 4 shows a diagram where data flows between the different processes and the two databases are shown.

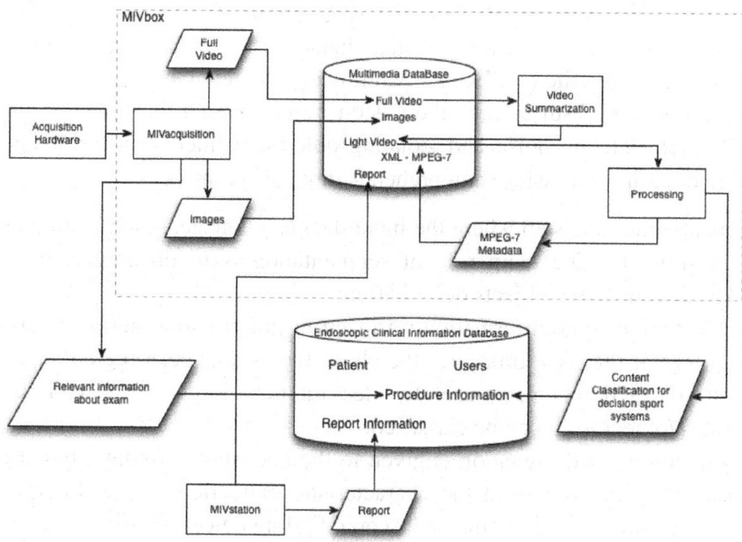

Fig. 4 Data flow between different processes and the two databases

Some of the processes that occur in figure 4 are part of the processing module that constitutes the *MIVbox*. This module aims to facilitate further analysis and processing of the video. In figure 5, the steps that occur in the processing module are detailed.

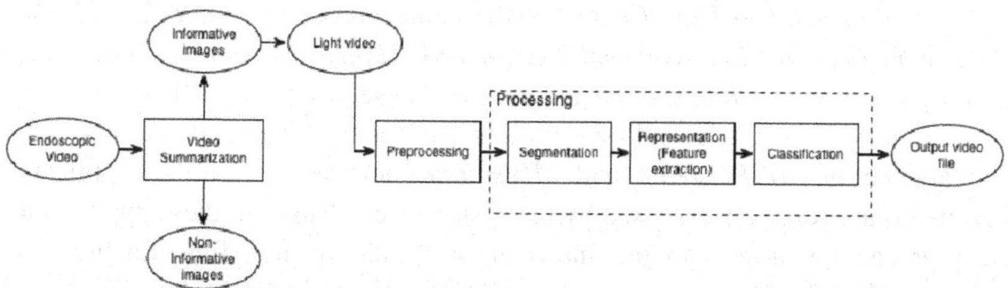

Fig. 5 Steps that occur on the endoscopic video processing module

Briefly, in the processing module, three principal steps take place: summarization, pre-processing and processing. Often, captured videos contain images with non-relevant information, e.g. images with only one predominant color, blurry. In order to eliminate these images, the video goes through a process called *video summarization*, which results in a light version of the video (that will be stored in the same directory as the rest of the exam resulting material).

After being stored, the light video undergoes a number of processing operations aiming the automatic extraction of visual characteristics. These characteristics are stored in a structured file (MPEG-7 – Multimedia Content Description Interface).

The first operation suffered by the video (pre-processing) is applied to improve contrast level, deleting noise and isolating objects of interest in the image. The second operation (processing) comprehends three different steps:

- Segmentation – the step where the input data (e.g. image, video) is divided into its constituents. The main role of segmentation is to distinguish the image background from the objects delimitation;
- Representation – the information retrieved in the previous step is transformed into a format that best translates the characteristics of each region/object. The Feature Extraction is a step used to select the most important characteristics of on image/object, so it can be classified;
- Classification – a designation is given to the endoscopic finding, based on the information retrieved from the characteristic extraction. In the Interpretation step, a meaning is given to the endoscopic finding detected earlier.

The purpose of the direct link between the *Multimedia DataBase* and the *Endoscopic Clinical Information Database* is to enable, during the evening of the exam day, the exchange of multimedia information (*light* video, images, MPEG-7 files) to be stored in the *Data Center*.

The operations described so far, occur in each *MIVbox* device. The system comprises various *MIVboxes* that can be part of different entities, but at the same time are parts of the overall system.

The *Mainframe* and *Data Center*, as the name suggests, includes a processing component (*Information Retrieval System* and *Application Server*) and storage (*Endoscopic Clinical Information Database*). These computational resources belong exclusively to the overall system and are not linked directly to any entity. The applications (*MIVstream* and *MIVstation*) are housed in the Application Server and are remotely accessed by the system users. Through these applications, the user can get access to the Information Search System functionalities and see/manipulate the stored information in the *Data Center* or *MIVboxes*.

The *MIVstation* application is the interface that allows professionals to consult the patient process, prescribe exams, schedule video streaming, elaborate reports and search similar cases, among other tasks.

Since search for information is one of the most important processes in sharing information between different entities and/or professionals, the existence of a sturdy search engine, that allows a search for multimedia content and annotations, is one of the most important characteristics so the professional can search similar cases or investigate rare cases.

In order to ensure compatibility of lexical information shared in the system, a standardized endoscopic vocabulary, the Minimal Standard Terminology (MST) was used [7].

3 Discussion

The architecture used in planning *MyEndoscopy* was based on the elements of a system found in [8]. Normally this kind of system is used in medical imaging departments with modalities such as, computer tomography, X Ray, ultrasound and magnetic resonance among others. Basically these systems comprise five basic steps: acquisition, storage, processing, communication and display. *MyEndoscopy*, despite being comprised by these steps, is considered a more complete distributed solution because it offers a greater scalability and number of functionalities in comparison with other systems in the market. For example the documentation system proposed by the *Richard Wolf Company* [9] is only compatible with the "HD ENDOCAM" camera, so the images and videos are storage on the system itself and there is no diffusion module to share information with other machines or entities (the information is copied to another USB, DVD or Blu-ray Disc). Most studies concerning endoscopy are focused on Wireless Capsule Endoscopy (WCE) [10], being noted the lack of studies directed to Upper GastroIntestinal (UGI) endoscopy [11]. Although the WCE is an innovative technique, it needs additional software to receive and process the videos, and usually there is no auxiliary information system to store and distribute the resulting material. At processing level, there are few studies related to Upper GastroIntestinal endoscopy, since it is a procedure in which human error in diagnosis are relatively low. Existing studies are mostly directed to help in the tumor and polyp detection with the Colonoscopy technique [12, 13] or with WCE videos [14].

It is noteworthy the compatibility between the *MIVbox* and the already existent devices (e.g. endoscopic, processor, light source).

The main improvements of the proposed solution are:

- Possible quality improvement of health care. This improvement relies on an efficient and global management of information, which allows the elaboration of a properly supported diagnosis and the avoidance of double endoscopic procedures;
- Quality and quantity improvement of information that health professionals have at their disposal, regardless where they do the consult/report. The possibility for a professional to remotely follow the realization of and endoscopic procedure, and give its personal opinion in real time is another main contribution, since the system presents a solution to the integration of additional specialists to the team (via streaming), if necessary;
- In the finance department and resource management, the *MIVbox* portability and versatility, as well as its extensive compatibility with the endoscopy equipment, may represent an improvement to some entities.

In general, the system was projected to connect different health care entities, through a cloud computing based service. This service contains not only centralized components but also components that are distributed across different entities.

The projected system is oriented to the Upper GastroIntestinal endoscopy. As a future work it can be easily extended to other endoscopy techniques, paying only special attention to the processing module, since it may be necessary to add algorithms to address some specific characteristics of each technique, as well as a new knowledge base.

It is our conviction that the proposed solution presents some unique characteristics at the levels of functionalities, integration, availability and scalability.

Acknowledgments. This work is funded by ERDF - European Regional Development Fund through the COMPETE Programme (operational programme for competitiveness) and by National Funds through the FCT - *Fundação para a Ciência e a Tecnologia* (Portuguese Foundation for Science and Technology) within project FCOMP-01-0202-FEDER-013853.

References

1. Shortliffe, E.H., Cimino, J.J. (eds.): Biomedical Informatics Computer Applications in Health Care and Biomedicine. Springer, New York (2006)
2. Jamal, A., McKenzie, K., Clark, M.: The impact of health information technology on the quality of medical and health care: a systematic review. The HIM Journal 38, 26–37 (2009)
3. Haux, R.: Medical informatics: past, present, future. International Journal of Medical Informatics 79, 599–610 (2010)
4. Direcção-Geral da Saúde, Direcção de Serviços de Epidemiologia e Estatísticas de Saúde, D. de E. de S.: Elementos estatísticos - Informação Geral: Saúde 2008. 159 (2010)
5. Cotton, P.B., Williams, C.B.: Practical Gastrointestinal Endoscopy the Fundamentals. Blackwell Publishing (2003)
6. Olympus: Endoscopy: Leading the Way to a New Future (2002)
7. Aabakken, L., Rembacken, B., LeMoine, O., Kuznetsov, K., Rey, J.-F., Rösch, T., Eisen, G., Cotton, P., Fujino, M.: Minimal standard terminology for gastrointestinal endoscopy - MST 3.0 (2009)
8. Sawhney, G.S.: Digital Image Acquisition and Processing. Fundamentals of Biomedical Engineering, p. 273. New Age International (P) Ltd. (2007)
9. Wolf, R.: VictOR HD for documentation in Full-HD Quality, http://www.richard-wolf.com/en/human-medicine/visualisation/documentation-systems/victor-hd.html
10. Muñoz-Navas, M.: Capsule endoscopy. World Journal of Gastroenterology 15, 1584 (2009)
11. Liedlgruber, M., Uhl, A.: Computer-aided decision support systems for endoscopy in the gastrointestinal tract: a review. IEEE Reviews in Biomedical Engineering 4, 73–88 (2011)
12. Vilariño, F., Lacey, G., Zhou, J., Mulcahy, H., Patchett, S.: Automatic Labeling of Colonoscopy Video for Cancer Detection. In: Martí, J., Benedí, J.M., Mendonça, A.M., Serrat, J. (eds.) IbPRIA 2007. LNCS, vol. 4477, pp. 290–297. Springer, Heidelberg (2007)
13. Alexandre, L.A., Casteleiro, J.M., Nobreinst, N.: Polyp Detection in Endoscopic Video Using SVMs. In: Kok, J.N., Koronacki, J., Lopez de Mantaras, R., Matwin, S., Mladenič, D., Skowron, A. (eds.) PKDD 2007. LNCS (LNAI), vol. 4702, pp. 358–365. Springer, Heidelberg (2007)
14. Chen, Y., Lee, J.: A Review of Machine Vision Based Analysis of Wireless Capsule Endoscopy Video

Texture Classification with Neural Networks

William Raveane and María Angélica González Arrieta

Universidad de Salamanca, Salamanca, España
{raveane,angelica}@usal.es

Abstract. Texture classification poses a well known difficulty within computer vision systems. This paper reviews a method for image segmentation based on the classification of textures using artificial neural networks. The supervised machine learning system developed here is able to recognize and distinguish among multiple feature regions within one or more photographs, where areas of interest are characterized by the various patterns of color and shape they exhibit. The use of an enhancement filter to reduce sensitivity to illumination and orientation changes in images is explored, as well as various post-processing techniques to improve the classification results based on context grouping. Various applications of the system are examined, including the geographical segmentation of satellite images and a brief overview of the model's performance when employed on a real time video stream.

1 Introduction

Texture classification is a specialized area within the field of pattern recognition. As such, other pattern classification techniques have successfully been applied in the past to this problem, such as statistical analysis [7], stochastic algorithms [4], geometric methods [1], and signal processing [8]. Each of these methods has its own advantages and they may be more or less appropriate depending on the particular scenario they are applied to.

Textures, in the sense portrayed in this paper, refer to patterns of colors and shapes formed by pixels in a digital image. Recognizing and distinguishing these textures tends to be a very complicated task, as small variations in scene illumination and view perspective can lead to drastic differences in the visual appearance of a texture, making it difficult for automated systems to successfully segregate them. However, the applications for such a system are many and diverse, making this an important research topic in computer vision. Some of the most typical uses of these texture analysis algorithms are the segmentation of aerial imagery, industrial surface inspection, biomedical image analysis, as well as the classification of textiles, minerals and even wood species.

S. Omatu et al. (Eds.): *Distrib. Computing & Artificial Intelligence*, AISC 217, pp. 325–332.
DOI: 10.1007/978-3-319-00551-5_40

In this paper, a different method is reviewed based on artificial neural networks. A simplified version of this method using a trivial neural network has already been proposed in the past [5]. This paper proposes a better implementation of this procedure involving additional steps dealing with image processing, neural network preparation and automated results correction. This allows for the successful application of the model on a much wider array of image types, and more importantly, to better generalize on new images that the system has not been trained with.

2 Classification Model

The proposed model is based on a supervised neural network. This section reviews each of the steps involved in its methodology, and then presents how the final system can be used to classify pre-defined texture classes.

2.1 Data Preparation

The application of a neural network to data that originates from photographic imagery renders the preparation step even more significant, as special care must be applied with many of the issues surrounding the handling of visual information in machine learning tasks.

It is customary to first submit images to an enhancement filter, for which there exist many options. Some of the best results observed are consistently obtained by a range of specialized processes such as the Retinex filter developed by NASA for the boosting of detail in satellite and aerial photography [6]. This filter enchances color information in the resulting image and normalizes the variations in illumination adaptively and locally. The neural network results are greatly improved when images are pre-processed in this fashion as it does not need to learn redundant lighting variations found within the shadows or highlights of the image, but can instead focus on the inherent image characteristics.

Alternatively, a histogram equalization process can also be applied independently on all three color channels in the image. This ensures that each data channel fed to the neural network is properly normalized and data is evenly distributed throughout the intensity level spectrum, a process akin to normalizing and re-distributing input data as done in most machine learning tasks.

Figure 1 compares both pre-processing methods and their resulting level histograms in each color channel.

To prepare the neural network system, several regions of interest in the input image are defined, each of which is formed by a number of pixels within a bounded area. A desired label is designated for each of these regions that will identify it as one of the target texture classes. This area of pixels should ideally form an adequate data sampling region including as much variety of the target texture as possible – as the pixels in these regions will become the training data for the neural network, and their labels the ideal output data. It is not necessary for a region of interest to outline a single continuous area, and in fact, it is often desirable to extract multiple regions

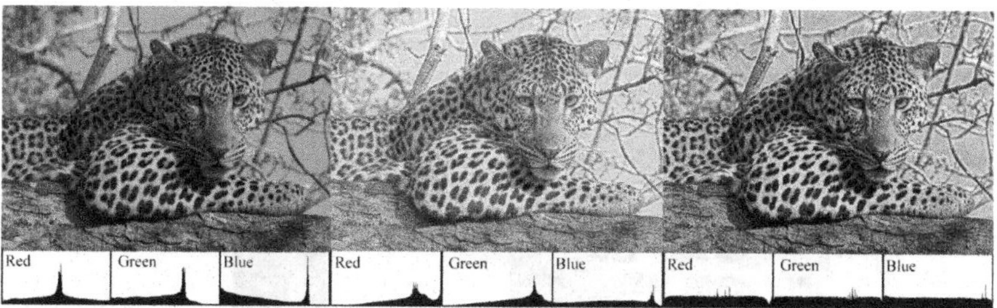

Fig. 1 Comparison of the original sample image (left), the result of the Retinex filter applied (middle), and the result of histogram equalization (right), along with their respective RGB histograms (bottom)

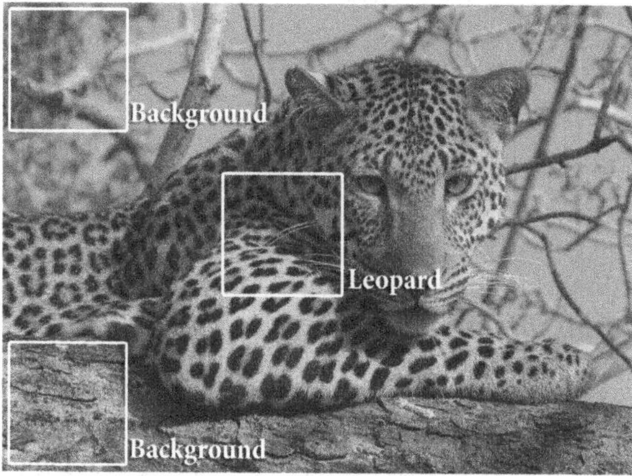

Fig. 2 Regions of interest defining the class labels for various textures in the input image

to provide a larger variation in the training data and in so doing, guarantee better generalization of the model. Figure 2 shows an example of the labeling process.

Once the image is ready for use and its target texture classes have been declared, the regions of interest extracted can be converted to data suitable for the neural network to train with. This is achieved by further subdividing these areas into a grid consisting of 5×5 pixels, thereby creating tiles of 25 pixels each. Figure 3 shows this process in detail.

The text labels assigned to each data sample must be converted to numerical values that the neural network can make sense of within a mathematical context. This is achieved by replacing each string label with an output vector of the same dimensionality as the total number of texture labels that have been assigned. Each output vector will then be an encoded representation that uniquely identifies each of the designated texture classes. With this end, a mapping mechanism is introduced that employs the following label replacement assignments:

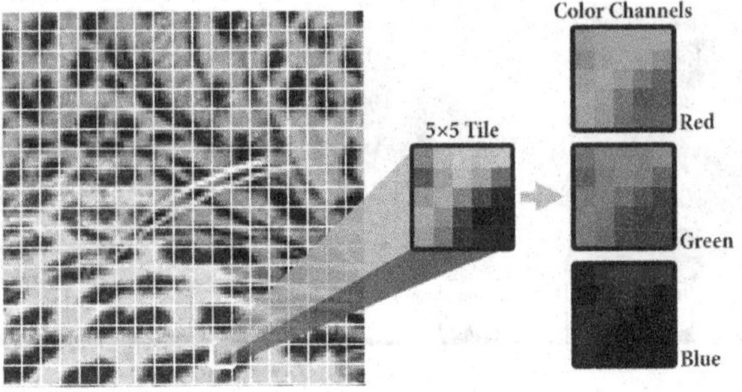

Fig. 3 Detail of a region of interest in the input image, subdivided into a 5×5 grid producing the sampling tiles for that particular texture; where for each tile, the collection of its 25 pixel values in each of the three color channels yields the $\mathbb{R}^{75}$ input vector for a training sample

$$label_0 = \begin{bmatrix} +1 \\ -1 \\ -1 \\ \vdots \\ -1 \end{bmatrix} \quad label_1 = \begin{bmatrix} -1 \\ +1 \\ -1 \\ \vdots \\ -1 \end{bmatrix} \quad \ldots \quad label_{n-1} = \begin{bmatrix} -1 \\ -1 \\ -1 \\ \vdots \\ +1 \end{bmatrix} \qquad (1)$$

2.2 *Texture Classification*

The neural network used is a simple three layer feed forward network. Although it is one of the most basic configurations available, it yields very good classification results for the purpose at hand and in return it is capable of training within an acceptable amount of time. The neural network layer structure is arranged as follows:

1. The input layer, with one neuron for each attribute in the input sampling data. Having 75 input neurons, this layer accomodates an input sampling vector in $\mathbb{R}^{75}$ consisting of the three color channel values for each of the 25 pixels in every sampling tile.
2. The hidden layer, with half as many neurons as in the input layer, which is a good rule of thumb to follow in configurations where the amount of samples considerably outnumbers the amount of attributes.
3. The output layer, with one neuron for each texture being classified. Each neuron in this layer corresponds to one of the class labels defined.

Once all of the training data has been laid out, the neural network is trained with it, using a learning algorithm such as Resilient-backpropagation (RPROP) [2], which in our tests has resulted as one of the most optimal methods when compared to other traditional algorithms such as simple Backpropagation.

Upon completion of the training process, the system can be used to classify the texture regions in an image. For this procedure, it is first necessary to subdivide the entire image into a similar 5×5 grid, and extract all of the individual tiles outlined by this grid. As before, each tile becomes a new sampling point that can be expressed

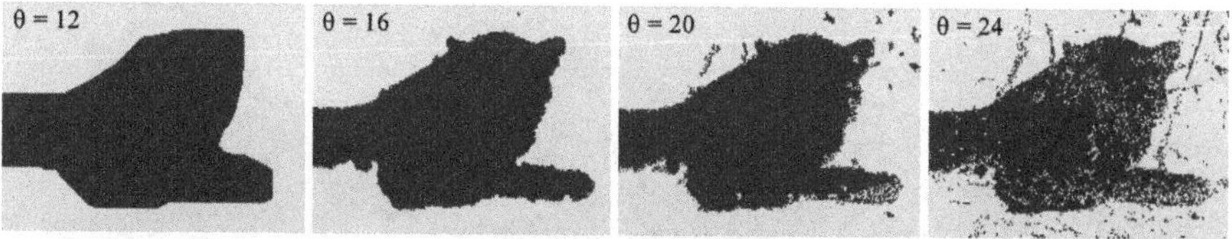

Fig. 4 The effect of varying θ in the final results, showing $\theta = 12, 16, 20$ and 24 respectively

as a vector in $\mathbb{R}^{75}$. The values for each sample tile in the testing image grid are sequentially fed to the trained neural network, and the output generated by each is then the predicted texture at that region. This process is repeated for every tile in the image, until all areas in the image have been classified.

For better results, a final post-processing procedure is applied to the resulting classified tiles. This consists of a simple nearest-neighbor search and replace process which seeks to correct any outlier tiles that usually represent misclassified data. Each tile is tested and compared against its neighbors up to a distance of 2 tiles apart, that is with the closest 24 neighboring tiles. A parameter θ is introduced as a threshold, where if the number of neighbors with a label different than that of the currently tested tile exceeds this parameter θ, then the current tile's label is changed to match that of the surrounding majority. This is analogous to a smoothing procedure as is usually applied at the end of many image processing algorithms to alleviate noise introduced by the system. Figure 4 shows the result of varying θ in the classification results.

3 Results

A comparison of the results obtained on the sample image when three different algorithms are applied is shown in Fig. 5. The first of these is a well known image processing algorithm known as Chan-Vese segmentation [3]. Although this algorithm has been widely proven to be succesful on images with a clear subject and background distinction, it is evident that in more complex images such as the one presented here, it fails to distinguish the textures, instead focusing on local variations such as the spots in the leopard's skin, while being quite sensitive to shadows.

The following result shown is that of the non-optimized neural network solution proposed by Natarajan et al.[5]. The results in this case are much closer to what would be expected of a texture classifier. Although it is able to distinguish the sky portion of the background properly, it is not capable of correctly segregating the branch region of the background.

The final image shows the result of applying the system proposed in this paper, where the silhouette of the leopard is fully distinguishable from the background, with very little noise remaining in the final classification.

The neural network can also be applied to other images having similar texture characteristics. Figure 6 shows an example of a continuous video stream where only the first frame of the stream was used to train the neural network model. Yet, the

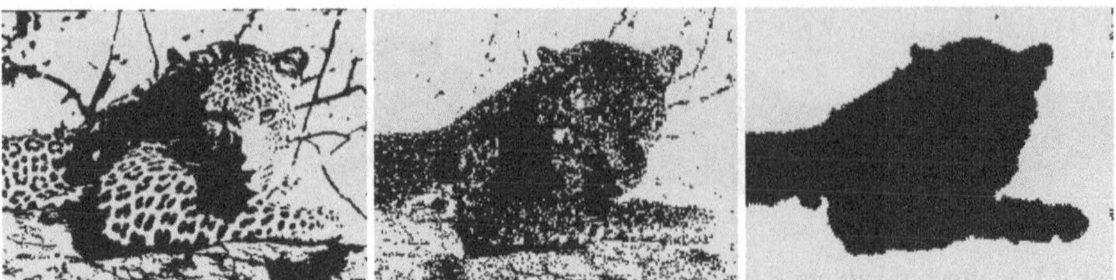

Fig. 5 Comparison of various systems, Chan-Vese segmentation [3] (left), a non-optimized neural network system [5] (middle), and the procedure proposed in this paper (right)

Fig. 6 Multiple frames of a video stream classified with a neural network model trained on a single frame of the video

Fig. 7 A single image with multiple texture labels classified by the proposed system

system is able to generalize the classification results to the rest of the frames in the video, properly identifying the walking leopard and following it throughout the rest of the video sequence.

The system can be extended to classify multiple textures by defining additional labeled classes in the training set, as can be seen in the example results of Fig. 7. Here, multiple regions in the image are defined for each of the textures of interest, each one with its own unique class label.

A common application for texture classification is in the topographical study of aerial photography. The proposed system can be successfully applied to this field as

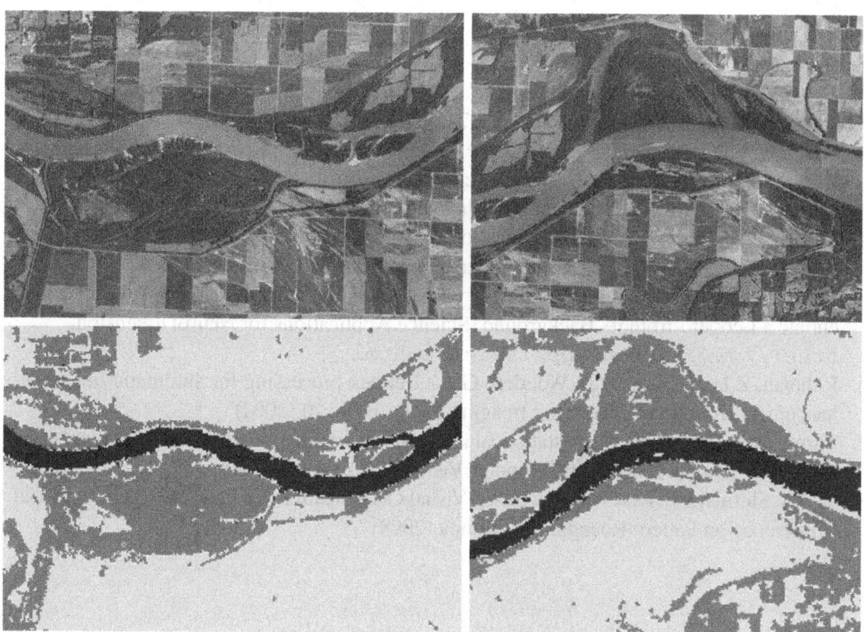

Fig. 8 An example of the system trained only on the left image for the classification of topographical terrain on satellite imagery of the Missouri River, where three types of surfaces are recognized by the model: croplands, woodlands, and the river surface

shown in Fig. 8. Here, the model is able to classify surface types by learning features in the various terrain types, even when such features may be complex and non-repeating such as in the segmentation of croplands shown in the example images.

4 Conclusions and Future Work

A new method was proposed for the classification of textures in an image through a supervised neural network model. The various steps involved in preparing and processing the image were described, and some results and applications of such a system were reviewed.

The current method gives reasonable results in a wide array of sample images, but improvements to the system's robustness can always be improved upon. In particular, enhancements to the context representation of neighboring tiles for better classification consistency would be of great benefit to this technique. Additionally, this neural network model could be a stepping stone to building an unsupervised classification system where no textures or labels are defined beforehand – but instead regions would automatically be clustered together by their features in a manner not unlike current color segmentation algorithms, but with all the improvements described here.

References

1. Chen, Y.Q., Nixon, M.S., Thomas, D.W.: Statistical geometrical features for texture classification. Pattern Recognition 28(4), 537–552 (1995)
2. Encog Online Documentation: Resilient Propagation. Heaton Research, http://www.heatonresearch.com/wiki/
3. Getreuer, P.: Chan-Vese Segmentation. Image Processing On Line (2012)
4. Liu, L., Fieguth, P.: Texture Classification from Random Features. IEEE Transactions on Pattern Analysis and Machine Intelligence 34(3), 574–586 (2012)
5. Natarajan, K., Subramanian, V.: Texture Classification, Using Neural Networks to Differentiate a Leopard from its Background. Science Applications for NeuroDimension, http://www.nd.com/apps/science.html
6. Rahman, Z.U., Jobson, D.J., Woodell, G.A.: Retinex processing for automatic image enhancement. Journal of Electronic Imaging 13(1), 100–110 (2004)
7. Varma, M., Zisserman, A.: A statistical approach to texture classification from single images. International Journal of Computer Vision 62(1), 61–81 (2005)
8. Yu, G., Slotine, J.J.: Fast Wavelet-Based Visual Classification. In: Proc. IEEE International Conference on Pattern Recognition, Tampa (2008)

A Context Aware Architecture to Support People with Partial Visual Impairments

João Fernandes[1], João Laranjeira[3], Paulo Novais[1], Goreti Marreiros[2,4], and José Neves[1]

[1] University of Minho
[2] Institute of Engineering – Polytechnic of Porto
[3] CCTC – Computer Science and Technology Center
[4] GECAD – Knowledge Engineering and Decision Support Group
pg20686@alunos.uminho.pt, {jopcl,mgt}@isep.ipp.pt, {pjon,jneves}@di.uminho.pt

Abstract. Nowadays there are several systems that help people with disabilities on their quotidian tasks. The visual impairment is a problem that affects several people in their tasks and movements. In this work we propose an architecture capable of processing information from the environment and suggesting actions to the user with visual impairments, to avoid a possible obstacle. This architecture intends to improve the support given to the user in their daily movements. The idea is to use speculative computation to predict the users' intentions and even to justify the reactive or proactive users' behaviors.

Keywords: decision support system, ambient intelligence, speculative computation.

1 Introduction

Nowadays there are many systems that assist/help the users on taking decisions in order to make their life more comfortable. People with visual impairments need to travel and move, in an independent way through places that often are not prepared for that. A simple task, as for example, go to a doctor's appointment at the hospital, can become transformed in a very difficult task due to the various obstacles that the user must overcome. In last years, several systems were developed that try to address this issue [1][2], some of them are designed for indoor movement, others for outdoor and there are also systems that consider both scenarios (outdoor and indoor). Despite the existent systems this research area is still an open problem, one of the major limitation lies in the fact that the information gathered from the environment is sometimes dotted with uncertainty or incomplete, leading to incorrect decisions. Other issue is the ability to recognize user intentions and therefore provide better assistance.

S. Omatu et al. (Eds.): *Distrib. Computing & Artificial Intelligence,* AISC 217, pp. 333–340.
DOI: 10.1007/978-3-319-00551-5_41

Decision Support Systems (DSS) are based on a knowledge base (user's characteristics, preferences and goals), with the aim of supporting the user [3]. This type of systems may be completely independent of human interaction and may also take advantage of humans to capture information and process it [3]. DSS may also be framed in ambient intelligence systems. The ambient intelligence (AmI) considers the user at the center of a digital ambient, i.e., AmI is based on an electronic environment that centers the user in a virtual reality that processes information in function of his presence and movements. DSS are sensible and adaptable, being able to meet the user's real needs, habits, attitudes and emotions [4].

The information collected by DSS about the environment will be the basis of all knowledge that will allow supporting users in their decision making processes. However, in a considerable number of situations, the information collected or inferred is incomplete or even uncertain. This may lead to wrong conclusions, which in turn will lead to "wrong support" and "bad decisions". The use of speculative computation to solve problems with missing information or weak conditions of communication has increased in recent years.

In this context, the aim of this work is to propose an architecture able to assist people with visual disabilities, helping those avoiding objects and falls. The ambient information captured by image sensors and devices makes it possible to determine the presence of obstacles (objects, furniture, walls, depth, etc.) in the ambient in which the user is. Through the use of speculative computation, we think it is possible to predict the users' intentions and even to justify the reactive or proactive users' behaviors.

The rest of the paper is organized as follow. First we present the state of the art. Next we present the environment modeling and after it we present a proposed architecture to be applied to a system. In the last section we conclude our work and present the future work.

2 State of the Art

Before starting the development process of this project some research of references had to be done in order to support all the project's specifications and all the decisions made. Regarding the research some areas and systems were taken as reference: Ambient Intelligence (AmI), Decision Support Systems (DSS), Speculative Computation and Multi-Agent Systems.

AmI defines a vision of a future in which environments support the people inhabiting them. We can find examples of AmI applications in several environments like smart homes, smart offices, intelligent meeting rooms, ambient healthcare, smart classrooms, etc [5][6][7]. However for the purpose of this proposal we will focus on our appliance domain of application, the smart offices and intelligent meeting rooms (IMR) context. The emergence of AmI concept contributes with a new perspective, a different way of viewing traditional offices and decision rooms, where it is expected to include key features like monitoring, prediction, autonomy, adaptation/learning and the ability to naturally communicate with humans. In addition the IMR are also defined as an environment that is able to adapt itself to the user needs by releasing the users from routine tasks they should perform to change the environment to suit to their preferences and to access services available at each moment by customized interfaces [7].

The DSS are computational solutions that might be used to assist on taking complex decisions and problem solving [5]. In this sense there are several systems that integrate decision support who predict the user's intention in order to facilitate his tasks and problems.

In the context of AmI we can find several projects that integrate the recognition of the user's intentions: military projects, used to detect terrorist activity [7] and anticipate the enemy's movements [8] and also to control urban riots [9]. Other projects were ambient support can be found are based on users that need medical care, such as elderly [10] and people with chronic diseases like Alzheimer [11]. Another concept of this type of systems is based on help deficient people like blind people that use a smart helmet with camera sensors for navigation [14].

The speculative computation is a provisional computation that uses propositions by default, i.e., uses assumptions to continue processing, when information is incomplete or even when there is no essential information to gain knowledge [12]. This type of computation was used for the first time in the 80's on parallel processing in order to improve the processing times [13]. At the beginning of the XXI century, the researcher Ken Satoh [12], started applying abductive logic on speculative computation with the objective of processing incomplete information [15][16]. To support these systems, Ken Satoh used a Japanese concept denominated "Kiga-Kiku" so that the system would understand a determined situation and take the most appropriate decision without being explicitly informed of what to do. This concept is applied when an individual is able to predict the others intention's and act proactively in function of pre-acquired knowledge [17]. Thus this concept, when applied in systems that incorporate speculative computation, aims to understand and know the diverse situations used to generate an answer, for example the perception of the user's intentions by learning what his preferences are. This concept is also able to manipulate incomplete information common to the environments in which the user is inserted [17].

During the past few years new projects to treat incomplete information's in multi-agent systems have appeared. The emergence of these systems is based on incomplete information and problem solving capacities that each agent has. This required the development of multi-agent systems with the goal of all working together in order to achieve results in solving problems [18]. This type of system can be applied at a domestic level in order to control the house and distribute tasks.

Overall, we can say that the speculative computation is very useful when we want to process incomplete information. Sometimes, in a decision making context, the decision does not allow a speculative processing. This type of computation can also be used in learning processes in order to generate useful information for future decisions.

3 Environment Modeling

The environment modulation is essential to understand the connection between the different models which constitute a particular system. In general, this system should be divided into layers (Fig. 1), where each layer processes the information according to its functionality.

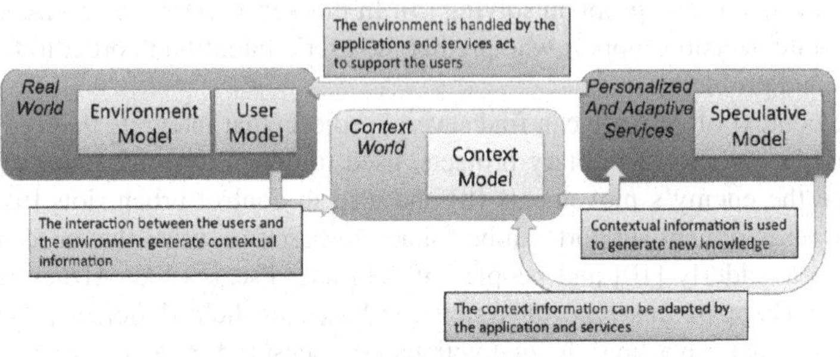

Fig. 1 Environment Modeling [adapted by [19]]

The first layer is named **Real World** and maintains the information acquired by the sensors. Through this model, the information is obtained through image capturing devices, like a video camera, that capture singular or sequence images. The environmental model is the first to interact with the information acquired by the sensors in order to maintain a real representation of the environment in a certain instant of time. The user model maintains specific information of a particular user. This information is very relevant to allow the system to generate the correct answer according to the user needs. Some examples of information that establish the user model are: age, height, weight, deficiencies, etc. Another type of information corresponds to the context in which certain system is operating in order to produce quality results according to the user's goals.

One of the user goals might be, for example, the movement from point A (current location) to a point B (destination). Part of this information is induced through the combination of the information obtained in the first layer, i.e., the environment information combined with the user information generates the context information for a certain situation. This information is maintained in the context model, on second layer, **Context World**.

The third layer, **Personalized and Adaptive Services**, incorporates the speculative model where all gathered information is interpreted and processed in order to produce answers. The answers that shall produce intend to help the user avoiding obstacles and falls in order to ensure a safe user movement. After the interpretation and processing of the information are produced results that will inform the user, through sonorous commands, about the physical actions that should be taken in order to move, thereby avoiding obstacles and possible falls.

The interconnection of all the layers begins at the first layer, **Real World**, where all the information is captured in a time instant. Afterwards, assuming that the user model already contains information, the combination of these with the information gathered on the first layer generates context information that is introduced on the second layer, **Context World**. When all the information is gathered, it's send to the third layer, **Personalized and Adaptive Services**, where, with the help of speculative computation, the parallel processing is performed in order to achieve a solution or a better answer to the problem. The processing result generates new context information that is send to the second layer in order to update the previous information that the system contained. Additionally, this information is send to the first information layer in order to interact with the user aiming to facilitate his movement.

4 Proposed Architecture

The definition of the architecture is fundamental because it will facilitate the understanding of the functionalities and the interconnections of all its components. In the environment modulation were identified all the layers and models that will now be integrated on architecture. At this section the goal is to specify the proposed architecture as well as the information flow and processing until the achievement of the results.

As it was previously mentioned, this architecture initiates its function through the acquisition of information by image capture devices. For that, our goal is to use a camera developed by Microsoft Corporation for the use on videogames on the console Xbox. This camera, denominated "Kinect", has the particularity of obtaining a depth matrix from the image captured, i.e., it's possible to know at what distance the objects and/or obstacles are. Through this functionality it is possible to obtain useful information for the following process. In Fig. 2 we can have a better perception of the proposed architecture, which starts with the image acquisition and ends at the point that the user is informed of the actions that he need perform.

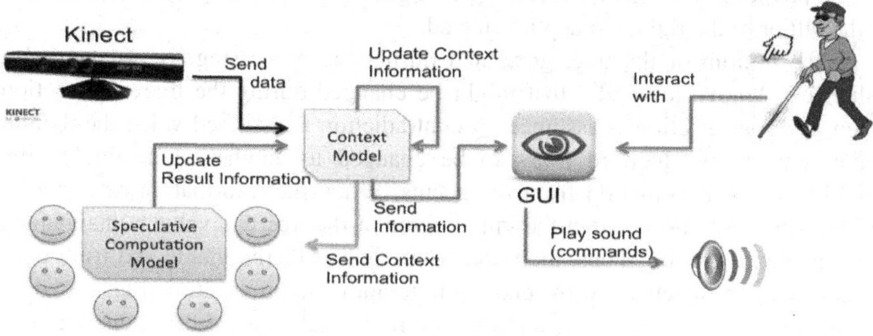

Fig. 2 Architecture for support blind people

The aim of this architecture is to predict the human behavior through speculative computation, covering all the possible cases of a determined problem. For this it is intended to include the multi-agents concept, where each agent is responsible of processing part of the information at a certain time instant in order to integrate the function pattern of the speculative computation and also reduce the processing load.

For a better understanding of the speculative module model function it will be explained how the information is processed. The information processing starts with the capture of information, through the camera, and ends with its processing by the speculative module model in order to obtain answers according to the user's needs. When the information is transmitted to each system agent, the process is initialized in order to generate a logic tree until a response is obtained from the corresponding scenario (Fig. 3).

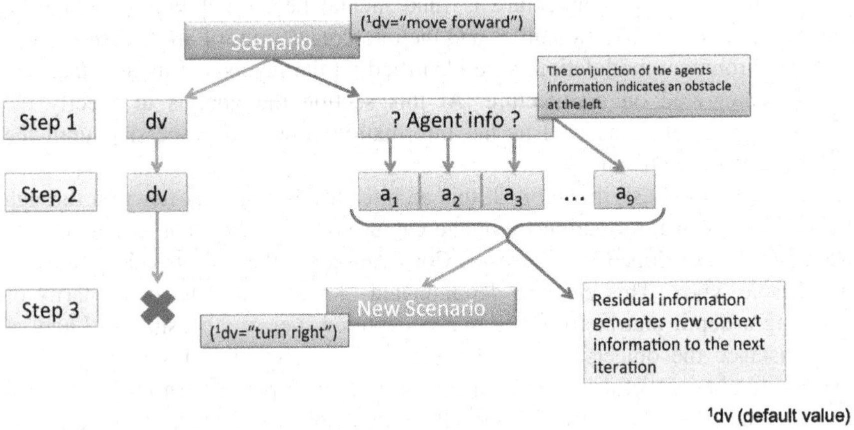

Fig. 3 Iteration process of speculative computation

The possible scenarios for this system might be: move forward, turn or deviate to the left or to the right and depth detected.

At the bottom of the tree, generated during the processing, is instantiated a value (e.g. "move forward") that might be changed during the tree construction when any contradiction is assumed. A contradiction is assumed when the default value (e.g. "move forward") has to be changed to another value due to the combination of information from the agents. When the information received by the agents does not contradict the initial value of the tree this value is maintained until the next iteration. Thus, the initial value is considered viable and truth until some agent contradicts it. At the end of an iteration the information transmitted by the agents, even if it doesn't contradict the initial value, is used to generate new context information for the next iteration.

5 Conclusions and Future Work

In this paper we propose a context aware architecture to support blind people. This architecture was made to be inserted in a multi-agent system. The focus of this architecture is support the blind people to execute some tasks in ambient intelligence context. This process collects information about the ambient intelligence by a sensorial device and the next step is to convert the information in knowledge to support the user decision making.

Our aim to the future work is to develop a multi-agent system based in the proposed architecture to be applied in an ambient intelligence context. The aim is to support the blind people to make better decisions.

Acknowledgments. This work is funded by National Funds through the FCT - Fundação para a Ciência e a Tecnologia (Portuguese Foundation for Science and Technology) within projects PEst-OE/EEI/UI0752/2011 and PTDC/EEI-SII/1386/2012. The work of João Laranjeira is also supported by a doctoral grant by CCTC – Computer Science and Technology Center (UMINHO/BI_1_2013_ENG_Agenda_2020).

References

[1] Pinedo, M., Villanueva, F., Santofimia, M., López, J.: Multimodal Positioning Support for Ambient Intelligence. In: 5th International Symposium on Ubiquitous Computing and Ambient Intelligence, pp. 1–8 (2011)

[2] Guerrero, L., Vasquez, F., Ochoa, S.: An Indoor Navigation System for the Visually Impaired. Sensors 12(6), 8236–8258 (2012)

[3] Shim, J., Warkentin, M., Courtney, J., Power, D., Sharda, R., Carlsson, C.: Past, present, and future of decision support technology. Decision Support Systems 33(2), 111–126 (2002)

[4] Preuveneers, D., Novais, P.: A survey of software engineering best practices for the development of smart applications in Ambient Intelligence. Journal of Ambient Intelligence and Smart Environments 4(3), 149–162 (2012)

[5] Sadri, F.: Ambient intelligence: A survey. ACM Computing Surveys (CSUR) 43(4), no. 36, 36–66 (2011)

[6] Ramos, C., Augusto, J., Shapiro, D.: Ambient Intelligence—the Next Step for Artificial Intelligence. IEEE Intelligent Systems 23, 15–18 (2008)

[7] Cook, D., Augusto, J., Jakkula, V.: Ambient Intelligence: applications in society and opportunities for AI. Pervasive and Mobile Computing 5(4), 277–298 (2009)

[8] Mao, W., Gratch, J.: A utility-based approach to intention recognition. In: AAMAS 2004 Workshop on Agent Tracking: Modelling Other Agents from Observations (2004)

[9] Suzic, R., Svenson, P.: Capabilities-based plan recognition. In: 9th International Conference on Information Fusion, Florence, Italy, pp. 1–7 (2006)

[10] Pereira, L., Anh, H.: Elder care via intention recognition and evolution prospection. In: 18th International Conference on Applications of Declarative Programming and Knowledge Managment (INAP 2009), Évora, Portugal (2009)

[11] Roy, P., Bouchard, B., Bouzouane, A., Giroux, S.: A hybrid plan recognition model for Alzheimer's patients: interleaved-erroneous dilemma. In: IEEE/WIC/ACM International Conference on Intelligent Agent Technology, California, USA, pp. 131–137 (2007)

[12] Satoh, K., Inoue, K., Iwanuma, K., Sakama, C.: Speculative Computation by Abduction under Incomplete Communication Environments. In: ICMAS 2000, pp. 263–270 (2000)

[13] Burton, F.: Speculative Computation, Parallelism and Functional Programming. IEEE Transactions on Computers c-34, 1190–1193 (1985)

[14] Mann, S., Huang, J., Janzen, R., Lo, R., Rampersad, V., Chen, A., Doha, T.: Blind navigation with a wearable range camera and vibrotactile helmet. In: Proceedings of the 19th ACM International Conference on Multimedia - MM 2011, USA, p. 1325 (2011)

[15] Satoh, K., Yamamoto, K.: Speculative computation with multi-agent belief revision. In: International Joint Conference on Autonomous Agents and Multiagent Systems (AAMAS), pp. 897–904. ACM Press, New York (2002)

[16] Satoh, K.: Speculative computation and abduction for an autonomous agent. IEICE Transactions 88-D(9), 2031–2038 (2005)

[17] Satoh, K.: Kiga-kiku computing and speculative computation. Awareness: Self-Awareness in Automic Systems (2012)

[18] Sycara, K.: Multiagent Systems. AI Magazine 19(2), 79–92 (1998)

[19] Garcia-Valverde, T., Serrano, E., Botia, J.: Combining the real worldwith simulations for a robust testing of ambient intelligence services. Artificial Intelligence Review, 1–24 (2010)

Automatic Prediction of Poisonous Mushrooms by Connectionist Systems

María Navarro Cáceres and María Angélica González Arrieta

University of Salamanca, Spain

Abstract. The research offers a quite simple view of methods to classify edible and poisonous mushrooms. In fact, we are looking for not only classification methods but also for an application which supports experts' decisions. To achieve our aim, we will study different structures of neural nets and learning algorithms, and select the best one, according to the test results.

1 Introduction

Fungus are well-known for centuries, at least in their most spectacular manifestations, the mushrooms. Old city Micenas might be called like this due to a fungus: *mykes* in Greek language. Micology, the science which studies fungus, derives from this term *mykes* as well [1].

Humanity interest in fungus is rather old. Our ancestors observed and used the fermentation process. Without this, we would have no bread, wine or beer, among others. About knowledge and use of mushrooms, we have very old evidence, like Pompeya where milk caps are shown in pictures, or the antique tradition of using hallucinogenic mushrooms in some ritual practices [2].

Nowadays, the support for mushroom and fungus is becoming more and more widespread due to the increasing interest in nature. A large amount of mycologic societies have appeared during the last years[3], as well as activities related to them (such as exhibitions or excursion to identify mushrooms). On the other hand, its utility in Biotechnology is also well-known, above all because they are the origin of antibiotics [4].

In this research we propose a way to automate mushrooms classification process. There are more and more people who wants to collect them, but lots of them have to consult an expert to check how many mushrooms are edible. Sometimes, this expert might be wrong, what might endanger health or even the human life [5]. In order to improve this identification system and reduce this error, we will use neural nets, MultiLayer Perceptron (MLP) and Base Radial Function Net(BRF), to classify mushrooms.

In related works about mushroom classification, the authors use data mining techniques due to the large amount of data [6, 7]. In other cases a new algorithm is

S. Omatu et al. (Eds.): *Distrib. Computing & Artificial Intelligence,* AISC 217, pp. 341–349.
DOI: 10.1007/978-3-319-00551-5_42

developed so that the classification can be done [8]. There is an interesting work about artificial vision to recognize different features of a mushroom although uses as well a data mining classification techniques [9]. The main problem of these ones is that the error is not always 0%. Then, we considered whether a neural net solution may give better results in classification problem, at least to classify them only in poisonous or edible ones.

1.1 Theoric Concepts

As we discuss, in this project we used two types of neural nets: supervised ones and hybrid ones. Next, we will describe both briefly.

1.1.1 Perceptron (MLP)

Multilayer Perceptron is a neural net with multiple layers. It can resolve non-linear separable problems (problem XOR), what is the main problem of simple perceptron. MLP can be totally or partially connected (among neurons)[10].

Backpropagation is an algorithm used to train these nets. Due to its properties, this net is one of the most used neural nets to solve problems.[11] Its output is function of the connection weights, a threshold value and input vectors.

Nonetheless, the problem of local minimums in error functions [12] makes difficult the training process, as once a minimum is reached, the algorithm always stops. A possible solution is change the topology, the initials weights or the order in which data training input was presented[10].

1.1.2 Base Radial Function Net (BRF)

BRF are a hybrid model which uses supervised and non-supervised learning. [10]It allows to make no lineal systems modelings. They always have three layers: Input (to transfer data), output (it uses a linear function) and hidden layer (uses Gaussian function to give an output.) The output of hidden layer depends on the distance between centroid and input vector [13, 14]. In this case, the response is not global, because neurons only give an output if distance between centroid and input vector is small enough.

However, topology of these nets depends on size and shape of input data. The number of neurons can be assigned randomly or using an algorithm to add neurons till a threshold error was achieved. Besides, we usually train first hidden layer to determine centroid values, using k-means algorithm [15], and after output layer, with a backpropagation algorithm.

2 Materials and Methods

2.1 BRF versus MLP

BRF and MLP are feedfowarding nets organised in layers, and both can aproximate any continuous function. As regards differences between BRF and MLP, we can note that BRF have one hidden layer, whereas MLP can have several of them. Neurons in BRF have no weights assigned in connections between input layer and hidden layer. Besides, the activation function of the output layer is always a linear function. However, the main difference is the function of activation in the hidden layer. In BRF we use a base radial function (like Gaussian function), which provokes that each neuron is activated in one zone of the input space data. On the other hand, MLP uses a sigmoid function, which permits each neuron was activated with each input. Due to this features, we can conclude that:

1. MLP gets global relationships between input and output. Because of this, learning is slower, [15] as the change of only one weight provokes changes in the output for all the input data.
2. BRF gets local relationships. Each element of the hidden layer specializes in a certain region of the outputs. So, this provokes a more quick learning, because the change of one weight only affects to the processing element associated to this weight [16], and also to a certain group of inputs, belonging to the class which the neural element affected is representing.

2.2 Data Processing

Data have been obtained from `http://archive.ics.uci.edu/ml/datasets/Mushroom` It is a database of 8124 registries, with 22 attributes of nominal type. The attributes are:

Table 1 Studied attributes

cap-shape	cap-surface
cap-color	bruises
odor	gill-attachment
gill-spacing	gill-size
gill-color	stalk-shape
stalk-root	stalk-surface-below-ring
stalk-surface-above-ring	stalk-color-above-ring
stalk-color-below-ring	veil-type
veil-color	ring-number
ring-type	spore-print-color
population	habitat

All the attributes were nominal type, so we have to convert it into numerical type so as the neural net could process them. The solution is to turn nominal data into binary data as we show in the table with the example: odor-attribute.

We have the next values for odor: *almond, anise, none, fishy, foul, spicy, musty, creosote* and *pungent*. If we have odor = *almond*, we turn it into a numerical value, as shown in Table 2.2.

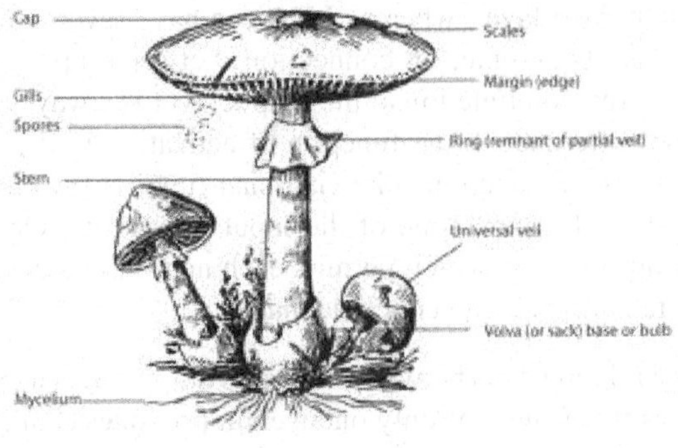

Fig. 1 Parts of a general mushroom

Table 2 Example of nominal data transformation

Odor-p	Odor-a	Odor-l	Odor-y	Odor-f	Odor-n	Odor-c	Odor-m	Odor-s
0	1	0	0	0	0	0	0	0

With this method, we got more than 50 values for each input registry, too much to process. Thats why we began to make tests to remove unnecessary attributes. After some inquiries and tests in MATLAB, we decided to use only 5 attributes: odor, stalk-surface, spore-color, bruises and gill-size.

2.3 *Process Modelling*

We used MATLAB to carry out this research. The simple is stored in an Excel file retrieved by MATLAB [17]and saved in two matrix: one for inputs and one for targets. We used rows from 1 to 6000 as training data, and from 6001 to 8124 as test data. We also implement a function which chose a certain number of random values from this matrix, in order to make different studies about the data and the results. Our MLP has one hidden layer with 60 neurons. On the other hand, BRF is automatically created by adding neurons in the hidden layer until threshold error was achieved. So, it is important not to have many training data to avoid the non-convergence of the threshold error established by the user.

3 Results

In order to study the different neural structures behavior, we put into practice them to predict edible or poisonous mushroom. We got different results, that we will discuss now. In both MLP and RBF cases, we carried out two experiments.

3.1 MLP Results

3.1.1 Training Data

Training sample size influences notably in the neural network precision. With a small sample, neurons are not capable of classifying test data right. As in Base Radial Network as in MLP, we used 100, 1000 and 6000 training registries. In MLP, 100 registries are not enough to achieve a minimal error. In fact, mean error obtained is 0.9135. The Figure 2 shows the classification results of the test sample. It is not an acceptable result at all.

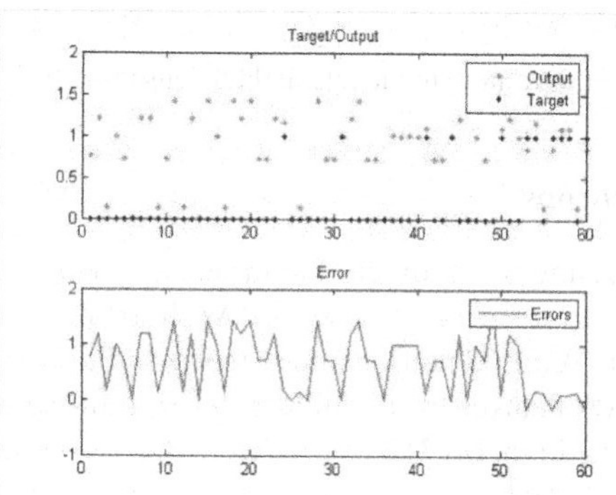

Fig. 2 MLP results with 100 training input registries

Lets see with 1000 data. Results shown in Figure 3 In this case, error is 0.7532,

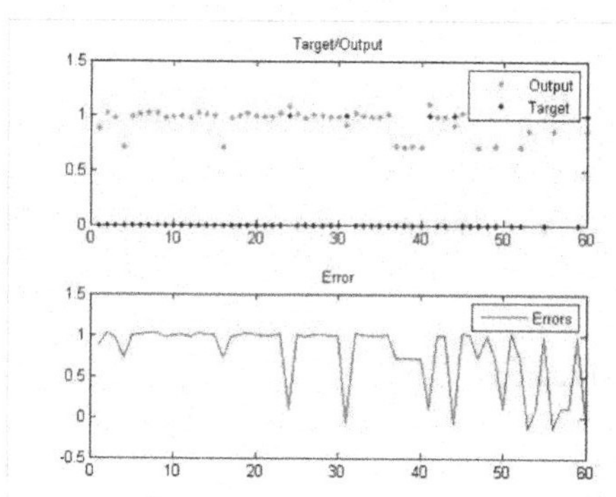

Fig. 3 MLP results with 1000 training input registries

not yet aceptable. However, with 6000 data, MLP network has a mean error of 0.0092. By inserting random data test, the results are quite positive, as is shown in the Figure 4.

In conclusion, the larger training simple, the better to train the neural net.

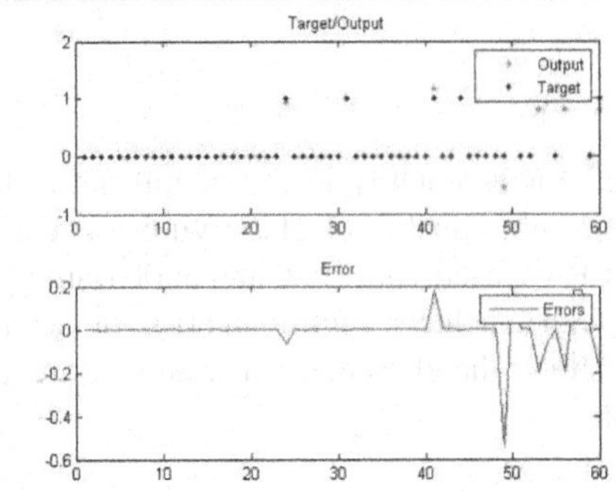

Fig. 4 MLP results with 6000 training input data and 60 neurons

3.1.2 Number of Neurons

We verify that by increasing the number of neurons, the error was reduced to some extent in which error can even increase again. With 10 neurons in hidden layer, error was about 0.7281. With 60 neurons, the error was 0.0094. (See Figure 4) On the other hand, with 200 neurons in the hidden layer, training time was increased and mean error was quite high, 0.7712.

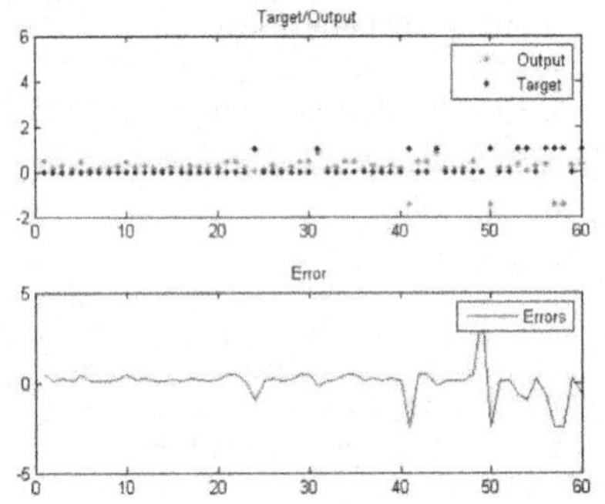

Fig. 5 MLP results with 10 neurons

3.2 *RBF*

3.2.1 Training Data

Threshold error was established in 10^{-3} to create the net. By inserting 100 training data, the network created has 99 neurons. The output and target data is shown in Figure 6.

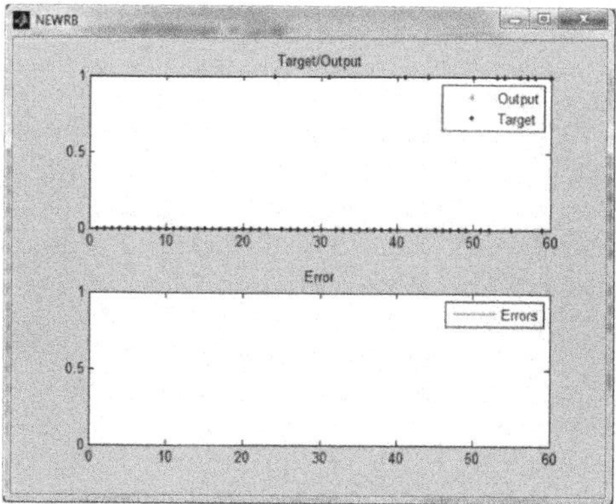

Fig. 6 RBF results with 100 training data (Threshold error is 10^{-3})

If we insert 500 values, the net never converge, because there are too data for the net to generalize, as we can see in Figure 7.

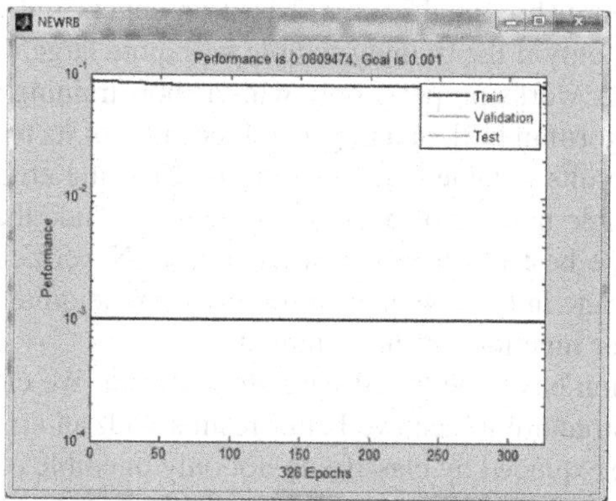

Fig. 7 RBF training process with 500 training data (threshold error is 10^{-3}) Convergence is not possible, at least in a reasonable period of time

With a simple of 100 values, this net predict the output very well. In fact, the error obtained in this case, was about 0.

3.2.2 Threshold Error

First of all, we put an error of 0.1. The number of neurons has been reduced (only 21). However, the difference between data output and target output was increased. The mean error is about 0.2472. The lower error we want, the more number of neurons we need. Perhaps, if we search a minimal error, the net may not ever converge, like in the excess of data situation.

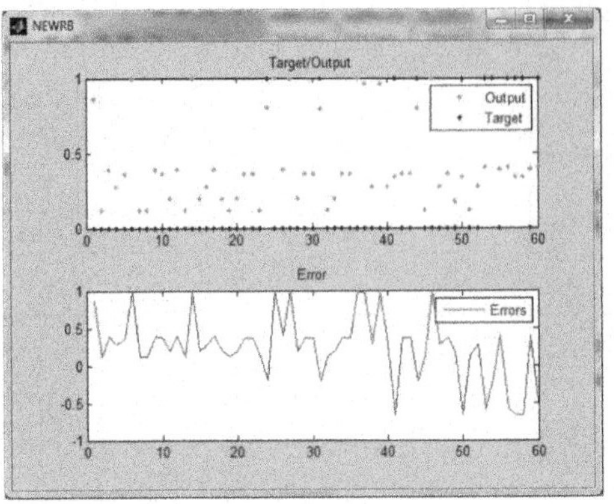

Fig. 8 RBF results. Threshold error set 0.1

4 Conclusions

In this research, two methods were studied in order to achieve the main purpose: recognize an edible mushroom. The first, based on supervised learning, gave acceptable results, but only if the training sample was quite large. On the other hand, Base Radial Network works properly only with a short training sample, due to its k-means training algorithm [15], and in general, because of its hybrid learning.

In view of the results obtained, and bearing in mind the error minimization in BRF with a reasonable number of neurons, we can conclude that BRF is a better model to achieve the best result in our studied case. Nevertheless, MLP will be much more appropriate in case we have a training sample quite large, although in our research the error function was not minimal.

Yet, the application has to be tested with another data. We can also consider to improve the MLP structure to achieve better results with an error of almost 0. Its functionality can be extended by classifying not only in edible or poisonous mushrooms, but also in its own species. In this hypothetical case, we may use another training algorithm, or even another learning mechanism which provokes a change in the neural structure. This new changes have to be chosen carefully, as the error has to be 0% in the majority of the situations in order to guarantee the reliability of this application. Also as a future work, it is important to visualize results in a properly way for the final user.

References

1. Alexopoulos, C.J.Y., Mins, C.W.: Introducción a la Micología. Omega, Barcelona (1985)
2. Ruiz Herrera, J.: El asombroso reino de los hongos. Avance y Perspectiva. Colección Micológica, Herbario de la Facultad de Biología (2001)
3. Micology experts and amateurs, `http://www.micologia.net/`
4. Palazó, F.: Setas para todos. Pirineo, Castellano (2001)

5. García Rollán, M.: Setas venenosas: intoxicaciones y prevención. Ministerio de Sanidad y Consumo, Madrid, Castellano (1990)
6. Chai, X., Deng, L., Yang, Q., Ling, C.X.: Test-Cost Sensitive Naive Bayes Classification (2003)
7. Bay, S.D., Pazzani, M.J.: Detecting Group Differences: Mining Contrast Sets. Data Min. Knowl. Discov. 5 (2001)
8. Li, J., Dong, G., Ramamohanarao, K., Wong, L.: DeEPs: A New Instance-based Discovery and Classification System. In: Proceedings of the Fourth European Conference on Principles and Practice of Knowledge Discovery in Databases (2001)
9. Matti, D., Vajda, P., Ebrahimi, T.: Mushroom Recognition (2010)
10. Hayking, S.: Neural Networks: A Comprehensive Foundation, 2nd edn. Prentice Hall (1998)
11. Bishop, M.: Neural Networks for Pattern Recognition. Oxford University Press (1995)
12. Jaynes, E.T.: Probability Theory: The Logic of Science. Cambridge University Press (2003)
13. Tao, K.K.: A closer look at the radial basis function (RBF) networks. In: Singh, A. (ed.) Conference Record of the Twenty-Seventh Asilomar Conference on Signals, Systems, and Computers. IEEE Comput. Soc. Press, Los Alamitos (1993)
14. Fine, T.L.: Feedforward Neural Network Methodology, 3rd edn. Springer, Nueva York (1999)
15. MacKay, D.J.C.: Information Theory, Inference and Learning Algorithms. Cambridge University Press (2004)
16. Alonso, G., Becerril, J.L.: Introducción a la inteligencia artificial. Multimedia Ediciones, S.A. Barcelona (1993)
17. Matlab Official Site, http://www.mathworks.es

Adaptive Learning in Games: Defining Profiles of Competitor Players

Tiago Pinto and Zita Vale

GECAD – Knowledge Engineering and Decision-Support Research Center, Institute of Engineering – Politechnic of Porto (ISEP/IPP), Porto, Portugal
{tmcfp,zav}@isep.ipp.pt

Abstract. Artificial Intelligence has been applied to dynamic games for many years. The ultimate goal is creating responses in virtual entities that display human-like reasoning in the definition of their behaviors. However, virtual entities that can be mistaken for real persons are yet very far from being fully achieved. This paper presents an adaptive learning based methodology for the definition of players' profiles, with the purpose of supporting decisions of virtual entities. The proposed methodology is based on reinforcement learning algorithms, which are responsible for choosing, along the time, with the gathering of experience, the most appropriate from a set of different learning approaches. These learning approaches have very distinct natures, from mathematical to artificial intelligence and data analysis methodologies, so that the methodology is prepared for very distinct situations. This way it is equipped with a variety of tools that individually can be useful for each encountered situation. The proposed methodology is tested firstly on two simpler computer versus human player games: the rock-paper-scissors game, and a penalty-shootout simulation. Finally, the methodology is applied to the definition of action profiles of electricity market players; players that compete in a dynamic game-wise environment, in which the main goal is the achievement of the highest possible profits in the market.

Keywords: Adaptive Learning, Artificial Intelligence, Player Profiles.

1 Introduction

Machine learning is a field of artificial intelligence that provides computers with the ability to learn without being explicitly programmed. Machine learning focuses on the development of computer programs that can teach themselves to adapt and expand when exposed to new data. Therefore, *"The goal of machine learning is to build computer systems that can adapt and learn from their experience"* [1].

The process of machine learning is similar to that of data mining. Both systems search through data to look for patterns. However, instead of extracting data for human comprehension - as is the case in data mining applications - machine learning uses that data to improve the program's own understanding. Machine

S. Omatu et al. (Eds.): *Distrib. Computing & Artificial Intelligence*, AISC 217, pp. 351–359.
DOI: 10.1007/978-3-319-00551-5_43

learning programs detect patterns in data and adjust program actions accordingly. Thus *"Learning denotes changes in a system that (...) enables a system to do the same task more efficiently the next time"* [2], this way making computer systems more capable of solving problems, by adapting and being more self-sufficient, with the purpose of supporting humans in their tasks.

This paper proposes an intelligent methodology with the purpose of providing adaptive learning capabilities to dynamic systems. The proposed methodology intelligently selects the best from a set of different machine learning algorithms, depending on their adequacy for each different context. For that, reinforcement learning algorithms are used, which choose the most appropriate from a set of different learning approaches. This is done through a constant update of the algorithms' confidence values, resulting from the algorithms' performance in each different case. These learning approaches have very distinct natures, from mathematical to artificial intelligence and data analysis methodologies, so that the methodology is prepared for very distinct situations.

The proposed methodology is applied to the popular rock-paper-scissors game, taking advantage on the easy comprehension of this game's procedures and consequent facilitation of the analysis of the results. The methodology is also used in a penalty-shootout simulation game-play, in which the challenge increases with the enlargement of the available actions that can be performed. Finally, the proposed methodology is applied to a realistic dynamic environment: the electricity markets. For that the Multi-Agent System for Competitive Electricity Markets (MASCEM) [3, 4] is used. MASCEM is an electricity markets simulator, which models a variety of market models and represents the most common entities found in this environment through software agents. The proposed methodology is used to provide decision support capabilities to a MASCEM's market negotiating agent, in the quest for the achievement of the highest possible profits.

2 Players Profiles Definition

The Rock-Scissors-Paper is a popular game in which two players face each other using one from three actions or plays: Rock, Scissors or Paper. The only three rules are: (i) Rock beats Scissors; (ii) Scissors beats Paper; (iii) Paper beats Rock. The core idea is that no strategy can win every time. Regardless of which "attack" you pick, you can either win or lose (or tie in the case of picking the same attack). Many game designs follow this pattern. This keeps players from finding one specific strategy that wins every time, encouraging players to play dynamically. To reach the full depth of game-play, a player must be able to predict opponent behavior, develop behavioral patterns, dynamically recognize them, intentionally trick, and make mistakes.

The second application of the proposed methodology is a penalty-shootout simulation of a football game. In this case, also considering a player versus computer scenario, each player must choose where to shoot when attacking, and which side to save when defending. In this simulation, six possible actions where defined, combining left, right and center with high, low and middle-height. In this game scenario the player must predict where the opponent will shoot to, in order to save the ball, and foresee where the opponent will defend, in order to choose where to shoot.

Finally, the proposed methodology is applied to a player acting in the electricity market. Electricity markets worldwide are complex and challenging environments, involving a considerable number of participating entities, operating dynamically trying to obtain the best possible advantages and profits [5]. The recent restructuring of these markets increased the competitiveness of this sector, leading to relevant changes and new problems and issues to be addressed [5]. Market players and regulators are very interested in foreseeing market behavior, thus a clear understanding of the impact of power systems physics on market dynamics and vice-versa is required. It is essential for these professionals to fully understand the market's principles and learn how to evaluate their investments in such a competitive environment [5]. For this, electricity markets simulators, such as the Multi-Agent System for Competitive Electricity Markets (MASCEM) [3, 4] are used. MASCEM uses real electricity markets data [6] to represent realistic scenarios, enabling its simulations to project the reality. Endowing MASCEM agents with adaptive learning capabilities enables a coherent and realistic study on whether the used methodologies are applicable and advantageous in a realistic environment.

In order to build suitable profiles of competitor players, it is essential to provide players with strategies capable of dealing with the constant changes in competitors' behavior, allowing adaptation to their actions and reactions. For that, it is necessary to have adequate forecasting techniques to analyze the data properly, namely the historic of other players past actions. The way each player's action is predicted can be approached in several ways, namely through the use of statistical methods, data mining techniques [4], neural networks (NN) [7], support vector machines (SVM), or several other methods [1, 2, 8]. However, since the competitor players can be able to adapt to the circumstances, there is no method that can be said to be the best for every situation, only the best for one or other particular case.

To take advantage of the best characteristics of each technique, we decided to create a method that integrates several distinct technologies and approaches. The method consists of the use of several forecasting algorithms, all providing their predictions, and, on top of that, a reinforcement learning algorithm that chooses the one that is most likely to present the best answer. This choice is done according to the past experience of their responses and also to the present scenario characteristics.

Three alternative reinforcement learning algorithms are used:

- A simple reinforcement learning algorithm, in which the updating of the values is done through a direct decrement of the confidence value C in the time t, according to the absolute value of the difference between the prediction P and the real value R. The updating of the values is expressed by (1).

$$C_{t+1} = C_t - |(R - P)| \quad (1)$$

- The revised Roth-Erev reinforcement learning algorithm [8] that, besides the features of the previous algorithm, also includes a weight value *W*, ranging from 0 to 1, for the definition of the importance of past experience. This version is expressed as in (2).

$$C_{t+1} = C_t \times W - |(R - P)| \times (1 - W) \quad (2)$$

- A learning algorithm based on the Bayes theorem of probability [9], in which the updating of the values is done through the propagation of the probability of each algorithm being successful given the facts of its past performance. The expected utility, or expected success of each algorithm is given by (3), being *E* the available evidences, *A* an action with possible outcomes *Oi*, *U(Oi|A)* the utility of each of the outcome states given that action *A* is taken, *P(Oi|E,A)* the conditional probability distribution over the possible outcome states, given that evidence *E* is observed and action *A* taken.

$$EU(A|E) = \sum_i P(O_i \mid E, A) \times U(O_i \mid A) \quad (3)$$

The algorithms used for the predictions are:

- Algorithms based on pattern analysis:
  - Sequences in the past matching the last few actions. In this approach are considered the sequences of at least 3 actions found along the historic of actions of this player. The sequences are treated depending on their size. The longer matches to the recent history are attributed a higher importance.
  - Most repeated sequence along the historic of actions of this player.
  - Most recent sequence among all the found ones.
- Algorithm based on history matching. Regarding not only the player actions, but also the result they obtained. This algorithm finds the previous time that the last result happened, i.e., what the player did, or how he reacted, the last time he performed the same action and got the same result.
- Algorithm returning the most repeated action of this player. This is an efficient method for players that tend to perform recurrent actions.
- Second-Guessing the predictions. Assuming that the players whose actions we are predicting are gifted with intelligent behavior, it is essential to shield this system, avoiding being predictable as well. So this strategy aims to be prepared to situations when the competitors are expecting the actions that the system is performing.
  - Second-Guess: if the prediction on a player action is *P*, and it is expecting the system to perform an action *P1* that will overcome its expected action, so in fact the player will perform an action *P2* that overcomes the

system's expected *P1*. This strategy prediction is the *P2* action, in order for the system to expect the player's prediction.

  - Third-Guess: this is one step above the previous strategy. If a player already understood the system's second guess and is expecting the system to perform an action that overcomes the *P2* action, than it will perform an action *P3* that overcomes the system prediction, and so, this strategy returns *P3* as the predicted player action.

- Self Model prediction. Once again if a player is gifted with intelligent behavior, it can perform the same historical analysis on the system's behavior as the system performs on the others. This strategy performs an analysis on its own historic of actions, to predict what itself is expected to do next. From that the system can change its predicted action, to overcome the players that may be expecting it to perform that same predicted action.
- Second-Guess the Self Model prediction. The same logic is applied as before, this time considering the expected play resulting from the Self Model prediction.

Additionally, for the electricity market players profiles definition, some specific approaches were added:

- A feed-forward neural network trained with the historic market prices, with an input layer of eight units, regarding the prices and powers of the same period of the previous day, and the same week days of the previous three weeks. The intermediate hidden layer has four units and the output has one unit – the predicted bid price of the analyzed agent for the period in question.
- Based on Statistical approaches. There are five strategies in this category:
  - Average of prices and powers from the agents' past actions database, using the data from the 30 days prior to the current simulation day, considering only the same period as the current case, of the same week day. This allows us to have a strategy based on the tendencies per week day and per period.
  - Average of the agent's bid prices considering the data from one week prior to the current simulation day, considering only business days, and only the same period as the current case. This strategy is only performed when the simulation is at a business day. This approach, considering only the most recent days and ignoring the distant past, gives us a proposal that can very quickly adapt to the most recent changes in this agent's behavior. And is also a good strategy for agents that tend to perform similar actions along the week.
  - Average of the data from the four months prior to the current simulation day, considering only the same period as the current case. This offers an approach based on a longer term analysis. Even though this type of strategies, based on averages, may seem too simple, they present good results when forecasting players' behaviors, taking only a small amount of time for their execution.

- Regression on the data from the four months prior to the current simulation day, considering only the same period of the day.
- Regression on the data of the last week, considering only business days. This strategy is only performed when the simulation is at a business day.

3 Experimental Findings

The first part of the experimental studies section considers the application of the proposed methodology to the Rock-Scissors-Paper game and the penalty shootout simulation. For both cases the results refer to the average of 50 repetitions, with the plays being executed against multiple test subjects.

Regarding the Rock-Scissors-Paper game, the results consider a game length of 150 plays, from which only winning and losing situations are considered (note that ties are possible in this game, when players choose the same action).

In order to test the scenario using the proposed methodology in the penalty shootout simulation, three cases were defined: (i) Penalty 1: considering only three possible actions: left, right and center; (ii) Penalty 2: considering the three actions from Penalty 1, and adding the possibility to shoot high or low; (iii) Penalty 3: considering the whole nine actions exposed in section 2. For all cases, the game length is 100 plays.

Table I presents the results of these tests, concerning the number of wins from both the human player and the computer, the percentage of the computer's methodology success, and the algorithm that achieved the higher confidence value in each case (i.e. the algorithm that presented the higher success for the specific scenario). The reinforcement learning algorithm chosen for all tests was the Roth-Erev algorithm, with a weight value for past events of 0.4; meaning a higher importance for the most recent events (success or failure of each of the different algorithms).

Table 1 Results of the proposed methodology when applied to the two games

Game	Success (%)	Computer wins	Player wins	Best Strategy
Rock-Scissors-Paper	71,03	76	31	History Matching
Penalty 1	69,00	138	62	Most Recent Pattern
Penalty 2	61,50	123	77	Longer Pattern (5 Moves)
Penalty 3	58,00	116	84	Longer Pattern (5 Moves)

From Table I it is visible that for all cases the proposed methodology presented a higher percentage of success than the human opponent, meaning that the computer won more often than the opposing player. This percentage, however, decreases as the number of possible actions increases. This fact is due to the increase in the actions search space, and exponential number of possible combinations when finding patterns in the competitor player's behavior.

Another interesting result is the variation of the approach that achieved the best results from case to case. In the Rock-Scissors-Paper game, the strategy that achieved the best results was the History Matching. Regarding the penalty shootout simulation, when considering a smaller amount of possible actions, the Most Recent Pattern was the most suitable strategy, while in the cases with a higher number of possibilities, the Longer Pattern achieved the best results. This is a very important result, for it supports the need and advantages in considering several distinct approaches, which are (or can be) more suitable to different cases.

The second part of this section regards the application of the proposed approach to the decision support of an electricity market negotiating agent. The tests were performed using a realistic test scenario, presented in [4], where the agents' actions are defined accordingly to real electricity market players from the Iberian electricity market – OMIE [6]. The MASCEM simulation concerns 61 consecutive days, starting on October 4th, and considers the negotiations in the day-ahead spot market (auction based) [6]. The test subject is a buyer of electric energy, which presents a market bid price based on the average prices of the previous four months, while requesting for a fix amount of power to buy regardless of the situation (note that one action in this type of market negotiation is composed by a bid price and an amount of power proposed to be sold or bought). Figure 1 presents the results of the test concerning the prediction of the player's bids along the 61 days.

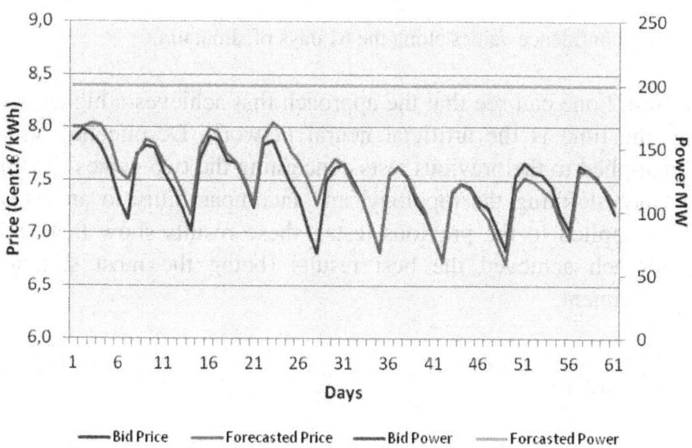

Fig. 1 Predicted and actual bid prices and powers of the subject player

Figure 1 shows that the proposed methodology was able to achieve very good prediction accuracy when faced with the actions of the electricity market player. The amount of power, being a fix value, caused no problem for the methodology, achieving 100% accuracy in the prediction. Concerning the bid price, the predictions are always very close to the actual value, being noticed a clear improvement as the time progresses. This improvement is due to the constant learning capabilities of the method, which allowed the predictions to adapt to the constant changes.

Figure 2 presents the comparison between the confidence values of the different considered algorithms that are used by the reinforcement learning algorithm.

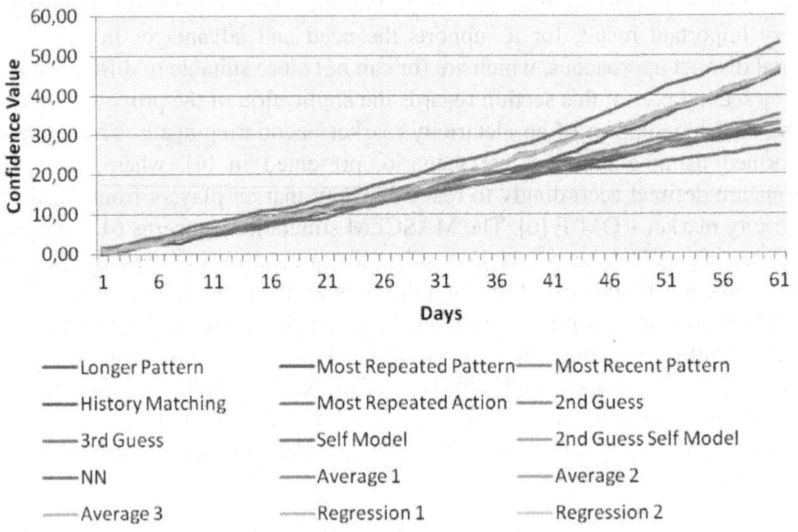

Fig. 2 Strategies' confidence values along the 61 days of simulation

From Figure 2 one can see that the approach that achieves a higher confidence value along the time is the artificial neural network. Despite this strategy not having been applied to the previous tests concerning the two games (because of its specificity when defining the topology, and incomparability to an eventual NN that could be applied to the previous tests), these results show that yet again a different approach achieved the best results (being the most suitable) for a different environment.

4 Conclusions

This paper proposed a methodology that provides adaptive learning capabilities to dynamic systems. The proposed methodology intelligently selects the best from a set of different machine learning algorithms, depending on their adequacy for each different context. For that, reinforcement learning algorithms are used. The proposed methodology is applied to the popular rock-paper-scissors game and to a penalty-shootout simulation. Finally, the proposed methodology is applied to a realistic dynamic environment: the electricity markets, using the MASCEM simulator.

The accomplished results show that this methodology is applicable to very distinct environments, having achieved good results in defining profiles of players of very distinct natures. Both when applied to simple software games and when applied to realistic situations as the electricity market negotiations, the proposed methodology showed adaptive learning capabilities which have originated advantageous results.

Another relevant conclusion is that the approaches that achieved the higher confidence values in the different cases were also different. This supports the need for considering different algorithms, of different natures, when dealing with dynamic environments. Considering such distinct approaches allows systems to be prepared for dealing with constantly changing scenarios, in which different views are required.

References

1. Dietterich, T.: Bridging the gap between specification and implementation. IEEE Expert 6(2), 80–82 (1991)
2. Simon, H.: Administrative Decision Making. IEEE Engineering Management Review 1(1), 60–66 (1973)
3. Praça, I., et al.: MASCEM: A Multi-Agent System that Simulates Competitive Electricity Markets. IEEE Intelligent Systems, Special Issue on Agents and Markets 18(6), 54–60 (2003)
4. Vale, Z., Pinto, T., Praça, I., Morais, H.: MASCEM - Electricity markets simulation with strategically acting players. IEEE Intelligent Systems. Special Issue on AI in Power Systems and Energy Markets 26(2) (2011)
5. Meeus, L., et al.: Development of the Internal Electricity Market in Europe. The Electricity Journal 18(6), 25–35 (2005)
6. OMIE – Operador del Mercado Iberico de Energia website, http://www.omie.es/ (acessed on January 2013)
7. Amjady, N., et al.: Day-ahead electricity price forecasting by modified relief algorithm and hybrid neural network. IET Generation, Transmission & Distribution 4(3), 432–444 (2010)
8. Erev, I., Roth, A.: Predicting how people play games with unique, mixed-strategy equilibria. American Economic Review 88, 848–881 (1998)
9. Korb, K., Nicholson, A.: Bayesian Artificial Intelligence. Chapman & Hall/CRC (2003)

Associative Learning for Enhancing Autonomous Bots in Videogame Design

Sergio Moreno[1], Manuel G. Bedia[1], Francisco J. Serón[2], Luis Fernando Castillo[1,2], and Gustavo Isaza[3]

[1] Universidad de Zaragoza, Spain
rhonellercy@hotmail.com
[2] Universidad de Zaragoza, Spain, Departamento de Ciencias de la Computación
{mgbedia,fjseron}@unizar.es
[3] Universidad de Caldas. Manizales, Colombia, Departamento de Sistema e Informática
{luis.castillo,Gustavo.isaza}@ucaldas.edu.co
[4] Universidad Nacional de Colombia Sede Manizales, Departamento de Ing. Industrial
lfcastilloos@unal.edu.co

Abstract. The Today's video games are highly technologically advanced, giving users the ability to step into virtual realities and play games from the viewpoint of highly complex characters. Most of the current efforts in the development of believable bots in videogames — bots that behave like human players — are based on classical AI techniques. Specifically, we design virtual bots using Continuous-Time Recurrent Neural Network (CTRNNs) as the controllers of the non-player characters, and we add a learning module to make an agent be capable of re-learning during its lifetime. Agents controlled by CTRNNs are evolved to search for the base camp and the enemy's camp and associate them with one of two different altitudes depending on experience. We analyze the best-evolved agent's behavior and explain how it arises from the dynamics of the coupled agent-environment system. The ultimate goal of the contest would be to develop a computer game bot able to behave the same way humans do.

1 Introduction

The use of Continuous-Time Recurrent Neural Network (CTRNNs) [1] to design controllers for mobile agents has excelled in Evolutionary Robotics [2]. While it is true that physical agents were used in the projects undertaken initially, in the last few years the trend of implementing models adapted to the design of synthetic agents in virtual simulation has dominated [3], such as in video games. The field of video games in addition to being a powerful entertainment industry is also a framework for experimentation and research to test Artificial Intelligence techniques in complex environments. This article aims to adapt a learning technique to the control of virtual characters or "bots" (synthetic agents with human behaviour) using CTRNNs for a current commercial video game. Associative learning requires

S. Omatu et al. (Eds.): *Distrib. Computing & Artificial Intelligence*, AISC 217, pp. 361–368.
DOI: 10.1007/978-3-319-00551-5_44

responses, which are paired with a particular stimulus. Organisms at several levels of 'complexity' provide evidence for this, including many extraordinarily simple ones (e.g. examples of the small nematode worm C. elegans, evidence for the formation of associations between temperatures and food has been known for quite some time). Unreal Tournament 2004, also known as UT2004 or UT2K4, is a first-person action video game, mainly oriented to the multi-player experience whose aim varies according to the chosen game mode, but always includes the use of violence (virtual, of course) to achieve the goal. The goal is to get a bot controlled by a CTRNN with the ability to learn in real time without synaptic plasticity, i.e. without making changes to the network parameters. In the first section, the experiment and desired behaviour for the bot is described. Then the focus shifts to how to achieve the desired behaviour using the evolution of CTRNN. In the section that follows, the capacity of the CTRNN to learn without synaptic plasticity is analysed. Section 4 includes an analysis of the dynamics of said CTRNN with its environment and formally analyses bot behaviour in depth from its bifurcation diagrams. Lastly, the conclusions drawn from the experiment are shown.

2 Theoretical Framework

2.1 Continuous-Time Recurrent Neural Networks (CTRNN)

Among the community that studies autonomous agents, there is a growing interest in using dynamic neural networks to control the behaviour of agents [4] along with the evolution of such networks [5,6,7]. These studies, among others, have shown that the combination of genetic algorithms and neural networks is an interesting technique to develop control structures in autonomous agents. However as a drawback, the required time for the evolution is the most important limit that seems to show the evolutionary focus. The main advantages are their noise resistance, interpolation ability, parallelism, its ability to model dynamic systems for dynamic and recurrent neural networks, and the response to the its evolution using Genetic Algorithms since the meaning of their actions "emerge" from both mechanisms. On the other hand, updating neurons is normally synchronous (appropriate for when the neuron network task involves the recognition and classification of patterns), but neural networks that use time as one of the multiple information processing supports are used more frequently each time. This is because one of the characteristics that should be drawn consists of the temporary relationship of other characteristics. Its performance is completely asynchronous with some interneuron connections that produce delays in the input of neurons. For the tasks outlined in this paper, a strategy based on the use of Continuous Time Recurrent Neural Networks (CTRNN) has been chosen [1]. In such networks cycles may exist in their structure and are mathematically equivalent to Dynamic Systems. In contrast with feed forward neural networks, which only support reactive behaviours, dynamic neuron networks allow the agent to start an action regardless of the immediate situation and organise its behaviour in anticipation of future events [8]. This type of neuron network also helps describe the agent as a Dynamic System coupled with the environment in which it is located, since it is shown that this type of neuron network is the simplest continuous dynamic neuron network nonlinear model. The neurobiological interpretation of the CTRNN has been shown and can be found in [1,10].

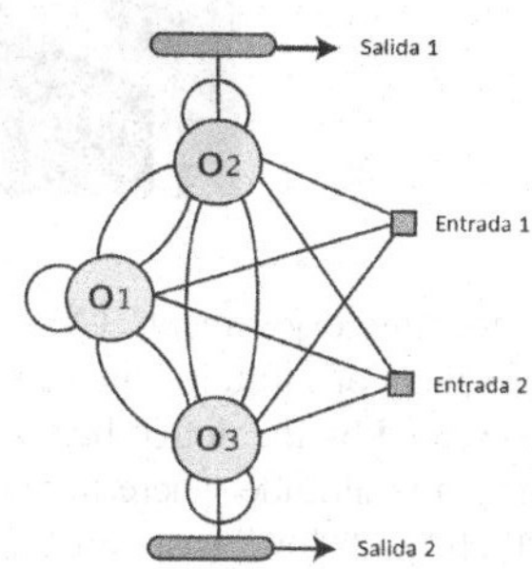

Fig. 1 Continuous-time Recurrent Neural Networks

2.2 Application of CTRNNs to the Bots Design

To implement bots, the Unreal Tournament 2004 was used with the Gamebots 2004 expansion (which provides a way to run bots in the videogame) and the Pogamut 3 platform (which allows for the programming of the virtual agent using Java (using JDK 6) programming language and connecting it and receiving information from the videogame using a plugin for Netbeans. When choosing certain behaviour for the bot, in which learning is required for its run time, it was decided to adapt the behaviour shown to the UT2004 environment by the Caenorhabditis elegans nematode [9]. Such behaviour consists of associating two stimuli (associative learning): temperature and food. This model was chosen because the C. elegans is a popular choice among researchers in the field of CTRNN evolution, since it shows behaviours simple enough to be modelled by small CTRNNs and complex enough to explore the memory capabilities of the CTRNNs [9]. When adapting this model to the UT2004 environment using Pogamut, a 2D environment with an altitude gradient is used along one of their dimensions, which is shown in figure 2A. In the model, there are two base types: "enemy" and "ally". Each base is located only in regions with a particular range of altitudes: "high" between [9,10] and "low" between [-10,-9]. The region where each base is located depends on the type of environment: in the A-ent the "enemy" base is located in the "high" region and the "ally" base is located in the "low" region, while in the B-ent the "enemy" base is located in the "low" region and the "ally" base in the "low" region as well. The altitude gradient extends throughout the environment, which is free of obstacles. To do this, an environment has been designed like the one shown in figure 2B.

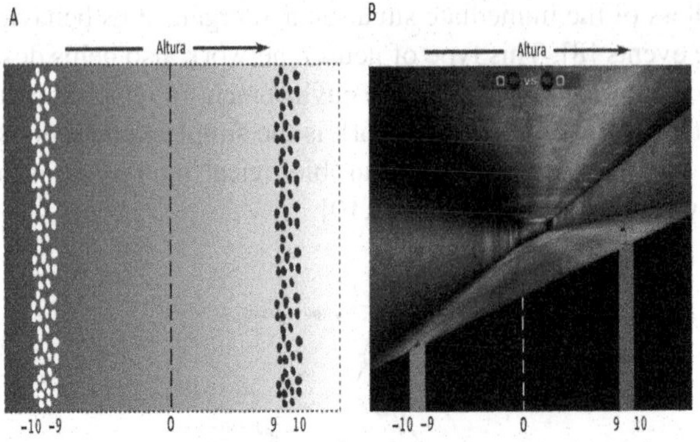

Fig. 2 Simulation environment for the experiment. (A) Two dimensional theoretical simulation environment with a gradient of altitudes, in which the "enemy" base can be located in one of two regions represented by the dotted bands. (B) Simulation environment in UT2004, in which the gradient is the altitude where the bot is located and the red and blue stripes represent where the "high" and "low" bases are located.

In the figure 2B The altitude sensor may have any real value. The "base" sensor returns a value of 0 unless the bot is located in one of the bases: it will return B=1 if it is located in the "enemy" base, and B=-1 if it is located in the "ally" base.

It is expected that the bot is capable of making associations during its runtime about altitude and "enemy" bases in each of the environments described and of memorising the altitude to return to it if the bot is relocated to the centre of the map. To accomplish this, the bot will appear in the position 0 of the altitude gradient in a random orientation. From this moment on, the bot will execute 200 cycles to move around the entire environment in search of the "enemy" base and to stay in that region as efficiently as possible. Once 200 cycles have been completed, the bot is repositioned to the 0 altitude with a random orientation, and should be able to go up or down in the altitude gradient depending on if during the previous execution it learned that it was in an A-ent or B-ent environment. If the type of environment is changed, the bot needs to be able to relearn and change its altitude preference.

3 Analysis of the Learning Behavior of the Bot

If the obtained CTRNN parameters are analysed (figure 3) three important characteristics can be seen.

First, one can see a high inverse bilateral symmetry: the way in which neurons 1 and 2 are connected to neurons 3 and 4 is by having similar weight but an opposite sign; both neurons are quickly activated; node 3 strongly excites neuron 1 and strongly inhibits neuron 2; both of their self-connections have small weights; the altitude sensor inhibits neuron 1 and excites neuron 2. Second, the base sensor has a large inhibitory force on the motor neurons, which will allow the bot to stop itself once it has found the enemy base camp. Lastly, and perhaps importantly, all neurons act as quickly as possible except for neuron 3, which is a lower magnitude order, which provides the CTRNN a multi-scale activity over time.

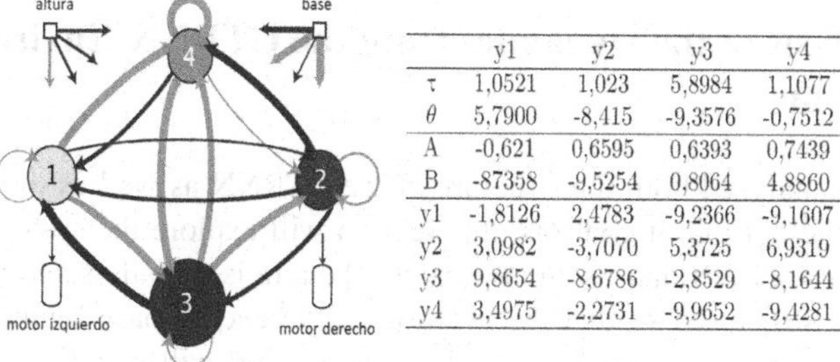

	y1	y2	y3	y4
τ	1,0521	1,023	5,8984	1,1077
θ	5,7900	-8,415	-9,3576	-0,7512
A	-0,621	0,6595	0,6393	0,7439
B	-87358	-9,5254	0,8064	4,8860
y1	-1,8126	2,4783	-9,2366	-9,1607
y2	3,0982	-3,7070	5,3725	6,9319
y3	9,8654	-8,6786	-2,8529	-8,1644
y4	3,4975	-2,2731	-9,9652	-9,4281

Fig. 3 Parameters for the best CTRNN with 4 completely interconnected and self-connected neurons. The nodes are shaded according to their bias. The excitatory (black) and inhibitory (gray) connections is proportional to the weight of the neurons. The time constants are represented by the size of the node, the slower nerons are the larger ones.

Fig. 4 Graphic illustration of the behaviour of a bot controlled by the CTRNN as a result of experiment 1 for the model

The obtained neural network is used to control the behaviour of the bot in a sequence of two environmental changes and three base searches per environment, like those used for the CTRNN evolution.

- At the beginning of the execution, the bot goes to the bottom of the map, but changes its behaviour and goes to upward before reaching the region where the base is located in the lower part of the map (between [-10,-9]). This seems to be part of the search strategy, since the bot still does not know in what kind of environment it is, and it is a phenomenon that has been observed in all executions of the bot.
- If the environment changes (the bot changes teams), the bot returns to the top part of the map, with the difference that the sensor receives a base value of B=-1 for being in the ally base. The bot continues going up beyond the base until its behaviour gradually changes to go down in the altitude gradient and it stops once it reaches the enemy base, this time located in the lower part of the

map. By placing it in the centre, the bot has learned where the new enemy base is located and goes toward it.

4 Analysis of the Dynamics from the CTRNN-Environment System

After having analyzed both the structure of the CTRNN as well as the observable behavior of the bot that it controls, this section will explore the system dynamics formed by the CTRNN and its environment. The aim is to understand how the bot dynamics are structured so that the location of the enemy base found in the past affects the direction of the altitude gradient in which it will move. To analyze the system dynamics, the bifurcation diagrams will be analyzed, which are received for all possible cases. The system dynamics change based on the altitude at which the enemy base is located and whether or not the enemy base is or is not found.

- *Analysis of the dynamics for B=0*

First, how the system equilibrium changes based on the altitude in the absence of based is checked. To do this, a case without bases, i.e. a case in which the sensor base is always B=0, is observed. Since the dynamic system defined by the CTRNN has four variables (activation of neurons) and the altitude is the parameter that changes, the bifurcation diagram is 5-dimensional. Figure 5 shows the four two-dimensional projections from the 5-dimensional diagram, one for each neuron activation value based on the altitude variable

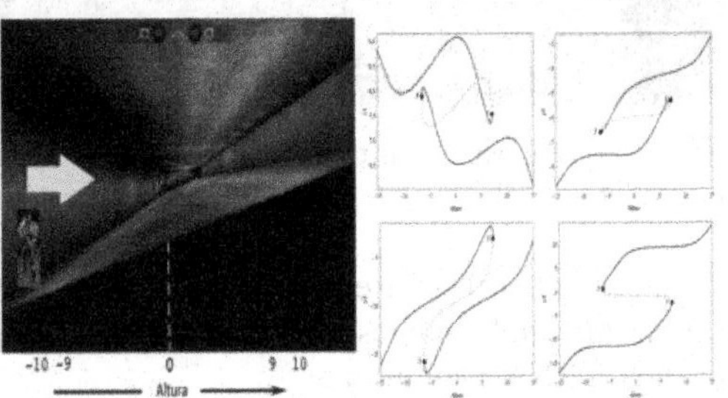

Fig. 5 (A) Graphic illustration of the behavior of a bot controlled by CTRNN. (B) Bifurcation diagram in the presence of the enemy base. Four two-dimensional projections from the 5-dimensional diagram, one for each of the neurons from the CTRNN. The solid lines represent stable equilibrium points, while the dashed lines represent unstable equilibrium points. The grey verticle lines show the altitude ranges where the enemy base could be located.

- *Analysis of the dynamics for B=1*

Second, the goal is to see how the system equilibrium changes based on the altitude enemy base presence, for which, although it is considered as a B=1 input in the total range of altitudes, only the shaded areas in grey are of interest, since the enemy base only can be found there. Figure 6 shows the four two-dimensional projections from the 5-dimensional diagram for this case- one for each neuron activation value based on the altitude variable. As is evidenced in the diagram, when the enemy base is located in the region corresponding to "high" altitudes, there is a single point of stable equilibrium. However, when the enemy base is located in "low" altitudes, the agent can be in one of two possible states, since the agent stays there long enough to reach the equilibrium point.

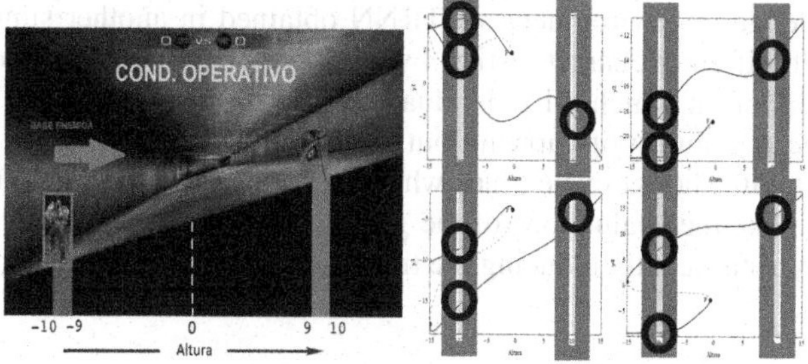

Fig. 6 (A) Graphic exmple of the behavior of a bot controlled by CTRNN (B) Bifurcation diagram in the presence of the enemy base. Four two-dimensional projections from the 5-dimentional diagram, one for each one of the CTRNN neurons. The solid lines represent stable equilibrium point while the dashed lines represent unstable equilibrium points. The vertical gray lines show the altitude ranges where you can find the enemy base.

Observations

The study of these diagrams suggests two main predictions:

- The first of which is that in the absence of any base, the system fall into an unlimited cycle in which the agent switches between going up and down the altitude gradient. Although the bot has not been trained for this scenario, this can be interpreted as a "high level" search behaviour of the enemy base, which comes from the two "low level" behaviours for which it has been trained. This behaviour supports the fact that it is not necessary to consider the base sensor when the value returned is -1.
- Second, as a result of the bistability observed in the graphics from figure 6, one could predict and confirm that, even after experimenting with environments with the enemy base in the "low" regions, if the agent is exposed to low altitudes in the presence of the enemy base for long enough, the agent could become reconditioned to navigate up the gradient altitudes.

5 Conclusions

This paper has proven how, thanks to the feature of the CTRNNs whose neurons work with different activation times. Once the agent dynamics have been analysed with their environment, it was noted that the agent had a behaviour in which it ignored preconceptions in its design: first, the CTRNN ignored the face that being in the ally base (indicated by a base value of =-1) and opted for ascending and descending behaviour in the altitude gradient in search of the enemy base; second, despite having designed the experiment as a model of associative learning with classical conditioning (associated altitude and enemy base), the CTRNN shows associative learning capacity with operative conditioning (associating stimuli with behaviour). As for the UT2004 simulation environment, Pogamut has proven to be of little use when the CTRNNs are evolving with many parameters (32 in this case). However, it has been shown that it is possible to successfully adapt to the UT2004 videogame environment a CTRNN obtained in another simulation environment with different characteristics, by using the Differential Evolution from an initial population in which all individuals are defined by the said CTRNN. This opens up the possibility of carrying out the evolutionary process in a simulation environment with shorter cycle times when executing the actions and in which it is possible to make the evaluations of the genetic algorithm individuals in a parallel way, thereby significantly reducing the time needed by the genetic algorithm.

References

1. Beer, R.D.: On the dynamics of small continuous-time recurrent neural networks. Adaptative Behavior 3(4), 459–509 (1995a)
2. Harvey, I., Di Paolo, E., Wood, R., Quinn, M., Tuci, E.A.: Evolutionary robotics: A new scientific tool for studying cognition. Artificial Life 11(1-2), 79–98 (2005)
3. Jakobi, N.: Minimal Simulations For Evolutionary Robotics. PhD thesis. COGS, University of Sussex (1998)
4. Collins, R.J., Jefferson, D.R.: Representations for artificial organisms. In: Meyer, J.-A., Roitblat, H., Wilson, S. (eds.) From Animals to Animats 1: Proceedings of the Second International Conference on the Simulation of Adaptive Behavior, pp. 382–390. MIT Press, Cambridge (1991)
5. Werner, G.M., Dyer, M.G.: Evolution of communication in artificial organisms. In: Langton, C.G., Taylor, C., Farmer, J.D., Rasmussen, S. (eds.) Artificial Life II, pp. 659–687. Addison-Wesley, Reading (1991)
6. Beer, R.D., Gallagher, J.: Evolving Dynamical Neural Networks for Adaptive Behavior. Adaptive Behavior 1(1), 91–122 (1992)
7. Miller, G.F., Cliff, D.: Protean behavior in dynamic games: Arguments for the coevolution of pursuit-evasion tactics. In: Cliff, D., Husbands, P., Meyer, J., Wilson, S. (eds.) From Animals to Animats 3: Proceedings of the Second International Conference on the Simulation of Adaptive Behavior, pp. 411–420. MIT Press, Cambridge (1994)
8. Funahashi, K., Nakamura, Y.: Approximation of dynamical systems by continuous time recurrent neural networks. Neural Networks 6, 801–806 (1993)
9. Xu, J.-X., Deng, X., Ji, D.: Study on C. elegans behaviors using recurrent neural network model. In: 2010 IEEE Conference on Cybernetics and Intelligent Systems (CIS), June 28-30, pp. 1–6 (2010), doi:10.1109/ICCIS.2010.5518591
10. Ruiz, S.M., Bedia, M.G., Castillo, L.F., Isaza, G.A.: Navigation and obstacle avoidance in an unstructured environment Videogame through recurrent neural networks continuous time (CTRNN). In: 2012 7th Colombian Computing Congress (CCC), October 1-5, pp. 1–6 (2012), doi:10.1109/ColombianCC.2012.6398004

An Integral System Based on Open Organization of Agents for Improving the Labour Inclusion of Disabled People

Alejandro Sánchez, Carolina Zato, Gabriel Villarrubia-González, Javier Bajo, and Juan Francisco De Paz

Departamento Informática y Automática, Universidad de Salamanca
Plaza de la Merced s/n, 37008, Salamanca, Spain
{sanchezyu,carol_zato,gvg,jbajope,fcofds}@usal.es

Abstract. This paper presents a system composed by a set of tools that facilitate the work of disabled people in their work environment. The PANGEA platform was used to build the base architecture of the system, where each tool is designed as a collection of intelligent agents that offer the services as Web-services. Moreover, all the system is implemented as an Open MAS. In this paper two tools are presented in detail, the proximity detection tool and the translator tool for people with hearing impairments.

Keywords: personalization workplace, disabled people, open MAS, agent platform, Zigbee, proximity detection, localization.

1 Introduction

Modern societies are characterized by two trends. The first is the rapid development of technologies, which has influenced our lives in many different ways. The second is the effort among governments, companies and associations toward enabling people with disabilities to have an independent life, which includes the possibility for remunerative employment. The effective integration of people with disabilities in the workplace is a huge challenge to society, and it presents an opportunity to make use of new technologies.

This paper presents a collection of tools that are being developed to formed an integral intelligent system. The different tools for the disabled people have been modelled with intelligent agents that use Web services. These agents are implemented and deployed within the PANGEA platform so they form an integral system that can be used regardless of their physical location or implementation. This project aims to develop new technologies that contribute to the employment of groups of people with visual, hearing or motor disabilities in office environments. Some of these tools are a head mouse to control the mouse with the eyes, a vibrator bracelet to send Morse messages, an avatar for the hearing impaired, a location system, etc. But due to space limitation, in this paper just an avatar and a proximity detection tool are presented.

S. Omatu et al. (Eds.): *Distrib. Computing & Artificial Intelligence*, AISC 217, pp. 369–376.
DOI: 10.1007/978-3-319-00551-5_45

The rest of the paper is structured as follows: The next section introduces the basis of open MAS and the PANGEA platform. Section 3 presents the proximity prototype tool. Section 4 explains the translator tool. Next, section 5 presents a case study. Finally, in section 6 some conclusions are presented.

2 Open MAS

Open MAS can be understood as the following step in multi-agent systems. These are systems in which the structure is able to change dynamically. Its components are not known a priori, change over time and may be heterogeneous. The open MAS must allow the participation of heterogeneous agents with different architectures and even languages [18]. This makes it difficult to rely on the agents' behavior, and necessitates controls based on societal norms or rules. The proposed system has been designed as an open MAS.

Nowadays, there are many multi-agent systems which help and facilitate the work with the agents [1,8,5]. The only inconvenience of these systems is that they are for general purpose. The architecture that will be used in this paper must be able to assume the tasks for the integration of the persons with disabilities to the workplace. In this line, the most known works are:

- The European project CommonWell [2] proposes an architecture to support European citizens with limited mobility, or a hearing or visual impairment. However, it focuses on the elderly and does not incorporate either advanced adaptive interfaces or identification and localization elements.
- The European project DTV4A11 [3] proposes the use of digital television to integrate persons with disabilities, but it relies on the television as the only mechanisms to provide services.
- The European project MonAMI [15] proposes a global framework to offer services to the elderly and handicapped, but it focuses on providing these individuals with a more independent lifestyle.

At a Spanish national level we can find:

- The DISCATEL project [4] aims to incorporate persons with disabilities to Contact Centers or allow them to telecommute from their home or residence.
- The INREDIS project (Interfaces for the Relationship between people with Disabilities) [11] is a CENIT project headed by Technosite, which is investigating the concept of using personal devices with interoperability and ubiquitous characteristics to strengthen accessibility of persons with disabilities..
- The eVia platform has the INCLUTEC [10] study group, which is oriented toward analysis, and promotes the use and development of mobility mechanisms, such as assisted wheelchairs and specialized vehicle, alternative and enhanced communication, manipulation, and cognition.

None of this multiagent platforms shown previously, adapt to our requirements because most of them focus on the elderly or on the social integration of people with disabilities, instead of our goal, which is the labor inclusion of this kind of people. PANGEA architecture whose novelty is a dynamic and adaptable architecture capable of integrating new services for incorporating persons with visual, hearing, or mobile impairments into the workforce.

2.1 *Description of PANGEA*

PANGEA [20] is a service oriented platform that allows the implemented open MAS to take maximum advantage of the distribution of resources. To this end, all services are implemented as Web Services. Due to its service orientation, different tools modeled with agents that consume Web services can be integrated and operated from the platform, regardless of their physical location or implementation. This makes it possible for the platform to include both a service provider agent and a consumer agent, thus emulating a client-server architecture. The provider agent (a general agent that provides a service) knows how to contact the web service, while the remaining agents know how to contact with the provider agent due to their communication with the ServiceAgent, which contains information about services.

Once the client agent's request has been received, the provider agent extracts the required parameters and establishes contact. Once received, the results are sent to the client agent. Using Web Services also allows the platform to introduce the SOA (Service-oriented Architecture) [12] into MAS systems. SOA is an architectural style for building applications that use services available in a network such as the web. It promotes loose coupling between software components so that they can be reused. Applications in SOA are built based on services.

3 Proximity Detection Tool

The proximity detection system is based on the detection of presence using the ZigBee Technology [19]. Every computer in the room must have a Zigbee router assigned, and the system have to know their exact positions at every moment. Furthermore, all the users have to carry a Zigbee tag (Figure 1), which is responsible for the identifying each of them. Once the Zigbee tag carried by the person has been detected and identified, its location is delimited within the proximity of the sensor that identified it.

Fig. 1 Zigbee tags

The agents that composed the tool are deployed in a specialized suborganization inside PANGEA, each one of these agents offer services (like a localization service) modeled as Web Services. The platform agents are implemented with Java, while the agents of the detection prototype are implemented in .NET and nesC.

Every user in the proposed system carries a Zigbee tag, which is detected by a ZigBeeReaderAgent located in each system terminal and continuously in communication with the ClientComputerAgent. Thus, when a user tag is sufficiently close to a specific terminal (within a range defined according to the strength of the signal), the ZigBeeReaderAgent can detect the user tag and immediately send a message to the ClientComputerAgent. The parameter RSSI is the responsible of measuring the receiving signal strength. The values fluctuate from an initial 0 to negative values. If values are close 0, the user tag is near a computer. If the user moves away the workplace, values begin to be negatives. For switching on the computers, the Wake-on-LAN protocol is used due to the system uses a LAN infrastructure [16] [13].

Next, this agent communicates the tag identification to the UsersProfileAgent, which consults the database to create the xml file that is returned to the ClientComputerAgent. The ClientComputerAgent then interacts with the ServiceAgent to invoke the Web Services needed to personalize the computer according to the user's profile.

In Figure 2 the interface of the tool is shown. In the upper part, the main controller has options to manage computers, events, sensors and users. And below, the information about the identification once the ClientComputerAgent has finished the communication process with the UsersProfileAgent.

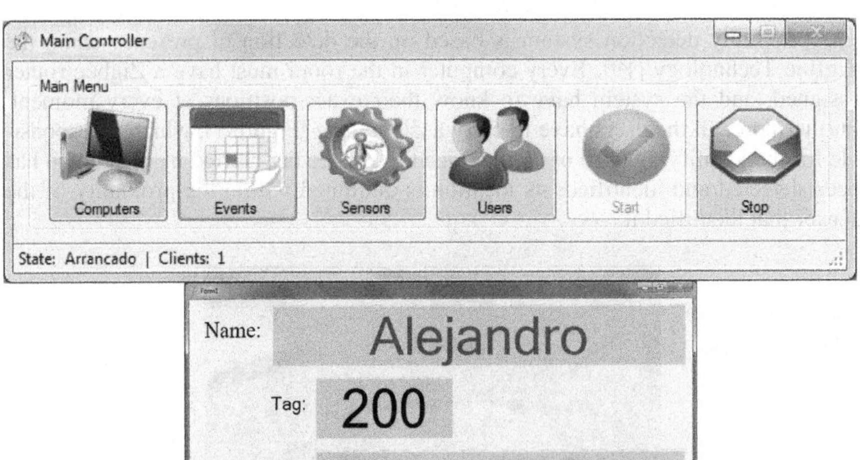

Fig. 2 Interface of the system

4 Translator Tool

This tool emerged as a result of the difficulty encountered by employers in communicating to their hearing impaired employees the actions that they have to do in their jobs. Given the ineffectiveness of avatar translators, the solution chosen was to study the main communication needs and provide some recorded videos with commands and explanations specifically related to the performance of a particular job.

Once the computer has been switched on and personalized thanks to the proximity detection tool, the worker with a hearing impairment will see automatically the avatar on his screen. As can be seen in the figure 3, the avatar appears on the screen to indicate to the user the tasks that must be performed on this workday. The boss can send the tasks through email or text message, and the worker will receive them on the computer screen.

Fig. 3 Avatar of the translator tool

The issuing agents, deployed on both Smartphones (Android or iPhone) or computers, will be responsible for playing the video required at each moment. Receptor agents, also available for Smartphones or computers, will be responsible for capturing by text or by voice, the command or instruction that the employer wishes to transmit to the disabled employee.

In PANGEA, the translator agent is called VideoTranslatorAgent and it is deployed within the suborganization TranslatorOrganization. VideoTranslatorAgent is responsible for receiving the instruction and mapping the specific video used by the emitter agent who is requesting the transfer. The figure 4 shows all the agents involved in the functioning of the tool.

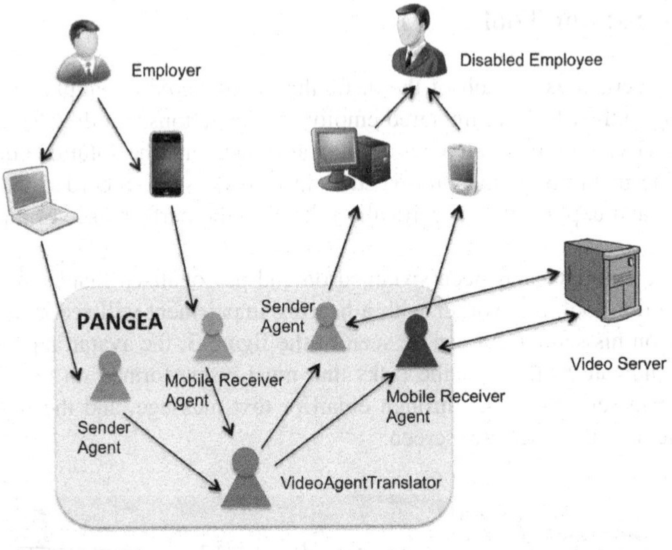

Fig. 4 Agents of the translator tool deployed in PANGEA

5 Case Study

All the tests have been centered on the final user. Tests have been carried out with individuals with different disabilities and a user profile exists for each person. User profiles stored data related to applications that are useful to users. These data can be classified according to the application, as shown in the following table.

Table 1 Data stored to personalize the computer

Screen		
Parameter	Value	Description
usage	Preferred/unpreferred	Field to describe if this tool is used.
invertColourChoice	1/0	Field to describe if the colors are inverted.
magnification	1-200	Field to describe the level of magnification the user want on the screen.
Avatar		
Parameter	Value	Description
usage	Preferred/unpreferred	Field to describe if this tool is used.
velocity	1-10	Field to describe the speed of the signs.
suze	1-10	Field to indicate the size of the avatar
localization	1-10	Field to where the avatar is displayed on the screen

Table 1 (*continued*)

Language		
Parameter	Value	Description
lenguage	ISO 3166-1 alfa-3	Field to describe the user's language
Virtual Keyboard		
Parameter	Value	Description
usage	Preferred/unpreferred	Field to describe if this tool is used.
Head Mouse		
Parameter	Value	Description
usage	Preferred/unpreferred	Field to describe if this tool is used.
cursorAcceleration	1-10	Field to describe the accelerating the cursor.
cursorSpeed	1-10	Field to describe the speed of the cursor

The tests have helped us to fix the correct values and to refine the functioning of these tools.

6 Conclusions

This system is specifically oriented to facilitate the integration of people with disabilities into the workplace. Thanks of the PANGEA platform, the system can be easily designed and deployed since the platform itself provides agents and tools for the control and management of any kind of open MAS or VO. Moreover, the platform makes it possible to deploy different agents, even those included in the mobile devices, and communicates with the agents embedded in the Zigbee sensors. On the other hand, due to the based-on-services PANGEA implementation, the system has a high scalability and more tools can be added easily.

The presented system offers a multiagent system which is able to communicate with a proximity detection system and to personalize the workspace to improve the adaptation to the company flow. This individual adaptation allows that, whatever the disability the person has, the workplace will be adapted automatically, facilitating his productivity and removing the existing barriers, as the case of turning on the computer with the proximity detection system. Moreover, the translator tool will facilitate the communication in case the worker has hearing impairments.

Acknowledgements. This research has been supported by the project OVAMAH (TIN2009-13839-C03-03) funded by the Spanish Ministry of Science and Innovation.

References

1. Agent Oriented Software Pty Ltd., JACK™ Intelligent Agents Teams Manual. Agent Oriented Software Pty Ltd. (2005)
2. CommonWell Project (2010), `http://commonwell.eu/index.php`

3. Digital Television for All project (2010), http://www.psp-dtv4all.org/
4. DISCATEL (2010), http://www.imsersounifor.org/proyectodiscatel/
5. Galland, S.: JANUS: Another Yet General-Purpose Multiagent Platform. Seventh AOSE Technical Forum, Paris (2010)
6. Giunchiglia, F., Mylopoulos, J., Perini, A.: The tropos software development methodology: Processes, models and diagrams. In: Giunchiglia, F., Odell, J.J., Weiss, G. (eds.) AOSE 2002. LNCS, vol. 2585, pp. 162–173. Springer, Heidelberg (2003)
7. Huang, Y., Pang, A.: A Comprehensive Study of Low-power Operation in IEEE 802.15.4. In: Proceeding of the 10th ACM Symposium on Modeling, Analysis and Simulation of Wireless and Mobile Systems, Chaina, Crete Island, Greece (2007)
8. Hübner, J.F.: J -Moise+ Programming organisational agents with Moise+ & Jason. Technical Fora Group at EUMAS 2007 (2007)
9. Ilyas, M., Dorf, R.C.: The handbook of ad hoc wireless networks. CRC Press Inc., Boca Raton (2003)
10. INCLUTEC (2011), http://www.idi.aetic.es/evia/es/inicio/contenidos/documentacion/documentacion_grupos_de_trabajo/contenido.aspx
11. INREDIS (2011), http://www.inredis.es/
12. Josuttis, N.M.: SOA in Practice. O'Reilly Media, Inc. (2007)
13. Lieberman, P.: Wake on LAN Technology, White paper (2011), http://www.liebsoft.com/pdfs/Wake_On_LAN.pdf
14. Martin, D., et al.: OWL-S: Semantic Markup for Web Services, W3C Member Submission (2004), http://www.w3.org/Submission/OWL-S/
15. Monami project (2010), http://www.monami.info/
16. Nedevschi, S., Chandrashekar, J., Liu, J., Nordman, B., Ratnasamy, S., Taft, N.: Skilled in the art of being idle: reducing energy waste in networked systems. In: Proceedings of the 6th USENIX Symposium on Networked Systems Design and Implementation, Boston, Massachusetts, pp. 381–394 (2009)
17. Razavi, R., Perrot, J.-F., Guelfi, N.: Adaptive modeling: An approach and a method for implementing adaptive agents. In: Ishida, T., Gasser, L., Nakashima, H. (eds.) MMAS 2005. LNCS (LNAI), vol. 3446, pp. 136–148. Springer, Heidelberg (2005)
18. Zambonelli, F., Jennings, N.R., Wooldridge, M.: Developing multiagent systems: The Gaia methodology. ACM Transactions on Software Engineering and Methodology 12(3), 317–370 (2003)
19. ZigBee Standards Organization: ZigBee Specification Document 053474r13. ZigBee Alliance (2006)
20. Zato, C., et al.: PANGEA – platform for automatic coNstruction of orGanizations of intElligent agents. In: Omatu, S., Paz Santana, J.F., González, S.R., Molina, J.M., Bernardos, A.M., Rodríguez, J.M.C. (eds.) Distributed Computing and Artificial Intelligence. AISC, vol. 151, pp. 229–240. Springer, Heidelberg (2012)

+Cloud: An Agent-Based Cloud Computing Platform

Roberto González, Daniel Hernández, Fernando De la Prieta, and Ana Belén Gil

University of Salamanca, Computer Science and Automatic Control Department,
Plaza de la Merced s/n, 37007, Salamanca, Spain
{rgonzalezramos,dan_her_alf,fer,abg}@usal.es

Abstract. Cloud computing is revolutionizing the services provided through the Internet, and is continually adapting itself in order to maintain the quality of its services. This study presents the platform +Cloud, which proposes a cloud environment for storing information and files by following the cloud paradigm. This study also presents Warehouse 3.0, a cloud-based application that has been developed to validate the services provided by +Cloud.

Keywords: Cloud Computing, cloud storage, agent-based cloud computing.

1 Introduction

The technology industry is presently making great strides in the development of the Cloud Computing paradigm. As a result, the number of both closed and open source platforms has been rapidly increasing [2]. Although at first glance this may appear to be simply a technological paradigm, reality shows that the rapid progression of Cloud Computing is primarily motivated by economic interests that surround its purely computational or technological characteristics [1].
The majority of well-known Cloud Computing platforms tend to underscore their ability to provide elastic infrastructure services (through virtualized hardware), without taking high level services, such as platform and software, into account.

Information storage is not performed in the same way today as it was in the past. During the incipient stages of computer sciences, information was stored and accessed locally in computers. The storage process was performed in different ways: in data files, or through the use of database management systems that simplified the storage, retrieval and organization of information, and were able to create a relationship among the data. Subsequently, data began to be stored remotely, requiring applications to access the data in order to distribute system functions; database system managers facilitated this task since they could access data remotely through a computer network. Nevertheless, this method had some drawbacks, notably that the users had to be aware of where the data were stored, and how they were organized. Consequently, there arose a need to create systems to facilitate information access and management without knowing the place or manner in which the information was stored, in order to best integrate information provided by different systems

S. Omatu et al. (Eds.): *Distrib. Computing & Artificial Intelligence,* AISC 217, pp. 377–384.
DOI: 10.1007/978-3-319-00551-5_46

This study proposes a Cloud architecture developed in the +Cloud system to manage information. +Cloud is a Cloud platform that makes it possible to easily develop applications in a cloud. Information access is achieved through the use of REST services, which is completely transparent for the installed infrastructure applications that support the data storage. In order to describe the stored information and facilitate searches, APIs are used to describe information, making it possible to search and interact with different sources of information very simply without knowing the relational database structure and without losing the functionality that they provide. Using a Cloud architecture and document manager facilitates the integration of information from different sources that is used by applications, thus simplifying the exchange of information among the different services offered by the Cloud. Finally, the study also presents Warehouse 3.0, which is a cloud storage application. This application has been developed with the aim to test and validate the functionality of the services proposed by the +Cloud platform.

This paper is structured as follows: the next section provides an overview of the +Cloud platform, paying special attention to PaaS layer; the Warehouse 3.0 is then presented; and finally the study finishes with the conclusion and future studies.2.1

2 +Cloud Platform

A complete cloud-computing environment was developed for this study. The system has a layered structure that coincides with the widely accepted layered view of cloud-computing [3]. This platform allows services to be offered at the PaaS (Platform as a Service) and SaaS (Software as a Service) levels.

The SaaS (Software as a Service) layer is composed of the management applications for the environment (control of users, installed applications, etc.), and other more general third party applications that use the services from the PaaS (Platform as a Service) layer. At this level, each user has a personalized virtual desktop from which they have access to their applications in the Cloud environment and to a personally configured area as well. The next section presents the characteristics and modules of PaaS Layer in +Cloud and +Cloud in greater detail. Both the PaaS and SaaS layers are deployed using the internal layer of the platform, which provides a virtual hosting service with automatic scaling and functions for balancing workload. Therefore, this platform does not offer an IaaS (Infrastructure as a Service) layer. The virtual and physical resources are managed dynamically. To this end, a virtual organisation of intelligent agents that monitor and manage the platform resources is used [4].

2.1 PaaS Layer

The Platform layer provides its services as APIs, offered in the form of REST web services. The most notable services among the APIs are the identification of users and applications, a simple non-relational database, and a file storage service that provides version control capabilities and emulates a folder-based structure.

The services of the Platform layer are presented in the form of stateless web services. The data format used for communication is JSON, which is more easily readable than XML and includes enough expression capability for the present case. JSON is a widely accepted format, and a number of parsing libraries are available for different programming languages. These libraries make it possible to serialize and de-serialize objects to and from JSON, thus facilitating/simplifying the usage of the JSON-based APIs.

2.1.1 File Storage Service (FSS)

The FSS provides an interface to a file container by emulating a directory-based structure in which the files are stored with a set of metadata, thus facilitating retrieval, indexing, searching, etc. The simulation of a directory structure allows application developers to interact with the service as they would with a physical file system. A simple mechanism for file versioning is provided. If version control is enabled and an existing file path is overwritten with another file, the first file is not erased but a new version is generated. Similarly, an erased file can be retrieved using the "restore" function of the API. In addition to being organized hierarchically, files can be organized with taxonomies using text tags, which facilitates the semantic search for information and makes the service more efficient.

The following information is stored for each file present in the system:

- Its virtual path as a complete name and a reference to the parent directory.
- Its length or size in bytes.
- An array of tags to organize the information semantically.
- A set of metadata.
- Its md5 sum to confirm correct transfers and detect equality between versions.
- Its previous versions.

Web services are implemented using the web application framework Tornado[1] for Python. While Python provides excellent maintenance and fast-development capabilities, it falls short for intensive I/O operations. In order to keep file uploads and downloads optimized, the APIs rely on the usage of the Nginx[2] reverse proxy for the actual reads and writes to disk. The actual file content is saved in a distributed file system so that the service can scaled, and the workload is distributed among the frontend servers by a load balancer. The structure of the service allows migrating from one distributed file system to another without affecting the client applications.

File metadata and folder structure are both stored in a MongoDB[3] database cluster, which provides adequate scalability and speed capabilities for this application. Web service nodes deploy Tornado and Nginx as well as the distributed file

---

[1] http://www.tornadoweb.org/

[2] http://nginx.org/

[3] http://www.mongodb.org/

Table 1 Restfull web services exposed by FSS

REST Web Call	Description
PutFile	creates a new file (or a new version of an existing file) in response to a request containing the file and basic metadata (path, name and tags) in JSON, structured in a standard multipart request.
Move	changes the path of a file or a folder
Delete	deletes a file. Can include an option to avoid the future recovery of the file, erasing it permanently
GetFolderContents	returns a JSON array with a list of the immediate children nodes of a specific directory.
GetMetadata	returns the metadata set of a file or directory providing its identifier or full path.
GetVersions	returns the list of all the recoverable versions of a file.
DownloadFile	returns the content of a file (a specific older version can be specified).
Copy	creates a copy of a file or a recursive copy of a folder.
CreateFolder	creates a new folder given its path.
DeleteVersion	permanently deletes a specific version of a file.
Find	returns a list of the children nodes of a folder (recursively).
GetConfiguration	retrieves the value of a configuration parameter for the application.
SetConfiguration	sets the value of a configuration parameter (e.g. enabling or disabling version control)
GetSize	retrieves the size of a file. If a folder path is passed, then the total size of the folder is returned.
RestoreVersion	sets an older version of a file as the newest.
Undelete	restores a file.

system clients (GlusterFS[4]/NFS), and the access to the MongoDB cluster that can be located either within or exterior to the nodes.

2.1.2 Object Storage Service (OSS)

The OSS is a document-oriented and schemaless database service, which provides both ease of use and flexibility. In this context, a document is a set of keyword-value pairs where the values can also be documents (this is a nested model), or references to other documents (with very weak integrity enforcement). These documents are grouped by collections, in a manner similar to how tuples are grouped by tables in a relational database. Nevertheless, documents are not forced to share the same structure. A common usage pattern is to share a subset of attributes among the collection, as they represent entities of an application model. By not needing to define the set of attributes for the object in each collection, the migration between different versions of the same application and the definition of the relationships among the data become much easier. Adding an extra field to a collection is as easy as sending a document with an extra key. A search on that

[4] `http://www.gluster.org/`

key would only retrieve objects that contain it. The allowed types of data are limited to the basic types present in JSON documents: strings, numbers, other documents and arrays of any of the previous types.

As with the FSS, the web service is implemented using Python and the Tornado framework. By not managing file downloads or uploads, there is no need to use the reverse proxy that manages them in every node; therefore Nginx is used only to balance the workload at the entry point for the service.

Table 2 Restfull web services exposed by OSS

REST Web Call	Description
Create	creates a new object inside a collection according to the data provided. It returns the created object, adding the newly generated identifier. If the collection does not exist, it is created instantly.
Retrieve	retrieves all objects that match the given query.
Update	updates an object according to the data provided (the alphanumeric identifier of the object must be provided).
Delete	deletes all objects that match the given query.

2.1.3 Identity Manager

The Identity Manager is in charge of offering authentication services to both customers and applications. Among the functionalities that it includes are access control to the data stored in the Cloud through user and application authentication and validation. Its main features are:

- Single sign-on web authentication mechanism for users. This service allows the applications to check the identity of the users without implementing the authentication themselves.
- REST calls to authenticate application/users and assign/obtain their roles in the applications within the Cloud.

3 Warehouse

Warehouse is the first non-native application that has been developed for the +Cloud platform. This tool makes intensive usage of both the file and object storage services and it serves the purpose of being the first real-application test for the developed APIs

3.1 Functionality

Using last-generation standards such as HTML5 and WebSockets[5], the tool allows storing and sharing information using the cloud environment. The user interface is shown at Fig. 1. The available mechanisms for file uploading include HTML5's drag&drop technique.

[5] http://www.websocket.org/

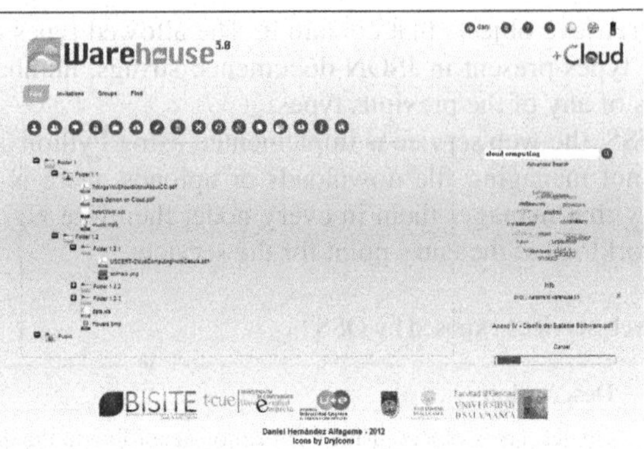

Fig. 1 Snapshot of the user interface

Every user has a root folder that contains all their data. The information stored by a user can be shared with other users through invitations for specific folders or using the user's "Public" folder. It is also possible to create groups of users, which work in a similar way to e-mail groups, in order to allow massive invitations. The user interface is updated asynchronously by using WebSockets. The changes made by a user over a shared resource are automatically displayed in the browsers of the other users. The application has syntactic and semantic search capabilities that are applied to different types of files (text, images or multimedia) due to the extraction and indexing of both the textual content and the metadata present in those files. Furthermore, the search results are presented next to a tag cloud that can be used to refine the searches even more. Finally, the application allows users to retrieve and manipulate different versions of their files. This function is powered by the mechanisms present in the underlying file storage API that has been previously described.

The contents of the files and the file system structure are stored in the FSS. Additional information is necessary to establish relationships between the data and to maintain the folder-sharing logic. This extra information is stored in the OSS. Due to the scalability and high-performance of the APIs, the application can execute tasks that will mainten the referential integrity of its model and the high number of recursive operations that are necessary to move and copy folders.

When a user creates a folder, three properties are assigned to it automatically: (i) Host user: keeps permissions over the folder. A number of operations are reserved for this user: move, delete, rename and cancel sharing; (ii) A list of invited users, initially empty; and finally, (iii) A list of users with access privileges to the file, initially containing only the host user.

The basic behaviour of the sharing algorithms is shown in the state diagram in Fig. 2. These algorithms are capable of multi-level folder sharing: a children folder of one shared folder can be shared with another list of users. This second group of users will only be allowed to navigate the most-nested folder.

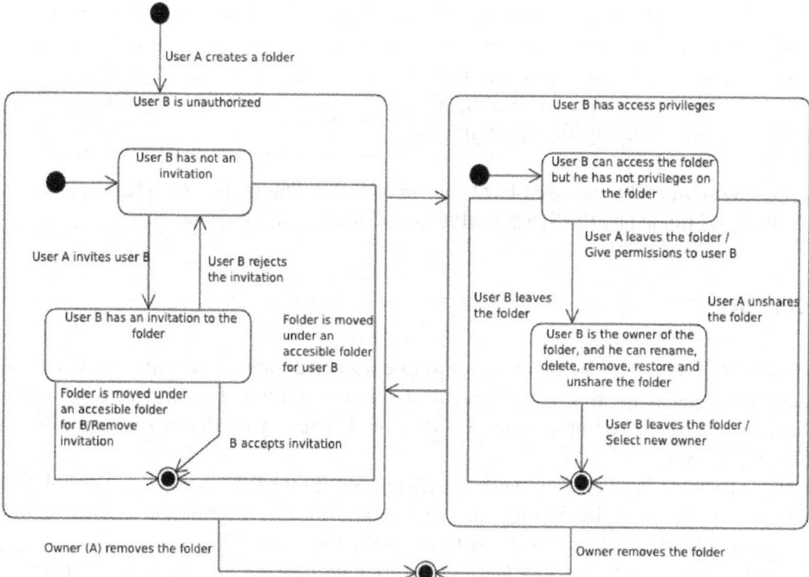

Fig. 2 State diagram for sharing

The actions related to folder sharing include:

- Invite: adds a user to the list of invited users.
- Accept or decline invitation: If the invitation is accepted, the user is added to the list of access-allowed users. Otherwise, the target user is removed from the list of invited users.
- Leave folder: the user that leaves the folder is removed from the list of access-allowed users. If the host user leaves the folder, the folder will be moved to another user's space and that user will be the new host. If there is more than one user remaining, the current host must choose which user will be the new host.
- Turn private: this operation can only be executed by the host user, and deleting all invitations and resetting the access list.
- Move: if the host moves the file, the other users will see a change in the reference to the shared folder. If the operation is done by another user, then only the reference of that user is modified (no move operation is performed).
- Delete: only the host can execute this operation. The shared folder can be moved to the space of another user, or be completely removed.

4 Conclusions

The cloud architecture defined in +Cloud has made it possible to transparently store information in applications without having previously established a data model. The storage and retrieval of information is done transparently for the applications, and the location of the data and the storage methods are completely transparent to the user. JSON can define information that is stored in the architecture, making it possible to perform queries that are more complete than those allowed by other cloud systems. This characteristic makes it possible to change the infrastructure layer of the cloud system, facilitating the scalability and inclusion of new storage systems without affecting the applications.

Acknowledgments. This research has been supported by the project *OVAMAH* (TIN2009-13839-C03-03) funded by the Spanish Ministry of Science and Innovation.

References

1. Buyya, R., Yeo, C.S., Venugopal, S.: Market-oriented cloud computing: Vision, hype, and reality for delivering it services as computing utilities. Department of Computer Science and Software Engineering (CSSE), The University of Melbourne, Australia, pp. 10–1016 (2008)
2. Peng, J., Zhang, X., Lei, Z., Zhang, B., Zhang, W., Li, Q.: Comparison of several cloud computing platforms. In: 2nd International Symposium on Information Science and Engineering, ISISE 2009, pp. 23–27. IEEE Computer Society (2009)
3. Mell, P., Grance, T.: The Nist Definition of Cloud Computing, pp. 1–3. NIST Special Publication 800-145, NIST (2011)
4. Heras, S., De la Prieta, F., Julian, V., Rodríguez, S., Botti, V., Bajo, J., Corchado, J.M.: Agreement technologies and their use in cloud computing environments. Progress in Artificial Intelligence 1(4), 277–290 (2012)

Representation of Propositional Data for Collaborative Filtering

Andrzej Szwabe, Pawel Misiorek, and Michal Ciesielczyk

Institute of Control and Information Engineering, Poznan University of Technology, M. Sklodowskiej-Curie Square 5, 60-965 Poznan, Poland
{Andrzej.Szwabe,Pawel.Misiorek, Michal.Ciesielczyk}@put.poznan.pl

Abstract. State-of-the-art approaches to collaborative filtering are based on the use of an input matrix that represents each user profile as a vector in a space of items and, analogically, each item as a vector in a space of users. When the behavioral input data have the form of *(userX, likes, itemY)* and *(userX, dislikes, itemY)* triples, one has to propose a bi-relational data representation that is more flexible than the ordinary user-item ratings matrix. We propose to use a matrix, in which columns represent RDF-like triples and rows represent users, items, and relations. We show that the proposed behavioral data representation based on the use of an element-fact matrix, combined with reflective matrix processing, enables outperforming state-of-the-art collaborative filtering methods based on the use of a 'standard' user-item matrix.

1 Introduction

When the purpose of the recommender system is to predict user choices rather than real-valued ratings, the entries of the input data matrix should have a form of binary numbers representing propositions of the form of *(userX, likes, itemY)*. Such a format is frequently regarded as the most convenient to model user actions [6], especially in on-line retailing and news/items recommendation scenarios. In such scenarios, the binary information about user actions (e.g., about purchase or page view) is the only data available to the system. An example is One-Class Collaborative Filtering system [7], for which only the data on positive user preferences are given and negative examples are absent.

In order to address a scenario of using binary behavioral data, we propose to use a new data processing framework that consists of: (i) a data representation method based on a binary element-fact matrix (for which rows represent users, items, and relations playing the roles of RDF subjects, objects, and predicates, respectively, and columns represent facts ,i.e., RDF-like triples), and (ii) the vector similarity quasi-measure based on the 1-norm length of the Hadamard product of the given tuple of vectors.

S. Omatu et al. (Eds.): *Distrib. Computing & Artificial Intelligence,* AISC 217, pp. 385–392.
DOI: 10.1007/978-3-319-00551-5_47

2 Related Work

Behavioral data processed by some of the recommender systems presented in the relevant literature have the form of propositional statements about the user-system interaction history. As such data have a natural relational representation, researchers working in the area of Machine Learning investigate collaborative filtering (CF) as one of the applications of Statistical Relational Learning (SRL) [9][8]. In order to enable the representation and processing of RDF triples of more than one relation, Singh *et al.* [8] proposed an approach based on the collective matrix processing. This approach was followed by proposals of similar models based on 3rd-order tensors [9][11].

The propositional nature of the algebraically represented data makes our proposal relevant to the challenge of unifying reasoning and search [4]. As far as algebraic transformation of graph data is concerned, the algorithms presented in the paper may be regarded as similar to RDF data search methods which are frequently based on Spreading Activation realized by means of iterative matrix data processing or, as it is in the case of Random Indexing (RI), by means of a single multiplication by a random projection matrix [3]. On the other hand, our method features a new kind of vector-space quasi-similarity measurement that allows us to estimate the likelihood of unknown RDF triples, rather than similarity measure [3].

3 Methodology

Apart from a few simplified recommendation scenarios, the proposed model has been investigated in a bi-relational scenario involving the use of two predicates: *Likes* and *Dislikes*. Such an investigation involved, among other steps, a special dataset preparation and a recommendation list generation. The evaluation of the proposed model has been realized with the use of state-of-the-art collaborative filtering data processing algorithms and the AUROC measure.

3.1 Proposed Data Representation Model

We propose to model the input data of a recommendation system (i.e., the information on all modeled elements: users, items, and predicates) as a set of RDF triples representing propositions (facts) stored in the form of an element-by-fact matrix referred to as A. It is worth noting that a fact-based data model can be used to represent graded user preference data since each level of a discrete-value rating scale may be modeled as a separate predicate. Taking into account the widespread adoption of the RDF technology, we believe that the proposed fact-based data representation model allows for flexible representation of highly heterogeneous data, leading to new successful applications of semantically-enhanced recommender systems [10].

The data representation model, which is commonly used in such a scenario, is based on a single classical user-item matrix;the rating scale is used in such a way that a high value indicates that a user likes an item, whereas a low value indicates the presence of the *Dislikes* relation. In our model, both the relations are modeled as separate predicates represented by separately trained vectors. Such an approach allows us to exploit the similarities between users and items that are observable within the set of triples including the *Likes* predicate or within the set of triples including the *Dislikes* predicate. Moreover, by showing a successful (i.e., recommendation quality improving) use of *Dislikes* triples as a secondary source of relational data (supporting the use of the primary *Likes* relational data), we feel encouraged to propose our model as the basis for future developments of recommender systems taking the advantage of using multi-relational data.

3.2 Data Representation Based on an Element-Fact Matrix

We introduce a data representation model based on the binary matrix A representing subject-predicate-object RDF-like triples. The rows of the matrix represent all the entities used to define triples, i.e. the elements playing the role of a subject, object or predicate, whereas columns represent propositions corresponding to the triples. We define a set $E = S \cup P \cup O$ as a set of elements referred to by the triples, where S is a set of subjects, P is a set of predicates, and O is a set of objects. We assume that $|S| = n$, $|O| = m$, and $|R| = l$. The model allows each modeled entity to freely play the role of a subject or an object, so in the general scenario we have $|S| = |O|$. In the case of the application scenario presented in this paper, sets S and O (i.e., sets of subjects and objects) are disjoint and correspond to sets of users and items, respectively. Additionally, we define set F as a set of all facts represented as the input dataset triples, such that $|F| = f$. Finally, we define the binary element-fact matrix $A = [a_{i,j}]_{(n+m+l)\times f}$. As a consequence of the fact that the columns of matrix A represent the triples, each column contains precisely three non-zero entries, i.e., for each j there are exactly three non-zero entries $a_{k_1,j}, a_{k_2,j}, a_{k_3,j}$, such that $1 \leq k_1 \leq n$, $n+1 \leq k_2 \leq n+m$, and $n+m+1 \leq k_2 \leq n+m+l$ – the entries corresponding to that correspond to the subject, object, and predicate of the modeled triple. At the same time, the number of non-zero entries in each row denotes the number of triples containing the element that corresponds to this row. Such a model is convenient to represent an RDF dataset, which consists of a finite number of predicates $l \geq 1$. Naturally, an element-fact matrix of the form proposed in the paper is much sparser than a typical user-item matrix. However, the fact that there are many zeros in the matrix is in fact neither an algorithmic nor an implementation obstacle, as the reflective data processing function may be effectively implemented by means of well known sparse matrix multiplication methods (featured by several widely used mathematical computation libraries).

3.3 Generation of Predictions

When using the element-fact matrix as the collaborative filtering data representation, one has to perform the prediction generation step (applied after processing input matrix A into its reconstructed form B) in a special way. We propose to calculate each of the predictions as the 1-norm length of the Hadamard product of row vectors corresponding to elements of the given RDF triples, i.e., the RDF triple forming the proposition which is the subject of probability estimation. More formally, the prediction value $p_{i,j,k}$ is calculated according to the formula:

$$p_{i,j,k} = \|b_i \circ b_j \circ b_k\|_1,$$

where b_i, b_j and b_k are the row vectors of the reconstructed matrix B corresponding to the subject, predicate, and object of the given RDF triple, and the symbol $b_i \circ b_j \circ b_k$ denotes the Hadamard product of vectors b_i, b_j and b_k. The proposed formula may be seen as a generalization of the dot product formula, as in the hypothetical case of measuring quasi-similarity of two (rather than three) vectors, the formula is equivalent to the dot product of the two vectors. The interpretation of the proposed formula as the probability of the joint incidence of two or more events represented as vectors is based on the quantum IR model [13].

3.4 Evaluation Scenarios

In this paper, we evaluate two matrix-based methods for the representation of RDF datasets that are applicable in the collaboration filtering scenario: the classical user-item matrix model and the novel element-fact matrix model. Both the models have been tested in two experimental scenarios: the one-class collaborative filtering scenario (with the use of RDF triples of the *Likes* predicate only) and the bi-relational collaborative filtering scenario (with the use of RDF triples of both the *Likes* and *Dislikes* predicates). In particular, the following four scenarios S1-4 have been investigated:

- S1 – the application of a binary user-item matrix $B = [b_{i,j}]_{n \times m}$ representing RDF triples of the *Likes* predicate, where n is the number of users, and m is the number of items,
- S2 - the application of a ternary $\{-1, 0, 1\}$ user-item matrix $B = [b_{i,j}]_{n \times m}$ representing RDF triples of the *Likes* predicate (denoted by positive numbers) and RDF triples of the *Dislikes* predicate (denoted by negative numbers),
- S3 - the application of a binary element-fact matrix $A_{i,j} = [a_{i,j}]_{(n+m+l) \times f}$ representing RDF triples of the *Likes* predicate, where n is the number of subjects, m is the number of objects, and $l = 1$ is the number of predicates (only the *Likes* predicate is represented),
- S4 - the application of a binary element-fact matrix $A_{i,j} = [a_{i,j}]_{(n+m+l) \times f}$ representing RDF triples of the *Likes* predicate or the *Dislikes* predicate, where the number of predicates is equal to 2 ($l = 2$).

In our experiments, we have used one of the most widely referenced data sets – the MovieLens ML100k set [2]. Each rating that is higher than the average of all ratings has been treated as an indication that a given user likes a given movie, i.e., that the RDF triple of the form (userA, *likes*, movieA) represents a true proposition. Analogically, each rating lower than the average, has been used as an indication that proposition (userB, *dislikes*, movieB) is true. We have generated five datasets

Table 1 Number of RRI/RSVD reflections used for each training ratio

x (training ratio)	Method							
	S1		S2		S3		S4	
	RRI	RSVD	RRI	RSVD	RRI	RSVD	RRI	RSVD
0.2	3	7	5	7	8	15	3	3
0.3	3	5	3	7	10	9	3	3
0.4	3	5	3	7	10	9	3	3
0.5	3	5	3	5	8	9	3	3
0.6	3	5	3	5	8	11	3	3
0.7	3	5	3	5	10	5	3	3
0.8	3	3	3	5	8	11	3	3

by randomly dividing the set of all known propositions into two groups – a train set and a test set according to the training *ratio*, denoted by x. To compensate for the impact that the randomness in the data set partitioning has on the results of the presented methods, each plot in this paper shows a series of values that represent averaged results of individual experiments.

We have compared several collaborative filtering data processing algorithms that are presented in the literature: popularity-based (favoring items having the higher number of ratings in the train set) [12][2], Reflective Random Indexing (RRI) [12], PureSVD [2], and Randomized SVD (RSVD) [1]. All these methods have been applied to the input data matrix (in the case of both matrix-based and triple-based data representations) [2], [1], [12]. When the classical matrix-based model is used, the input matrix is decomposed, and then reconstructed in order to generate predicted ratings as presented in [10]. We have tested each of the methods, using the following parameters (where applicable): vector dimension: 256, 512, 768, 1024, 1536, 2048; seed length: 2, 4, 8, SVD k-cut: 2, 4, 6, 8, 10, 12, 14, 16, 20, 24. The combinations of parameters that lead to the best recommendation quality (i.e., the highest AUROC value) were considered optimal, and used in experiments illustrated in this paper. Table 1 shows the number of reflections used in the RRI and RSVD algorithms for each training ratio.

Finally, we have compared the above-specified algorithms with a popularity-based as in [2]. In such baseline method, herein referred to as 'Popularity', the recommendation is based on the number of a given item's ratings in the train set.

We have compared the proposed data model with the standard one from the perspective of the application of each of these models within a complete collaborative filtering system. To obtain quantitative results of such an analysis, we have evaluated an ordered list of user-item recommendations by means of the AUROC (i.e., the Area Under the ROC curve) measure [5].

4 Experiments

Figures 1 and 2 show a comparison of the investigated recommendation algorithms (explicitly: popularity-based, RRI, PureSVD, RSVD), each using either the classical

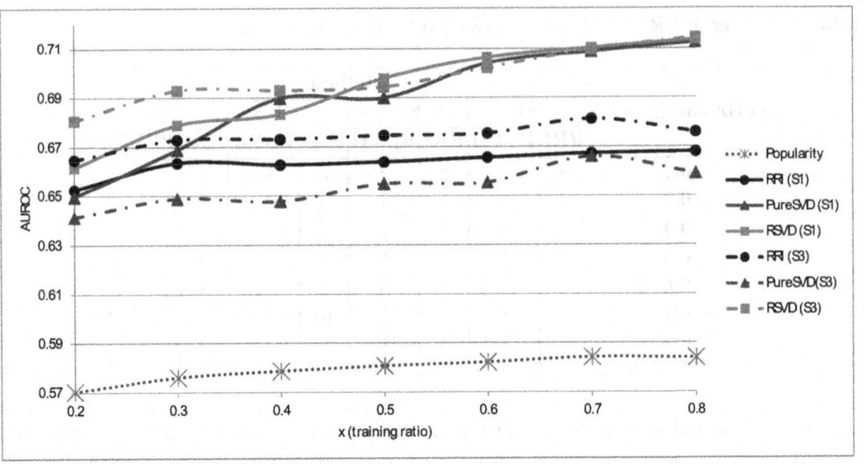

Fig. 1 Results achieved in the S1 and S3 scenarios

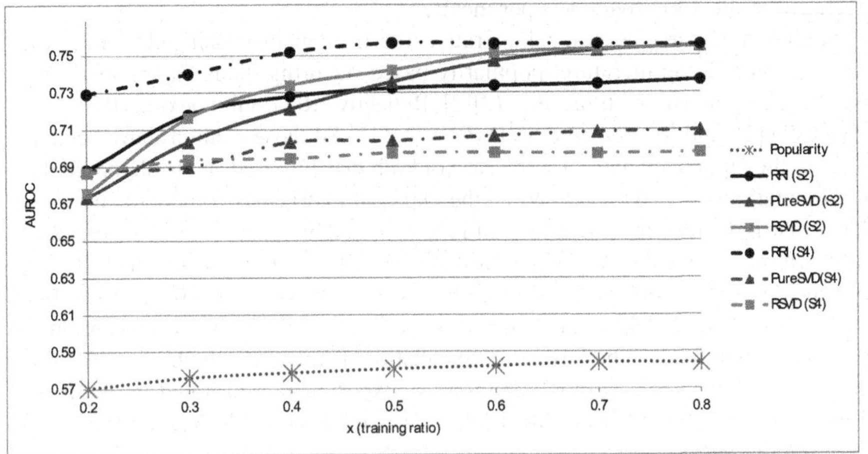

Fig. 2 Results achieved in the S2 and S4 scenarios

user-item or the element-fact matrix data representation. The comparison has been performed using datasets of various sparsity. Fig. 1 presents AUROC evaluation results obtained for the case of using the dataset containing only *Likes* predicates, whereas Fig. 2 presents the analogical results obtained for the case of using the full dataset, i.e., the one containing both the *Likes* and *Dislikes* predicates (i.e., both the positive and the negative feedback).

The results of the experiments presented in the paper indicate that the presence of the additional information about the *Dislikes* relation improves the recommendation quality. The results for S2 and S4 scenarios (see Fig. 2) are significantly better than the results obtained in S1 and S3 scenarios (see Fig. 1). However, the main conclusion from the experiments is that the best quality is observed in scenario S4 (in which we applied the proposed data representation and prediction method) for the case of the RRI-based data processing application.

The 1-norm length of the vector Hadamard product may be seen as a natural extension of the vector dot product (in our case - as a 'kind' of 'group inner product' of the three vectors representing the RDF subject, object, and predicate) whereas the dot product may be seen as an elemental step of the matrix multiplication, i.e., the basic operation used in reflective matrix processing (each cell of a matrix multiplication result is in fact a dot product). On the other hand, the prediction based on the Hadamard product does not suit well the data processing techniques based on SVD decomposition. This explains the relatively weak results of the dimensionality reduction methods in the scenarios in which the proposed data modeling method is used.

5 Conclusions

We have shown that using the proposed bi-relational matrix data representation together with reflective data processing enables the researcher to design a collaborative filtering system outperforming systems based on the application of the dimensionality reduction technique.

While taking the perspective of related areas of research (such as SRL and web scale reasoning), one may find particularly interesting to investigate our proposal of using the 1-norm length of the Hadamard product (a 'quasi-similarity 'of a given triple's constituents) as an unknown triple likelihood measure. We believe that, as a result of realizing probabilistic reasoning as a vector-space technique, our solution provides a basic means for extending the capacity for reasoning on RDF data beyond the boundaries imposed by widely-known non-statistical methods.

Acknowledgement. This work is supported by the Polish National Science Centre under grant DEC-2011/01/D/ST6/06788, and by Poznan University of Technology under grant 45-085/12 DS.

References

1. Ciesielczyk, M., Szwabe, A.: RSVD-based Dimensionality Reduction for Recommender Systems. International Journal of Machine Learning and Computing 1(2), 170–175 (2011)
2. Cremonesi, P., Koren, Y., Turrin, R.: Performance of Recommender Algorithms on Top-n Recommendation Tasks. In: Proceedings of the Fourth ACM Conference on Recommender Systems (RecSys 2010), New York, NY, USA, pp. 39–46 (2010)
3. Damljanovic, D., Petrak, J., Lupu, M., Cunningham, H., Carlsson, M., Engstrom, G., Andersson, B.: Random Indexing for Finding Similar Nodes within Large RDF Graphs. In: García-Castro, R., Fensel, D., Antoniou, G. (eds.) ESWC 2011. LNCS, vol. 7117, pp. 156–171. Springer, Heidelberg (2012)
4. Fensel, D., van Harmelen, F.: Unifying reasoning and search to web scale. IEEE Internet Computing 11(2), 94–95 (2007)

5. Herlocker, J.L., Konstan, J.A., Terveen, L.G., Riedl, J.T.: Evaluating Collaborative Filtering Recommender Systems. ACM Trans. Information Systems 22(1), 5–53 (2004)
6. Pan, R., Zhou, Y., Cao, B., Liu, N.N., Lukose, R., Scholz, M., Yang, Q.: One-Class Collaborative Filtering. Technical Report. HPL-2008-48R1, HP Laboratories (2008)
7. Sindhwani, V., Bucak, S.S., Hu, J., Mojsilovic, A.: A Family of Non-negative Matrix Factorizations for One-Class Collaborative Filtering Problems. In: Proceedings of the ACM Recommender Systems Conference, RecSys 2009, New York (2009)
8. Singh, A.P., Gordon, G.J.: Relational Learning via Collective Matrix Factorization. In: Proceeding of the 14th ACM SIGKDD International Conference on Knowledge Discovery and Data Mining, pp. 650–658 (2008)
9. Sutskever, I., Salakhutdinov, R., Tenenbaum, J.B.: Modelling Relational Data Using Bayesian Clustered Tensor Factorization. Advances in Neural Information Processing Systems 22 (2009)
10. Szwabe, A., Ciesielczyk, M., Janasiewicz, T.: Semantically enhanced collaborative filtering based on RSVD. In: Jędrzejowicz, P., Nguyen, N.T., Hoang, K. (eds.) ICCCI 2011, Part II. LNCS, vol. 6923, pp. 10–19. Springer, Heidelberg (2011)
11. Szwabe, A., Misiorek, P., Walkowiak, P.: Reflective Relational Learning for Ontology Alignment. In: Omatu, S., Paz Santana, J.F., González, S.R., Molina, J.M., Bernardos, A.M., Rodríguez, J.M.C. (eds.) Distributed Computing and Artificial Intelligence. AISC, vol. 151, pp. 519–526. Springer, Heidelberg (2012)
12. Szwabe, A., Ciesielczyk, M., Misiorek, P.: Long-Tail Recommendation Based on Reflective Indexing. In: Wang, D., Reynolds, M. (eds.) AI 2011. LNCS, vol. 7106, pp. 142–151. Springer, Heidelberg (2011)
13. van Rijsbergen, C.J.: The geometry of IR, The Geometry of Information Retrieval, pp. 73–101. Cambridge University Press, New York (2004)

Face Identification by Real-Time Connectionist System

Pedro Galdámez and Angélica González

University of Salamanca, Plaza de los Cados, 37008 Salamanca
{peter.galdamez,angelica}@usal.es

Abstract. This document provides an approach to biometrics analysis which consists in the location and identification of faces in real time, making the concept a safe alternative to Web sites based on the paradigm of user and password. Numerous techniques are available to implement face recognition including the principal component analysis (PCA), neural networks, and geometric approach to the problem considering the shapes of the face representing a collection of values. The study and application of these processes originated the development of a security architecture supported by the comparison of images captured from a webcam using methodology of PCA, and the Hausdorff algorithm of distance as similarity measures between a general model of the registered user and the objects (faces) stored in the database, the result is a web authentication system with main emphasis on efficiency and application of neural networks.

Keywords: Neural networks, eigenfaces, Hausdorff distance, Face Recognition.

1 Introduction

In social relationships our principal target is the face, playing an important role in the perception of identity and emotions. Although the ability to relate intelligence or character from facial appearance is mistaken, the human ability to recognize faces is remarkable. Humans are able to recognize thousands of faces learned throughout their life and identify familiar faces at a glance. It is an amazing ability, pondering the great changes in the visual stimulus, given the conditions of observation, expression, aging, and distractions such as glasses, or changes in the hair style.

Face recognition has become a major issue in several applications, ranging from security systems, credit card verification to criminal identification; where the ability to modeling a face and distinguish it from a large number of stored models greatly improve criminal identification processes. During the 80s facial recognition work was inert. And later on in the 90s, a significant growth and interest in the investigation was started. Although there are different approaches to the problem of face recognition, there are two basic methods of which most approaches build their base. The first one is based on the concepts of information theory, that means, in the

S. Omatu et al. (Eds.): *Distrib. Computing & Artificial Intelligence,* AISC 217, pp. 393–400.
DOI: 10.1007/978-3-319-00551-5_48

methods of principal component analysis. In this approach, the most relevant information that best describes a face image is derived from the entire face. Based on the Karhunen Loeve expansion in pattern recognition M. Kirby and L. Sirovich Have shown [5] that any particular face could be mathematically represented in terms of a coordinate system that is called "eigenfaces". The second method is based on extracting feature vectors of the basic parts of a face such as eyes, nose, mouth and chin. In this method, with the help of deformable templates and math, key information of the basic parts of a face meets and then becomes a feature vector [14].

2 Facial Recognition Using Eigenfaces

The face recognition algorithm with eigenfaces is described basically in the figure 1. First, the original images of the training set are transformed into a set of eigenfaces E, Then, weights are calculated for each image on the (E) set, and then are stored in the (W) set. Observing an image X unknown, weights are calculated for that particular image, and stored in the vector W_X. Subsequently, W_X compared to the weights of images, which is known for sure that they are faces (the weights of the training set W).

2.1 *Classification of a New Face*

The process of classifying a new face in the Γ_{new} to another category (known faces) is the result of two steps. First of all, the new image is transformed into its eigenface components. The resulting weights forms the weight vector Ω^T_{new}.

$$\omega_k = u_k^T(\Gamma_{new} - \Psi) \;\; k = 1,\ldots,M' \qquad \Omega^T_{new} = [\omega_1 \; \omega_2 \; \ldots \; \omega_{M'}] \tag{1}$$

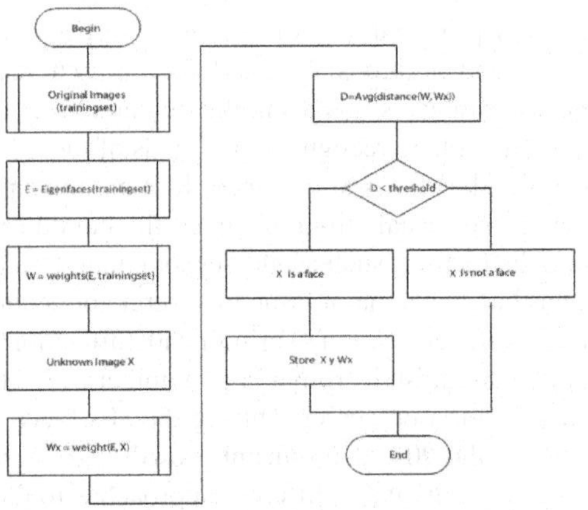

Fig. 1 Facial recognition algorithm based on eigenfaces

The Euclidean distance between two vectors of weights $d(\Omega_i, \Omega_j)$ provides a measure of similarity between the corresponding images i and j. If the Euclidean distance between Γ_{new} and the rest of images on average exceeds a certain threshold value, through this can be assumed that Γ_{new} is not a recognizable face [8].

2.2 *The Core System*

The previous section discussed the procedure for obtaining the eigen vectors of a collection of images, in order to perform the classification. The flow of the algorithm based on this technique is explained below. The following method provides a comprehensive procedure applied, it describes the methods of preprocessing and flow catches ranging from the acquisition of the images to their identification, which subsequently will make emphasis in the latest techniques applied in the heart of the system to achieve the goal of face recognition.

Data Acquisition

The first activity to do is to take pictures of users to recognize them later. The project uses a general purpose webcam, this represents the first contact with the users logon screen, where he will decide whether or not to register a user. If so, when trying to login, the system will identify him. If the user is not recognized by the system, he or she can be registered on a screen dedicated for that purpose. Captured images are processed and stored in a database as an array of bytes, the original image is stored along with the grayscale image and the extracted face is capture.

Image Processing

Once users have been registered in the system, it is possible to make identifications, for which we proceed to capture the image of the login screen, the image is resized

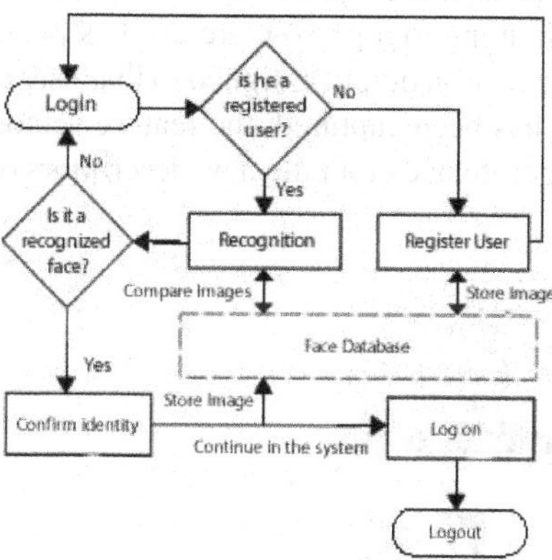

Fig. 2 Basic flow system

to 320×240 pixels. The captured image is converted to grayscale, in this step the face is extracted and adjusts its size to 100×100 pixels. Both actions use the library EmguCV [3]. For better results, the method needs to remove the background applying a mask that is just a black ellipse around the identified face, finally the procedure equalize the image, standardizing their lighting. The process is performed for all images in the database and for the image captured at the time of the login attempt.

User Recognition

With all the pictures processed, an array is created with labels indicating to whom belongs each image, the eigenfaces and eigenvalues are calculated with the PCA algorithm. Subsequently, it obtains the Euclidean distance of the comparison of the weight vectors obtained. Defining a threshold value of 1000 representing a similarity between two images of 90%, under the assumption that a smaller distance means a greater similarity between the sets.

In other words, the eigenvectors are obtained from the set of images stored in the database, the eigenfaces and unknown image are compared both with the distance measurement, resulting in a distance vector from the input image regarding the collection of the database. The lowest value is obtained and compared with the threshold, being the label with the lower value the identified user.

3 Neural Network for Face Recognition

Neural networks have been trained to perform complex functions in various fields of application including pattern recognition, speech, vision and control systems. In this project, there is a neural network that identifies each person in the database. After calculating the eigenfaces, the feature vectors of the faces are stored in the database. These vectors are used as inputs to train the network. In the training algorithm, the vectors of values belonging to a person, are used as positive for returning said individual neuron 1 as the neuron output assigned to that user and 0 in other neurons.

When the new image has been captured, the feature vectors are calculated from the eigenfaces obtained before, we compute new descriptors of the unknown image.

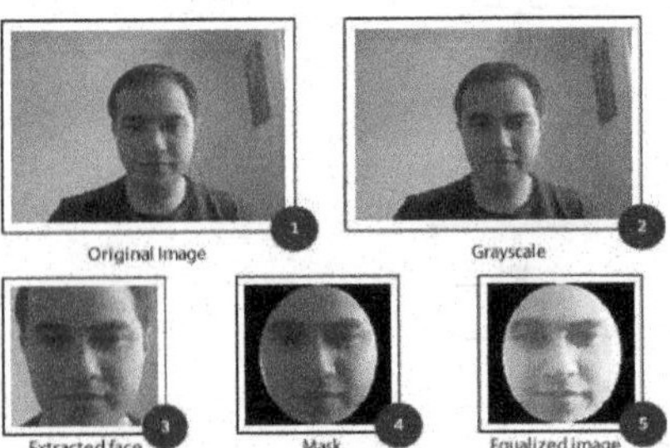

Fig. 3 Processing Flow of an Image

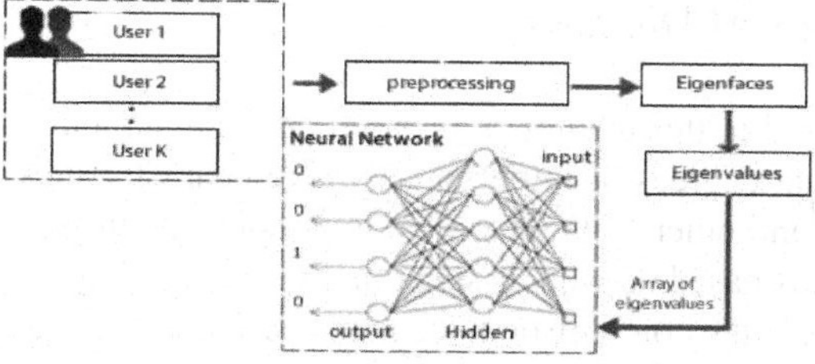

Fig. 4 Neural Network Training

These descriptors are entered into the neural network, the outputs of individual neurons are compared, and if the maximum output level exceeds the predefined threshold, then it is determined that the user belongs to the face assigned to the neuron with the index activated.

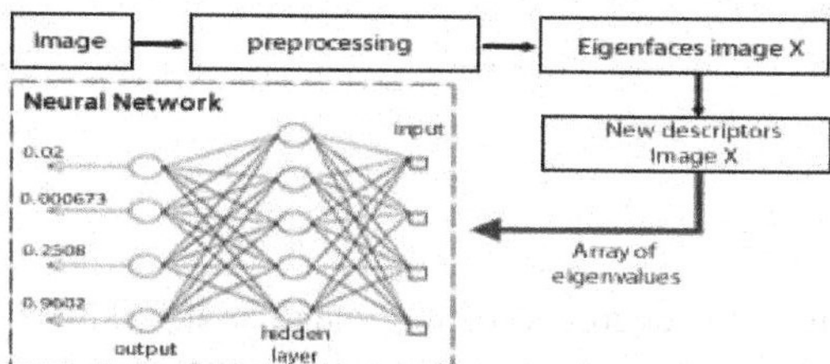

Fig. 5 Neural Network Simulation

3.1 Algorithm Implementation Summary

The approach of face recognition using eigenfaces and neural networks can be summarized in the following steps:

1. Build a library (database) of faces.
2. Choose a training set (M) that includes a number of images for each person with a variation in expression and lighting.
3. Construct the matrix L of $M \times M$, find its eigenvalues and eigenvectors, and choose the vectors M' with the highest values.
4. For each image in the database, calculate and store its feature vector.
5. Create a neural network with one output neuron per registered user.
6. Train the network, using the images of each user as positive examples of neurons and negative examples for the rest.
7. For each new image calculate its eigenvector and get their eigenvalues.
8. Use the new vector calculated as input to the network.
9. Select the neuron with the maximum value. If the output of the selected neuron passes a predefined threshold, is presented as the recognized face.

4 Application of the Hausdorff Distance

The Hausdorff distance measure used in this document is based on the assumption that the facial regions have different degrees of importance, where characteristics such as eyes, mouth, face contour and others; play the most important role in face recognition. The algorithm applied is based on what is stated in [6]. In applying the Hausdorff distance, operates basically the comparison of edge maps. The advantage of using edges to match two objects, is that this representation is robust to illumination change. Accordingly, the edge detection algorithm used will have a significant effect on performance.

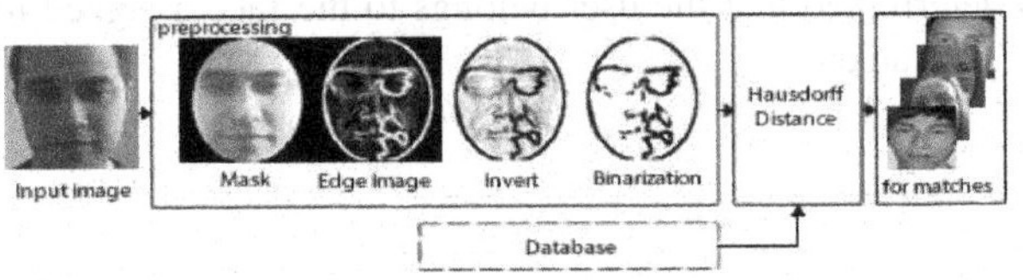

Fig. 6 Flow applying the Hausdorff distance

The procedure involves removing the background of the image as it was performed in the preprocessing original, added some steps after image masking, we proceed to obtain the edges using the Sobel filter, the image is reversed to operate with a white background, then the face is binarized, similar procedure is applied to each image stored in the database.

With the obtained objects, as if they were geometric figures performing a comparison process, calculating the Hausdorff distance, we compare pixels to get how similar are the two figures, resulting in a collection of values that contain the distance of the input image with respect to each item in the database. The object having the smaller relative distance can be presented as an option, if not exceeds the minimum threshold value, identifies the user, otherwise the problem is considered as an unsolved. In the developed system, the Hausdorff algorithm is presented as an alternative to the neural network and recognition using eigenfaces, if the three procedures identify that the user is the same, even without exceeding the thresholds defined in each process, the image is accepted belongs to user input identified by all three techniques.

5 Results and Experimental Set-Up

The proposed method has been tested on 10 different users. Each user has more than one image with different conditions (expression, lighting). For the PCA algorithm is defined a threshold of 90%. Regarding neural network the number of output neurons used is equal to the number of people in the database, the parameters are defined:

- Type: FeedFoward, resilient backpropagation.
- Number of layers: 3 (input, hidden layer, output).
- Input layer neurons: M (number of eigen values obtained to describe faces).
- Neurons in the hidden layer: 10
- Neurons in the output layer: one per registered user in the database.
- Recognition threshold: 90%

With respect to the algorithm Hausdorff, we defined a threshold value of 90%, the result is conditioned by the response of the neural network and PCA algorithm. Only the user shall be deemed recognized by the Hausdorff distance if the three procedures have agreed in their response. In the experiment, we developed a Web application using EmguCV library, that is a wrapper for OpenCV to implement the PCA algorithm and face detection using ViolaJones cascade classifier. We decided to replace the traditional security mechanism of passwords by using HTML5 canvas object that was received from a video which it was generated by a web cam with the facial recognition information. When the algorithm was implemented, we got a number of false positives the result was about two in ten attempts for the login process, this situation could increase to three if is modified the worst lighting conditions. By including the neural network, sending as input the calculated Eigen vectors, the recognition process was improve in nine out of ten attempts. Finally using the Hausdorff distance was observe that the system became robust to different background conditions and lighting changes, but even with that algorithm the face recognition was not perfect and we got nine of ten positives results in face recognition process.

6 Conclusions and Future Work

In this research, we studied three approaches for face recognition. The first based on an approximation Eigenfaces using Euclidean distance for comparison of results, the second approach use the eigenvalues of images as input to a neural network that performs recognition and finally we applied another measure, the Hausdorff distance on edges maps of images of registered users. The main advantage of working in this format is the independence over the lighting problem. The result is a web site with a login using the user's face. The PCA technique allows to obtain the most important components of a face, its strength on feature based approaches is its simplicity and speed, that combined with the application of neural networks provides a learning potential acceptable for use in web applications. This methodology can be extended by replacing the Euclidean distance for a more robust approach to consider the image obtained, and distinguish between sections within it, assigning weights. Finally in a system that is robust and secure, you should consider the changes in lighting, the presence of details in the face such as glasses, beard, and more variables considering the background of the image. Project Experimental results show that the orientation of the face and background of image, represent a major drawback, as future work, the system could try detecting rotation angle of the face.

References

1. Goldstein, A.J., Harmon, L.D., Lesk, A.B.: Identification of human faces. Proc. IEEE 59, 748–760 (1971)
2. Intel, Intel Open Source Computer Vision Library, v2.4.1 (2006), http://sourceforge.net/projects/opencvlibrary/
3. Intel, EmguCV Envoltorio de la biblioteca OpenCV, v2.4.0 (2012), http://www.emgu.com
4. Kerin, M.A., Stonham, T.J.: Face recognition using a digital neural network with self-organizing capabilities. In: Proc. 10th Conf. on Pattern Recognition (1990)
5. Kirby, M., Sirovich, L.: Application of the Karhunen-Loeve procedure for the characterization of human faces. In: IEEE PAMI, vol. 12 (1990)
6. Lin, K.-H., Lam, K.-M., Siu, W.-C.: Spatially eigen-weighted Hausdorff distances for human face recognition. Polytechnic University, Hong Kong (2002)
7. Manjunath, B.S., Chellappa, R., Malsburg, C.: A feature based approach to face recognition. Trans. of IEEE, 373–378 (1992)
8. Dimitri, P.: Eigenface-based facial recognition (Diciembre 2002)
9. Rowley, H., Baluja, S., Kanade, T.: Neural network face detection, San Francisco, CA (1996)
10. Smith, L.I.: A tutorial on principal components analysis (February 2002), http://www.cs.otago.ac.nz/cosc453/student_tutorials/principal_components.pdf (accessed on April 27, 2012)
11. Terrillon, J., David, M., Akamatsu, S.: Automatic detection of human faces in natural scene images by use of a skin color model, Nara, Japan (1998)
12. Turk, M., Pentland, A.: Eigenfaces for recognition (1991a), http://www.cs.ucsb.edu/mturk/Papers/jcn.pdf (accessed on April 27, 2012)
13. Viola, P., Jones, M.J.: Robust real-time face detection. International Journal of Computer Vision (2004)
14. Yuille, A.L., Cohen, D.S., Hallinan, P.W.: Feature extraction from faces using deformable templates. In: Proc. of CVPR (1989)

QoS Synchronization of Web Services: A Multi Agent-Based Model

Jaber Kouki[1], Walid Chainbi[2], and Khaled Ghedira[3]

[1] High Institute of Human Sciences of Tunis/SOIE, Tunisia
jaber.kouki@hotmail.com
[2] Sousse National School of Engineers/SOIE, Tunisia
Walid.Chainbi@gmail.com
[3] Higher Management Institute of Tunis/SOIE, Tunisia
Khaled.Ghedira@isg.rnu.tn

Abstract. From the last decade, Web services technology has witnessed a great adoption rate as a new paradigm of communication and interoperability between different software systems. This fact, has led to the emergence of Web services and to their proliferation from outside the boundary of the UDDI business registry to other potential service resources such as public and private service registries, service portals, and so on. The main challenge that arises from this situation is the fact that for the same service implementation, several service descriptions are published in different service registries. Accordingly, if the service implementation is updated all of its descriptions have to be updated too over all of these registries. Otherwise, the service user may not bind to the suitable Web service if its descriptions are inaccurate or outdated. To address the above challenge, we propose in this paper a multi agent-based model that focuses on synchronizing the description of Web services, especially their quality of service, to maintain their consistency and sustainable use.

1 Introduction

To support interoperable machine-to-machine interaction over a network, Web services are emerged as a promising technology that delivers application functionalities as services which are language and platform independent. One of the major building blocks of Web services technology is the UDDI (Universal Description Discovery and Integration) [1] business registry (UBR) where service providers and service consumers (or users) publish and discover Web services respectively.

To use the UBR service consumers can only provide functional descriptions of Web services they need (e.g. keywords, inputs, outputs, etc.), however to select the best suitable Web service nonfunctional descriptions (e.g. QoS) are required but, they are missing in UBR.

Focusing on this limit, other service resources start to be deployed such as public business registries (PBRs), and service portals (e.g. Xmethods.net, webservicelist.com, webservicesx.net, etc). Nevertheless, the main issue with these resources is the fact that they do not adhere with Web service standards (e.g. UDDI) and therefore to be potential service resources like UBR does. Furthermore, as Web services substantially increase in number all over the Web, the number of service

S. Omatu et al. (Eds.): *Distrib. Computing & Artificial Intelligence,* AISC 217, pp. 401–408.
DOI: 10.1007/978-3-319-00551-5_49

ressources is also increased. Accordingly, service users will not be able to browse all of them separately to select Web services of interest.

To cope with these facts, we have proposed in a previous work [2] a local repository-based framework (LRBF) which provides service users with local Web service repositories (LWSRs) to collect service information including binding details and QoS descriptions. This information is used to bind locally to Web services that interest service users without the need to browse heterogeneous service resources independently.

However, since service providers may change the implementation of their published Web services, the service information collected in LWSRs may become inaccurate and outdated. Due to this change, Web services that are locally bound may no more interest service users. To deal with this fact, we extend the LRBF framework with a multi agent-based model to synchronize service information (especially QoS) of the updated Web services over different service resources, and therefore to ensure that the locally bound Web services are always the best suitable ones according to the service users need.

The remainder of this paper is organized as follows. Section 2 outlines the background material of the QoS synchronization issue within the LRBF framework. The extension to this framework with a multi agent-based model for QoS synchronization is discussed in Sect. 3.The implementation issues are presented in Sect. 4. Prior to the conclusion and future work in Sect. 6, related work are reviewed in Sect. 5.

2 QoS Synchronization Background

Considering the need to optimize the binding to relevant Web services from heterogeneous environments, the LRBF framework provides service users with a local access point to bind to Web services of interest without having to browse separately hundreds if not thousands of different service resources. The architecture of this framework involves three levels of service repositories where two levels of binding optimization are performed.

The first level of optimization, which is public, is performed by a public Web service crawler engine (PWSCE) that collects service information from the first level of service repositories that encompasses UBRs, PBRs and service portals into the second level which is represented by a public Web service repository (PWSR). Having collected service information of all Web services in this repository, is very practical for service users to see Web service candidates from different service resources grouped under one roof, and then to select easily the best suitable one. However, what is impractical in this level of optimization is the fact that service users have to access the PWSR each time they need to re-bind to the same Web services which may be a useless work and a waste of time. In fact, as the PWSR is hosted on the administrator server host and as it contains a huge number of service information, to respond to an excessive amount of service requests may take a considerable amount of time.

To deal with these facts, a local Web service crawler engine (LWSCE) performs the second level of binding optimization which is local. In this level of optimization, the LWSCE collects service information of the best suitable Web services from the PWSR into the third level of service repositories which is represented by a local Web service repository (LWSR). In this repository, service users can find locally (within their local hosts) service information required to bind to Web services of interest without using the PWSR.

However, having collected service information in more than one repository within the LRBF framework, means that such information is duplicated over the involved service repositories. According to this fact, service information of Web services and especially their QoS descriptions need to be synchronized for two reasons. The first one is to ensure that the different levels of service repositories, involved within the LRBF framework, see all the changes that have been originated by the service providers, and the second one is to ensure that the locally bound Web services still always fit the service user interest. To meet these two objectives, we introduce in the next section a multi agent-based model for dynamic synchronization of Web service QoS descriptions.

3 QoS Synchronization Proposed Model

The need to agent paradigm in our proposed synchronization model is basically argued by the need of both the autonomy and pro-activeness properties of software agents to synchronize dynamically QoS descriptions of Web services over service repositories with the minimum of human intervention (either by service providers or service users).

As defined by Wooldridge, software agent is "a computer system that is situated in some environment, and that is capable of autonomous actions in this environment in order to meet its delegated objectives" [3]. In our proposed model, the actions to carry out by software agents within LRBF are first to synchronize QoS descriptions of Web services over the involved service repositories once new updates are delivered by service providers and then to ensure that, upon each update in the service information, the locally bound Web services always rank better than their candidates.

From the above considerations and given the infrastructure of LRBF framework which encompasses three levels of service repositories, the architecture of the proposed synchronization model (see Fig. 1) involves three software agents including PISA (Public Information Synchronization Agent), UISA (Universal Information Synchronization Agent), and LISA (Local Information Synchronization Agent). Each class of software agents operates at one level of service repositories.

As depicted in Fig. 1, the processing of QoS synchronization module within LRBF starts when new updates in the service QoS description are delivered by the service provider to the PISA agent that operates at the first level of service repositories which includes UBRs, PBRs and service portals. To interact with this agent, the service provider uses a GUI by which he can provide new updates in the QoS description (e.g. response time) of a particular Web service. Upon receiving these updates, the PISA agent saves them first in the repository (e.g. PBR) of the service provider, and then it sends to the UISA agent a message incorporating the updated QoS description.

As soon as the UISA agent receives this message from the PISA agent, it saves first the included QoS description in the PWSR, the second level of service repositories on which this agent operates, and then it starts listening for new update requests from the LISA agent. When new requests are received, the UISA agent replies the requestor agent with a message incorporating the updated QoS description received from the PISA agent. Upon receiving this message and before saving the included data in the LWSR (the third level of service repositories), the LISA agent checks first whether the locally bound Web service with its new QoS description still match better the predefined requirements of the service user or not. Accordingly, if this Web service always ranks better than its candidates, the LISA agent will save its new QoS description in the LWSR. Otherwise, this agent selects from the list of Web service candidates the service that best fits requirements of the service user and then it collects the service information of the new selected Web service from the PWSR into the LWSR to replace the existing one.

To conduct a reliable synchronization of QoS information over the different levels of service repositories, software agents have to carry out suitable behaviors. According to the Fipa-compliant agent platform, JADE (Java Agent Development Framework), software agents can carry out One-shot behavior that is executed only once and then completes immediately, Cyclic behavior that is executed forever, and Generic behavior that is executed when a given condition is met [4]. Besides, further behaviors may be executed at given points in time such as Waker behavior that is

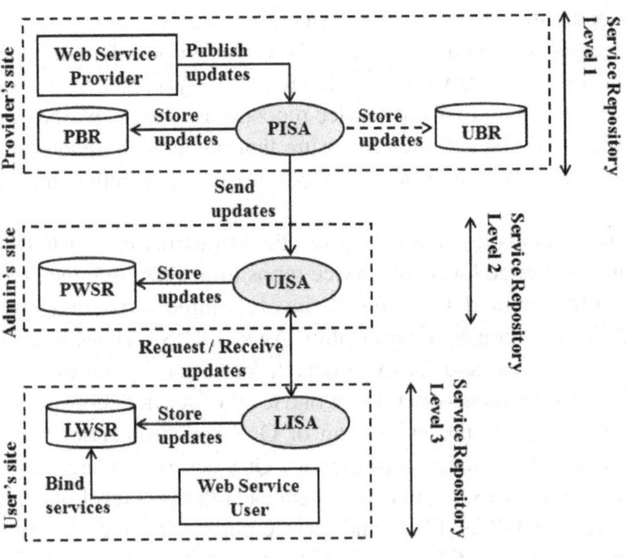

Fig. 1 Architecture of the QoS synchronization model

executed only once just after a given timeout is elapsed and Ticker behavior that is executed periodically waiting a given period after each execution.

According to the tasks to carry out by each agent within the proposed synchronization model and given the types of agent behaviors described above, software agents involved in this model have to carry out the following behaviors:

- PISA behaviors: the PISA agent executes one-shot behavior for each updates reception from the service provider and another one-shot behavior for each updates delivery to the UISA agent.
- UISA behaviors: the UISA agent executes two cyclic behaviors: one dedicated to serve requests for updates reception from the PISA agent and the other dedicated to serve requests for updates delivery to the LISA agent.
- LISA behaviors: the LISA agent carries out a ticker behavior that deals with the update requests to the UISA agent and one-shot behavior that deals with the updates reception from this agent. The last behavior is executed only if the updated QoS descriptions belong to the locally bound Web services. Upon updating these services or their potential candidates the LISA agent carries out another ticker behavior to select periodically the new suitable Web service that fits better the service user needs.

4 Implementation

The QoS synchronization model is implemented using JADE framework to develop software agents involved in this model such as PISA, UISA and LISA. Furthermore, we have integrated these agents in ATAC4WS[1] (Agent Technology and Autonomic Computing for Web Services) which implements the LRBF framework that we have developed to create the three levels of service repositories on which software agents synchronize QoS information of Web services. Upon running ATAC4WS, both PWSCE and LWSCE (the two Web service crawler engines) start collecting service information from UBRs, PBRs and other potential service resources into the PWSR for all Web services and then from PWSR into LWSR for only Web services that interest service users. To launch the processing of the QoS synchronization module within LRBF, service providers use a GUI to deliver to the PISA agent new updates in the QoS description (e.g. cost) of a particular Web service.

Upon receiving these updates, the PISA agent incorporates them first into the repository (e.g. PBR) of the service provider within the first level of service repositories and then it sends them to the UISA agent. To synchronize the updates in QoS description over the second and third levels of service repositories, the UISA and LISA agents exchange a message incorporating the new Qos description to be included in the PWSR and LWSR respectively.

Finally, upon each updates reception in QoS description, the LISA agent computes the QoS ranking of the locally bound Web services compared to their candidates according to the QoS preferences of the service user. The focus of this computation is to select Web services that interest more the service user and therefore to

[1] ATAC4WS provides an environment to create, register, discover and use Web services.

collect their service information in the LWSR where to be bound locally in subsequent use.

Details about the running of LISA agent in response to the update of the cost attribute of two Web services (e.g. GlobalWeather and WeatherForecast) are shown in the NetBeans IDE outputs of Fig. 2. These Web services are used to get up-to-date weather conditions for all major cities around the world.

In the first output (see the top of Fig. 2), although the cost of GlobalWeather (a locally bound web service) is raised to 0.4 $, this service still always the best suitable one assuming that the service user is only interested in the cheapest Web service. However, in the second output (see the buttom of Fig. 2), upon decreasing the cost of WeatherForecast (a Web service candidate) to 0.3 $, this service becomes cheaper than GlobalWeather and in this case the LISA agent will collect its service information from the PWSR into the LWSR to replace the existing service (GlobalWeather).

Note that, before each request of new QoS updates, the LISA agent tries to find new UISA agents since they may dynamically appear and disappear in the system.

5 Related Work

As Web services start to expand across the Internet, most of them are still inaccessible or have duplicated descriptions that do not match the new implementation of the original Web services. Therefore, several works focused on addressing QoS issues to guarantee the reliability of published Web services while others are interested in addressing synchronization concern of Web services in different aspects.

```
LISA-agent Lisa@JABER-DELL:1099/JADE trying to find UISA agents:
The following UISA agents are found:
Uisa@JABER-DELL:1099/JADE
LISA-agent Lisa@JABER-DELL:1099/JADE requesting new updates
New updates are sent to LISA-agent
GlobalWeather Updated in LWSR
New Cost = 0.4 ($/tr)
GlobalWeather is always the best Web service
QoS rating = 0.8461538
-------------------------------------------------------------
LISA-agent Lisa@JABER-DELL:1099/JADE trying to find UISA agents:
The following UISA agents are found:
Uisa@JABER-DELL:1099/JADE
LISA-agent Lisa@JABER-DELL:1099/JADE requesting new updates
New updates are sent to LISA-agent
WeatherForecast doesn't exist in LWSR!
WeatherForecast is a Web service candidate
New Cost = 0.3 ($/tr)
WeatherForecast is become the best Web service
QoS rating: 0.875
```

Fig. 2 LISA running outputs

To address the QoS issue of Web services, many researchers have extended the service oriented architecture (SOA) [5] to incorporate nonfunctional properties of Web services including QoS. For example, in [6] the author proposed a regulated UDDI model by adding a new data structure type where to represent QoS information of Web services such as availability, reliability, etc.

The authors in [7] did not extend the UDDI model but they used the existing data structure type tModels. In tModels, QoS are represented as a KeyedReference element where the KeyName attribute contains the QoS name and the KeyValue contains the QoS value. In the same effort, Hollunder in [8] and the authors in [9] used WS-Policy [10] and OWL-S [11] respectively instead of tModels to advertise QoS information of Web services.

However, since service providers may change the QoS description of their published Web services, it is important to synchronize the new description over different services ressources. Although it is hard to find works addressing the synchronization issue of QoS description, some other works focused on the validity of service information in service registries in general.

For example, the authors in [12] extended WS-Policy as WS-TemporalPolicy to describe service properties with time constraints. That is, to define the validity of a policy or its included service properties a set of time attributes are used such as startTime, endTime and expires.

The authors in [13] extended UDDI as UDDIe where Web services hold a lease which defines how long the service information should remain registered in a registry. If a lease expires, the service provider should renew it otherwise the registered information will be deregistered.

Finally, the authors in [14] proposed an agent-based model for dynamic synchronization of Web services. However the synchronization aspects of this model focus only on service binding details.

6 Conclusion

In this work, we have proposed a multi agent-based model for QoS synchronization of Web services which is integrated within LRBF framework to synchronize the QoS description of updated Web services over different levels of service repositories (PBRs, PWSR, and LWSR respectively). This model has been implemented using the JADE platform.

In Future work, we envision to deal with the trustworthy of the delivered QoS information since the service provider may greatly influence how QoS metrics are generated to obtain the most suitable results, and therefore may provide inaccurate information about QoS.

References

1. Clement, L., Hately, A., Riegen, V.C., Rogers, T.: Universal description, discovery, and integration (UDDI 3.0.2). Technical committee draft, OASIS (2004), http://uddi.org/pubs/uddi_v3.htm/

2. Kouki, J., Chainbi, W., Ghedira, K.: Binding optimization of web services: a quantitative study of local repository-based approach. In: ICWS 2009: Proceedings of the IEEE International Conference on Web Services, pp. 646–647. IEEE computer society (2012)
3. Wooldridge, M.: An introduction to multiagent systems. John Wiley & Sons Ltd., UK (2009)
4. Bellifemine, F., Caire, G., Trucco, T., Rimassa, G.: Jade programmers guide. Technical report, TILab (2010), http://jade.tilab.com/doc/programmersguide.pdf/
5. Krafzig, D., Banke, K., Slam, D.: Enterprise soa: service-oriented architecture best practices. Prentice-Hall Inc., New Jersey (2005)
6. Ran, S.: A model for web services discovery with qos. ACM SIGecom Exchanges 4(1), 1–10 (2003)
7. Rajendran, T., Balasubramanie, P.: An optimal agent-based architecture for dynamic web service discovery with qos. In: ICCCNT 2010: International Conference on Computing Communication and Networking Technologies, pp. 1–7 (2010)
8. Hollunder, B.: Ws-policy: on conditional and custom assertions. In: ICWS 2009: Proceedings of the IEEE International Conference on Web Services, pp. 936–943. IEEE Computer Society (2009)
9. Lakhal, R.B., Chainbi, W.: A multi-criteria approach for web service discovery. In: MobiWIS 2012: Proceedings of the 9th International Conference on Mobile Web Information Systems, vol. 10, pp. 609–616. Elsevier PCS (2012)
10. Vedamuthu, A., Orchard, D., Hirsch, F., Hondo, M., Yendluri, P., Boubez, T., Yalcinalp, U.: Web services policy (1.5). W3C recommendation, W3C (2007), http://www.w3.org/TR/ws-policy/
11. Martin, D., Burstein, M., Hobbs, J., Lassila, O., McDermott, D., McIlraith, S., Narayanan, S., Paolucci, M., Parsia, B., Payne, T., Sirin, E., Srinivasan, N., Sycara, C.: Owl-s: semantic markup for web services. W3C member submission, W3C (2004), http://www.w3.org/Submission/OWL-S/
12. Mathes, M., Heinzl, S., Freisleben, B.: Ws-temporalpolicy: a ws-policy extension for describing service properties with time constraints. In: COMPSAC 2008: Proceedings of the 32nd Annual IEEE International Computer Software and Applications, pp. 1180–1186 (2008)
13. ShaikhAli, A., Rana, O.F., Al-Ali, R., Walker, D.W.: UDDIe: an extended registry for web services. In: SAINT-W 2003: Proceedings of the 2003 Symposium on Applications and the Internet Workshops, pp. 85–89. IEEE Computer Society (2003)
14. Kouki, J., Chainbi, W., Ghedira, K.: An agent-based approach for binding synchronization of web services. In: AWS 2012: Proceedings of the 1st International Workshop on the Adaptation of Web Services, pp. 921–926. Elsevier PCS (2012)

Comparing Basic Design Options for Management Accounting Systems with an Agent-Based Simulation

Friederike Wall

Alpen-Adria-Universitaet Klagenfurt, Universitaetsstrasse 65-67, 9020 Klagenfurt, Austria
friederike.wall@uni-klu.ac.at

Abstract. The paper applies an agent-based simulation to investigate the effectiveness of basic design options for management accounting systems. In particular, different settings of how to improve the information base by measurement of actual values in the course of adaptive walks are analyzed in the context of different levels of complexity and coordination modes. The agent-based simulation is based on the idea of NK fitness landscapes. Results provide broad, but no universal support for conventional wisdom that lower inaccuracies of information lead to more effective adaptation processes. Furthermore, results indicate that the effectiveness of improving judgmental information by actual values subtly depends on the complexity of the decisions and the coordination mode applied.

1 Introduction

An important function of management accounting is to provide decision-makers with judgmental information for evaluating options [1]. For deciding whether to change the status quo in favor of an alternative, a decision-maker requires information on the pay-offs of both options. Information on the status quo may result from (possibly imperfect) measurements of actual values (i.e., "weighting", "counting") within accounting systems. In contrast, alternative options, in principle, are subject to ex-ante evaluations by decision-makers who suffer from cognitive limitations [2]. However, even ex-ante evaluations might be based on measurements, i.e., actual values received on basis of decisions made in former periods and used to "learn" for future decisions. For instance, plan cost accounting often relies on cost functions which are built from actual costs realized in former periods [3].

Moreover, management accounting systems are embedded in an organizational structure which in turn affects imperfections of judgmental information. In particular, the overall decision problem is segmented into partial decisions which are delegated to decentral decision-makers (e.g. [4], [5]). With delegation further difficulties occur: For example, partial decisions may be interdependent and decision-makers likely have different knowledge and information. To avoid losses with respect to the organization's performance, coordination is required, though, according to Ackoff [6], more intense coordination not necessarily increases organizational performance.

S. Omatu et al. (Eds.): *Distrib. Computing & Artificial Intelligence,* AISC 217, pp. 409–418.
DOI: 10.1007/978-3-319-00551-5_50

Against this background the paper investigates the question *in which settings of organizational and "accounting" design it is effective to use measured actual values by management accounting systems for improving judgmental information*. For investigating the research question, we apply an *agent-based simulation* since this method allows mapping interacting heterogeneous agents which is in the very center of our research question [7]: One reason for delegating decisions to subordinates is to benefit from their specialized knowledge. Thus, it is rather likely, that different decision-makers in the organization have different areas of expertise and, hence, given management accounting numbers are interpreted differently in the organization. Obviously, these interrelated issues would be particularly difficult to control in empirical research and would induce intractable dimensions in formal modeling.

The paper contributes to research in management science since, to the best of the author's knowledge, for the first time different settings of memorizing actual values and dynamic adjustments through actual values in accounting are investigated in interaction with major organizational design variables.

The remainder of this article is organized as follows: section 2 introduces the architecture of the simulation model. In the third part we present and discuss results of the simulations.

2 Simulation Model

The simulation model is based on the NK model introduced by Kauffman [8], [9] in the context of evolutionary biology and successfully applied in management research (e.g. [10], [11], [12], for an overview [7]). The NK model allows representing a multi-dimensional decision problem where N denotes the number of dimensions and K the level of interactions among these dimensions.

We adopt an advanced version of the NK model with noisy fitness landscapes, as introduced by Levitan and Kauffman [13] and, in an even modified form, employed to analyze decision-making with imperfect information [14], [15]. Our model consists of three major components: (1) the organizational structure which is mapped similar to Siggelkow and Rivkin [11]; (2) a representation of imperfect judgmental information that corresponds to organizational segmentation and specialization; (3) different options of management accounting systems. The components (2) und (3) are regarded to be the distinctive features of the model.

2.1 Organizational Structure

The artificial organizations face a ten-dimensional binary decision problem, i.e., they have to make decisions $i, i = 1, \ldots 10$ (hence, $N = 10$ in terms of the NK model), with states $d_i \in \{0,1\}$ leading to $\mathbf{d} = (d_1, \ldots, d_{10})$. Each single state d_i of decision i provides a contribution C_i with $0 \leq C_i \leq 1$ to overall performance $V(\mathbf{d})$. A decision i might interact with K other decisions. K can take values from 0 (no interactions) to $N-1$ (maximum interactions). Thus, performance contribution C_i may not only depend on the state of d_i but also on the state of K other decisions so that

$$C_i = f_i(d_i; d_i^1, ..., d_i^K). \tag{1}$$

In line with the NK model, we assume that for each possible vector $(d_i; d_i^1, ..., d_i^K)$ the value of C_i is randomly drawn from a uniform distribution over the unit interval, i.e., $U[0,1]$. Hence, given equation 1, whenever one of the states $d_i; d_i^1, ..., d_i^K$ is altered, another (randomly chosen) performance contribution C_i becomes effective. The overall performance $V(\mathbf{d})$ is given as normalized sum of performance contributions C_i with

$$V(\mathbf{d}) = \frac{1}{N} \sum_{i=1}^{N} f_i(d_i; d_i^1, ..., d_i^K). \tag{2}$$

Our organizations consist of a main office and three departments $r \in \{1,2,3\}$. The organizations segment their ten-dimensional decision problem $\mathbf{d}$ into three partial problems and delegate each of these to one of the r departments. Hence, each department has primary control over a subset of the ten decisions (department 1 over decisions 1 to 3, department 2 over decisions 4 to 7 and department 3 over decisions 8 to 10). Department r perceives the overall problem $\mathbf{d}$ to be partitioned into a partial vector $\mathbf{d_r^{own}}$ related for those single decisions which are in the "own" responsibility and into $\mathbf{d_r^{res}}$ for the "residual" decisions that other departments are in charge of. However, in case of cross-departmental interactions, choices of a certain department may affect the contributions of decisions other departments are responsible for and vice versa.

In each period of the adaptive walk a department head seeks to identify the best configuration for the "own" subset of choices assuming that the other departments do not alter their prior subsets of decisions. In particular, head of department r randomly discovers two alternative partial configurations $\mathbf{d_r^{own}}$: an alternative $a1$ differing in one dimension and another alternative $a2$ differing in two dimensions from the status quo. Hence, each department head has three options to choose from (status quo and two alternatives). According to economic literature, a department head favors that option which he/she perceives to promise the highest value base for compensation. In our model department heads are compensated on basis of the overall performance $V(\mathbf{d})$ according to a linear incentive scheme so that we can ignore conflicts of interests between organizational and departmental objectives. However, due to specialization our department heads have different knowledge about the organization's decision problem $\mathbf{d}$ (we return to that point in the section 2.2). In consequence, even though in our model no conflicts of interests occur, departments can have different preferences which might evoke a need for coordination. We analyze two different modes of coordination ([11], [16] for these and other modes):

1. In the "decentral" mode, in fact, there is no coordination: each department autonomously decides on $\mathbf{d_r^{own}}$ and the overall configuration $\mathbf{d}$ results as a combination of the departmental choices without any intervention of the main office. Hence, the function of the main office is limited to (perhaps inaccurately) observing the overall performance achieved.
2. In the "proposal" mode each department proposes two alternative configurations $\mathbf{d}$ to the main office which, among all proposals received, chooses that one which

promises the highest overall performance. Hence, by their proposals the departments shape the search space of the main office.

2.2 *Informational Structure*

To represent inaccurate judgmental information we add noise according to segmentation, delegation and specialization: A common idea of many organizational theories is that decision-makers in organizations dispose of information with different levels of imperfections (e.g. [4], [5]). For example, departmental decision-makers are assumed to have relatively precise information about their own area of competence, but limited cross-departmental knowledge whereas the main office might have rather coarse-grained, but organization-wide knowledge.

We assume that departments decide on basis of the perceived value base for compensation, i.e., the perceived overall performance $\widetilde{V}_r(\mathbf{d})$ rather than the actual $V(\mathbf{d})$. In particular, the perceived value base for compensation $\widetilde{V}_r(\mathbf{d})$ is computed as normalized sum of the actual own performance and actual residual performance, each distorted with an error term as

$$\widetilde{V}_r(\mathbf{d}) = [\widetilde{V}_r^{own}(\mathbf{d}_r^{own}) + \widetilde{V}_r^{res}(\mathbf{d}_r^{res})]/N \tag{3}$$

$$\text{where } \widetilde{V}_r^{own}(\mathbf{d}_r^{own}) = V_r^{own}(\mathbf{d}_r^{own}) + e_r^{own} \tag{4}$$

$$\text{and } \widetilde{V}_r^{res}(\mathbf{d}_r^{res}) = V_r^{res}(\mathbf{d}_r^{res}) + e_r^{res}. \tag{5}$$

Likewise, in the "proposal" mode the main office chooses of the proposals on basis of the *perceived* overall performance $\widetilde{V}_r^{main}(\mathbf{d})$ computed as the sum of the true overall performance $V(\mathbf{d})$ and an error term e^{main}. With respect to accounting systems [17], it is reasonable to assume that high (low) true values of performance come along with high (low) distortions. Hence, we reflect distortions as relative errors (for other functions see [13]), which, for simplicity, follow a Gaussian distribution $N(\mu;\sigma)$ with expected value $\mu = 0$ and standard deviations σ_r^{own}, σ_r^{res} and σ^{main}. We differentiate the standard deviations according to specialization of departments and the main office as mentioned above (see also notes in Table 2).

2.3 *Design Options of Management Accounting Systems*

For comparing the contributions of alternative design options of management accounting systems, as is in the center of this paper, we distinguish five settings of measurement and usage of actual values in the adaptive walk (Table 1):
(1) In case that "no measurement" is used, the evaluation of the status quo cannot be based on measured values. Hence, in a way, this reflects an organization which does not have any accounting system.
(2) In the setting "measurement only" the accounting systems allow departments to perfectly determine the performance achieved with the current configuration c of the decisional vector $\mathbf{d}^c$ in the last period. Hence, throughout each adaptive walk, department heads perfectly know the status quo, but they suffer from inaccurate

knowledge of the performance contributions of alternatives $\mathbf{d}_r^{a1}$ and $\mathbf{d}_r^{a2}$. The accounting system does not provide any tracking or memory about the configurations that have been implemented before.

(3) In "stepwise refinement" the status quo can be perfectly estimated like in setting (2); additionally, the measured actuals are used for some rather simple kind of "learning": whenever a certain configuration has been implemented, decision-makers receive information about the related contributions to performance. In future periods, this information will be partially memorized leading to an improved evaluation of that configuration.[1] For each configuration $\mathbf{d}_n, n = 2^N$ (i.e., $n = 1024$ since $N = 10$) a counter $count_n$ is introduced which is incremented by 1 whenever this configuration is implemented. If configuration $\mathbf{d}_n$ is evaluated again in a later period, the corresponding errors (e_r^{own}, e_r^{res} and e^{main}) are divided by $count_n$. Thus, for example, in the "proposal" mode the main office perceives the performance of configuration $\mathbf{d}_n$ as

$$\widetilde{V}_{main}(\mathbf{d}_n) = V(\mathbf{d}_n) \cdot (1 + \frac{1}{count_n} \cdot e^{main}(\mathbf{d}_n)). \tag{6}$$

(4) In the case "immediate adjustment" the accounting systems provide perfect memorizing and immediate correction of ex-ante evaluations due to measured actual values. Hence, whenever in the adaptive walk a configuration is considered, which has already been implemented at least once, the decision-makers get perfect information about the level of performance. For example, the main office evaluates the overall performance of a configuration $\mathbf{d}_n$ as

$$\widetilde{V}_{main}(\mathbf{d}_n) = V(\mathbf{d}_n) \cdot (1 + e^{main}(\mathbf{d}_n)) \text{ for } count_n = 1 \tag{7}$$

$$\text{and } \widetilde{V}_{main}(\mathbf{d}_n) = V(\mathbf{d}_n) \text{ for } count_n \geq 2. \tag{8}$$

Table 1 Settings of measurement and usage of actuals

Name of Setting	Measurement of actuals for status quo	Adjustment of inaccuracies while adaptive walk
(1) No measurement	no	no
(2) Measurement only	yes	no
(3) Stepwise refinement	yes	stepwise
(4) Immediate adjustment	yes	immediately at once
(5) Perfect evaluation	yes	(not necessary)

[1] For example, this reflects a cost planning system where cost functions are adjusted with each measurement of the performance that a certain configuration of cost drivers provides: with each determined combination of cost drivers and cost measures the statistical basis is broadened from which a cost function could be derived (for example by regression analysis).

(5) Perfect evaluations in our simulations serve as a "benchmark" so that performance differences due to imperfect evaluations can be determined. Neither the estimation of the status quo nor the evaluation of the alternative options suffer from any noise, i.e., all error terms are set to zero.

3 Results and Interpretation

For simulating an adaptive walk, after a "true" performance landscape is generated, distortions are added as described in section 2.2. Next the organizations are placed randomly in the landscape and observed for 300 periods while searching for higher levels of performance. The search process is modelled as a steepest-ascent hill-climbing algorithm [7]. Two interaction structures of decisions were simulated which, in a way, represent two extremes [12]: in the low complexity case intra-departmental interactions are maximally intense with no cross-unit interdependencies; in the high complexity case all decisions affect the performance contributions of all other decisions, i.e., complexity is raised to maximum. Empirical findings report errors of judgmental information between 5 up to 30 percent [18]. The results presented here relate to errors around 10 percent, though differentiated due to specialization (see section 2.2 and note to table 2).[2]

Table 2 reports three measures: "Speed" (ΔV_{5-1}) captures the performance enhancements achieved in the first 5 periods which most purely allow for analyzing the effects of refinements (settings 3 and 4). The performance in the last observation period, V_{300}, serves as indicator for the effectiveness of the search process as well as the frequency of how often the global maximum in the performance landscape was finally found. Furthermore, Figures 1A to 1D reflect the performance differences in the course of the adaptive walks of the noisy against the perfect evaluations.

3.1 Effectiveness of Various Settings of Management Accounting Systems

We discuss results in two steps. Firstly, we focus on comparing the different settings of measuring and using actuals against each other and afterwards we discuss the moderating effects of complexity and coordination (section 3.2).

Obviously, evaluating options with imperfect information can result in a choice which *appears* favorable, whereas, in fact, it reduces performance compared to the status quo ("false positive" decision) [13]. Underestimating the status quo level of performance might foster the false estimation. Vice versa, with "false negative" decisions an alternative is rejected because its marginal contribution to performance

[2] The results were subject to robustness analyses, especially with respect to the magnitude of errors and the spread between knowledge about the "own" area of competence and the rest of the organization. We found that the results appear robust in a range up to a magnitude of overall error around 22 percent and with several levels of spread according to specialization of decision-makers.

Table 2 Condensed Results

setting	decentral mode speed ΔV_{5-1}	final perf. V_{300}	frequ. global max.	proposal mode speed ΔV_{5-1}	final perf. V_{300}	frequ. global max.
low complexity						
(1) No measurement	0.04419	0.83689	1.66%	0.06878	0.83506	1.46%
(2) Measurement only	0.05211	0.85251	1.94%	0.07277	0.85222	2.52%
(3) Stepw. refinemt.	0.05734	0.89748	8.24%	0.08473	0.87681	4.84%
(4) Immed. adjustmt.	0.05771	0.89381	7.76%	0.08601	0.87724	4.56%
(5) Perfect evaluation	0.07321	0.89730	10.16%	0.09781	0.89518	9.54%
high complexity						
(1) No measurement	0.12479	0.84374	1.94%	0.05739	0.83541	1.76%
(2) Measurement only	0.12488	0.86941	3.16%	0.05754	0.83665	1.90%
(3) Stepw. refinemt.	0.12818	0.86738	3.00%	0.06387	0.85041	2.30%
(4) Immed. adjustmt.	0.12121	0.86519	3.46%	0.06742	0.84803	1.78%
(5) Perfect evaluation	0.14536	0.86466	2.44%	0.06510	0.86716	2.40%

Notes: Each entry represents results of 5,000 adaptive walks: 1,000 distinct fitness landscapes with 5 adaptive walks on each over 300 periods. Confidence intervals for V_{300} at a confidence level of 0.001 range between ±0.003 and ±0.004. Common parameters in settings (1) to (4): $\sigma_r^{own} = 0.05$, $\sigma_r^{res} = 0.15$ and $\sigma_{main} = 0.1$ (in (5) all set to 0); all errors with expected value $\mu = 0$.

compared to the (possibly overestimated) status quo appears worse than it actually is and, thus, the status quo is perpetuated [13]. This lets us hypothesize the following:

With increasing levels of measurement and usage of actuals for improving judgmental information (1) the speed of performance enhancements increases and (2) higher levels of organizational performance are achieved.

The settings of accounting systems displayed in Table 1 incorporate an order of increasing information accuracy. We find that the speed measure ΔV_{5-1} (Table 2) in most cases is increasing with the more advanced accounting systems. Furthermore, as Figures 1A to D show at a glance, the more advanced settings tend to have lower performance losses againt the perfect system. However, the results provide no universal support for the hypothesis stated above and some of the results deserve a closer analysis.

First of all, in setting (1) where no actual numbers are available at all, the performance achieved is lowest in all of the four scenarios and in three scenarios with remarkable performance losses even against the "measurement only" setting (we discuss the "high complexity-proposal mode" scenario below). Apparently, over- or underestimating the status quo leads to severe losses of speed and level of

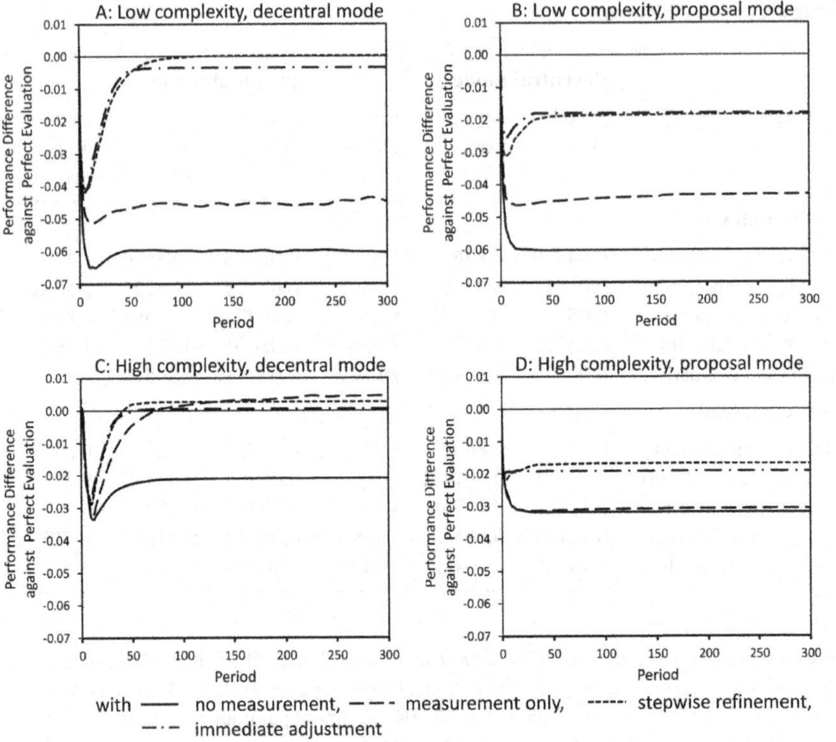

Fig. 1 Performance differences against perfect evaluations

performance enhancements. Hence, this indicates that using an accounting system, at least, to track the status quo (e.g., an actual cost system) is effective.

Second, settings "stepwise refinement" (3) and "immediate adjustments" (4) show rather similar results for all scenarios. An obvious reason is that with our "stepwise refinement" the decision-makers get better knowledge of the fitness landscape. Slower learning might yield other results. Third, these settings (3) and (4) bring performance to levels higher than achieved with "measurement only" – except for the case of high complexity and decentral coordination which is discussed below. This indicates that accounting systems, which allow for memorizing actual values, contribute to higher performance levels.

3.2 *Effects of Complexity and Coordination Mode*

We start the discussion with the "low complexity-decentral mode" case (Figure 1A). Here no cross-unit interactions exist, and, thus, no cross-unit coordination is required: with imperfect information departments may decide in favor of a suboptimal partial option, but there are no external effects in terms of reducing the performance of the other departments' decisions. The accounting systems 3 and 4, after around

75 periods in average reach the level of perfect information while systems 1 and 2 induce a rather high, nearly constant distance to perfect evaluations.

To a certain extent, things seem to change in case of high complexity: With decentral coordination even the "measurement only" setting leads to performance levels beyond that achieved with perfect evaluation. After around 50 to 75 periods all noisy accounting systems with measurements of actuals (i.e. settings (2), (3) and (4) in Table 1) exceed the performance achieved with perfect information (Figure 1C). This "beneficial" effect of inaccuracies might be caused by "false positive" evaluations: "False positives" might take an organization down a "wrong road" in short-term, but with the chance to discover superior configurations in a longer term ([14], [15]). In particular, imperfect knowledge may afford the opportunity to leave a local peak in the fitness landscape and this effect becomes more likely with higher complexity: as is well examined, with higher levels of complexity more local maxima exist, and, hence, the search process is more likely to stick to a local peak (e.g. [9], [10]). Inaccuracies induce diversity in the search process, and provide the chance to discover superior levels of performance and, eventually, the global maximum. Results provide support for this intuition: In the "high complexity-decentral mode" scenario the global maximum is found less frequent with perfect evaluations than with noisy accounting systems as far as they measure the status quo.

In the "proposal" mode of coordination differences between the settings of management accounting systems decline. Hence, in a way, with introducing the information-processing power of the main office the relevance of the setting of the accounting system tends to be reduced. Moreover, our results (Figure 1A versus 1B and 1C versus 1D) suggest that in the proposal mode organizations with inaccurate judgmental information by far cannot achieve performance levels that can be reached with perfect evaluations.

In order to provide an explanation we find it helpful to remember that in the proposal mode the status quo only is abandoned if two conditions are met. First, at least, one department has to discover a partial vector that promises a higher compensation to the respective department head (otherwise he/she would not propose the alteration); second, the main office has to accept the proposal. Hence, for being implemented each proposal has to pass an additional instance - with inaccurate evaluation, i.e., an additional source of error is induced. By that, the "false positives" are less likely to do their beneficial work, "false negative" evaluations by the main office might occur and the organization is more likely to suffer from inertia compared to the decentral mode. With more inertia the fitness landscape is less likely to be "experienced" which reduces benefits of the "stepwise refinement" and "immediate adjustment" accounting systems.

4 Conclusion

The results provide broad support for the intuition that improving judgmental information by measured values of accounting systems increases overall performance. However, the results might throw some new light on management accounting systems: apparently, the contributions of improving information accuracy subtly interfere with coordination need and mode. In particular, when decision problems are highly complex, better accounting systems not necessarily are more beneficial. Furthermore, apparently inaccuracies might have their positive sides for complex decisions - given that inaccuracies are accompanied by decentral coordination. Hence, taking into account that improving management accounting systems usually is not costless, these findings put claims for extensive investments in perspective.

References

1. Demski, J.S., Feltham, G.A.: Cost Determination: A Conceptual Approach. Iowa State Univ. Press, Ames (1976)
2. Simon, H.A.: A behavioral model of rational choice. Q. J. Econ. 69, 99–118 (1955)
3. Horngren, C.T., Bhimani, A., Datar, S.M., Foster, G.: Management and cost accounting, 3rd edn. Pearson, Essex (2005)
4. Galbraith, J.: Designing complex organisations. Addison-Wesley, Reading (1973)
5. Ginzberg, M.J.: An Organizational Contingencies View of Accounting and Information Systems Implementation. Account Organ. Soc. 5, 369–382 (1980)
6. Ackoff, R.L.: Management misinformation systems. Manag. Sci. 14, B-147–B-156 (1967)
7. Chang, M., Harrington, J.E.: Agent-based models of organizations. In: Tesfatsion, L., Judd, K.L. (eds.) Handbook of Computational Economics, vol. 2. Elsevier, Amsterdam (2006)
8. Kauffman, S.A.: The origins of order: Self-organization and selection in evolution. Oxford Univ. Press, Oxford (1993)
9. Kauffman, S.A., Levin, S.: Towards a general theory of adaptive walks on rugged landscapes. J. Theor. Biol. 128, 11–45 (1987)
10. Rivkin, J.W., Siggelkow, N.: Balancing search and stability: Interdependencies among elements of organizational design. Manag. Sci. 49, 290–311 (2003)
11. Siggelkow, N., Rivkin, J.W.: Speed and search: Designing organizations for turbulence and complexity. Organ. Sci. 16, 101–122 (2005)
12. Rivkin, J.W., Siggelkow, N.: Patterned interactions in complex systems: Implications for exploration. Manag. Sci. 53, 1068–1085 (2007)
13. Levitan, B., Kauffman, S.A.: Adaptive walks with noisy fitness measurements. Mol. Divers 1, 53–68 (1995)
14. Knudsen, T., Levinthal, D.A.: Two faces of search: Alternative generation and alternative evaluation. Organ. Sci. 18, 39–54 (2007)
15. Wall, F.: The (beneficial) role of informational imperfections in enhancing organisational performance. In: LiCalzi, M., Milone, L., Pellizzari, P. (eds.) Progress in Artificial Economics. Springer, Berlin (2010)
16. Dosi, G., Levinthal, D., Marengo, L.: Bridging contested terrain: Linking incentive-based and learning perspectives on organizational evolution. Ind. Corp. Chan. 12, 413–436 (2003)
17. Labro, E., Vanhoucke, M.: A Simulation Analysis of Interactions among Errors in Costing Systems. Account Rev. 82, 939–962 (2007)
18. Redman, T.C.: Data Quality for the Information Age. Artech House, Boston (1996)

Potential Norms Detection in Social Agent Societies

Moamin A. Mahmoud[1], Aida Mustapha[2], Mohd Sharifuddin Ahmad[1], Azhana Ahmad[1], Mohd Zaliman M. Yusoff[1], and Nurzeatul Hamimah Abdul Hamid[3]

[1] College of Information Technology, Universiti Tenaga Nasional, Kajang, Selangor, Malaysia
[2] Faculty of Computer Science & Information Technology. Universiti Putra Malaysia, Serdang, Selangor, Malaysia
[3] Faculty of Computer and Mathematical Sciences, Universiti Teknologi MARA, Shah Alam, Selangor, Malaysia
moamin84@gmail.com,
{sharif,azhana,zaliman}@uniten.edu.my,
aida@fsktm.upm.edu.my, nurze@fkstm.uitm.edu.my

Abstract. In this paper, we propose a norms mining algorithm that detects a domain's potential norms, which we called the Potential Norms Mining Algorithm (PNMA). According to the literature, an agent changes or revises its norms based on variables of local environment and amount of thinking about its behaviour. Based on these variables, the PNMA is used to revise the norms and identify the new normative protocol to comply with the domain's norms. The objective of this research is to enable an agent to revise its norms without a third party enforcement unlike most of the work on norms detection and identification, which entail sanctions by an authority. We demonstrate the execution of the algorithm by testing it on a typical scenario and analyse the results on several issues.

Keywords: Intelligent Software Agents, Norms Detection, Norms Mining, Normative Protocol.

1 Introduction

Norms detection and identification has increasingly become a significant research area [1, 2]. In a multi-agent system, the concept of norms detection entails a visitor agent identifying the norms and normative protocol that are prevailing within the interactions of the agent society. The need to do so is motivated by the importance of the visitor agent to enact acceptable behaviours that conform to the norms of the local agent society to avoid sanctions or sustained repudiation, thus preventing the agent from achieving its goals.

A norm has several definitions based on the area of study such as social science, game theory, psychology and legal theory [3]. According to Webster's

S. Omatu et al. (Eds.): *Distrib. Computing & Artificial Intelligence*, AISC 217, pp. 419–428.
DOI: 10.1007/978-3-319-00551-5_51

Dictionary, norm is "a principle of right of action binding upon the members of a group and serving to guide, control, or regulate proper and acceptable behaviour" (www.webster.com). Boella et al. [4] defined norms in a multi-agent system as "a multi-agent system organized by means of mechanisms to represent, communicate, distribute, detect, create, modify, and enforce norms, and mechanisms to deliberate about norms and detect norm violation and fulfilment". A normative protocol, on the other hand, is a set of executable norms that are imposed on some multi-agent community [5]. Therefore, norms are often considered as constraints on behaviour [3]. Other researchers defined it as "the order of occurrence of events or protocols that are related to a set of norms [6], for example, arrive, order, eat, pay, tip and then depart could be the norms of dining in a restaurant.

The fundamental norms are represented by injunctive norms, which refer to a group's beliefs about what have to be done [7] and descriptive norms, which refer to beliefs about what is actually done by the majority in a social group [8]. For example, in a formal meeting, since a majority of attendees are silent and attentive (descriptive norms), others usually act in a similar manner and that they will incur social sanctions such as frowning or given silent gestures if they do not comply (injunctive norms) [8].

Villatoro et al. [9] classified norms on two different kinds, which are conventions and essential norms. Conventions are naturally emerging norms without any enforcement. Young [10] defined conventions as "a pattern of behaviour that is customary, expected, and self-enforcing. Everyone conforms, everyone expects others to conform, and everyone wants to conform given that everyone else conforms". Conventions fix one norm among a set of norms and is always efficient as long as each one in the population practice the same norm i.e. greetings, driving side of the road [11]. Essential norms facilitate collective action problems, e.g., when there is a conflict between an individual and the collective interests [9, 11].

To influence an agent to adhere to the domain's norms, the norms are enforced by sanctions when an agent fails to comply. Usually, a third-party enforcement agent implements the sanctions [12]. But a non-compliant norm is also influenced by emotion of shame or guilt even in the absence of the third party enforcement [13]. This fact is specifically useful in large-scale communities, where it may be difficult to monitor adherence to equilibrium behaviour that needs sanctions by a third party [14].

Due to variability of norms in a normative agent society, e.g. with or without authority and presence or absence of sanctions, we propose a norms mining algorithm to detect a domain's potential norms, which we called the Potential Norms Mining Algorithm (PNMA). We define a potential norm as a behaviour that is practised by a majority of members in a multi-agent community. According to the literature [5, 15, 16], an agent changes or revises its norms based on variables of local environment and the amount of thinking or reflection about its behaviour. Thus, any change on these variables, the PNMA revises the norms and identifies the normative protocol that complies with the domain's norms. The

PNMA entails the process of data formatting, filtering and extracting the potential norms and normative protocol.

The objective of this work is to enable an agent to revise its norms even in the absence of a third party enforcement unlike most of the work on norms detection and identification, which entail sanctions by a third party [6, 15, 16, 17, 18, 19, 20].

The next section dwells upon the related work on norms identification and emergence. In Section 3, we present our work in potential norms mining algorithm. Section 4 illustrates the application of the mining process on a typical scenario to explain the algorithm. In Section 5, we present the analysis of the example and Section 6 concludes the paper.

2 Related Work

A social norm could be described as the rules of a specific social group of agents [21]. Numerous theories of norms have been proposed by researchers. For example, in [22], if a group, *g*, which is represented by members, *m*, entails social norm, *s*, which is a pattern of behaviour, (a) *m* within *g* regularly conform with *s*, (b) *m* consider *s* as a standard that judges whether the behaviours of *m* are acceptable or otherwise, (c) *m* believe, (i) all must equally conform to *s*, and (ii) they are justified in forcing each other to conform to *s*, (d) a deviant member will be pressured to conform with *s*, such as sanction.

In social norms learning within a social group [23], every agent in the group learns simultaneously from repeated interactions with random neighbours [9]. In norm adoption and diffusion [15], Epstein [24] discussed two macro phenomena of normative behaviour. Firstly, norms are locally stable, i.e., a group of agents near to each other adhere to the same norms. Secondly, norms are globally diverse, i.e., norms changes are based on two factors, which are the local environment and the amount of thinking or reflection about its behaviour. An agent is said to think less about its behaviour when it conforms to nearby agents' behaviours but thinks more otherwise. This means that the agent checks its conformity to the norms with its surrounding neighbours within a personal vision radius. If it conforms, it keeps the norm and thinks less about the norm. If it does not conform, then it thinks more about the norm and changes its behaviour to conform to the norm. It is assumed that in such scenario, the agent automatically adopts the majority norm, not thinking when it is in a stable norm environment and the norms are not enforced by any third party agent [15].

Savarimuthu et al. [6] proposed a norm identification technique that infers the norms of an agent community without the norms being explicitly imposed on the agents. They used association rule mining to identify the *tip* norm in an agent-based simulation of a virtual restaurant in which agents are located in the restaurant where other agents entering the restaurant may not be aware of the protocol associated with ordering and paying for food items and the associated norms. The agent situated in the environment observes these actions and records them in its belief base. The norm inference mechanism consists of two sub-components. The first sub-component is Obligation Norm Inference (ONI) algorithm to generate candidate obligation norms, and the second sub-component is the norm verification component, which verifies whether a candidate norm can be identified as a norm in the society. The Mechanism of ONI algorithm exploits the sanctioning action in the environment to identify the obligation norms.

In the EMIL Project [25], a theory of norm innovation is developed in coping with particular kinds of complicated systems i.e. a social system. The theory consists of two processes: emergent processes (from interaction between agents) and emergent effects (emergence of norms into the agents' minds). The EMIL project entails three main components which are

- EMIL-M fulfils a general model of norm-innovation, a complicated social dynamics between intelligent social agents to be verified by tools of agent-based simulation in a specific context.
- EMIL-S is a simulation platform for experiments and has been developed to test EMIL-M by running simulations and compare results with available data, specifically in the area of Open Source.
- EMIL-T assesses the achievement of the model by comparing the results of the simulations with the experiential data documented in the open-source.

3 The Potential Norm Mining Algorithm (PNMA)

In social norms learning within a normative system, every agent in the community learns simultaneously from repeated interactions with randomly selected neighbours [23]. Accordingly, to detect the potential norms, an agent needs to collect data from its neighbours for analysis. Two situations that change the agent's belief and force the agent to think more about its norms are its non-conformity with the neighbours or if the agent is in a new domain.

To detect a domain's norms and to check the conformity of norms by other agents, an agent needs to be able to observe other agents' behaviours. Savarimuthu et al. [6] in their research on norms identification, assume that an agent is able to sense its surrounding events via some signalling actions. Adopting the same assumption in a typical train scenario, an agent is able to sense the surrounding events such as *arrive, wait, enter, litter, depart*.

The agent starts collecting data and detects the norms when it observes its behaviours do not conform to those of the nearby agents or it is in a new domain. The agent collects data by gathering the related events and adding data about these events in its Record file, which stores structured data about the events. It stops collecting data when repeated events are encountered. Figure 1 shows the architecture of the PNMA.

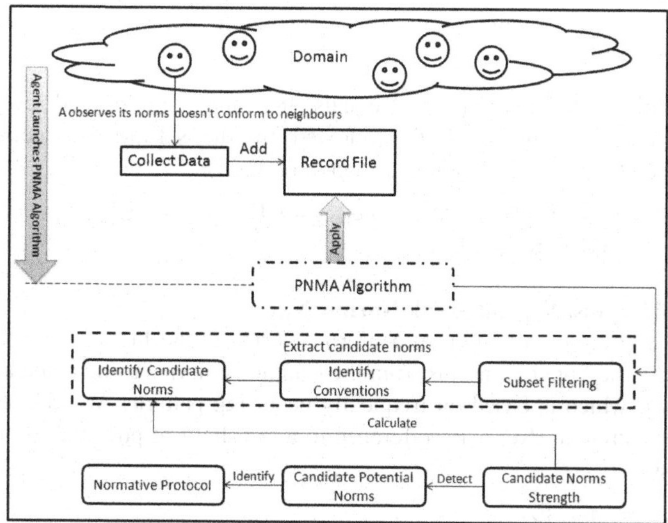

Fig. 1 The PNMA Algorithm Architecture

The objective of the Potential Norm Mining Algorithm (PNMA) is to extract the potential candidate norms and the subsequent normative protocol. Based on Figure 1, the PNMA algorithm steps proceed as follows.

- **Extract Candidate norms from the Record file**

A Candidate norm is a behaviour that is practised by some members of an agent community, which may or may not be a potential norm. Extracting the Candidate norms involves the following processes:

Subset Filtering: This is applied on events in the Record file by removing the subset events, which have the same sequence of events of a bigger set. For example, if we have two sets P = {a, b, c, d} and Q = {b, c}, we recognize that the second set, Q, is a subset of first set, P, i.e., $Q \subset P$. In subset filtering, we remove all events, which are subset of a bigger set or subset of itself, if they have the same sequence of the bigger set. In the above example, we remove the set Q = {b, c}.

Filtered events, E_F equals the set-theoretic complement between all events, E_A and subset events, E_S, i.e.,

$$E_F = E_A \setminus E_S$$

Identify Conventions, E_C: According to Conventions and Essential Norms in the literature [9, 10, 11], conventions equal the intersection set of E_F. If event, e_f is a filtered event, then,

$$\text{If } E_F = e_{f1},\ e_{f2}, \ldots, e_{fr}$$

$$E_C = e_{f1} \cap e_{f2} \cap \ldots, \cap e_{fr} \Rightarrow E_C = \cap_{r=1}^{k} e_{fr} \text{ where } k = 1, 2, 3, \ldots, r$$

The above formula means that conventions equal the intersections of filtered events.

Candidate Norms, N_C: N_C equals the union of filtered events, e_{fr}, represented as the set U-E_F followed by the set-theoretic complement between U-E_F and the candidate events set, E_C, i.e.,

$$U - E_F = e_{f1} \cup e_{f2} \cup \ldots, \cup e_{fk} \Rightarrow U - E_F = \cup_{r=1}^{k} e_{fr}$$
$$N_c = (U - E_F) \setminus (E_C)$$

- **Potential Norms, N_P, and Weak Norms, N_W,**

According to Adoption of Majority Norms in the literature [19], we can identify the Potential norms from the Candidate norms using association rule mining, which is based on frequently occurring patterns. We calculate each norm's strength (S_N), with reference to a threshold of potential norms, T, as follows:

$$N_C = n_{C1}, n_{C2}, \ldots, n_{Cr}$$
$$S_N(n_{Cr}) = \text{Number of events which include } n_{Cr} \text{ in Record file} \\ / \text{Total events number}$$
$$\text{If } S_N(n_{Cr}) \geq T \Rightarrow n_{Cr} \in N_P$$
$$\text{Otherwise, if } S_N(n_{Cr}) \leq T \Rightarrow n_{Cr} \in N_W$$

From the above equations, we notice that the potential norms are identified based on the candidate norm's strength. If the norm strength is greater than a threshold, then this candidate norm is a potential norm otherwise it is a weak norm. The norm strength is calculated by dividing the occurrence of Candidate norms in the Record file by the total number of events in the Record file.

- **Identify the Normative Protocol (N-P),**

The agent identifies the N-P by finding the set-theoretic complement between $U - E_F$ and N_W, i.e.,

$$N - P = (U - E_F) \setminus (N_W)$$

The equation means that the normative protocol is the union of filtered events but excluding the weak norms.

We submit that the morality of potential norms is beyond the scope of this research and will not be deliberated in this paper.

4 The PNMA Example (A Train Scenario)

To illustrate the algorithm and to test its validity, we present an example of agents' actions in taking a train. The aim is to identify the potential norms in taking a train.

In this scenario, we assume an agent, A, moved from its local domain D_L to a new domain D_N. In the new domain, there is a train and agent A with its own set of normative protocol from its previous experience in the domain, which is *arrive, a*; *wait, w*; *enter, e*; *depart, d*. In this example, the protocol is represented as (a, w, e, d).

Since the agent is in a new domain, it thinks more about its behaviours and the prevailing norms. Hence, it launches the PNMA to detect the new domain's (train) normative protocol.

We assume a set of norms that are commonly practised in the new domain, which are *arrive, a*; *wait, w*; *enter, e*; *litter, l*; *excuse, x*; *depart, d*. We also assume the following events are in the Record file, which have been collected by the agent, represented only as symbols, and A_1-A_{10} are agents:

$E_A = (A_1)$ = (a, w, e, x, d)
(A_2) = (w, e, x, d)
(A3) = (w, e, l, x)
(A4) = (a, w, e, x, d)
(A5) = (a, w, e, d)
(A6) = (w, e, x)
(A7) = (a, w, e, l, x, d)
(A8) = (w, e, l)
(A9) = (a, w, e)
(A10) = (w, e, d)

- **Extract Candidate Norms**

Subset Filtering (E_F). The agent filters all observed events by removing the subset events, which have the same sequence of events as the bigger set.

$E_F = E_A \setminus E_S$

Es = (w, e, x, d); (w, e, l, x); (a, w, e, x, d); (w, e, x); (w, e, l); (a, w, e); (w, e, d)

$E_F =$ (a, w, e, x, d); (a, w, e, d); (a, w, e, l, x, d)

Identify the Conventions (E_C). Using the formula,

If $E_F = e_{f1}, e_{f2}, \ldots, e_{fr}$

$E_C = e_{f1} \cap e_{f2} \cap \ldots, \cap e_{fk} \Rightarrow E_C = \bigcap_{r=1}^{k} e_{fk}$ where k = 1, 2, 3,…., r, then

$E_C = (a, w, e, d)$

Identify the Candidate norms (N_C). Using the formula,

$U - E_F = e_{f1} \cup e_{f2} \cup \ldots, \cup e_{fk} \Rightarrow U - E_F = \bigcup_{r=1}^{k} e_{fr}$

$N_c = (U - E_F) \setminus (E_C)$

Then, $U - E_F = (a, w, e, l, x, d)$

From above, $E_C = (a, w, e, d)$

The result, $N_C = (a, w, e, l, x, d) \setminus (a, w, e, d) \Rightarrow N_C = (l, x)$

- **Identify Potential Norms, N_P, and Weak Norms, N_W**

Using the formula,

$N_C = n_{C1}, n_{C2}, \ldots, n_{Cr}$

$S_N(n_{Cr}) =$ Number of events which include n_{Cr} in record file / Total events number

If $S_N(n_{Cr}) \geq T \Rightarrow n_{Cr} \in N_P$

Otherwise, $S_N(n_{Cr}) \leq T \Rightarrow n_{Cr} \in N_W$

In this example, we assume the threshold, T, of potential norms of 0.5.

From the above,

$N_C = (l, x)$

$n_{C1} = l \Rightarrow S_N(l) = 3/10 = 0.33$

$n_{C2} = x \Rightarrow S_N(x) = 6/10 = 0.6$

If $S_N(n_{Cr}) \geq 0.5 \Rightarrow n_{Cr} \in N_p$

Otherwise $S_N(n_{Cr}) \leq 0.5 \Rightarrow n_{Cr} \in N_W$

$S_N(l) = 0.33 \leq 0.5 \Rightarrow l \in N_W$

$S_N(x) = 0.6 \geq 0.5 \Rightarrow x \in N_P$

Consequently, $N_P = (x)$, $N_W = (l)$

- **Identify Normative Protocol (N-P)**

Using the formula,

$N - P = (U - E_F) \setminus (N_W)$

From the above, $U - E_F = (a, w, e, l, x, d)$

But, $N_W = (l)$,

Thus, $N - P = (a, w, e, l, x, d) \setminus (l) \Rightarrow N - P = (a, w, e, x, d)$

5 Analysis of Results

Although the given example is somewhat trivial, it does illustrate the potential solution that could be applied using the PNMA algorithm. From the example, we observe the following points:

1. It clearly shows that the algorithm succeeded in identifying the potential norms which is *excuse*. This makes the usage of the algorithm very useful and can be used to detect other norms in general.
2. The algorithm succeeded in identifying multiple norms. The example shows the algorithm detected one potential norm and one weak norm.
3. The algorithm successfully identified the normative protocol in the domain by excluding the weak norms. This gives the agent the new knowledge to practise the acceptable norms of the new domain. The agent's initial knowledge of a normative protocol is (arrive, wait, enter, depart) and after detecting the norms of the new domain, the agent revises the normative protocol to become (arrive, wait, enter, excuse, depart).

6 Conclusion and Further Work

In this research, we present the potential norms mining algorithm as a new tool to identify the potential norms of a domain. It could provide a novel solution for software agent and robot communities to adapt to various environments and learn new behaviours, which lead to improved capabilities.

Potential norms mining algorithm entails the process of data formatting, filtering, and extracting the potential norms and normative protocols. The results show that the PNMA succeeded in detecting the potential norms in general. It overcomes the problem of a third party agent to enforce the norms and sanctions to change the beliefs.

In our further work, we shall build a virtual environment that has several domains with a number of agent societies to further explore and test the PNMA for discovering other related aspects such as the detection time, the location of agents in the domain, and the domain size. The success of detecting norms will be measured on relations between the time and the location of agents, between the domain size and locations and other similar relations.

Acknowledgements. This research is funded by Fundamental Research Grants Scheme by Ministry of Higher Education Malaysia in collaboration with the Centre of Agent Technologies at University Tenaga Nasional, Malaysia.

References

1. Hollander, C., Wu, A.: The Current State of Normative Agent-Based Systems. Journal of Artificial Societies and Social Simulation 14(2), 6 (2011)
2. Alberti, M., Gomes, A.S., Gonçalves, R., Leite, J., Slota, M.: Normative systems represented as hybrid knowledge bases. In: Leite, J., Torroni, P., Ågotnes, T., Boella, G., van der Torre, L. (eds.) CLIMA XII 2011. LNCS, vol. 6814, pp. 330–346. Springer, Heidelberg (2011)
3. Hexmoor, H., Venkata, S., Hayes, D.: Modeling social norms in multi-agent systems. J. Exp. Theor. Artif. Intell. 18(1), 49–71 (2006)
4. Boella, G., van der Torre, L., Verhagen, H.: Introduction to the special issue on normative multiagent systems. Autonomous Agents and Multi-Agent Systems 17(1), 1–10 (2008)
5. Mahmoud, M., Ahmad, M.S., Ahmad, A., Mohd Yusoff, M.Z., Mustapha, A.: A Norms Mining Approach to Norms Detection in Multi-agent Systems. In: International Conference on Computer & Information Sciences (2012)
6. Savarimuthu, B.T.R., Cranefield, S., Purvis, M., Purvis, M.: Obligation Norm Identification in Agent Societies. Journal of Artificial Societies and Social Simulation (2010)
7. Cialdini, R.B., Reno, R.R., Kallgren, C.A.: A focus theory of normative conduct: Recycling the concept of norms to reduce littering in public places. Journal of Personality and Social Psychology 58, 1015–1026 (1990)

8. Lapinski, M.K., Rimal, R.N.: An Explication of Social Norms. Communication Theory 15(2), 127–147 (2005)
9. Villatoro, D., Sen, S., Sabater-Mir, J.: Of social norms and sanctioning: A game theoretical overview. International Journal of Agent Technologies and Systems 2, 1–15 (2010)
10. Young, H.P.: The evolution of conventions. Econometrica 61(1), 57–84 (1993)
11. Villatoro, D.: Self-organization in decentralized agent societies through social norms. In: The 10th International Conference on Autonomous Agents and Multiagent Systems, AAMAS 2011, vol. 3, pp. 1373–1374 (2011)
12. Grossi, D., Gabbay, D., van der Torre, L.: Book Title: The Norm Implementation Problem in Normative Multi-Agent Systems, Specification and Verification of Multi-agent Systems, pp. 195–224. Springer US (2010)
13. Elster, J.: Alchemies of the Mind. Cambridge University Press, Cambridge (1999)
14. Young, H.P.: Social Norms. The New Palgrave Dictionary of Economics, 2nd edn. Palgrave Macmillan (2008)
15. Xenitidou, M., Elsenbroich, C.: Construct validity and theoretical embeddedness of agent-based models of normative behaviour. Int. J. Interdiscip. Soc. Sci. 5(4), 67–80 (2010)
16. Axelrod, R.: An evolutionary approach to norms. American Political Science Review 80, 1095–1111 (1986)
17. Vázquez-Salceda, J., Aldewereld, H., Dignum, F.: Norms in multiagent systems: from theory to practice. Computer Systems: Science & Engineering 20(4) (2005)
18. Villatoro, D., Sen, S., Sabater-Mir, J.: Of social norms and sanctioning: A game theoretical overview. International Journal of Agent Technologies and Systems 2, 1–15 (2010)
19. Savarimuthu, B.T.R., Cranefield, S., Purvis, M., Purvis, M.: Norm Identification in Multi-agent Societies. Discussion Paper, department of Information Science, University of Otago (2010)
20. Ahmad, A., Yusoff, M.Z.M., Ahmad, M.S., Ahmad, M., Mustapha, A.: Resolving Conflicts between Personal and Normative Goals in Normative Agent Systems. In: The Seventh International Conference on IT in Asia 2011 (CITA 2011), Kuching, Sarawak (2011)
21. Detel, W.: On the concept of basic social norms. Analyse & Kritik 30, 469–482 (2008)
22. Hart, H.L.A.: The Concept of Law, Oxford (1961)
23. Sen, S., Airiau, S.: Emergence of norms through social learning. In: Proceedings of IJCAI 2007, pp. 1507–1512 (2007)
24. Epstein, J.: Learning to be thoughtless: social norms and individual computing. Center on Social and Economic Dynamics. Working Paper (2000)
25. Andrighetto, G., Conte, R., Turrini, P., Paolucci, M.: Emergence in the Loop: Simulating the Two Way Dynamics of Norm Innovation. In: Normative Multi-agent Systems, Dagstuhl Seminar Proceedings. Internationales Begegnungs-und Forschungszentrum für Informatik (IBFI), Schloss Dagstuhl, Germany (2007)
26. Symeonidis, A.L., Mitkas, P.A.: Agent Intelligence Through Data Mining. In: The 17th European Conference on Machine Learning and the 10th European Conference on Principles and Practice of Knowledge Discovery in Databases (2006)

Collision Avoidance of Mobile Robots Using Multi-Agent Systems

Angel Soriano, Enrique J. Bernabeu, Angel Valera, and Marina Vallés

Instituto U. de Automática e Informática Industrial,
Universitat Politècnica de Valencia, Camino de Vera s/n
46022 Valencia, Spain
{ansovi,ebernabe,giuprog,mvalles}@ai2.upv.es

Abstract. This paper presents a new methodical approach to the problem of collision avoidance of mobile robots taking advantages of multi-agents systems to deliver solutions that benefit the whole system. The proposed method has the next phases: collision detection, obstacle identification, negotiation and collision avoidance. In addition of simulations with virtual robots, in order to validate the proposed algorithm, an implementation with real mobile robots has been developed. The robots are based on Lego NXT, and they are equipped with a ring of proximity sensors for the collisions detections. The platform for the implementation and management of the multi-agent system is JADE.

1 Introduction

The area of artificial intelligence (AI) has expanded considerably in recent years. It not only dominates the area of games versus computers, but nowadays it applies in many sectors like databases management or web pages. As it is well known, the main topic of AI is the concept of intelligent agent defined as an autonomous entity which observes through sensors and acts upon an environment using actuators [20]. This definition is very close to services that a robot can provide, so the concept of agent often is related with robots, [4], [27], [16].

On the other hand, detecting and avoiding a collision is a previous step for overcoming the motion planning problem. In fact, collision detection has been inherently connected with the motion-planning algorithms from the very beginning. Current planning algorithms require the collision detection of mobile and nondeterministic obstacles.

Continuous collision detection (CCD) techniques are the most effective when dealing with multiple mobile agents or robots. CCD algorithms basically make a return if a collision between the motion of two given objects is presented or not; and if a collision is going to occur then, the instant in time of the first contact is returned, [2], [3], [7], [19], and [26].

In this paper, collision detection strategies of autonomous mobile robots based on [3] are combined with strategies based on artificial intelligence to offer a new method of collision avoiding management.

S. Omatu et al. (Eds.): *Distrib. Computing & Artificial Intelligence*, AISC 217, pp. 429–437.
DOI: 10.1007/978-3-319-00551-5_52

The method is divided into three basic concepts which are merged in this paper: obstacle detection by a mobile robot, the concept of abstraction robotic agent as a software agent within MAS, and distributed artificial intelligence as a method of communication and negotiation between these software agents.

Nowadays, there are many sensors on the market that allow robots to know if there is an obstacle that stands between them and its trajectory, and where is that obstacle. This process is usually local, i.e. it is performed inside a robot. In the case of two mobile robots at the same scenario, each one represents an obstacle to the other, but neither is aware of it because it is handled as a local process. The concept of robotic agent in a multi-agent robotic system is proposed as a next level or upper layer to fix it and to manage a more intelligent solution.

Multi-agent robot systems (MARS) represent a complex distributed system, consisting of a large number of agents-robots cooperating for solving a common task. In this case, each agent of MARS represents a real physical mobile robot that informs its software agent of all it perceives. The ability of communication, cooperation and coordination between the agents, allows conversations, negotiations and agreements, which are the basis of the algorithm is presented.

2 Avoiding Collision Method

The aim of this section is reviewing a CCD methodology in [3] for obtaining the instant in time when two robots or agents in motion will be located at their maximum-approach positions while they are following straight-line trajectories.

The mentioned maximum approach is also calculated. Therefore, if the involved robots do not collide while they are following their respective motions, then their minimum separation is returned. Otherwise, their maximum penetration is computed as a minimum translational distance [5]. A remarkable aspect is that both the instant in time and the corresponding minimum separation or maximum approach are computed without stepping any involved trajectory.

Some collision avoiding configurations for the involved robots or agents are directly generated from the computed instant in time and maximum penetration. These collision-free configurations are determined in accordance with a given coordination between the robots or agents.

2.1 *Obtaining the Instant in Time and the Maximum Approach*

Consider two robots or agents in motion each one enveloped or modeled by a circle. Let A be a circle in motion whose start position at time t_s is $A(t_s)=(c_A(t_s),r_A)$. Where $c_A(t_s)\in\Re^2$ is the A's center at t_s and $r_A\in\Re$ is its radius. A is following a straight-line trajectory whose final position at t_g is given by $A(t_g)=(c_A(t_g),r_A)$. Let $v_A\in\Re^2$ be the A's velocity for the time span $[t_s,t_g]$.

Let B be a second circle in motion whose start and goal positions at the respective instants in time t_s and t_g are $B(t_s)=(c_B(t_s),r_B)$ and $B(t_g)=(c_B(t_g),r_B)$. The B's velocity for the time span $[t_s,t_g]$ is $v_B\in\Re^2$.

All the infinite intermediate positions of the mobile circle A for $t\in[t_s,t_g]$ while A is in motion is parameterized by λ with $\lambda\in[0,1]$, as follows:

$$A(\lambda)=\{(c_A(\lambda),r_A): c_A(\lambda)=c_A(t_s)+\lambda\cdot(c_A(t_g)-c_A(t_s))\ ;\ t=t_s+\lambda(t_g-t_s);\ \forall\lambda\in[0,1]\} \qquad (1)$$

Note that the positions $A(\lambda)$ and $A(t)$, with $t=t_s+\lambda(t_g-t_s)$, are equal for all $t\in[t_s,t_g]$ and $\lambda\in[0,1]$. All the infinite intermediate positions of the mobile circle B are analogously parameterized for $\lambda\in[0,1]$ as indicated in (1).

Observing equation (1) is easy to conclude that the maximum approach d_M between in-motion circles A and B will be obtained by finding the parameter $\lambda_c\in[0,1]$ that minimizes $\|c_A(\lambda)-c_B(\lambda)\|-(r_A+r_B)$

Let $\lambda_c\in[0,1]$ be the parameter that minimizes (2). λ_c is obtained by minimizing by computing the distance from the origin point O to the straight-line $c_A(\lambda)-c_B(\lambda)$.

Note that $c_A(\lambda)-c_B(\lambda)$ for all $\lambda\in[0,1]$ is really a segment whose extreme points are respectively $c_0=c_A(t_s)-c_B(t_s)$ and $c_1=c_A(t_g)-c_B(t_g)$. Then, the parameter $\lambda_c\in[0,1]$ is obtained by projecting O onto mentioned segment, $O^\perp$, as

$$\lambda_c=-\frac{c_0\cdot(c_1-c_0)}{\|c_1-c_0\|^2} \quad \text{with } \lambda_c\in[0,1]. \text{ Then } O^\perp=c_0+\lambda_c\cdot(c_1-c_0). \qquad (2)$$

Once λ_c is obtained, d_M and the associated instant in time t_M are computed as

$$d_M=\|c_A(\lambda_c)-c_B(\lambda_c)\|-(r_A+r_B) \qquad t_M=t_s+\lambda_c(t_g-t_s). \qquad (3)$$

If d_M is negative, then d_M holds a penetration distance and, then A and B will collide with the maximum penetration d_M at t_M.

2.2 Determining Avoiding Collision Configurations

In case of collision, the positions where A and B present their maximum penetration are, as mentioned, $c_A(\lambda_c)$ and $c_B(\lambda_c)$ respectively with $\lambda_c\in[0,1]$, $d_M<0$, $t_M\in[t_s,t_g]$. One of these positions can be minimally translated in order to bring both circles into contact by using the unit vector $\hat{v}_{MTD}=O^\perp/\|O^\perp\|$, with $\|\hat{v}_{MTD}\|=1$,

Let $A_f(t_M)$ and $B_f(t_M)$ be the mention collision-free configurations,

$$\begin{aligned} A_f(t_M)&=(c_{Af}(t_M),r_A):\ c_{Af}(t_M)=c_A(t_M)-\delta\cdot\alpha\cdot d_M\cdot\hat{v}_{MTD} \\ B_f(t_M)&=(c_{Bf}(t_M),r_B):\ c_{Bf}(t_M)=c_B(t_M)+\delta\cdot(1-\alpha)\cdot d_M\cdot\hat{v}_{MTD}. \end{aligned} \qquad (4)$$

where $c_A(t_M)=c_A(t_s)+\lambda_c(c_A(t_g)-c_A(t_s))$ and $c_B(t_M)=c_B(t_s)+\lambda_c(c_B(t_g)-c_B(t_s))$. Parameter $\delta\geq1$ is a safety threshold. If $\delta=1$, then configurations $c_{Af}(t_M)$ and $c_{Bf}(t_M)$ will be in contact. Finally, parameter $\alpha\in[0,1]$ configures the degree of motion modification applied to each mobile robot or agent. In this way, if $\alpha=1$, then $c_B(t_M)$ and $c_{Bf}(t_M)$ are equal and, consequently, mobile robot or agent B do not change its current motion. A graphical example is shown in Fig. 1.

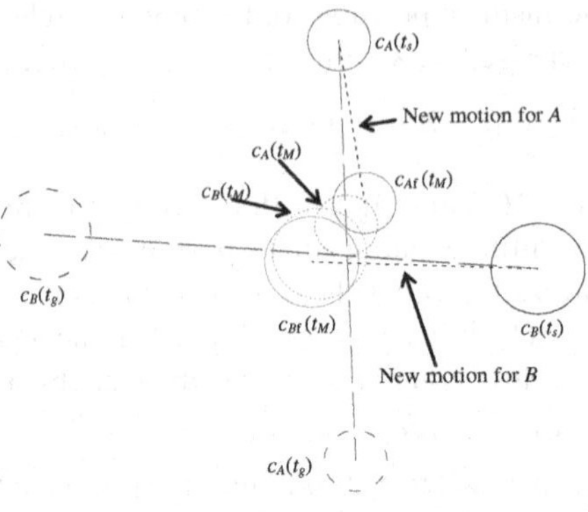

Fig. 1 Avoiding collision configurations with α=0.7 and δ=1.03

3 Hybrid Control Collision Avoidance

The implementation of the collision avoidance proposed methodology has six phases (see Fig. 2). A scenario where multiple robots follow a path infinite straight line between two target points is considered. These two points are alternated when they are achieved. All robots have their representation as a software agent in the MAS which encompasses the whole system, so there is no moving object within the scene that is not a software agent.

In phase 1 (**detection**), the local system (each robot) has defined a detection object area. If the local system detects an obstacle that may be a threat of collision (from now threat-object), it calculates the position of threat-object in the global scenario and it sends to the agent who represents the local system in MAS to manage it.

When an agent receives the position of a threat-object (from now threat-position) by the local system, it must identify what kind of threat it is. To know this, in the **obstacle identification** phase, the agent detects the threat (from now detector-agent), consults the other agents to know who is located within that area of threat. If there is not any agent within that area it is identified as a static object threat and directly the collision detection phase is performed. Otherwise, the threat-agent is identified through communication among agents and the next phase starts.

When the two involved agents in a possible threat have been identified, the communication between them is used to obtain the information needed to apply the detection algorithm presented in 2.1. In the **time to talk, negociate and resolve phase**, the inputs of the algorithm are four: the positions of each of the agents involved in that instant ($c_A(t_s)$, $c_B(t_s)$) and the target positions where they will be at time t_g ($c_A(t_g)$, $c_B(t_g)$). This time t_g must be the same for the two robots therefore, to calculate it, the agents communicate to each other to know which one reaches its destination before. Who plans to take more time to reach their destination calculates an intermediate destination from its current trajectory and

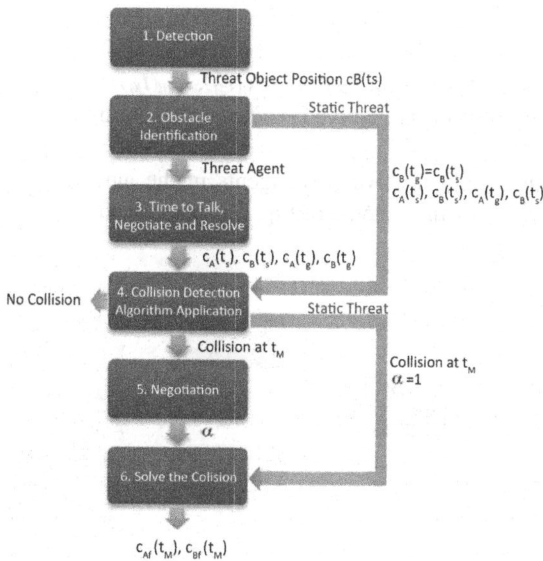

Fig. 2 Phases of the proposed methodology

the arrived time of the other agent to its destination. In this way the two agents shared the time it takes to reach their destination.

In the **collision detection phase**, the input requirements to implement collision detection algorithm are: current position coordinates of detector-agent ($c_A(t_s)$), its destination, ($c_A(t_g)$), current position of threat-object ($c_B(t_s)$) and its destination ($c_B(t_g)$). If in the Phase 2, the threat-object was identified as a static-threat, the target is the same as the initial position ($c_B(t_s)= c_B(t_g)$). Therefore the inputs are applied to the algorithm and it returns the probability of collision with the threat-object. In case there is no collision, threat is discarded and the method ends but if a collision is detected, the method informs to detector-agent the time of maximum penetration (t_M). The next step is to avoid the collision by the method described in section 2.2. If the object is a static-threat, the detector- agent should take over the entire cost of the collision avoidance (α=1) and jump to Phase 6. Otherwise the negotiations between the two agents involved are opened to decide how much charge is allocated to each.

To decide the load percentage (α) that each robot will have in the collision avoidance, in the **negotiation phase** the two agents communicate with each other and exchange parameters such as priority, the difficulty of maneuvering, maximum speed, etc., which define the easiness or availability that each agent offers to change its trajectory and avoid collision. Once each agent agreed with the selection of α, the detector-agent runs the last method described below.

The detector-agent, by the method 2.2, computes the two new positions that the robot should be achieve at time t_M to avoid collision in the **solve the collision** phase. The threat-agent receives, from the detector-agent, the avoidance position ($c_{Bf}(t_M)$) and the time in which must be achieve. Both change their trajectories to go to the new destination partial ($c_{Af}(t_M)$, $c_{Bf}(t_M)$) at the right time. Once it's reached, the collision is resolved, each robot continues its original path and the method ends.

The method only involves two agents in the agreement. When a third object intervenes, for now the scheme of Fig. 3 is performed.

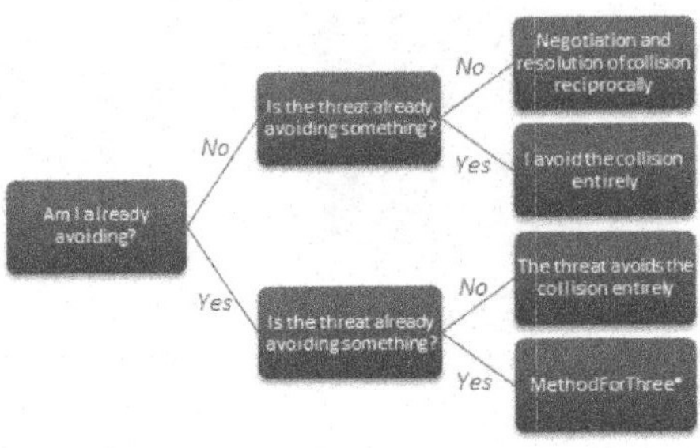

Fig. 3 Performance binary tree

The critical point occurs when an agent (a1) which is already avoiding a collision with other agent (a2) detect that exists a new collision with a third agent (a3), who is already avoiding other agent too. MethodForThree* is a provisional method that progressively reduces the speed of the agents a1 and a2, and may even stop them completely until a3 is no longer a threat to a1. Collision detection by agreements involving more than two agents is also being developed.

In order to test the effectiveness of this method, different scenarios with mobile robots have been simulated. Figures 4 shows the execution obtained in two instants (t=3s, t=10s). There are six robots (circles of different colors) that they must arrive to the opposite location (marked by the same color star). The figure also shows the detection area (a trapezoid in front of each robot), three static objects (black squares) and the path described by each robot for the first seconds of the simulation.

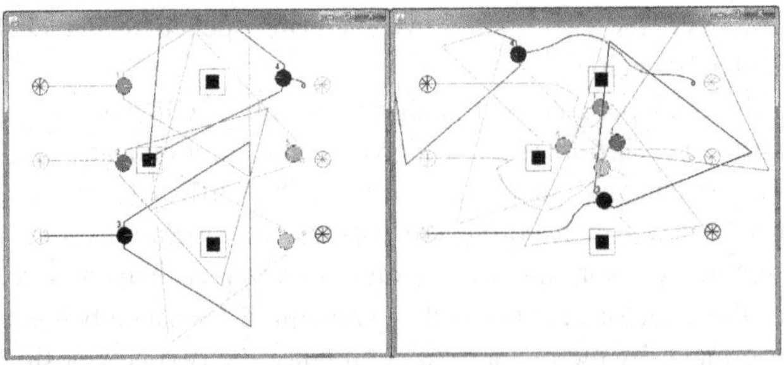

(a) t=3s (b) t= 10s

Fig. 4 Collision avoidance simulation with six robots

4 Practical Implementation with Mobile Robots

A practical implementation with mobile robots has been developed in order to test the robustness of the presented algorithm. The mobile robots used are LEGO Mindstorms NXT [12] and the platform for the management of MAS chosen was JADE [11].

Two LEGO differential wheeled mobile robots have been built [9]. Each robot has defined two destinations points. In order to achieve the trajectory, a control strategy based on a pure pursuit algorithm [28] was implemented in the robots. The robots have been equipped with a ring of proximity sensors to detect possible obstacles.

Robots are connected to their software agents (computers) via Bluetooth and those computers are part of a network that forms the overall MAS through JADE. The connection diagram is presented in Fig. 5a. Each robot carries a triangle to detect its position from an overhead camera located at the top of the scenario. This camera is also used to monitoring and minimizing odometry problems.

These robots have multiple threads running different functional modules. Each of them has one module to control the robot trajectory, a second one for detection that manages the IR sensors and a third one for communication that receives and sends information to the software agent (see Fig. 5b).

While IR sensors does not detect anything, the robot follows its fixed trajectory, but when something is detected by IR, the communication module informs to the software agent and expects a solution to the possible collision from MAS. If the solution leads to a new destination for the robot, the communication module receives the new destination and sends it to the control module for change the path.

The management of the agents in JADE is simple. When a software agent receives the position of a detected threat, the agent asks everyone if anyone is located in the threat area. Thus, if other robot is the threat, it's identified as the threat agent and they exchange their destinations and speeds to verify if there will be a collision or not. If finally there is it, they negotiate the way to avoid it and send the new destinations and speeds to their respective robots.

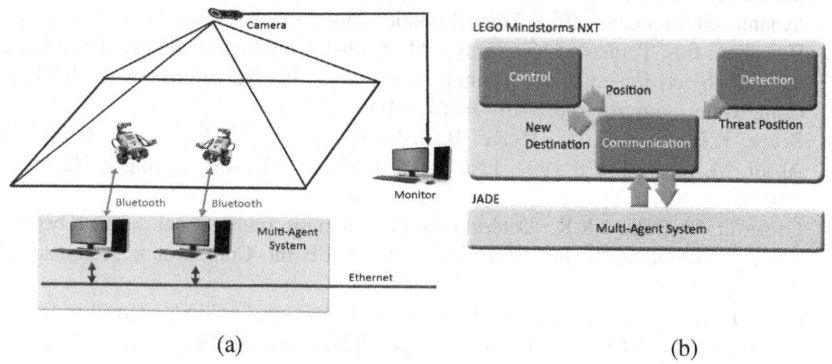

(a) (b)

Fig. 5 Control architecture and modules connection scheme

In http://idecona.ai2.upv.es, a video demonstration of practical experiment with Lego robots (*Robots Móviles* folder, at videos multimedia gallery) and two compiled versions of the platform that allow the simulation with robots (*Desarrollos de Software* folder, at Results option, Project menu) can be obtained.

5 Conclusions

A collision avoidance method that takes advantages and benefits of MAS has been presented in this work. This method is located one level above the traditional methods of obstacle avoidance where the management is performed locally and the possible communications between the local systems are solved functionally. The application of techniques provided by the area of artificial intelligence to the robotic area opens a wide range of possibilities that offers more natural results and gives human characteristics of communication like negotiation between robots. This work has succeeded in unifying concepts of agent theory with concepts from the area of mobile robotics, providing more intelligence to robots and offering solutions that otherwise cannot be provided. The methodology has been tested both in simulations and in real executions with mobile robots.

The kinematic configuration of the used agents is holonomic, then considering only linear trajectories might be acceptable. However, as a future work, the collision detection using another kind of movements, like natural Splines and Bezier curves are being considered.

Acknowledgements. This work has been partially funded by the Ministerio de Ciencia e Innovación (Spain) under research projects DPI2011-28507-C02-01 and DPI2010-20814-C02-02.

References

1. Austin, J.L.: How to Do Things With Words. Oxford University Press, Oxford (1962)
2. Bernabeu, E.J.: Fast generation of multiple collision-free and linear trajectories in dynamic environments. IEEE Trans. Robotics 25(4), 967–975 (2009)
3. Bernabeu, E.J., Tornero, J., Tomizuka, M.: Collision prediction and avoidance amidst moving objects for trajectory planning applications. In: Proceedings of the IEEE Int. Conf. Robot. Automat., pp. 3801–3806 (2001)
4. Bruce, K.B., Cardelli, L., Pierce, B.C.: Comparing Object Encodings. In: Ito, T., Abadi, M. (eds.) TACS 1997. LNCS, vol. 1281, pp. 415–438. Springer, Heidelberg (1997)
5. Cameron, S., Culley, R.K.: Determining the minimum translational distance between two convex polyhedra. In: Proceeding of the IEEE Int. Conf. Robot. Automat., pp. 591–596 (1986)
6. Choi, Y.-K., Wang, W., Liu, Y., Kim, M.-S.: Continuous collision detection for two moving elliptic disks. IEEE Trans. Robotics 22(2), 213–224 (2006)

7. Campion, G., Bastin, G., Dandrea-Novel, B.: Structural properties and classification of kinematic and dynamic models of wheeled mobile robots. IEEE Transactions on Robotics and Automation 12(1), 47–62 (1996)
8. Java Agent Development Framework, `http://jade.tilab.com`
9. LEGO Mindstorms home page, `http://mindstorms.lego.com`
10. Michalewicz, Z.: Genetic Algorithms + Data Structures = Evolution Programs, 3rd edn. Springer, Heidelberg (1996)
11. Redon, S., Kheddar, A., Coquillart, S.: Fast continuous collision detection between rigid bodies. Computer Graphic Forum 21(3), 279–288 (2002)
12. Russell, S.J., Norvig, P.: Artificial Intelligence: A modern approach (2003)
13. Searle, J.: Speech acts: An essay in the philosophy of language. Cambridge University, Cambridge (1969)
14. Van den Bergen, G.: Continuous collision detection of general convex objects under translation. In: Game Developers Conf. Morgan Kaufmann (2005), `http://www.dtecta.com/interesting`
15. van Leeuwen, J. (ed.): Computer Science Today. LNCS, vol. 1000. Springer, Heidelberg (1995)
16. Wallace, R., Stentz, A., Thorpe, C., Moravec, H., Whittaker, W., Kanade, T.: First Results in Robot Road-Following (1985)

Multi-agent Architecture for Intelligent Insurance Systems

Egons Lavendelis

Riga Technical University,
Faculty of Computer Science and Information Technology,
Department of Systems Theory and Design,
1 Kalku Street, Riga, LV 1658, Latvia
egons.lavendelis@rtu.lv

Abstract. Modern insurance information systems need intelligence to provide new functions that till now as a rule have been carried out by humans. Introduction of intelligent mechanisms into information systems allows the insurance companies to automate processes in the insurance business and achieve two benefits. Firstly, the amount of work done by humans is reduced and secondly more services can be provided to customers electronically, which increases the level of customer service. Additionally, insurance information systems need to communicate with many other systems to get the needed data. These demands fit the characteristics of intelligent agents. Thus the paper proposes to implement an insurance information system as a multi-agent system using intelligent agents to realize the modules of insurance information systems. A novel multi-agent architecture for insurance information system development is proposed.

Keywords: Insurance systems, multi-agent systems, multi-agent architecture.

1 Introduction

Modern insurance information systems (IISs) need wide functionality, including both traditional tasks of information systems like data processing and storing and more advanced functions that have been done by humans so far, for example, risk evaluation and insurance premium calculation. These functions are more and more needed to automate the tasks in the insurance business and so decrease the response time for customers and reduce the amount of work done by employees of the insurance company. Nowadays it is possible to monitor and predict the situation in the insurance market automatically, to create offers for clients and make insurance deals online, to process insurance cases automatically and to automate many other tasks. For the system to be capable to do these tasks as well as humans do, the system must have many characteristics of intelligent systems like capabilities to react on the changes in the environment, capabilities of goal directed behaviour and social capabilities. To automate many processes the IIS must communicate with many other systems to retrieve data. It is unclear how to build such

S. Omatu et al. (Eds.): *Distrib. Computing & Artificial Intelligence*, AISC 217, pp. 439–447.
DOI: 10.1007/978-3-319-00551-5_53

systems using traditional software development techniques, because they lack natural methods to implement intelligence.

Intelligent agents are widely used to implement intelligent and distributed systems that need to communicate with other systems in many domains [1]. Agents can be used in domains where the tasks of the system can be decomposed and allocated to individual agents. The author sees that it is the case in the insurance domain. As a consequence multi-agent approach can be used to implement intelligence in the IISs. Moreover, holonic multi-agent systems [2] offer hierarchical architecture with high modularity and openness for new functionality. High modularity and openness simplify the change implementation into already working systems that is usually needed in IISs, because of high dynamism of the business and as a consequence frequent demands of new functionality. The paper addresses the issues of intelligent IISs development by the usage of the multi-agent systems. It proposes the conceptual approach of a multi-agent architecture for IIS development. Additionally, the benefits of introducing holonic multi-agent systems into the IIS development are analysed and as a consequence holonic multi-agent IIS architecture is proposed.

The remainder of the paper is organized as follows. The Section 2 describes the traditional architecture of IISs used as a background in the paper. The Section 3 describes IISs as intelligent systems. The Section 4 proposes the use of intelligent agents for implementation of IIS's modules and outlines the basic idea of the novel multi-agent architecture for IIS development. The Section 5 gives hierarchical and holonic view of the proposed architecture. The Section 6 concludes the paper.

2 Traditional Insurance Information System Architecture

Nowadays IISs are usually built using modular architecture that simplifies changing and extending the system's functionality during the maintenance phase. IIS has to provide the following business functions [3]: contract management, claims management, reinsurance management, coinsurance management, broker and commission management, accounting, calculating of reserves. Uhanova and Novitsky offer to integrate the main modules of IIS (Sales, Brokers/Commissions, Contract management, Reinsurance, Coinsurance and Rewards/Liabilities) with the main modules of information management system (Accountancy and Calculating of reserves). It enables efficient information flows between separate business processes: front office, back office and accountancy. So the full modular architecture of IIS consists of the following main components (see Figure 1):

- Policy and insurance contract management module, including the functionality for sales agents and brokers;
- Reinsurance and coinsurance modules;
- Loss/reward regulations module;
- Reserve calculations and recalculations module;

- Financial analysis, monitoring and management module;
- Accounting module.

Such an architecture implements all functions of the information system needed for the insurance business. So there is no need for multiple information systems in the same company. At the same time the system consists of loosely coupled modules that can be developed and maintained separately. As a consequence the architecture simplifies the development of the IIS. Still, there is a question how to implement the intelligent mechanisms into these components. The next section analyses what intelligent mechanisms are required in the IIS and how to implement them.

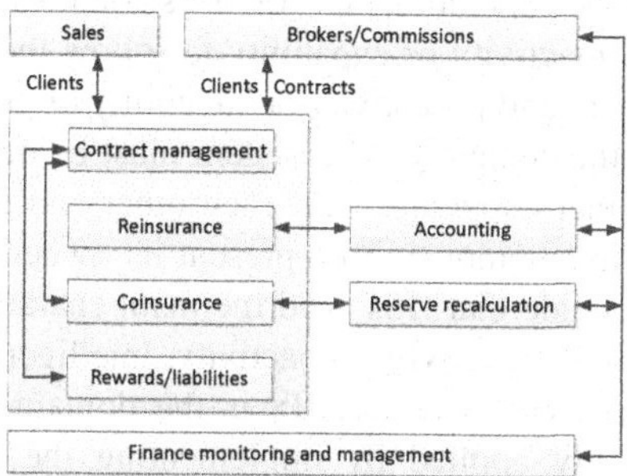

Fig. 1 Traditional architecture of IISs (modified from [3])

3 Insurance Systems as Intelligent Systems

Modern IISs are intelligent systems due to several reasons. Firstly, these systems have to carry out such tasks as market prediction, reserves recalculation and other tasks that require the ability to predict the evolvement of the environment [4]. Intelligent models and mechanisms are needed to implement such a proactive or goal driven prognosis.

Secondly, the deals that the company should offer to its clients should vary based on the actions taken (offers made) by other insurance companies that are direct competitors. Thus, the IIS has to monitor its reserves, market situation and react accordingly to different changes by adapting the offers.

Thirdly, during the processing of different insurance cases, the IIS needs to get various data from other systems, for example, to process a road traffic accident data the insurance company needs data from other insurance companies whose clients are involved in the accident. Such information is stored in the information systems of corresponding companies and some information is maintained in some centralized database provided by the government. Nowadays, at least in Latvia, retrieval of such information is manual due to different structures of data and lack of automatic data retrieval mechanisms. Similar problems exist in processing of many other types of insurance cases, like medical insurance, travel insurance, etc.

To simplify communications, the IIS should be capable to automate communications with other systems so shortening the time elapsed to get the required data. In fact there is a need to create a distributed system inside which the IIS can exchange data with other systems.

Finally, nowadays insurance companies try to automate many actions related to the sales and marketing, for example, selling insurance policies can be done by the IIS on the Internet. In this case the IIS is autonomous and it represents its user in the market. Various intelligent mechanisms are used by autonomous intelligent computer systems to represent their users in the electronic marketplace [5]. Usage of such an approach in the IIS simplifies the process of making various deals, for example the policy issuing can be simpler for both the company and its clients.

To summarize, it can be concluded that IIS or its components need such characteristics as intelligence, reactivity or capability to follow the changes in the environment and react on them and proactivity or capability of goal directed behaviour to achieve the goals of the company. The system must be distributed and it has to consist of components that cooperate with each other and with other systems. It must be capable to act autonomously to represent its owner – the insurance company. Thus one can conclude that IISs need the main characteristics of intelligent agents given in [6], namely, reactivity, proactivity, intelligence, social capabilities and distributedness. As a consequence, IIS is the domain where the intelligent agents may be effectively applied by implementing the IIS as a multi-agent system.

4 Multi-agent Architecture

Previous research has proven that intelligent software systems can be successfully implemented using intelligent agents [1, 7, 8, 9]. Moreover, if the system has a modular architecture, then these modules can be preserved in the agent oriented system. In this case each module is implemented as one or several agents [7]. These agents are named as higher level agents. To increase modularity and implement openness higher level agents can be implemented as multiple lower level agents or multi-agent systems that are seen as a single agent from outside and called holons [2]. The current section describes the proposed basic multi-agent architecture for IISs, while the concept of holonic multi-agent architecture is analysed in the following section.

In the basic multi-agent architecture the modules making up the IIS are implemented as one or several agents. Agents realize the functionality of corresponding module and communications with other modules of the system, information systems of other insurance companies and third party systems, whose data are used in the functionality of the IIS. The functionalities of particular agents are the following:

- Contract management agent is responsible for risk management, premium and insurance case accounting;

- Policy selling agent carries out insurance premium calculation, insurance rules determination and is responsible for deal making;
- The brokering agent provides as precise as possible insurance offer calculations. In some insurance cases the agent asks human experts to provide their evaluations;
- The accounting agent implements the accounting system of the company;
- Reinsurance agent maintains reinsurance contracts in larger insurance companies and makes the necessary changes as soon as the resources of the company become too small for all insurance deals made;
- Reserve recalculation agent monitors the insurance deals and makes calculations with the goal to keep track if the company has enough reserves. The agent informs other agents in case some action is required to ensure the needed reserves;
- Coinsurance agent defines the conditions of coinsurance deals;
- Finance management agent ensures monitoring of companies finance resources and informing the CEO about the financial condition of the company;
- Insurance case processing agent ensures the processing of insurance cases.

Agents must implement all algorithms that are needed to realize the functions of the IIS, for example, contract management agent classifies insurance deals and calculates amounts of insurance premiums using models of mathematical regression and classification [10, 11]. The basic multi-agent architecture for IISs is given in Figure 2.

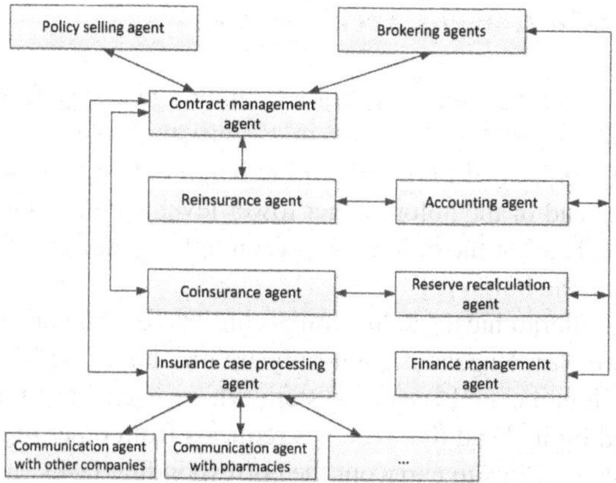

Fig. 2 Basic multi-agent architecture for IISs

The proposed architecture is open in the sense that it allows to implement such systems, whose functionality can be extended by adding new agents and without changing existing code. The openness can be effectively used, for example, in implementation of insurance case processing. To process different insurance cases the system needs data from various other systems, including databases of other companies and centralized databases of the country, for example, the vehicle insurance registry. These databases vary from each other. Firstly, each company has its own database with unique access mechanism and data structure. Secondly, the data and as a consequence databases vary from one type of insurance case to another. So the IIS must be capable to access very different databases with different access mechanisms. This problem can be at least partly solved by implementing each access mechanism in separate agent, allowing adding new agents at any moment, so adding interactions to new systems. As a consequence the architecture of the system consists of multiple levels. There is one main agent that is responsible for any communications with other systems, but it directly does not communicate to any external system. To communicate with certain types of systems it uses particular lower level agents, for example, agent for road traffic accidents, agent for medical insurance, etc. Additionally, to the agents for each type of the insurance, separate agents can be used for each particular system needed to communicate with. These agents are used by the corresponding insurance type agents. For example, to process a road traffic insurance case the corresponding agent requests to carry out the communication the agents that are capable to communicate to other road traffic insurance companies.

The proposed architecture allows implementing any higher level agent as a multi-agent system. Similarly, in case of simpler systems some agents can be joined into one agent. Therefore the architecture given in the Figure 2 is a high level framework and can be customized to meet the requirements of the particular company.

5 Holonic Multi-agent Architecture

The multi-agent system is said to be holonic multi agent system if (some) agents consist of smaller agents. The agents that consist of other agents are named holons. They are represented to outside of the holon only by one lower level agent that is named the head of the holon. Other lower level agents communicate only to each other and the head of the holon. Thus each holon is seen as one agent for any other agent outside the holon [2].

The proposed hierarchical architecture consists of higher and lower level agents. Each lower level agent is used only by one higher level agent. Thus the architecture is a holonic multi-agent system where each higher level agent is a holon represented by its head that receives requests from other higher level agents and uses lower level agents to carry out the corresponding tasks.

To enable building open IISs the notion of open and closed holons are introduced [8]. Closed holons consist of predefined sets of agents, while open holons contain predefined types of agents, but the exact instances and exact number of agents can change during the lifetime of the system. Higher level agent is an open holon if it is expected that the corresponding functionality frequently will change during the lifetime of the system.

One particular design of the hierarchy is given in Figure 3. This example shows how the architecture can be customized to meet the requirements of particular system. If the system needs various reinsurance options and communications to other systems to process all insurance cases, then two open holons are created, in particular, the reinsurance agent and the insurance case processing agent are open holons. Each body agent of the reinsurance holon is responsible for separate reinsurance option while each agent of insurance case processing holon is responsible for one type of insurance cases. One of the lower level agents of this holon is a multi-agent system on its own. It allows implementing communications to particular insurance company systems in separate third level agents to implement openness for additions of new interaction mechanisms with the systems of other insurance companies.

6 Conclusions

The paper proposes to implement IISs as multi-agent systems using the holonic multi-agent architecture. The author sees the following benefits from the usage of intelligent agents in the IIS development. Agents can act autonomously to represent the interests of the company during the deal making process. In the insurance business agents can automate deal making among the stakeholders of the insurance market. Firstly, upon receipt of the information about new insurance deals made by contract management agent, the reinsurance agent can find bigger insurance company to reinsure these deals. During this process the reinsurance agent can also find the most suitable reinsurance offers. Various agent interaction mechanisms can be used for this purpose. Secondly, if the insurance companies have autonomous agents then an auction based market mechanism where agents of insurance companies compete for the deal with particular client, can be created. In other words, introduction of agents enables usage of new market mechanisms.

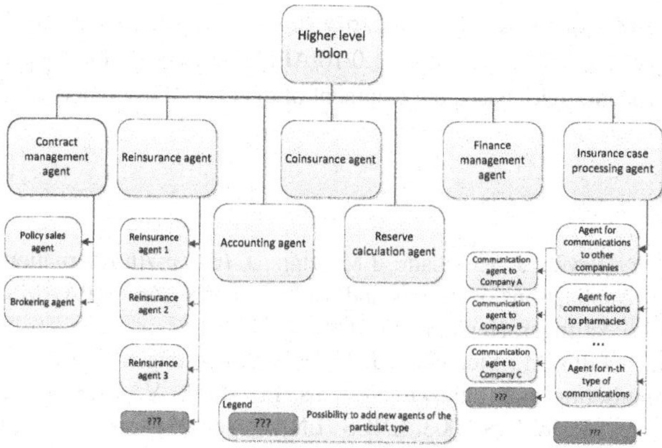

Fig. 3 Customized holonic architecture for particular system

Introduction of intelligent agents into IISs opens new opportunities for communications between insurance companies and clients. Nowadays many insurance companies offer to buy policies on the Internet. Still at the moment it is done only for the insurance types where all the data about costumers and insurance objects are precisely known and it is possible to use standard price calculation. The author believes that usage of agent technologies can make IISs capable to acquire data about the customers and insurance objects from various data sources and at the same time to analyse situation in the market, so gaining possibilities to offer new types of insurance electronically. Lastly, if a significant number of insurance companies agree, it becomes possible for client agents to carry out negotiations to different insurance companies and make a deal with one of them completely autonomously.

The proposed IIS architecture uses the results of research of other open intelligent systems, in particular, intelligent tutoring systems [8]. As it is proven for the intelligent tutoring systems, it allows extending system's functionality by adding new agent(s) without changing the existing code. This is true for new functionality of certain types that corresponds to open holons.

The main direction of the future work is to implement a case study of agent based IIS using the proposed architecture. The case study will provide more practical implementation details of the proposed architectural framework. The proposed architecture can be used together with some agent oriented software engineering methodology. In particular, the MASITS methodology [9] is built for the systems with similar characteristics to the IISs – intelligent tutoring systems. The multi-agent architecture is one of the artefacts needed to successfully use this methodology for other systems then intelligent tutoring systems. So the other direction of the future work is to extend the MASITS methodology to make it usable also for IIS development.

Acknowledgement. The work has been partly supported by European Regional Development Fund project 2011/0008/2DP/2.1.1.1.0/10/APIA/VIAA/018 "Intelligent agents, modelling and web Technologies based development of distributed insurance software".

References

1. Demazeau, Y., Pavón, J., Corchado, J.M., Bajo, J. (eds.): 7th International Conference on Practical Applications of Agents and Multi-Agent Systems (PAAMS 2009). AISC, vol. 55, 589 p. Springer, Heidelberg (2009)
2. Fischer, K., Schillo, M., Siekmann, J.H.: Holonic Multiagent Systems: A Foundation for the Organisation of Multiagent Systems. In: Mařík, V., McFarlane, D.C., Valckenaers, P. (eds.) HoloMAS 2003. LNCS (LNAI), vol. 2744, pp. 71–80. Springer, Heidelberg (2003)
3. Uhanova, M., Novitsky, L.: Application of Modeling and Internet Technologies in Marine Insurance Business Processes. In: Proceedings of 4th Int. Conference on Computer and IT Applications in the Maritime Industry, COMPIT 2005, Hamburg, pp. 34–43 (2005)

4. Novickis, L., et al.: Intelligent agents, modelling and web Technologies based development of distributed insurance software. ERDF Project report, 70 p. (2011) (in Latvian)
5. Wooldridge, M.: An Introduction to MultiAgent Systems, 2nd edn., 365 p. John Wiley & Sons (2009)
6. Russell, S., Norvig, P.: Artificial Intelligence. A Modern Approach, 2nd edn., 1112 p. Pearson Education, Upper Saddle River (2003)
7. Grundspenkis, J., Anohina, A.: Agents in Intelligent Tutoring Systems: State of the Art. In: Scientific Proceedings of Riga Technical University, Computer Science. Applied Computer Systems. 5th series, vol. 22, pp. 110–121. RTU Publishing, Riga (2005, 2009)
8. Lavendelis, E., Grundspenkis, J.: Open Holonic Multi-Agent Architecture for Intelligent Tutoring System Development. In: Proceedings of IADIS Int. Conference, Intelligent Systems and Agents 2008, Amsterdam, The Netherlands, pp. 100–108 (2008)
9. Lavendelis, E., Grundspenkis, J.: MASITS – A Multi-Agent Based Intelligent Tutoring System Development Methodology. In: Proceedings of IADIS International Conference, Intelligent Systems and Agents 2009, Algarve, Portugal, pp. 116–124 (2009)
10. Jekabsons, G.: Adaptive Basis Function Construction: an approach for adaptive building of sparse polynomial regression models. In: Zhang, Y. (ed.) Machine Learning, pp. 127–156. In-Tech (2010)
11. Jekabsons, G., Lavendels, J.: Polynomial regression modelling using adaptive construction of basis functions. In: Proceedings of IADIS International Conference, Applied Computing 2008, Mondragon unibertsitatea, Algarve, Portugal, pp. 269–276 (2008)

Evaluation Framework for Statistical User Models

Javier Calle, Leonardo Castaño, Elena Castro, and Dolores Cuadra

Computer Science Department, Carlos III University of Madrid, Spain
{jcalle,lcastano,ecastro,dcuadra}@inf.uc3m.es

Abstract. This paper analyzes the main barriers that user model developers have to face when evaluating a statistical user model. Main techniques used to evaluate statistical user models, mostly borrowed from the areas of Machine Learning and Information Retrieval, are examined. Then an evaluation methodology for statistical user models is proposed together with a set of metrics to specifically evaluate statistical user models. Finally, a benchmark for statistical user models is proposed, thus making possible to compare and replicate the evaluations. Thus, main contribution of this paper is to enable that several user model evaluations were comparable.

Keywords: Statistical user model, evaluation methodology, benchmarking.

1 Introduction

One of the most difficult tasks of user modeling is the evaluation. This is a non-standardized and time consuming task. In fact, it is one of the hardest barriers for user model developers. For several years there had been a lack of standardized tools (methods and frameworks) to evaluate user models. In addition, there is a need of empirical evaluations to make results comparable and experiments replicable [11]. Designing experiments that can be tested separately is the basic to achieve high-quality empirical evaluations.

This paper describes a complete framework to evaluate and measure the performance of any statistical user model. The main advantages of having a complete evaluation framework are that evaluations could be replicable and that evaluation results could be comparable with other results. These advantages are important for user model developers, but these advantages become even more important taking into account that most user models are still research prototypes. Therefore, user model research requires a framework to test its user models and compare with the results yielded by other similar user models. As stated afore, there is a lack of benchmark for statistical user models. This proposal aims to fulfill this need by introducing a complete evaluation framework comprising a database of users and several batteries of experiments, which's statistical significance is properly justified.

S. Omatu et al. (Eds.): *Distrib. Computing & Artificial Intelligence*, AISC 217, pp. 449–456.
DOI: 10.1007/978-3-319-00551-5_54

After a brief summary of the related work, the proposed methodology and its metrics are explained. Then, the benchmark is described as a useful resource to attain such sort of evaluation.

2 State of the Art

According to Larson [12] statistical models are intended to perform predictions of a dependent parameter, departing from observed samples. Regarding user modeling, such parameter represents an aspect of a user future behavior, likes or preferences. The usage of statistical and probabilistic user models has been manifested by the increasing publication of research in these lines. However, for several years it is claimed that those approaches are still far from being applicable and effective in some fields [15], and therefore there is still room for research in this area. Statistical user models can be classified according to their approach to estimate ratings. According to [3] statistical user models can be classified into the following categories, content-based systems, collaborative systems and hybrid approaches. In content-based approach, the behavior of a user is predicted from his or her past behavior, content-based approach has its roots in Information Retrieval [2]. Collaborative filtering relies on other like-minded people to provide predictions. Finally, hybrid approaches combine collaborative filtering with another method (usually content-based) to improve its performance.

One of the most difficult tasks concerning statistical user models refers to evaluation. Two main issues must be faced when evaluating a statistical user model. The first, involves the selection of the dataset [6] and the second one involves selecting proper metrics to evaluate. Some statistical user models are designed for a specific dataset and taking into account the structure of this dataset [13]. This dataset-dependent user models deviates from the genericity desired in these systems. In addition, the selection of proper metrics might also depend on the dataset or the domain [1]. Most important barrier when selecting proper metrics to evaluate are in the lack of standardization. New works in the area of user modeling introduce new metrics challenging the comparison between two different user models [10]. The same work classified the metrics used to evaluate statistical user models into three main typologies, predictive accuracy metrics, classification accuracy metrics and rank accuracy metrics. Predictive accuracy metrics are based on comparison between true user ratings and system's predictions to measure the distance between them [5]. Secondly, classification metrics are suitable for recommenders aimed to classify items as good or bad for the user. Both metrics measure the ability of any recommender to make correct or incorrect decisions [9, 4]. Finally, rank accuracy metrics are suitable for systems providing an ordered list of relevant items to the user [5, 8].

On the other hand, Cosley et al. [7], identify some considerations that affect the evaluation process. They study how sensitive are users to predicted ratings, concluding that predicted ratings can influence current ratings given by the user. In addition, this work identifies that providing information about ratings together with ratings increases the user confidence in the system.

Summarizing, statistical user models must deal with the selection of the dataset and proper metrics to be evaluated. Even though, these are the most important issues when evaluating a statistical user model, some other considerations referring users confidence and user ratings must be taken into account.

3 Evaluation Methodology

The area of statistical user models requires its own evaluation techniques. These techniques can be defined adapting and extending those techniques used in Machine Learning and Information Retrieval [14]. First step is to define the baseline experiment needed to evaluate a statistical user model. Then, the evaluation process should be described. Figure 1 depicts a general proposal for the evaluation methodology process.

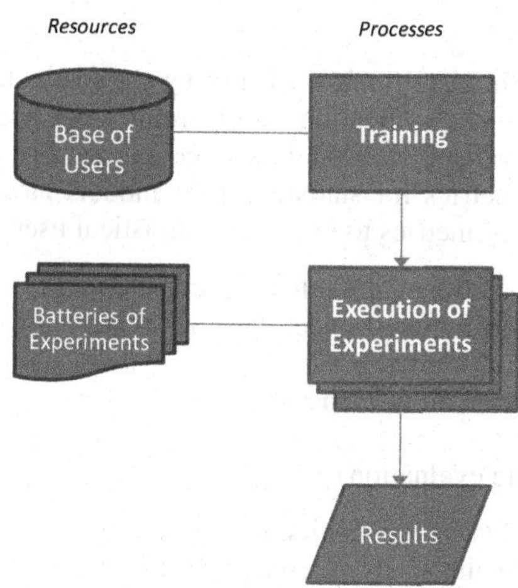

Fig. 1 Schema of the Evaluation Methodology

As shown in figure 1, the evaluation process is supported by some resources: a base of users (for both training and evaluating) and a battery of experiments. Once the model is trained, the chosen batteries of experiments are executed providing the correspondent results, which can be later analyzed and compared to the obtained through other model evaluation.

3.1 Experiments Definition

The base of users should comprise two sets of samples; one for training and the other one just for evaluation. A sample is defined as a user description including its full characterization. That is, a set of facts each of them regarding a feature and the value it takes for that user. Notice that a given (multivalued) feature could appear several times (regarding several concurrent values for the same user). The baseline experiment involves taking a sample, which will be the "current user" and selecting one of its facts to be the inference goal. The process of evaluation involves feeding the model with incremental knowledge and inferring the goal each time a new fact is acquired. Therefore, an isolated experiment consists of as many iterations 'learn fact' and 'infer' as required till the inference is successful (or until there are no more available facts for that user, except for the goal that will be never fed). An alternative experiment could complete always all the iterations, detecting the eventuality of losing the correctness of the inferences when new facts are fed. In addition a battery of experiments is needed to complete the evaluation process as depicted in figure 1. Any battery of experiments is characterized by its type regarding the sample set (within training set or out of it), and the statistical significance intervals and levels it provides. It could also be characterized by other problem specific features, such as those in which the facts are fed or the subset of features from which the goal is selected.

4 Metrics

In order to analyze the results obtained during the evaluation process, some metrics are required to make evaluations comparable. As previously explained, most techniques used to evaluate statistical user models are borrowed from other areas where non-specific metrics for statistical user models have been defined. Therefore the main proposed metrics to evaluate a statistical user model are:

1. inference success (if the model infers the correct fact);
2. number of iterations required to get the correct fact;
3. certainty of successful inferences;
4. certainty of unsuccessful inferences.
5. response time
6. success rate (whole evaluation)

To preserve the validity of the results through metric (2), the model should not be fed back with the results of its inferences (either success or fail). Thus, at next iteration the model can infer again the same (unsuccessful) value unless the newly fed fact about current user makes it to change its consideration. Nevertheless, another method and metric can be posed by feeding back the model and checking how many iterations requires to success. Finally, the success rate is observed as the number of correct predictions divided by the number of users in the test set. The response times for the inference process will also be observed for each iteration (response time for a single inference).

5 Evaluation Framework Description

Benchmark corpus has been obtained from a real domain used in Cadooh (TSI-020302-2011-21) research project, where $4{\cdot}10^4$ samples (user descriptions) were acquired. Even though some of the features were biometric, the data domains of the user features were discretized in order to ease its application to any statistical user model. The set of samples was split into a *training set* (28.000 randomly chosen samples) for feeding the statistical user model before its evaluation, and a *test set* (the remaining 12.000 samples) containing enough samples as to support different confidence and interval levels that assure the significance of most experiments.

The following Figure 2 depicts main issues concerning benchmark samples including cardinality distribution, features per user distribution and rows per user of the proposed benchmark.

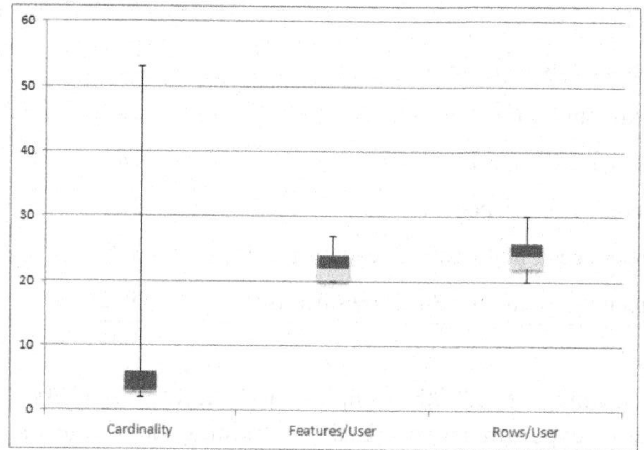

Fig. 2 Benchmark statistical distribution

In addition, some issues related to benchmark data structure will be indicated. Table 1 includes the characterization of the benchmark domain (definition of user description). On the other hand, table 2 provides a brief description of the training set.

Table 1 Domain characterization

	Benchmark domain
Number of users	40000
Number of features	29
Maximum feature cardinality	53
Minimum feature cardinality	2
Average feature cardinality	7

Even though, this benchmark has been obtained departing from a real domain used in Cadooh (TSI-020302-2011-21) research project, it has been standardized and can be used with other user models. However, most user models designed for a specific domain and application could not provide good results when performing over this benchmark. However, statistical user models are domain-independent and application-independent. Thus, all statistical user models can be tested with this benchmark.

Table 2 Samples characterization

	Benchmark domain
Number of users	$28 \cdot 10^3$
Maximum rows per user	33
Minimum rows per user	18
Average rows per user	25
Maximum features per user	27
Minimum features per user	18
Average features per user	22
Percentage of multi-valued features per user	3.33%
Percentage of multi-valued rows per user	5.42%

Once the training set (28000 samples) and the test set (12000 samples) were formed, a set of experiments with both the training and the test set samples were prepared. In order to obtain these experiments, a statistical significance analysis has been achieved, setting two confidence levels and two confidence intervals. The proposed confidence levels are 95% and 99%, on the other hand, the confidence levels are 2% and 5%. In addition, two different batteries of experiments are proposed. First one comprises experiments in which for each sample a random fact is selected (to be inferred), and the rest of the facts (to be fed) are also randomly ordered. The experiments in the second battery are domain dependent (from a real case). In them, the fact is selected from a subset of inferable features, and the rest (non inferable features) are fed in a given order. Table 3 provides some information on those batteries.

Table 3 Battery sets description

Data set	Confidence level	Confidence interval	Sample size
Training set	95%	2%	2250
Training set	95%	5%	400
Training set	99%	2%	3650
Training set	99%	5%	650
Test set	95%	2%	2050
Test set	95%	5%	400
Test set	99%	2%	3100
Test set	99%	5%	650

This benchmark is for public usage and is currently available at http://labda.inf.uc3m.es/doku.php?id=en:labda_lineas:um_1, along with its detailed description, and some examples of use.

6 Conclusions and Future Work

Current evaluations with statistical user models are done using techniques inherited from the Information Retrieval and Machine Learning areas. These techniques have been extended and adapted to the case of statistical user model evaluations.

This work has presented a benchmark for user modeling for supporting evaluations not only comparable but also replicable. In addition, it has been standardized departing from a real domain, easing its application for any statistical user model.

Besides, two issues have been regarded: unifying methodological aspects of statistical user model evaluations; and providing a framework to make all evaluation replicable and comparable.

Finally, according to current state of the art, this proposal tries to fill the latent emptiness of a complete framework to evaluate this type of user models. Providing, a methodology of evaluation, a set of specific metrics and a benchmark to accomplish the evaluation.

Acknowledgements. This proposal development belongs to the research projects Thuban (TIN2008-02711), MA2VICMR (S2009/TIC-1542) and Cadooh (TSI 020302-2011-21), supported respectively by the Spanish Ministry of Education and the Spanish Ministry of Industry, Tourism and Commerce.

References

1. Amento, B., Terveen, L., Hix, D., Ju, P.: An empirical evaluation of user interfaces for topic management of web sites. In: Proceedings of the Conference on Human Factors in Computing Systems (CHI 1999), pp. 552–559. ACM, New York (1999)
2. Baeza-Yates, R., Ribeiro-Neto, B.: Modern Information Retrieval. Addison-Wesley (1999)
3. Balabanovic, M., Shoham, Y.: Fab: Content-based, collaborative recommendation. Commun. ACM 40, 66–72 (1997)
4. Basu, C., Hirsh, H., Cohen, W.: Recommendation as Classification: Using Social and Content-Based Information in Recommendation. In: Proceedings of the 15th National Conference on Artificial Intelligence, Madison, WI, pp. 714–720 (1998)
5. Breese, J., Heckerman, D., Kadie, C.: Empirical analysis of predictive algorithms for collaborative clustering. In: Proc. of the Fourteenth Annual Conference on Uncertainty in Artificial Intelligence, San Francisco, vol. 43 (1998)

6. Bollen, D., Knijnenburg, B.P., Willemsen, M.C., Graus, M.: Understanding choice overload in recommender systems. In: Proceedings of the Fourth ACM Conference on Recommender Systems, pp. 63–70. ACM, New York (2010), http://doi.acm.org/10.1145/1864708.1864724, doi:10.1145/1864708.1864724
7. Cosley, D., Lam, S.K., Albert, I., Konstan, J.A., Riedl, J.: Is seeing believing? How recommender interfaces affect users' opinions. CHI Lett. 5 (2003)
8. Heckerman, D., Chickering, D.M., Meek, C., Rounthwaite, R., Kadie, C.: Dependency networks for inference, collaborative filtering, and data visualization. J. Mach. Learn. Res. 1, 49–75 (2000)
9. Herlocker, J.L., Konstan, J.A., Riedl, J.: An empirical analysis of design choices in neighborhood-based collaborative filtering algorithms. Inf. Retr. 5, 287–310 (2002)
10. Herlocker, J.L., Konstan, J.A., Terveen, L.G., Riedl, J.T.: Evaluating collaborative filtering recommender systems. ACM Transactions on Information Systems 22(1), 5–53 (2004)
11. Hernández, F.O., Gaudioso, E.: Evaluation of recommender systems: A new approach. Expert Syst. 35(3), 790–804 (2008), http://dx.doi.org/10.1016/j.eswa.2007.07.047, doi:10.1016/j.eswa.2007.07.047
12. Larson, H.J.: Introduction to Probability Theory and Statistical Inference. Wiley International Edition (1969)
13. Miller, B., Albert, I., Lam, S.K., Konstan, J., Riedl, J.: MovieLens Unplugged: Experiences with a Recommender System on Four Mobile Devices. In: Proceedings of the 17th Annual Human-Computer Interaction Conference (HCI). British HCI Group (September 2003)
14. Pu, P., Chen, L., Hu, R.: A user-centric evaluation framework for recommender systems. In: Proceedings of the Fifth ACM Conference on Recommender Systems, pp. 157–164. ACM, New York (2011), http://doi.acm.org/10.1145/2043932.2043962, doi:10.1145/2043932.2043962
15. Wade, W.: A grocery cart that holds bread, butter and preferences. NY Times (January 16, 2003)

Agent Mediated Electronic Market Enhanced with Ontology Matching Services and Emergent Social Networks

Virgínia Nascimento, Maria João Viamonte, Alda Canito, and Nuno Silva

GECAD – Knowledge Engineering and Decision Support Research Center,
Institute Of Engineering – Polytechnic of Porto (ISEP/IPP), Porto, Portugal
{vilrn,mjv,alrfc,nps}@isep.ipp.pt

Abstract. Agent technology has been applied in e-commerce to help coping with problems that arise due to its rapid growth. However, despite the amount of research in this area, the level of automation achieved is still limited. This is mainly due to the natural diversity existent in e-commerce environments, where agents may possess different conceptualizations about their needs and capabilities, giving rise to interoperability issues. In this paper we approach this problem and present the AEMOS system as a possible solution. AEMOS is an agent-based e-commerce platform that includes ontology matching services facilitating the interoperability between agents that have different conceptualizations about the same domain of knowledge. The system also explores emergent social networks in order to improve its efficiency by enhancing its ontology matching services and supporting agents in their decisions.

1 Introduction

The rapid growth of e-commerce has led to an increasing demand for automated processes to support both customers and suppliers in buying and selling products. In this context, the automated negotiation carried by software agents has been receiving an increasing attention by the scientific community [1]. However, the diversity of information existent on the web, which is mainly represented for human comprehension only, turns the development of fully automated systems into a challenge [2].

In order to overcome this problem, ontology centered approaches have been proposed (e.g. [2-5]). However, given the natural diversity existent in e-commerce environments, usually each agent has its own conceptualization about its needs and capabilities, giving rise to a semantic heterogeneity problem that is seen as a corner stone for the agents' interoperability.

Although most frequent approaches for AMEC systems consider simplified and limited solutions in order to avoid semantic problems (e.g. [2,3]), there are already some where semantic problems are being dealt with. For example in [4] and [5] the exploitation of the ontology matching paradigm [6] is proposed in order to

S. Omatu et al. (Eds.): *Distrib. Computing & Artificial Intelligence*, AISC 217, pp. 457–464.
DOI: 10.1007/978-3-319-00551-5_55

facilitate agents' interoperability. However, ontology matching is a natural ambiguous and subjective process, which can be performed using different techniques, leading to different alignments that may influence the negotiations' outcomes. The quality and adequacy of an ontology alignment is very important in the negotiation since it may determine the efficiency of the interaction. However, given the multiple variables that may contribute for the negotiations' success, detecting and quantifying the alignment's influence in the negotiations is not a trivial task.

In order to provide an answer for this problem we developed AEMOS, which is an AMEC platform that provides ontology matching services improved with the exploitation of emergent social networks (SN). In this paper we present the AEMOS system model (Section 2) including a brief description of the implemented ontology matching services (Section 3) and SN-based support (Section 4). In order to assess our proposal we describe some experiments and analyze the achieved results (Section 5). Finally we dray some conclusions (Section 6).

2 The AEMOS System Overview

The AEMOS (Agent-based Electronic Market with Ontology Services) [7-10] system is an innovative project (PTDC/EIA-EIA/104752/2008) supported by the Portuguese Agency for Scientific Research (FCT). It is based on the ISEM system [3], which is an agent-based simulation system for e-commerce that aims to study agents' market strategies. In reality, the AEMOS system is an evolution of the ISEM system, keeping all its original functionalities, but allowing agents' to use different ontologies to represent their domain of knowledge.

The main goal of this system is to enable an efficient and transparent negotiation between agents even when they use different ontologies to represent the same domain, ensuring that the agents are able to understand each other and correctly assess the terms and conditions of each transaction.

For that, the system follows an ontology-based information integration approach, exploiting the ontology matching paradigm [6], selecting and suggesting possible alignments (i.e. sets of correspondences) between the agents' ontologies and letting the agents choose which one should be used to translate the subsequent exchanged messages.

In order to overcome issues related to how the chosen ontology alignment may influence the business negotiation efficiency, the system includes a support component based in emergent SN, capable of improving the ontology alignments recommendations and supporting the agents' decisions about which alignment to choose.

2.1 Multi-agent Model

The multi-agent model includes several types of agents classified in two main categories: the business agents and the supporting agents.

The business agents represent real world entities whose behavior is intended to be simulated and studied. Currently we consider two types of business agents: Buyer (B) representing a consumer, and Seller (S) representing a supplier.

The supporting agents are those who provide services that allow business agents to communicate with each other. A thorough description of these agents can be found in [9]. In this paper we introduce only the most relevant actors in the interaction protocol, namely:

- Market Facilitator (MF) – coordinates the interaction between business agents, being responsible for ensuring that the communicating agents are able to understand each other. Normally there are multiple agents of this type per marketplace, being initialized by the MM when necessary. When a B is registered, a MF is associated such that, from that moment on, all messages related to the business negotiation process pass through the associated MF;
- Ontology Matching intermediary (OM-i) – responsible for the ontology matching services, recommends possible ontology alignments for each business negotiation and transforms the exchanged messages according to the approved alignment. Normally there are multiple agents of this type per marketplace, being initialized by the MM when necessary. When a MF is initiated, an OM-i is associated, such that, from that moment on, all the requests related to ontology matching services are sent to the associated OM-i;
- Social Network intermediary (SN-i) – responsible for the SN-based support, provides advice about the adequacy of the alignments to each business negotiation. Normally there are multiple agents of this type per marketplace, being initialized by the MM when necessary. When an OM-i is initiated, or a business agent registers in the market, a SN-i is associated, such that, from that moment on, all requests related to SN-based support are sent to the associated SN-i.

2.2 General Interaction

To participate in the market, the business agents must register first, providing information about the ontologies that they use, and sharing (parts of) the profile of the entity they represent. This information is stored by MF and SN-i agents.

Once registered, the agents are allowed to negotiate. For that, B agents start announcing their buying products and wait for S agents to formulate proposals. Figure 1 illustrates the main interaction that may occur during a business negotiation.

When the negotiation starts, the responsible MF selects the S agents that might be able to satisfy the B agent's request. For that it follows an ontology-based approach, selecting: (i) the S agents that use the same ontology as the B; and, (ii) supported by an OM-i, the ones that use ontologies which can be aligned with it. Therefore, the business negotiations may occur in two different scenarios: (i) a scenario where both agents use the same; and (ii) a scenario where the agents use different ontologies.

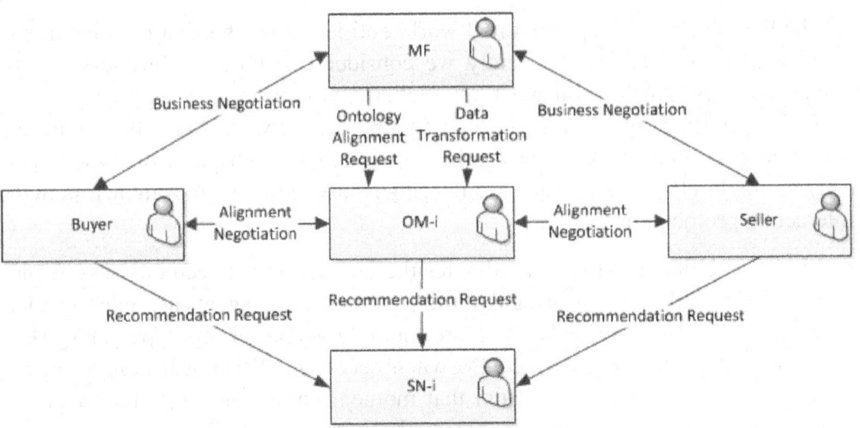

Fig. 1 Main Agents' Interaction during a Business Negotiation

In the first scenario, the MF acts as a proxy between B and S (simply receiving and forwarding messages), while in the second, it is necessary to find an agreement about the alignment between the respective ontologies that should be used to translate the exchanged messages. For that the MF requests an OM-i to mediate an ontology alignment negotiation between B and S. If an agreement is achieved, the subsequent exchanged messages are sent to the OM-i, which translates their content according to the agreed alignment, ensuring that the message receiver will be able to understand it.

During the business negotiation the involved agents, B and S, exchange proposals and counterproposals, terminating the negotiation when an agreement is achieved or when they have no more proposals to formulate.

When a business agent satisfies all its business goals, or its deadlines are reached, it must terminate its activity, notifying the market and declaring the achieved results.

3 Ontology Matching Services

In AEMOS the ontology services are provided by the OM-i agents. An OM-i is responsible for (i) providing information about ontologies and alignments, (ii) proposing alignments for negotiation, (iii) coordinating alignment negotiations, and (iv) transforming messages' content (ontology's instances) when requested.

In order to improve performance, currently, the ontology matching process is performed externally to the negotiation process. It is then considered a registry of the ontologies that are recognized by the agent and a repository of possible alignments between them. This information can be updated any time, as new ontologies are discovered and ontology alignments are created.

When the alignment negotiation is requested, the OM-i selects from its alignments repository the ones that involve both indicated ontologies. It then ranks, sorts and filters the alignments either by (i) requesting a SN-i to rank the alignments for the business negotiation, or, (ii) analyzing their coverage of the ontology's concepts and properties used by the B to describe the requested product. This process results in a set of possible alignments and their respective score, which is sent to both B and S, who analyze the recommended alignments taking into account their preferences and/or requesting advice of a SN-i, replying to the OM-i with the list of the alignments that they consider acceptable.

The OM-i analyzes both replies and checks if there is an agreement, i.e., if some alignment was selected by both agents. If there is no agreement, depending on the system configuration, the negotiation may terminate, or proceed, with the OM-i refining its list of recommended alignments and asking agents to reconsider their options and criteria. Otherwise, if there is an agreement, the OM-i notifies both agents and the MF about the agreement and proceeds with the transformation of the request made by the B.

From that moment on, all the subsequent exchanged messages between the agents are forward to the OM-i for transformation. The transformation of a message's content (i.e. ontology's instance) is performed using the agreed alignment. This process is provided by information integration tools such as MAFRA Toolkit [11] and it's transparent to the agents.

4 Social Network Based Support

During the market activity, the SN-i agents collect information about its participants and their interactions. Then it builds and maintains the relationship graph, applying SNA techniques [12] in order to capture proximity relations between agents, and adequacy relations from alignments to agents, which emerge during the agents' activities in the market. By combining this information, the SN-i is able to evaluate the adequacy of the alignments to each business negotiation. A detailed description of the SN-i agent's model, as well as the fundaments behind it, can be found in [7]. This paper addresses only the key aspects of its responsibilities. There are as follows.

Collect information through the Market: the SN-i receives information from the other agents on the market that will allow it to, in return, support them in their tasks: (i) business agents provide information about the profile of the entities they represent, about ontologies' usage and preferences; (ii) MF agents provide information about the business negotiations between the agents (e.g. both agents' identification, the used alignments, the negotiation outcome, the satisfaction of B agent with the deal); and (iii) OM-i agents provide information about ontologies, alignments and previous alignment negotiations.

Evaluate Agent-To-Agent Proximity: based in theories supported in literature, the SN-i combines a series of factors in order to capture proximity relations between agents. These factors are: (i) the similarity between the agents' profiles and ontologies usage and preferences; (ii) the similarity between their interactions with other agents; (iii) the success rate of their own previous business negotiations; and (iv) the average satisfaction of B about purchased products from S.

Evaluate Alignment-To-Agent Adequacy: to determine the adequacy of an alignment to an agent, the SN-i evaluates: (i) the alignment's coverage of the agent's used ontologies' concepts and properties (considering their respective relevance); (ii) the agent's success rate in business negotiations using the alignment; and (iii) the agent's average satisfaction in deals using the alignment.

Evaluate Alignment-To-Business-Negotiation Adequacy: the adequacy of an alignment to a business negotiation will depend on many factors, namely: (i) the coverage of the alignment in relation to the requested product's description; (ii) the general success rate in negotiations using the alignment; (iii) the general average satisfaction in deals using the alignment; (iv) the adequacy of the alignment to each of the involved agents; and (v) the adequacy of the alignment to the agents closest (i.e. with high proximity relations) to the involved agents.

5 Experiments and Results

In order to test and validate our model, several experiments were performed. We consider a simple marketplace where agents negotiate the same type of product, although using different ontologies to describe it. To demonstrate the usefulness of our system we include situations where agents that use different ontologies could provide better deals. Since we are focusing in the ontology dimension of the negotiation, the satisfaction of a B with a transaction will correspond to the similarity between the purchased product and the desired one, weighted by the relevance that B attributes to each attribute. More details on this function can be found in [7].

We consider three different ontologies for the same domain. For each pair of ontologies we create two alignments: (i) one with all correct correspondences; and (ii) another with a higher coverage but with less correct correspondences.

In order to demonstrate the usefulness of our system four different scenarios were tested, namely: (i) a scenario without both ontology services and SN support; (ii) a scenario with ontology services (including only the semantically correct alignments) but not SN support; (iii) a scenario similar to the previous, but including all created alignments; and (iv) a scenario with both ontology services and SN support. We ran each scenario several times. The average results are presented in Table 1.

Table 1 Average results in each scenario

	Satisfaction in Deals	Alignments' Adequacy
Scenario 1	0.55	-
Scenario 2	0.65	-
Scenario 3	0.52	0.14
Scenario 4	0.62	0.30

By comparing the first two scenarios we are able to demonstrate how the system benefits by the inclusion of ontology services by itself. However, this second scenario represents an unrealistic situation by considering that the alignments are always semantically correct and equally adequate. When we ran the third scenario we observe that the average satisfaction is significantly reduced. This is due to the fact that, in the considered scenario, the agents will continue choosing the inadequate alignments (have a higher coverage) which will cause a severe impact on their business satisfaction. When the SN support is included the agents tend to choose more adequate alignments achieving, a higher business satisfaction.

6 Conclusions

In this paper we present AEMOS, an innovative project that combines in a sole model multiple technologies that enable and improve the automated negotiation in e-commerce platforms.

The experimentation results validate the AEMOS model suggesting that the AMEC systems can highly benefit from the inclusion and combination of ontology services and emergent SN. The ontology services allow agents to negotiate with a larger amount of agents increasing the probability of reaching more satisfactory deals. On the other hand, the SN support can highly improve the communications' efficiency as it allows to progressively recommend the most adequate alignments, keeping agents from constantly choosing inadequate alignments.

Acknowledgments. This work is supported by FEDER Funds through the "Programa Operacional Factores de Competitividade - COMPETE" program and by National Funds through FCT "Fundação para a Ciência e Tecnologia" under the projects: FCOMP-01-0124-FEDER-PEst-OE/EEI/UI0760/2011; PTDC/EIA-EIA/104752/2008.

References

1. Zhang, L., Song, H., Chen, X., Hong, L.: A simultaneous multi-issue negotiation through autonomous agents. Eur. J. Oper. Res. 210(1), 95–105 (2011)
2. Cui-Mei, B.: Combining Intelligent Agent with the Semantic Web Services for Building An e-Commerce System. Paper presented at the 2009 IEEE International Conference on E-Business Engineering (ICEBE 2009), Macau, China, October 21-23 (2009)
3. Viamonte, M.J., Ramos, C., Rodrigues, F., Cardoso, J.: ISEM: A Multi-Agent System That Simulates Competitive Electronic MarKetPlaces. International Journal of Engineering Intelligent Systems for Electrical Engineering and Communications: Special Issue on Decision Support 15, 191–199 (2007)
4. Malucelli, A., Palzer, D., Oliveira, E.: Ontology-based Services to help solving the heterogeneity problem in e-commerce negotiations. Electron. Commer. Res. Appl. 5, 29–43 (2006)

5. Wang, G., Wong, T.N., Wang, X.A.: Negotiation Protocol to Support Agent Argumentation and Ontology Interoperability in MAS-Based Virtual Enterprises. In: 2010 Seventh International Conference on Information Technology: New Generations (ITNG), April 12-14, pp. 448–453 (2010), doi:10.1109/ITNG.2010.39
6. Euzenat, J., Shvaiko, P.: Ontology matching. Springer-Verlag New York, Inc., Secaucus (2007)
7. Nascimento, V., Viamonte, M.J., Canito, A., Silva, N.: Enhancing ontology alignment recommendation by exploiting emergent social networks. Paper presented at the 2012 IEEE/WIC/ACM International Conference on Intelligent Agent Technology (WI-IAT 2012), Macau, China (2012)
8. Viamonte, M.J., Nascimento, V., Silva, N., Maio, P.: AEMOS: An Agent-Based Electronic Market Simulator With Ontology-Services And Social Network Support. Paper presented at the 24th European Modeling & Simulation Symposium (Simulation in Industry) (EMSS 2012), Viena, Austria (2012)
9. Viamonte, M.J., Silva, N., Maio, P.: Agent-Based Simulation of Electronic Marketplaces With Ontology-Services. Paper presented at the 23rd European Modeling & Simulation Symposium (Simulation in Industry) (EMSS 2011), Rome, Italy (2011)
10. Silva, N., Viamonte, M.J., Maio, P.: Agent-Based Electronic Market With Ontology-Services. Paper presented at the 2009, IEEE International Conference on e-Business Engineering (ICEBE 2009), Macau, China (2009)
11. Maedche, A., Motik, B., Silva, N., Volz, R.: MAFRA – A mApping fRAmework for distributed ontologies. In: Gómez-Pérez, A., Benjamins, V.R. (eds.) EKAW 2002. LNCS (LNAI), vol. 2473, pp. 235–250. Springer, Heidelberg (2002)
12. Wasserman, S., Faust, K.: Social Network Analysis: Methods and Applications (Structural Analysis in the Social Sciences). Cambrige University Press, Cambrige (1994)

Semantic Multi-agent Architecture to Road Traffic Information Retrieval on the Web of Data

José Javier Samper-Zapater[1], Dolores M. Llidó Escrivá[2], Juan José Martínez Durá[1], and Ramon V. Cirilo[1]

[1] Computer Sciences Department, University of Valencia, Spain
{jose.j.samper,Juan.Martinez-Dura,Ramon.V.Cirilo}@uv.es
[2] Computer Languages and Systems Department, Universitat Jaume I, Spain
dllido@uji.es

Abstract. In this paper, we describe a system based on FIPA standards to help the process of advertisement, discovery, invocation and reuse of traffic information on the web of data. The use of semantic web services (SWS) can be exploited to improve the outcomes in the discovery process, allowing end users to specify their need using concepts not keywords. Most of the traffic information is generally recovered by end users through web forms that specify their requirements, and must refill each time the same parameters to obtain the updated value from the web sites. Using agents besides Service Oriented Architecture (SOA), we will achieve interoperability between systems and also automatize the process to obtain the data updated periodically. The amount of SWSs and the development of domain ontologies is increasing. However, in our domain (Intelligent Transport Systems) they are not very extended, so we had the need to develop our own ontologies. Our ontologies will allow to annotate semantically web services and also to translate the data provided by web forms to web services. To provide semantic search and retrieval information in real time we store the semantic WS profiles and our domain ontologies in a knowledge repository(KB) .This framework should improve the Human-Machine and the Machine-to-Machine interaction with the web of data, thanksto agents and the use of semantically annotated web service profiles.

1 Introduction

The need of making road Traffic safer and more efficient implies the use of the advanced technologies to develop intelligent systems. These systems, called Intelligent Transport Systems (ITS) allow users, both drivers and road managers, to receive and use Traffic information. This information is supplied by different providers and multiple web interfaces or APIs. Pre-trip systems requires data from diferent sources, and this is still an obstacle to overcome and requires human interaction. Generally, the search information process is based on purely syntactic criteria and it is restricted to find the occurrence of keywords in the text. So although it is useful, it is quite limited, and could not be integrated with other systems.

This scenario requires the use of a specific system that allows the exchange of Traffic information between heterogeneous Traffic Information Centers. We propose

S. Omatu et al. (Eds.): *Distrib. Computing & Artificial Intelligence,* AISC 217, pp. 465–472.
DOI: 10.1007/978-3-319-00551-5_56

to use an architecture with a Multi-Agent platform, in which clients and service providers interact with each other. The agents replicate the web sites and maintain a repository of profiles semantically annotated. This architecture also has a user interface to allow semantic queries to help the user to ask and retrieve data using ontology concepts and not only keywords. This paper is organized as follows: next section describes the state of the art of semantic applied in Road Traffic systems. Section 3 introduces the required ontologies. Section 4 presents the architecture approach and the life cycle . Finally some conclusions are exposed.

2 Semantic in the Area of Road Traffic Information

Although there is a lot of work and development using XML for information diffusion, data exchange and Traffic modelling, there aren't semantic specifications for Traffic information. For example, regarding road information diffusion by a management center, there are some initiatives that make use of XML language for information diffusion like CARS (Condition Acquisition and Reporting System) [1] in the United States, where TMDD (Traffic Management Data Dictionary) is used. Another one is the project of the Research and Development Department of the Hokkaido Development Bureau within the ITS/Win Research programme where they propose a language that uses RWML (Road Web Mark-up Language)[2] in order to manage Traffic information. As other experiences of mark-up languages for road information using XML we can mention : Traffic Data Mark-up Language (TDML)[1], and also in Europe TPEG (Transport Protocol Experts Group) with Road Traffic Message Application ML[2]

3 Road Traffic Information Ontologies

We have developed two kinds of ontologies[3]: a **Road Traffic Ontology** for specifying the road information concepts and relations, and a **Road Traffic Service Categorization Ontology** for the service category specification and other non functional parameters for WS.

The **Road Traffic Ontology** is a domain ontology with concepts related to traffic information. It has three functionalities: 1) services matching, 2) description of Traffic information services and, 3) specification assertional KB instance. For Services matching similarity degree we need to compare the different functional parameters, and to do this they are specified using the Road Traffic Ontology on the WS and in the user request profile. Also this ontology is used to annotate semantically the traffic data extracted from the Traffic web sites and to build instances. We have described in detail this ontology in [3]. To tackle the problem of a road Traffic Ontology development we have planned the definition of sub-domains. Establishing sub-domains is a very important aspect, since supporting independent ontologies

[1] http://www.travelerinformation.com
[2] http://www.tpeg.org/pdf/standardisation

means that each one can be modified independently. Considering the analysis of road Traffic information, these are the selected sub-domains: Road classification (Motorways, dual carriageway, etc.), Vehicle classification (Truck, car), Location (Area, Point, Section, etc.), Geography (Towns, Countries, etc.), Events (Accidents, Incidents, Measures), People (Driver, Passenger etc.) and Routes (Urban, interurban etc.) .

The second ontology **Road Traffic Service Categorization Ontology** will allow to decrease the processing time in the discovering process because it allow to filter firstly Traffic by non functional categories required on real time traffic services. Non functional categories make distinctions between services that do the same thing but in different ways or with different performance characteristics. For no functional parameters there are two classification ontologies NAICS[3] and UNSPSC[4] but the are too wide for our domain. As pointed out by Paolucci[7] the use of service categories may have little meaning if the categories allowed in the ontology are too wide. So we have defined a Road Traffic Service Categorization Ontology more specific taking the ISO standardization of WG TC 204 as a start point. Then our ontology will cover the taxonomy necessary to classify Road Traffic information. This ontology will allow us to improve the cost time on the discovery process filtering the desired service by one of these categories.

4 Traveller Information Architecture

In this section the specific architecture multi-agent platform of a traveller information system is presented (see figure 1). This architecture allows the development of the complete cycle of advertisement, request, discovery, invocation and execution of within it. In our prototype Client and Provider Agents uses OntoService[5]. OntoService is a web client interface that lets us to annotate semantically a WS using ontologies and specify the request profiles. So, this tool helps the providers to annotate semantically on a domain and the clients to query using the OWL-S profile. The following Agents components can be identified in a Traveller information architecture:

Provider Agent: It represents the WS within the platform, the provider has to annotate semantically its WS, and register its profile on the repository.

Client Agent: Is the representation of the client within the platform. It helps the user on the information search process allowing to specify the user requirements on a semantic way, and specify it as an OWL-S profile using OntoService. When the user select a result WS, this agent also helps the user to retrieve the selected information from the WS.

Matchmaker Agent: This agent implements the matchmaking algorithm to retrieve the best profiles that fits the user requirements. The matching process[4] requires a reasoner to compare the similarity between the profile specified by the user and all the profiles advertised on the repository.

[3] `http://www.census.gov/eos/www/naics/`

[4] `http://www.unspsc.org`

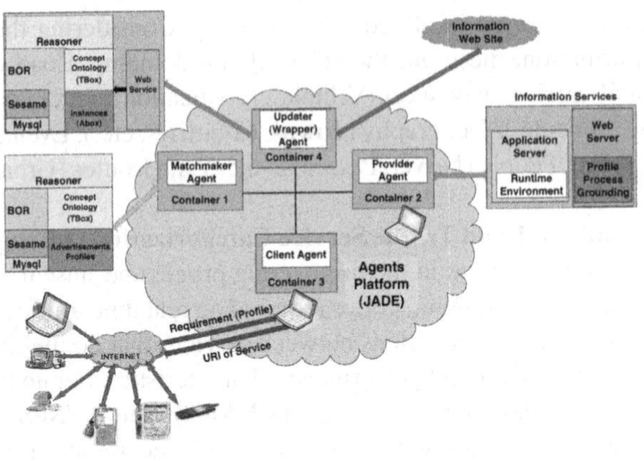

Fig. 1 Multiagent Architecture

Updater Agent: This wrapper agent replicates a list of web pages as a semantic WS, and wraps the information from the web periodically.

Directory Facilitator: This agent is necessary in any agent platform, can be considered the yellow pages of the system. Agents register their services on this agent or search for other agents.

The multi-agent platform has been implemented using JADE[11] which follows FIPA standards as a matchmaking model. The Client Agent queries the provider, and the Matchmaker Agent is an intermediary between the client and the provider. The complete life cycle of our platform would be as follows:

- **Advertising a WS:** When a Provider User wants to advertise a new WS, she/he interacts with the Provider Agent, and using 'OntoService' annotates semantically the service, generating a OWL-S profile. The Provider agent registers it on the Directory Facilitator agent and then the Directory Facilitator in turns, informs the Matchmaker Agent about a new WS giving its profile.
- **User favourites web pages:** The user can initialize the Updater agent with a list of URL's (favourites web pages). For each URL, the Updater Agent first checks in the KB if there is a concept (representation schema) that fits the information on the page. If it is necessary a new WS for this URL is created and advertised registering its profile on the Directory Facilitator. The wrapper of this Updater agent extracts periodically the new data on the web page to create new instances of this concept on the KB to allow to retrieve the data from the web pages updated without the need to access and fill any web form.
- **Search Traffic Information:** The user to retrieves road traffic information has to launch a Client agent, and using the OntoService gives a description of the required information like a profile of a SWS. The Client agent asks the Matchmaker agent to find a similar service to the request profile. The reasoner contrasts the description of requested service, with all of the available providers on the repository and returns the closest WS profile to the Client agent (see figure 2).

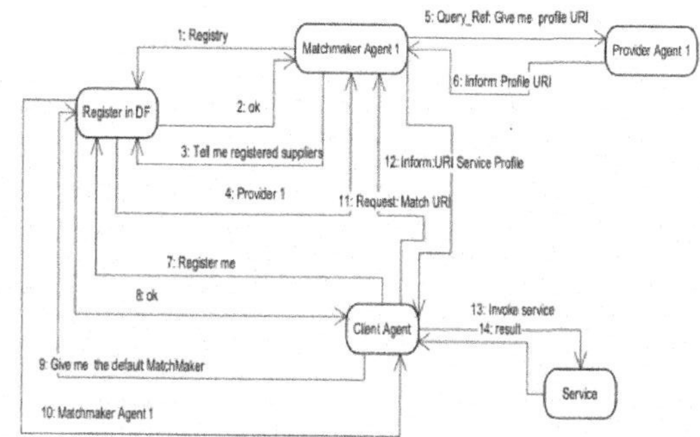

Fig. 2 Agent Collaboration Diagram: The matching process

- **Retrieve Traffic Information:** Once the Client Agent dynamically knows the description that closely resembles the user requirements, it generates a graphic interface to obtain the input data necessary to make a correct invocation of the service. When the user fills out the form the Client Agent invokes directly the WS and the user requested information is returned to the user.

4.1 Automatic WS Discovery: MatchMaking Agent

Automatic WS Discovery is an automated process to find WS[8] that can provide a particular class of service capabilities, while adhering to some client-specified constraints. To achieve this objective is very important to match appropriately the different properties or capabilities of the provided information , taking into the account their OWL-S descriptions. The main problem searching for and obtaining information in keeping with the users requirements, is still the main obstacle to overcome.

To automatic WS discovery in [4] we presented an algorithm for the matchmaking system based on Paolucci's matchmaking system[7]. However, it has some differences, which are concentrated on four main aspects:

- Filtering the providers profiles used for matchmaking, depending on the request profile category to improve the response time.
- Using non functional parameters. To improve the precision.
- Obtaining the matchmaking measure through the use of new degrees and modifying others as exact, to improve the recall and to differentiate between different grades of matchmakings.
- The count of the failures to avoid false positives/negatives.
- Ordering results (newSortRule) to adapt to the defined measures.

Figure 3 shows the diagram protocol of the Client Process to find a WS Provider:The Client Agent is started, and it queries to the Directory Facilitator Agent for a matchmaker Agent to retrieve a desired service. Figure 4 presents the Matchmaking process: the MatchMaker Agent is started, queries the Directory Facilitator Agent to

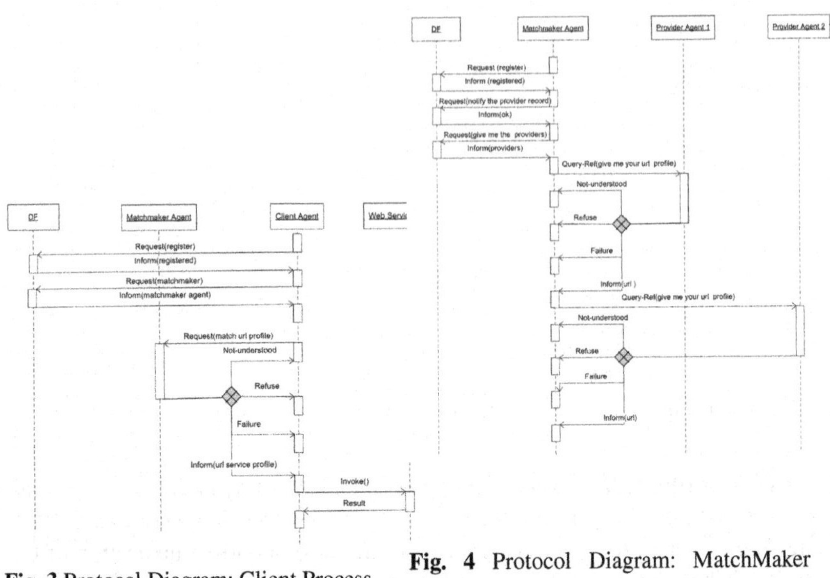

Fig. 3 Protocol Diagram: Client Process

Fig. 4 Protocol Diagram: MatchMaker Process

find all the available Provider Agents and contacts with them to match the more suitable service.

4.2 *From Web Sites to semantic WS*

The KB) [8] was stored in the Sesame Repository . Sesame supports the query languages SeRQL[6] (Sesame RDF Query Language) to access to the knowledge. It allows interoperability with a description logic reasoner BOR [9] and Racer. Road Traffic concepts ontology, Traffic categorization ontology, the Web pages schema representation and the definition of OWL-S[5] was stored as TBox(terminological box), and the instances of the schema representation and profiles are stored on the ABox(assertional box).

Then using the same representation schema of our KB we can create a SWS in our system that offers the same information as the web pages as a proxy. To extract information from the web page and store it as an instance of a concept of the ABox we used WebL libraries[10](see figure 1). Then using a WS semantic skeleton and the schema repository the user administrator specifies the query to access to the schema representation instances and creates using Apache Axis a new WS. A web page will be stored in the KB as a concept on the TBox and the data from the web page will be instances of this model, and will be updated periodically and stored as instances of this concept in the ABox. Each time the information of a web site change we are going to have a new instance of this term on our ABox. The Road Traffic Ontology is going to be the base knowledge to model terms and relation of the representation schema of a web page.

[5] `http://www.ai.sri.com/daml/services/owl-s/1.2/overview/`

For each web page a new updater Agents is created. The Agent first tries to find the best concept of our ontology that fits this web page information, if it doesn't exist it has to create a new concept on the Tbox as a representation schema of the web page and a WS. Then periodically the Wrapper access to the web pages to update the information of the site as description of its representation schema in the Abox. The creation of the WS is a semi-automatic process, because we have defined a skeleton of a WS that allows to query and retrieve the instances of the concepts on the Knowledge-Base, so when the Updater Agent have to create a new WS, the user administrator has to specify the query to retrieve the instances of the web pages and to annotate semantically this new WS using OntoService. When this new service is created using the Provider Agent this new WS is registered on the DF like a new provider resource.

5 Conclusions

The work we have presented here has served us as a proof of concept for the use of ontologies and in the Road Traffic domain. Their use allows making semi-automatic treatments of the data from the Web. Although the chosen domain for testing, was the Road Traffic information domain, it could be applied to other domains, by adapting the ontology. In fact, nowadays the results of this research are being applied in an e-Health project funded by the University of Valencia. For the elaboration of the work, two different points of view have been taken: (i) the need of an information interchange among the different organizations (or Traffic administrations), as well as the decision-taking according to the data obtained and, on the other hand, (ii) the point of view of the driver or traveller who might need to have information from different sources, in different formats. In this last sense it is our intention to make the user's consultation much more precise so the pertinent information will be achieved with the use of ontologies. Now, there is a new technology on the Web, the user of linked data. We need to extend our system architecture to integrate also this kind of knowledge and Data. Probable we also can improve our system using the DBPedia Ontology and adapting all the system to a multilingual environment. Improvements of the system: The Updater Agent is useful if we have no access to the system, but probably a better solution bill be to create a SWS that query to the web site without replicate the information.

References

1. Hamwi, B., Choudhry, O.: Condition Acquisition and Reporting System (CARS). In: ITS Canada Conference (2005)
2. Kajiya, Y., Yamagiwa, Y., Kudo, Y.: Road Web Markup Language-XML for Road-related Information Distribution. Transportation Research Board (2004)
3. Javier Samper, J., Carrillo, E., Cervera, A.: Semantic Modelling of Road Traffic Information elements. Revista Colombiana de Computacin 7(1), 76–91 (2006) ISSN 1657 - 2831

4. Javier Samper, J., Javier Adell, F., van den Berg, L., José Martinez, J.: Improving Semantic Web Service Discovery. Journal of Networks 3(1) (January 2008) ISSN : 1796-2056
5. Javier Samper, J., Tomás, V.R., Carrillo, E., do Nascimento, R.P.C.: Visualization of ontologies to specify semantic descriptions of services. IEEE Transactions on Knowledge and Data Engineering, ISSN: 1041-4347
6. Aduna, B.V., Sirma AI Ltd.: The SeRQL query languag, rev. 1.1 from User Guide for Sesame, http://www.openrdf.org/doc/sesame/users/ch06.html
7. Paolucci, M., Kawamura, T., Payne, T.R., Sycara, K.: Semantic Matching of Web Services Capabilities. In: Horrocks, I., Hendler, J. (eds.) ISWC 2002. LNCS, vol. 2342, pp. 333–347. Springer, Heidelberg (2002)
8. Baader, F., Nutt, W.: The Description Logic Handbook. Part I: Theory. Basic Descriptions Logics, Cambridge, p. 46 (2003) ISBN 0-521-78176-0
9. Simov, K., Jordanov, S.: BOR: a Pragmatic DAML+OIL Reasoner. In: Deliverable 40, On-To-Knowledge project (June 2002)
10. Hannes, M., Tom, R.: Automating the Web with WebL. Compaq Systems Research Center (1999)
11. Yu, W.H.: Application of Java Serialization into JADE Agent Communication. Journal Advanced Materials Research 433-440, 7357–7361 (2012), www.scientific.net/AMR.433-440.7357, doi:10.4028

Personalised Advertising Supported by Agents

Bruno Veloso[1,2], Luís Sousa[1], and Benedita Malheiro[1,2]

[1] Instituto Superior de Engenharia do Porto, Instituto Politécnico do Porto, Porto, Portugal
[2] INESC TEC – INESC Technology and Science (formerly INESC Porto)
{bruno.miguel.veloso,letvsousa}@gmail.com, mbm@isep.ipp.pt

Abstract. This paper reports the development of a B2B platform for the personalization of the publicity transmitted during the program intervals. The platform as a whole must ensure that the intervals are filled with ads compatible with the profile, context and expressed interests of the viewers. The platform acts as an electronic marketplace for advertising agencies (content producer companies) and multimedia content providers (content distribution companies). The companies, once registered at the platform, are represented by agents who negotiate automatically the price of the interval timeslots according to the specified price range and adaptation behaviour. The candidate ads for a given viewer interval are selected through a matching mechanism between ad, viewer and the current context (program being watched) profiles. The overall architecture of the platform consists of a multiagent system organized into three layers consisting of: (*i*) interface agents that interact with companies; (*ii*) enterprise agents that model the companies, and (*iii*) delegate agents that negotiate a specific ad or interval. The negotiation follows a variant of the Iterated Contract Net Interaction Protocol (ICNIP) and is based on the price/s offered by the advertising agencies to occupy the viewer's interval.

Keywords: Multiagent system, B2B, Multimedia, Brokerage, Profile Matching, Fixed ICNIP, Web Services.

1 Introduction

Media content personalisation has been addressed by several projects such as MiSPOT, NoTube, LinkedTV or HBB-NEXT. MiSPOT proposes a non-invasive and fully personalized form of advertising, using semantic reasoning techniques to select advertisements suited to the preferences, interests and needs of each viewer [1]. NoTube defines a flexible end-to-end architecture for the personalised creation, distribution and consumption of TV content [2]. LinkedTV focuses on integrating TV and Web contents based on user personalisation and contextualisation [3]. HBB-NEXT goal is to mix broadcast, Internet and user-generated content by adopting user-centric technologies to enrich the TV-viewing experience with social networking, multiple device access and group-tailored recommendations [4].

S. Omatu et al. (Eds.): *Distrib. Computing & Artificial Intelligence*, AISC 217, pp. 473–481.
DOI: 10.1007/978-3-319-00551-5_57

The work described in this paper reports on the Business-to-Business (B2B) MultiMedia Brokerage (MMB) platform for media content personalisation under development [5,6]. The personalisation is based on the profiles of viewers and media components (ads and intervals). The application domain is the personalization of advertising intervals, i.e., the content of viewer intervals will be negotiated to ensure an ad alignment compatible with the current context, expressed interests and previous interactions of the viewer. It is expected that, in the future, advertising agencies and media content distributors use the platform to personalize advertising and perform niche segmentation.

The platform acts as an electronic market that supports the automated trading between the advertising agencies (producers) and the content providers (distributors). The companies, once registered at the platform, are represented by agents that negotiate items automatically according to the specified negotiation behaviour, i.e., the price range and adaptation tactics. Distributors act as sellers and the producers as buyers of interval timeslots. The negotiation follows a variant of the Iterated Contract Net Interaction Protocol (ICNIP) [10] – the Fixed ICNIP – and is based on the price/s offered by producers to occupy the interval timeslots.

This paper is organised in four sections: the introductory section, the development section, covering the system architecture and functionalities, the tests and results section and the conclusions section.

2 Multimedia Brokerage Platform

The brokerage platform is a multiagent system organized into three layers: interface, enterprise and market layers. Fig. 1 displays an overview of the system architecture.

The agents of the platform are divided into four types: (*i*) interface agents to interface with businesses; (*ii*) enterprise agents that model the businesses; (*iii*) market delegate agents dedicated to specific ad or interval negotiations on behalf of enterprise agents; and (*iv*) layer manager agents (interface, enterprise and market layer agents). Each business (producer or distributor enterprise) is represented at the platform by: (*i*) an enterprise interface agent, which exposes a Web service with a set of interface operations, located in first layer, (*ii*) the enterprise agent that models the enterprise at platform, which exposes a Web service with a set of operations for the agents of the other layers, residing in the intermediate layer; and (*iii*) an undetermined number of delegates involved in specific negotiations. Table 1 shows the set of Web service operations exposed by the platform agents.

The domain knowledge is represented by three ontologies developed with the Protégé ontology editor: the MMB platform ontology, the viewer and program profiles ontology and the ad profile ontology. The MMB platform ontology is a Protégé frame ontology with the following main concepts: (*i*) AgentAction that contains all actions concerning the operations of Web services; (*ii*) AgentType

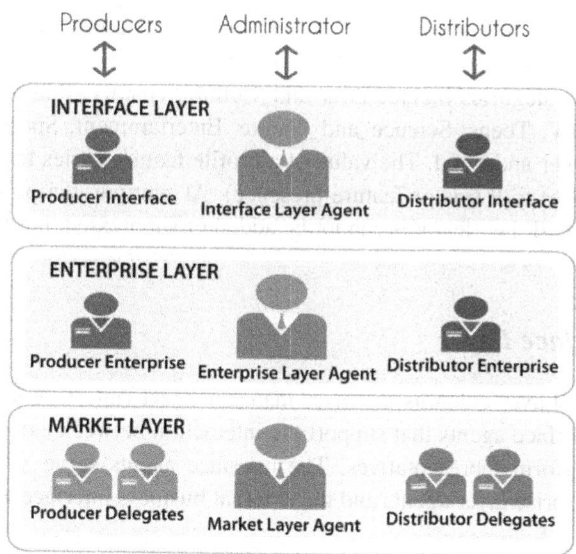

Fig. 1 MMB platform architecture

Table 1 Platform agent Web services and operations

Interface Layer	**Service Type**	**Producer**	**Distributor**	**Layer**
	Operations	SetAd SetAdProfile GetAdResults RemoveAd	SetInterval SetIntervalProfile SetViewerProfile GetIntervalResults RemoveInterval	CreateAgent KillAgent
Enterprise Layer	**Service Type**	**Producer**	**Distributor**	**Layer**
	Operations	GetAdProfile RemoveAd SetAdProfile SetAdResult SetAd GetAd GetAdResults GetProduct	GetIntervalProfile SetViewerProfile SetIntervalResult RemoveInterval SetInterval GetIntervalResults GetInterval SetIntervalProfile	CreateAgent KillAgent
Market Layer	**Service Type**	**Producer**	**Distributor**	**Layer**
	Operations			CreateProducerDelegate CreateDistributorDelegate SetMarketProtocol

that includes all types of agents used in the platform; and (*iii*) AgentData that holds the ads, intervals and viewers, including the corresponding profiles. The viewers and programs ontology is an OWL ontology inspired in the BBC program categories [6] and the ads ontology is an OWL ontology based on the Yellow Pages classified ads categories [8].

A viewer, program or ad profile is a vector of 15 features, where each feature corresponds to a category specified in the corresponding ontology. For example, the programs categories include Arts_and_Culture, History, Radio, News, Learning, Music, TV, Teens, Science_and_Nature, Entertainment, Sport, Health, Gardening, Weather and Food. The value of a profile feature varies from 0 (complete feature absence) to 9 (major feature presence). At runtime, the profile dimension cannot be altered, i.e., features cannot be added or suppressed, but features can be substituted.

2.1 *Interface Layer*

The Interface Layer contains the layer manager agent (interface layer agent) and dedicated interface agents that support the interaction between external businesses and their platform representatives. The interface agents serve as intermediaries between enterprise layer agents and the external business interface applications.

2.2 *Enterprise Layer*

The Enterprise Layer holds the layer manager agent (enterprise layer agent) and the agents that model the businesses (distributor and producer enterprises) within the platform. The distributor agents are continuously trying to find and invite producer agents with ads matching the upcoming viewer intervals. Interval profiles are based on the context (the program being watched) and on the viewer profiles.

The matching between interval and ad profiles is performed by the producer agents in order to select the ads to negotiate in the market layer and, thus, decide whether to accept the distributor agent invitations. Since ads and intervals use distinct ontologies, it is necessary, before applying any algorithms, to map the ads and programs ontologies. The adopted mapping is straightforward two step mechanism: a search for ad features that are identical to program features followed by a final search for ad sub-features that are identical to any unmapped program features. The matching mechanism is supported by a set of dedicated rules which use two distinct similarity algorithms to match ads with intervals profiles to rank the candidate ads. The first algorithm determines the similarity using the cosine similarity and the Euclidean distance [9]. The second algorithm computes the similarity based on the dominant characteristic of the interval. Depending on the resulting similarity ranking, the producer agents decide whether to accept the invitation to negotiate an ad and, consequently, launch a producer delegate in the market layer. The rules outcome is a similarity ranking between 1 (lowest) and 4 (highest).

2.3 *Market Layer*

The Market Layer contains the layer manager agent (market layer agent) and the enterprise delegate agents that represent the ads and intervals (distributor and producer delegates) under negotiation. Each delegate negotiates a single ad or interval on behalf of the corresponding producer or distributor agent according to the defined ad or interval negotiation behaviour: price range and adaptation tactic.

The market implements the Fixed ICNIP (FICNIP) negotiation protocol which is a variant of the Foundation for Intelligent Physical Agents (FIPA) Iterated Contract Net Interaction Protocol (ICNIP) [10]. While the FIPA-ICNIP stops as soon as there is an offer that matches the seller's target price, the FICNIP iterates for a fixed number of times regardless of the values of the buyers' offers received so far.

3 Tests and Results

Three types of tests were conducted to assess the operation of the MMB platform: (*i*) the ad selection mechanism; (*ii*) the negotiation of an interval involving a distributor and producers with different negotiation tactics and equal price variation ranges; and (*iii*) the negotiation of an interval involving a distributor and producers with different negotiation price variation ranges and equal adaptation tactics.

3.1 *Ad Selection*

This scenario is intended to illustrate the application of the similarity metrics implemented to select advertisements (the cosine similarity and dominant characteristic similarity) and involves a producer and a distributor enterprise. The producer enterprise submits two different ads. The distributor enterprise uploads to the platform an upcoming interval together with the corresponding viewer and context (current program) profiles. Table 2 displays the ad profiles.

Table 2 Ad characteristics

Product	Enterprise	Timeslot (s)	Profile	Ref. Price/s (€)
Ferrari	Prod001	30	897864156494888	10
Jaguar	Prod001	30	987489496848499	10

Table 3 presents the viewer, context (program being watched) and resulting interval profiles held by the distributor agent.

Table 3 Viewer, context and resulting interval profiles

Viewer	Enterprise	Viewer Profile	Channel	Program	Program Profile
1	Dist001	104351267334794	Discovery	MythBusters	826492673411245

Interval	Enterprise	Timeslot (s)	Ref. Price/s (€)	Interval Profile
1	Dist001	60	25	415371465322464

Table 4 holds the results of applying to both ads the matching rules based on the cosine similarity and on the interval dominant characteristic. In the first case, the producer agent chooses to negotiate the Ferrari ad and, in the latter case, chooses to negotiate the Jaguar ad. The ad ranking, which varies between a minimum of 1 and a maximum of 4, is determined through defined matching rules. If a tie results from the application of a similarity metric, the producer chooses randomly one of the tied ads.

Table 4 Matching rules results

Product	Similarity	Distance	Ranking	Int. Dom. Char.	Ad Dom. Char.	Ranking
Ferrari	0.431	1.066	1	7	6	2
Jaguar	0.379	1.114	1	7	8	3

These results demonstrate the proper functioning of matching mechanism.

3.2 *Price Adaptation Tactics*

This scenario involves one distributor and five producer enterprises. The producer enterprises submit the ads presented in Table 5. All ads have equal price variation ranges, but adopt different price adaptation tactics during the negotiation. The negotiation protocol is FICNIP.

Table 5 Ads characteristics

Product	Ferrari	Porsche	Toyota	LandRover	Mazda
Enterprise	Prod001	Prod002	Prod003	Prod004	Prod005
Price Tactic	Quadratic	Exponential	Linear	Random	Fixed
Ref. Price/s (€)	10	10	10	10	30
Max. Price/s (€)	50	50	50	50	50
Timeslot (s)	30	30	30	30	30

The distributor enterprise uploads the interval features and corresponding viewer and context (current program) profiles. The viewer profile, viewer context profile and the resulting interval profile are presented in Table 6.

Table 6 Viewer, context and resulting interval characteristics

Viewer	Enterprise	Viewer Profile	Channel	Program	Program Profile
2	Dist001	950567777851928	Discovery	MythBusters	826492673411245

Interval	Enterprise	Timeslot (s)	Ref. Price/s (€)	Interval Profile
2	Dist001	150	25	833474675631536

The results of this interval negotiation are shown in Table 7 and correspond to the final interval content. They demonstrate the negotiation of a full interval involving producers with equal price ranges and diverse price adaptation tactics.

Table 7 Interval 2 negotiation results

Product	Ferrari	Porsche	Toyota	LandRover	Mazda
Negotiated Price/s (€)	50.0	50.0	45.0	49.3	30.0

3.3 *Price Ranges*

This scenario illustrates the FICNIP negotiation behaviour with ads with the same adaptation tactics and different price ranges. It involves one distributor and two producer enterprises. The producer ad parameters are presented in Table 8.

Table 8 Ads characteristics

Product	Jaguar	Porsche
Enterprise	Prod001	Prod002
Ref. Price/s (€)	15	10
Max. Price/s (€)	90	50
Price Tactic	Exponential	Exponential
Timeslot (s)	60	60

The distributor enterprise uploads the viewer and viewer context (current program) profiles and upcoming viewer interval features. Table 9 presents the uploaded data and the resulting interval profile.

Table 9 Viewer, context and resulting interval profiles

Viewer	**Enterprise**	**Viewer Profile**	**Channel**	**Program**	**Program Profile**
2	Dist001	950567777851928	Discovery	MythBusters	826492673411245

Interval	**Enterprise**	**Timeslot (s)**	**Ref. Price (€)**	**Interval Profile**
3	Dist001	60	25	833474675631536

The negotiation results are shown in Table 10. These results demonstrate a negotiation involving two producers with the same price adaptation tactics and different price ranges. The Porsche ad was not added to the interval because it was full after the successful negotiation of the Jaguar ad.

Table 10 Interval 3 negotiation results

Product	Jaguar
Negotiated Price/s (€)	90.0

4 Conclusions

This paper presents the application domain, the developed platform functionalities as well as three different test scenarios that illustrate the platform operation.

In terms of achievements, the developed MMB platform prototype is able to trade timeslots between content producers and distributors based on the viewer, program, interval and ad profiles. The architecture is organized into three layers: (*i*) interface layer that is responsible for interacting with the external businesses; (*ii*) enterprise layer where distributor invite producer agents to negotiate and producer agents select ads for negotiation based on the profile matching mechanism; and (*iii*) market layer where delegate agents of the intermediate layer agents meet to negotiate interval timeslots according to the ascribed interval and ad negotiation behaviour (price range and adaptation tactic). The communication between agents of different layers and between the platform and the external entities is performed through Web services. The market agents communicate by exchanging FIPA-ACL messages and implement the Fixed ICNIP (FICNIP) negotiation protocol. All Web services exposed by the agents are published in a UDDI registry for discovery and consumption.

Concerning future developments, the mapping between the programs and ads profile ontologies can be refined. Currently it implements a direct mapping between the corresponding features of programs and ads and between the remaining unmatched program features and the corresponding ad sub-features. Different weights should be attributed to features and sub-features and, in the case of several matching sub-features, the mean value should be attributed instead of the highest value currently used. The matching mechanism between ad and interval profiles can be enhanced. The producer agents select the ads based on the similarity between ad and interval profiles. The similarity based on the interval dominant characteristic can use, instead of a single dominant characteristic, the top five interval characteristics (one third of the profile features). Furthermore, unexpected ads should be regularly chosen, provoking a sense of novelty and unpredictability on the viewer, i.e., introducing serendipity. Finally, the platform evaluation needs to be carried out with real users and data.

Acknowledgment. This work is financed by the ERDF – European Regional Development Fund through the COMPETE Programme (operational programme for competitiveness) and by National Funds through the FCT – Fundação para a Ciência e a Tecnologia (Portuguese Foundation for Science and Technology) within project «FCOMP - 01-0124-FEDER-022701».

References

1. López-Nores, M., et al.: MiSPOT: dynamic product placement for digital TV through MPEG-4 processing and semantic reasoning. Knowledge and Information Systems 22(1), 101–128 (2010)
2. NoTube, NoTube (2012), http://notube.tv/ (accessed in February 2013)
3. LinkedTV, LinkedTV (2012), http://www.linkedtv.eu/ (accessed in February 2013)
4. HBB-NEXT, HBB-NEXT Next Generation Media (2012), http://www.hbb-next.eu/index.php (accessed in February 2013)

5. de Sousa, L.V., Malheiro, B., Foss, J.: Negotiation platform for personalised advertising. In: de Strycker, L. (ed.) Proceedings of the Fifth International European Conference on the Use of Modern Information and Communication Technologies (ECUMICT 2012), pp. 361–373 (2012)
6. Foss, J.D., Malheiro, B., Burguillo, J.C.: Personalised placement in networked video. In: Proceedings of the 21st International Conference Companion on World Wide Web, WWW 2012 Companion, pp. 959–968. ACM, New York (2012)
7. British Broadcasting Corporation. Program categories, `http://www.bbc.co.uk/a-z/` (accessed in September 2012)
8. Yellow Pages. Classified ads categories, `http://www.yellowpages.com/` (accessed in September 2012)
9. Madylova, A., Oguducu, S.: A taxonomy based semantic similarity ofdocuments using the cosine measure. In: Proceedings of the 24th International Symposium on Computer and Information Sciences (ISCIS 2009), pp. 129–134 (2009)
10. Foundation for Intelligent Physical Agents, FIPA iterated contract net interaction protocol specification, FIPA TC Communication, Standard 30 (2002), `http://www.fipa.org/specs/fipa00030/SC00030H.pdf` (accessed in September 2012)

Two Approaches to Bounded Model Checking for a Soft Real-Time Epistemic Computation Tree Logic*

Artur Męski[1,**], Bożena Woźna-Szcześniak[2], Agnieszka M. Zbrzezny[2,**], and Andrzej Zbrzezny[2]

[1] University of Łódź, FMCS, Banacha 22, 90-238 Łódź, Poland,
ICS PAS, J. Kazimierza 5, 01-237 Warsaw, Poland
meski@ipipan.waw.pl

[2] Jan D ługosz University, IMCS, Armii Krajowej 13/15,
42-200 Częstochowa, Poland
{b.wozna,a.zbrzezny,agnieszka.zbrzezny}@ajd.czest.pl

Abstract. We tackle two symbolic approaches to bounded model checking (BMC) for an existential fragment of the soft real-time epistemic computation tree logic (RTECTLK) interpreted over interleaved interpreted systems. We describe a BDD-based BMC method for RTECTLK, and provide its experimental evaluation and comparison with a SAT-based BMC method. Moreover, we have attempted a comparison with MCMAS on several benchmarks.

1 Introduction

Multi-agent systems (MASs) are distributed systems in which many intelligent agents, representing autonomous entities such as software programs, interact (communicate, negotiate, cooperate, etc.) with each other to solve problems that are beyond the individual capacities or knowledge of a single agent. A plethora of verification techniques of MASs that use temporal-epistemic logics, among others [5, 13, 16], have been proposed and refined over the years. Many of them are based on the model checking method [17], the practical applicability of which is strongly limited by the state explosion problem. To avoid this problem a number of state reduction techniques [2, 3, 9] and symbolic model checking approaches (symbolic storage methods based on binary decision diagrams (BDDs) [16], bounded (BMC) [6, 10, 14, 19] and unbounded [7] model checking) have been developed. Recent research in

* Partly supported by National Science Center under the grant No. 2011/01/B/ST6/05317.

** The study is co-funded by the European Union, European Social Fund. Project PO KL "Information technologies: Research and their interdisciplinary applications", Agreement UDA-POKL.04.01.01-00-051/10-00.

S. Omatu et al. (Eds.): *Distrib. Computing & Artificial Intelligence*, AISC 217, pp. 483–491.
DOI: 10.1007/978-3-319-00551-5_58

the verification of MASs tries to complete the picture of BMC methods by using BDDs to represent the reachable state space truncated up to a specific depth, rather than translating it to the SAT-problem, and by performing BDD-based verification on a truncated model [6].

So far, the SAT-based BMC approaches developed for MASs, which are modelled by interleaved interpreted systems (IIS) [9] (i.e., a special class of interpreted systems (IS) [4], in which only one action at a time is performed in a global transition), deal with the properties expressed in ECTLK (i.e., the existential fragment of CTL extended with epistemic components) [14], RTECTLK (i.e., the existential fragment of soft-real time CTL extended with epistemic components) [19], ELTLK (i.e., the existential fragment of LTL extended with epistemic components) [15], and EMTLK (i.e., the existential fragment of discrete MTL extended with epistemic components) [18]. The BDD-based BMC techniques developed for MASs modelled by IS deal with the properties expressed in ECTLK [6] and ELTLK [12].

The main objective of this paper is to introduce a BDD-BMC method for RTECTLK and MASs modelled by IIS. We implemented this BDD-BMC algorithm into VerICS [8], and we compare it with the SAT-based BMC method for RTECTLK [19], which is also implemented in VerICS. For a constructive evaluation of our two BMC methods we use two scalable scenarios: *Train Controller system* and *Generic Pipeline Paradigm*. In addition we compare SAT-based BMC method for ECTLK [14] with the BDD-BMC method for ECTLK [6] that is implemented in the MCMAS tool [11].

The rest of the paper is organised as follows. We begin in Section 2 by presenting IIS and the RTECTLK language. In Section 3 we present the BMC methods for RTECTLK. In Section 4 we discuss our experimental results. In the last section we conclude the paper.

2 Preliminaries

Interleaved Interpreted Systems (IIS). We assume that a MAS consists of n agents, and by $Ag = \{1, \ldots, n\}$ we denote the non-empty set of agents; note that we do not consider the environment component. This may be added with no technical difficulty at the price of heavier notation. We assume that each agent $c \in Ag$ is in some particular local state at a given point in time, and that a set L_c of local states for agent $c \in Ag$ is non-empty and finite.

With each agent $c \in Ag$ we associate a finite set of *possible actions* Act_c such that a special "null" action (ϵ_c) belongs to Act_c; as it will be clear below the local state of agent c remains the same, if the null action is performed. We do not assume that the sets Act_c (for all $c \in Ag$) are disjoint. Moreover, we define the following two sets: $Act = \bigcup_{c \in Ag} Act_c$ and $Agent(a) = \{c \in Ag \mid a \in Act_c\}$.

Next, with each agent $c \in Ag$ we associate a *protocol* that defines rules, according to which actions may be performed in each local state. The protocol for agent $c \in Ag$ is a function $P_c : L_c \to 2^{Act_c}$ such that $\epsilon_c \in P_c(\ell)$ for any $\ell \in L_c$, i.e., we insist on the null action to be enabled at every local state.

Further, for each agent c, there is defined a (partial) *evolution function* $t_c : L_c \times Act_c \to L_c$ such that for each $\ell \in L_c$ and for each $a \in P_c(\ell)$ there exists $\ell' \in L_c$ such that $t_c(\ell,a) = \ell'$; moreover, $t_c(\ell,\epsilon_c) = \ell$ for each $\ell \in L_c$.

For a given set of agents Ag and a set of propositional variables $\mathcal{PV}$, an IIS is a tuple: $(\{\iota_c, L_c, Act_c, P_c, t_c, \mathcal{V}_c\}_{c \in Ag})$, where $\iota_c \in L_c$ is an initial state of agent c, and $\mathcal{V}_c : L_c \to 2^{\mathcal{PV}}$ is a valuation function for agent c. With each IIS it is possible to associate a *Kripke model* as defined below.

Definition 1. *A* model for IIS *(or* a model*) is a tuple $M = (\iota, S, T, \mathcal{V})$, where (1) $\iota = \iota_1 \times \ldots \times \iota_n$ is the initial global state; (2) $S = L_1 \times \ldots \times L_n$ is a set of all possible global states. $l_c(s)$ denotes the local component of agent $c \in Ag$ in a global state $s = (\ell_1, \ldots, \ell_n)$; (3) Let $t : S \times Act_1 \times \cdots \times Act_n \to S$ be a global interleaved evolution function defined as follows: $t(s, a_1, \ldots, a_n) = s'$ iff there exists an action $a \in Act \setminus \{\epsilon_1, \ldots, \epsilon_n\}$ such that for all $c \in Agent(a)$, $a_c = a$ and $t_c(l_c(s), a) = l_c(s')$, and for all $c \in Ag \setminus Agent(a)$, $a_c = \epsilon_c$ and $t_c(l_c(s), a_c) = l_c(s)$. In brief we write the above as $s \xrightarrow{a} s'$. $T \subseteq S \times Act \times S$ is a transition relation defined by the global interleaved evolution function as follows: $(s, a, s') \in T$ iff $s \xrightarrow{a} s'$. We assume that the relation T is total, i.e., for any $s \in S$ there exists $s' \in S$ and an action $a \in Act \setminus \{\epsilon_1, \ldots, \epsilon_n\}$ such that $s \xrightarrow{a} s'$; (4) $\mathcal{V} : S \to 2^{\mathcal{PV}}$ is the valuation function defined as $\mathcal{V}(s) = \bigcup_{c \in Ag} \mathcal{V}_c(l_c(s))$. $\mathcal{V}$ assigns to each state a set of propositional variables that are assumed to be true at that state.*

A *path* in M is an infinite sequence $\pi = (s_0, s_1, \ldots)$ of states such that $(s_j, s_{j+1}) \in T$ for each $j \in \mathbb{N}$. For such a path, and for $m \in \mathbb{N}$, by $\pi(m)$ we denote the m-th state s_m. By $\Pi(s)$ we denote the set of all the paths starting at $s \in S$; note that since T is total $\Pi(s)$ is never empty.

Given an IIS and $\Gamma \subseteq Ag$, we recall definitions of the following relations: $\sim_c \subseteq S \times S$ is an indistinguishability relation for agent c defined by: $s \sim_c s'$ iff $l_c(s') = l_c(s)$, $\sim_\Gamma^E \stackrel{def}{=} \bigcup_{c \in \Gamma} \sim_c$, $\sim_\Gamma^C \stackrel{def}{=} (\sim_\Gamma^E)^+$ (the transitive closure of $\sim_\Gamma^E$), and $\sim_\Gamma^D \stackrel{def}{=} \bigcap_{c \in \Gamma} \sim_c$.

RTECTLK. In the syntax of RTECTLK we assume the following: $p \in \mathcal{PV}$ is an atomic proposition, $\top$ denotes the Boolean constant true, $c \in Ag$ is an index of an agent, $\Gamma \subseteq Ag$ is a subset of indices of agents, and I is an interval in $\mathbb{N} = \{0, 1, 2, \ldots\}$ of the form: $[a, b)$ where $a \in \mathbb{N}$, $b \in \mathbb{N} \cup \{\infty\}$, and $a \neq b$.

The RTECTLK formulae are built according to the following grammar:

$$\varphi ::= \top \mid p \mid \neg p \mid \varphi \wedge \varphi \mid \varphi \vee \varphi \mid \mathrm{EX}\varphi \mid \mathrm{E}(\varphi \mathrm{U}_I \varphi) \mid \mathrm{EG}_I \varphi \mid \overline{\mathrm{K}}_c \varphi \mid \overline{\mathrm{D}}_\Gamma \varphi \mid \overline{\mathrm{E}}_\Gamma \varphi \mid \overline{\mathrm{C}}_\Gamma \varphi$$

The derived basic temporal modalities are defined as follows: $\mathrm{E}(\alpha \mathrm{R}_I \beta) \stackrel{def}{=} \mathrm{E}(\beta \mathrm{U}_I (\alpha \wedge \beta)) \vee \mathrm{EG}_I \beta$, and $\mathrm{EF}_I \alpha \stackrel{def}{=} \mathrm{E}(\top \mathrm{U}_I \alpha)$.

Note that the fragment of RTECTLK in which all the intervals are of the form $[0,\infty)$ is the existential fragment of CTL augmented to include epistemic modalities (i.e., ECTLK). Therefore, for simplicity, we write $\mathrm{E}(\alpha \mathrm{U} \beta)$ and $\mathrm{EG}\alpha$ instead of $\mathrm{E}(\alpha \mathrm{U}_{[0,\infty)}\beta)$ and $\mathrm{EG}_{[0,\infty)}\alpha$, respectively.

An RTECTLK formula φ is *true* in the model M (in symbols $M \models \varphi$) iff $M,\iota \models \varphi$ (i.e., φ is true at the initial state of the model M). For every $s \in S$ the relation $\models$ is defined inductively as follows:

$$
\begin{array}{ll}
M,s \models \top,\ M,s \models p & \text{iff } p \in \mathcal{V}(s),\ M,s \models \neg p \text{ iff } p \notin \mathcal{V}(s),\\
M,s \models \alpha \wedge \beta & \text{iff } M,s \models \alpha \text{ and } M,s \models \beta,\\
M,s \models \alpha \vee \beta & \text{iff } M,s \models \alpha \text{ or } M,s \models \beta,\\
M,s \models \mathrm{EX}\alpha & \text{iff } (\exists \pi \in \Pi(s))(M,\pi(1) \models \alpha),\\
M,s \models \mathrm{E}(\alpha \mathrm{U}_I \beta) & \text{iff } (\exists \pi \in \Pi(s))(\exists m \geq 0)(m \in I \text{ and } M,\pi(m) \models \beta\\
 & \quad \text{and } (\forall j < m) M,\pi(j) \models \alpha),\\
M,s \models \mathrm{EG}_I \alpha & \text{iff } (\exists \pi \in \Pi(s))(\forall m \in I)[M,\pi(m) \models \alpha],\\
M,s \models \overline{\mathrm{K}}_c \alpha & \text{iff } (\exists \pi \in \Pi(\iota))\ (\exists i \geq 0)(s \sim_c \pi(i) \text{ and } M,\pi(i) \models \alpha),\\
M,s \models \overline{Y}\alpha & \text{iff } (\exists \pi \in \Pi(\iota))(\exists i \geq 0)(s \sim \pi(i) \text{ and } M,\pi(i) \models \alpha),\\
 & \quad \text{where } \overline{Y} \in \{\overline{\mathrm{D}}_\Gamma, \overline{\mathrm{E}}_\Gamma, \overline{\mathrm{C}}_\Gamma\}, \text{ and } \sim \in \{\sim_\Gamma^D, \sim_\Gamma^E, \sim_\Gamma^C\}.
\end{array}
$$

The *model checking problem* asks whether $M \models \varphi$.

3 Bounded Model Checking (BMC)

The SAT-based BMC is a verification technique whose basic idea is to consider only finite prefixes of paths (i.e., prefixes whose length are bounded by some integer k) that may be a witnesses to an "existential" model checking problem. If no error is found, then one increases k until either an error is found, or the problem becomes intractable. In this paper we use the SAT-based BMC technique for RTECTLK that was introduced in [19]. Therefore, we do not present the technique here. The crux of BDD-based BMC is to interleave verification with the construction of the reachable states.

BDD-based BMC for RTECTLK. Let $M = (\iota, S, T, \mathcal{V})$ be a model. By S_T we denote the set of all reachable states of the model. For $X \subseteq S_T$, we define the set $pre(X) = \{s \in S_T \mid (\exists s' \in X)(\exists a \in Act)\ (s,a,s') \in T\}$ of all the reachable states from which a transition into a state in X is possible.

By $[\![M,\varphi]\!]$ we denote the set of all the reachable states of the model M at which φ holds; we write $[\![\varphi]\!]$ if M is implicit. If φ is an ECTLK formula, then $[\![\varphi]\!]$ is defined as in [16]. We now define this set for the real-time modalities in a similar way to that of [1]. Let α, β be RTECTLK formulae, then:

- $$[\![\mathrm{EG}_{[a,b)}\alpha]\!] = \begin{cases} pre([\![\mathrm{EG}_{[a-1,b-1)}\alpha)]\!]) & \text{if } a > 0 \text{ and } b > 1,\\ pre([\![\mathrm{EG}_{[a-1,\infty)}\alpha]\!]) & \text{if } a > 0 \text{ and } b = \infty,\\ [\![\alpha]\!] \cap pre([\![\mathrm{EG}_{[0,b-1)}\alpha]\!]) & \text{if } a = 0 \text{ and } b > 1,\\ [\![\mathrm{EG}\alpha]\!] & \text{if } a = 0 \text{ and } b = \infty,\\ [\![\alpha]\!] & \text{if } a = 0 \text{ and } b = 1. \end{cases}$$

- $$[\![\mathrm{E}(\alpha \mathrm{U}_{[a,b)}\beta)]\!] = \begin{cases} [\![\alpha]\!] \cap pre([\![\mathrm{E}(\alpha \mathrm{U}_{[a-1,b-1)}\beta)]\!]) & \text{if } a > 0 \text{ and } b > 1, \\ [\![\alpha]\!] \cap pre([\![\mathrm{E}(\alpha \mathrm{U}_{[a-1,\infty)}\beta)]\!]) & \text{if } a > 0 \text{ and } b = \infty, \\ [\![\beta]\!] \cup ([\![\alpha]\!] \cap pre([\![\mathrm{E}(\alpha \mathrm{U}_{[0,b-1)}\beta]\!])) & \text{if } a = 0 \text{ and } b > 1, \\ [\![\mathrm{E}(\alpha \mathrm{U}\beta)]\!] & \text{if } a = 0 \text{ and } b = \infty, \\ [\![\beta]\!] & \text{if } a = 0 \text{ and } b = 1. \end{cases}$$

Algorithm 1. The algorithm for computing $[\![\mathrm{EG}_{[a,b)}\alpha]\!]$

```
1: X := S_T; X_p := ∅; i := (b−1)−a
2: while X ≠ X_p and i > 0 do
3:     X_p := X;   X := [[α]] ∩ pre(X);   i := i−1
4: end while
5: i := a; X_p := ∅
6: while X ≠ X_p and i > 0 do
7:     X_p := X;   X := pre(X);   i := i−1
8: end while
9: return X
```

Algorithm 2. The algorithm for computing $[\![\mathrm{E}(\alpha \mathrm{U}_{[a,b)}\beta)]\!]$

```
1: X := ∅; X_p := S_T; i := (b−1)−a
2: while X ≠ X_p and i > 0 do
3:     X_p := X;   X := [[β]] ∪ ([[α]] ∩ pre(X));   i := i−1
4: end while
5: i := a; X_p := S_T
6: while X ≠ X_p and i > 0 do
7:     X_p := X;   X := [[α]] ∩ pre(X);   i := i−1
8: end while
9: return X
```

Algorithms 1 and 2 are similar to those implemented in NuSMV (`http://nusmv.fbk.eu/`) supporting RTCTL model checking; Note that the algorithms only support intervals of the form $[a,b)$. In the case of $[a,\infty)$, the first *while* loops (Lines 2–6) that appear in both algorithms should not depend on i. To perform BMC, we use Algorithm 3, which depends on a notion of a submodel defined as follows:

Definition 2. *Let $M = (\iota, S, T, \mathcal{V})$ be a model and $U \subseteq S$ be a subset of the set of states such that $\iota \in U$. The* submodel *induced by U is a tuple $M|_U = (\iota, U, T', \mathcal{V})$, where $T' = T \cap (U \times Act \times U)$.*

Given a model M and an RTECTLK formula α, Algorithm 3 checks if α is valid in the model, i.e., if $M, \iota \models \alpha$. For any $X \subseteq S$ by $X_{\leadsto} = \{s' \in S \mid (\exists s \in X)(\exists a \in Act)\ (s,a,s') \in T\}$ we mean the set of all immediate successors of the states in X. The algorithm starts with the set *Reach* of reachable states that initially contains only the state ι (Line 1). In each iteration of the loop, the formula α is verified against the current submodel induced by *Reach* (Line 3), and the set *Reach* is extended with the immediate successors (Line 6). The algorithm operates on submodels $M|_{Reach}$ generated from the set *Reach*, and

Algorithm 3. The algorithm for checking α in the model $M = (\iota, S, T, \mathcal{V})$

1: $Reach := \{\iota\}$, $Reach_p := \emptyset$
2: **while** $Reach \neq Reach_p$ **do**
3:   **if** $\iota \in [\![M|_{Reach}, \alpha]\!]$ **then**
4:     **return** *true*
5:   **end if**
6:   $Reach_p := Reach$; $Reach := Reach \cup Reach_{\rightsquigarrow}$
7: **end while**
8: **return** *false*

checks if ι is in the set of states where α holds. The computation continues until the fixed point is reached. If α is valid in any submodel induced by the set *Reach*, then the the algorithm terminates and returns *true* (Line 4). If we reach a fixed point in calculating the reachable state space, and if α is not valid in any of the generated submodels then the algorithm terminates with *false*. We note that if the fixed point is reached, then $M|_{Reach}, \iota \models \alpha$ is the same as checking $M, \iota \models \alpha$. The correctness of the results obtained by Algorithm 3 is summarised by the following theorem:

Theorem 1. *Let $M = (\iota, S, T, \mathcal{V})$ be a model, and φ an RTECTLK formula. Then, $M, \iota \models \varphi$ iff there exists $S' \subseteq S$ such that $\iota \in S'$, and $M|_{S'}, \iota \models \varphi$.*

Proof. Follows easily by induction on the length of the formula.

4 Experimental Results

Both the presented in the paper BMC algorithms have been implemented in C++. The BDD-based method uses the CUDD library for ROBDD operations while the SAT-based technique uses CryptoMiniSat for testing satisfiability. We consider two standard scalable benchmarks which we use to evaluate the performance of our SAT- and BDD-based BMC algorithms, as well as the BDD-based BMC algorithm of the tool MCMAS for the verification of several properties expressed in RTECTLK and ECTLK. The memory used is given in MB and the time in seconds. For the tests we have used a computer equipped with two Intel Xeon 2.00GHz processors and 4 GB of RAM, running Ubuntu Linux with kernel version 2.6.35-23-server, and we have set the timeout to 3600 seconds.

Generic Pipeline Paradigm (GPP) consists of three parts: Producer producing data, Consumer receiving data, and a chain of n intermediate Nodes that transmit data produced by Producer to Consumer. The evaluation of our both BMC algorithms for RTECTLK with respect to the GPP system has been done by means of the following RTECTLK specifications:

$\varphi_1 = \mathrm{EF}\,(Send \wedge \mathrm{EG}_{[a,\infty)}\overline{\mathrm{K}}_C\overline{\mathrm{K}}_P Rec)$, where $a = 2n+1$, $n \geq 1$;
$\varphi_2 = \mathrm{EF}\overline{\mathrm{K}}_P(Send \wedge \mathrm{EF}_{[0,4)} Rec)$.

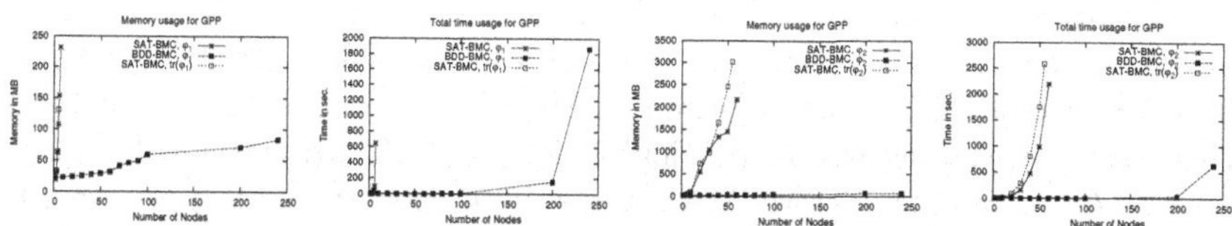

Fig. 1 Performance evaluation of GPP for RTCTLK fromulae φ_i $(i = 1,2)$ and equivalent ECTLK formulae $tr(\varphi_i)$ $(i = 1,2)$. The methods implemented in VerICS.

The evaluation with respect to the GPP system of our SAT-BMC algorithm for ECTLK and BDD-based BMC algorithm for ECTLK implemented in MCMAS has been done by means of the following ECTLK specifications: $tr(\varphi_1) = \mathrm{EF}(Send \wedge tr(\mathrm{EG}_{[a,\infty)}\overline{\mathrm{K}}_C\overline{\mathrm{K}}_P Rec))$, where $a = 2n+1, n \geq 1$; $tr(\varphi_2) = \mathrm{EF}\overline{\mathrm{K}}_P(Send \wedge (Received \vee \mathrm{EX}(Rec \vee \mathrm{EX}(Rec \vee \mathrm{EX}Rec))))$.

A train controller system (TC) consists of n trains (for $n \geqslant 2$) and a controller. Each train uses its own circular track for travelling in one direction. At particular part of the track, all trains have to pass through a tunnel that can only accommodate a single train. There are signalling lights on both sides of the tunnel, which can be either red or green. The trains notify the controller, which controls the access to the tunnel, to request access to the tunnel or when they leave the tunnel.

Let n be the number of considered trains. The evaluation of our both BMC algorithms for RTECTLK with respect to the TC system has been done by means of the following properties:

$$\varphi_1 = \mathrm{EF}(\overline{\mathrm{K}}_{T_1}(InTunnel_1 \wedge \mathrm{EG}_{[1,\infty)}(\neg InTunnel_1)))$$
$$\varphi_2 = \mathrm{EF}(InTunnel_1 \wedge \overline{\mathrm{K}}_{T_1}(\mathrm{EG}_{[1,n+2)}(\textstyle\bigwedge_{i=1}^{n}(\neg InTunnel_i))))$$

The evaluation with respect to the TC system of our SAT-BMC algorithm for ECTLK and BDD-based BMC algorithm for ECTLK implemented in MCMAS has been done by means of the following ECTLK specifications:

$$tr(\varphi_1) = \mathrm{EF}(\overline{\mathrm{K}}_{T_1}(InTunnel_1 \wedge \mathrm{EXEG}(\neg InTunnel_1)))$$
$$tr(\varphi_2) = \mathrm{EF}(InTunnel_1 \wedge \overline{\mathrm{K}}_{T_1}(tr(\mathrm{EG}_{[1,n+2)}(\textstyle\bigwedge_{i=1}^{n}(\neg InTunnel_i)))))$$

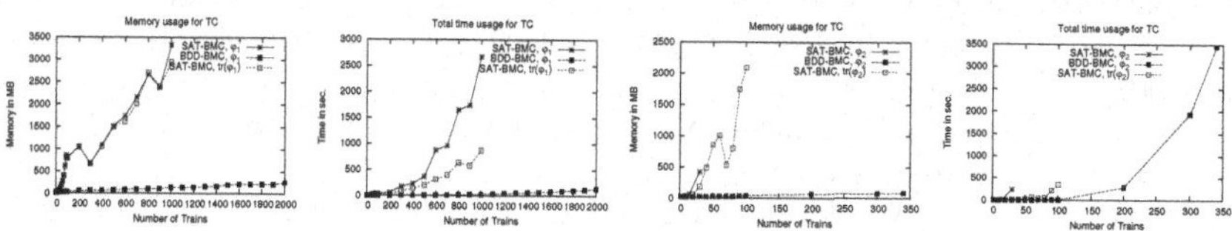

Fig. 2 Performance evaluation of the TC system for RTCTLK fromulae φ_i $(i = 1,2)$ and equivalent ECTLK formulae $tr(\varphi_i)$ $(i = 1,2)$. The methods implemented in VerICS.

Performance Evaluation. The comparison shows that for the GPP and the TC system our SAT-BMC is superior to BDD-BMC implemented in MCMAS for most of the tested formulae. Therefore we decided not to include these results in the figures. In particular, for the GPP system and ECTLK formulae $tr(\varphi_1)$ and $tr(\varphi_2)$ MCMAS was able to check 8 and 9 processes (it has consumed all the available memory), respectively, while our SAT-BMC was able to check 5 and 55 processes, respectively. For the TC system and ECTLK formulae $tr(\varphi_1)$ and $tr(\varphi_2)$ MCMAS was able to check 10 processes only (it has consumed all the available memory), while our SAT-BMC was able to check 1000 and 100 processes, respectively. When comparing our SAT- and BDD-based methods for (see Fig. 1–2), the BDD-based approach is clearly superior. The reason for the inferiority of MCMAS in most of our results most likely follows from the fact that MCMAS is designed for IS. Moreover, the version of MCMAS that we use has garbage collection and reordering of the variables disabled. Therefore it is possible that the results are not comparable.

5 Conclusions

We have proposed, implemented, and experimentally evaluated a BDD-BMC method for RTECTLK interpreted over IIS. Moreover, we have experimentally demonstrated that although RTECTLK model checking is not harder than ECTLK (RTECTLK, is the extension of ECTLK where temporal modalities may carry numerical time bounds), the BMC for RTECTLK has superior performance in comparison to the BMC for ECTLK on equivalent formulae. In our future work, we would like to test our methods with more complex and irregularly structured systems.

References

1. Campos, S.: A quantitative approach to the formal verification of real-time systems. Ph.D. thesis, School of Computer Science, Carnegie Mellon University, USA (1996)
2. Cohen, M., Dam, M., Lomuscio, A., Qu, H.: A data symmetry reduction technique for temporal-epistemic logic. In: Liu, Z., Ravn, A.P. (eds.) ATVA 2009. LNCS, vol. 5799, pp. 69–83. Springer, Heidelberg (2009)
3. Cohen, M., Dam, M., Lomuscio, A., Russo, F.: Abstraction in model checking multi-agent systems. In: Proc. of AAMAS 2009, vol. 2, pp. 945–952. IFAAMAS (2009)
4. Fagin, R., Halpern, J.Y., Moses, Y., Vardi, M.Y.: Reasoning about Knowledge. MIT Press, Cambridge (1995)
5. van der Hoek, W., Wooldridge, M.J.: Model checking knowledge and time. In: Bošnački, D., Leue, S. (eds.) SPIN 2002. LNCS, vol. 2318, pp. 95–111. Springer, Heidelberg (2002)
6. Jones, A.V., Lomuscio, A.: Distributed BDD-based BMC for the verification of multi-agent systems. Proc. of AAMAS 2010, 675–682. IFAAMAS (2010)

7. Kacprzak, M., Lomuscio, A., Lasica, T., Penczek, W., Szreter, M.: Verifying multi-agent systems via unbounded model checking. In: Hinchey, M.G., Rash, J.L., Truszkowski, W.F., Rouff, C.A. (eds.) FAABS 2004. LNCS (LNAI), vol. 3228, pp. 189–212. Springer, Heidelberg (2004)
8. Kacprzak, M., Nabialek, W., Niewiadomski, A., Penczek, W., Półrola, A., Szreter, M., Woźna, B., Zbrzezny, A.: VerICS 2007 - a model checker for knowledge and real-time. Fundamenta Informaticae 85(1-4), 313–328 (2008)
9. Lomuscio, A., Penczek, W., Qu, H.: Partial order reductions for model checking temporal-epistemic logics over interleaved multi-agent systems. Fundamenta Informaticae 101(1-2), 71–90 (2010)
10. Lomuscio, A., Penczek, W., Woźna, B.: Bounded model checking for knowledge and real time. Artificial Intelligence 171, 1011–1038 (2007)
11. Lomuscio, A., Qu, H., Raimondi, F.: McMAS: A model checker for the verification of multi-agent systems. In: Bouajjani, A., Maler, O. (eds.) CAV 2009. LNCS, vol. 5643, pp. 682–688. Springer, Heidelberg (2009)
12. Męski, A., Penczek, W., Szreter, M.: BDD-based Bounded Model Checking for LTLK over Two Variants of Interpreted Systems. Proc. of LAM 2012, 35–50 (2012)
13. van der Meyden, R., Shilov, N.: Model checking knowledge and time in systems with perfect recall. In: Pandu Rangan, C., Raman, V., Sarukkai, S. (eds.) FST TCS 1999. LNCS, vol. 1738, pp. 432–445. Springer, Heidelberg (1999)
14. Penczek, W., Lomuscio, A.: Verifying epistemic properties of multi-agent systems via bounded model checking. Fundamenta Informaticae 55(2), 167–185 (2003)
15. Penczek, W., Woźna-Szcześniak, B., Zbrzezny, A.: Towards SAT-based BMC for LTLK over interleaved interpreted systems. Fundamenta Informaticae 119(3-4), 373–392 (2012)
16. Raimondi, F., Lomuscio, A.: Automatic verification of multi-agent systems by model checking via OBDDs. Journal of Applied Logic 5(2), 235–251 (2005)
17. Wooldridge, M.: An introduction to multi-agent systems. John Wiley, England (2002)
18. Woźna-Szcześniak, B., Zbrzezny, A.: SAT-based BMC for Deontic Metric Temporal Logic and Deontic Interleaved Interpreted Systems. In: Baldoni, M., Dennis, L., Mascardi, V., Vasconcelos, W. (eds.) DALT 2012. LNCS, vol. 7784, pp. 170–189. Springer, Heidelberg (2013)
19. Woźna-Szcześniak, B.z., Zbrzezny, A., Zbrzezny, A.: The BMC method for the existential part of RTCTLK and interleaved interpreted systems. In: Antunes, L., Pinto, H.S. (eds.) EPIA 2011. LNCS (LNAI), vol. 7026, pp. 551–565. Springer, Heidelberg (2011)

Towards an AOSE: Game Development Methodology

Rula Al-Azawi[1], Aladdin Ayesh[1], Ian Kenny[1], and Khalfan Abdullah AL-Masruri[2]

[1] DMU, UK
rulaalaazawi@yahoo.com, aayesh@dmu.ac.uk, ikenny@dmu.ac.uk
[2] Higher College of Technology, Muscat, Oman
khalfan.almasruri@hct.edu.om

Abstract. Over the last decade, many methodologies for developing agent based systems have been developed, however no complete evaluation frameworks have been provided. Agent Oriented Software Engineering (AOSE) methodologies enhance the ability of software engineering to develop complex applications such as games; whilst it can be difficult for researchers to select an AOSE methodology suitable for a specific application. In this paper a new framework for evaluating different types of AOSE, such as qualitative and quantitative evaluations will be introduced. The framework assists researchers to select a preferable AOSE which could be used in a game development methodology. Furthermore the results from this evaluation framework can be used to determine the existing gaps in each methodology.

1 Introduction

Within the last few years, with the increase in complexity of projects associated with software engineering, many AOSE methodologies have been proposed for development[1]. Nowadays, intelligent agent-based systems are applied to many domains including robotics, networks, security, traffic control, and commerce. This paper will focus on the application of AOSE to game domains.

In the evaluation framework literature, the majority of the researchers [2][3] have used qualitative evaluation which has been dependent on the author's viewpoint or with regards to the questionnaire. Furthermore some of those evaluations have been presented by the same author of the methodology in question which makes those evaluations subjective.

In this paper a framework is presented which is based on two criteria: firstly some adopted from existing qualitative frameworks; secondly new quantitative evaluation frameworks. Furthermore the results derived from this particular framework guided us in the selection of the most suitable methodology for the game development domain.

The paper is structured in the following manner: Section 2 presents methodologies in AOSE and explains why we have selected MaSE and Tropos for the purpose of evaluation in our framework. Section 3 presents an overview of game development. Section 4 presents common evaluation frameworks introduced by other

S. Omatu et al. (Eds.): *Distrib. Computing & Artificial Intelligence,* AISC 217, pp. 493–501.
DOI: 10.1007/978-3-319-00551-5_59

authors. Section 5 presents our own framework where both qualitative and quantitative evaluations are explained in detail. In section 6, the critical evaluation of our own framework is presented and is compared with the results of other authors. Furthermore an analysis of the weaknesses in AOSE methodologies is also presented. Section 7 contains the conclusion followed by future works in Section 8.

2 Choosing AOSE Methodologies

There are several methodologies in AOSE. Each one has its own life cycle. However, some of them are precise only for analysis and design such as Gaia, while others cover the complete life cycle such as Trops, MaSE and Prometheus.

In this paper we perform comparisons of these well-known methodologies. These methodologies were selected based on the following criteria: a) they were described in the most details and had a complete life cycle methodology. Most of the existing AOSE only focus on analysis and design; whilst MaSE, Prometheus and Tropos have a full life cycle. b) were influenced by software engineering root. c) were perceived as significant by the agent community [4].

Depending on these criteria we decided to take in account only the MaSE .[5] [6] and Tropos[7] [8] methodologies (recent references are available for more details) for evaluation in our own evaluation framework.

3 Overview of Game Development

Game development has evolved to have large projects employing hundreds of people and development time measured in years. Unlike most other software application domains, game development present unique challenges that stem from the multiple disciplines that contribute to games. Until now some game development companies are still using the waterfall methodology but with modifications. A major issue leveled against the game development industries is that most adopt a poor methodology for game creation[9].

The relationship between games and AOSE is clear given that software agent or intelligent agents are used as virtual players or actors in many computer games and simulations. The development process is very close to the process of game development[10].

The goal of evaluating AOSE methodologies is to find the most convincing methodology that could be adopted for game development with modifications.

4 Common Evaluation Frameworks

The majority of the existing work found in the literature is based on qualitative evaluations by using different techniques such as feature based analysis, surveys, case studies and field experimented. In [11]with regard to qualitative evaluation are dealt with four main criteria: Concepts and properties; Modeling techniques;

and Process and pragmatics. His evaluation technique is based on a feature based analysis. Each criterion contains different attributes whereby Yes or No have been used to represent the criteria in each of the methodologies. The criteria with the same definition have also been used in [12].

Another interesting evaluation of methodologies is provided by [13] in which he attempted to cover ten of the most important methodologies by using different criteria. Thus instead of using Yes or No as in the case of the previous author, 'H' has been used for high, 'L' for low and 'M' for medium.

Some of the authors like [14] have presented a quantitative evaluation which evaluated the complexity of diagrams for MESSAGE and Prometheus using case study methods to measure magnitude and diversity and to determine the final complexity. Increasing magnitude and diversity increases the complexity of the model.

Another effort has been performed in [2] through measuring the complexity of the methodology using different attributes. Lower software complexity provides advantages such as lower development and maintenance time and cost, less functional errors and increased re-usability. Therefore, this is commonly used in software metrics research for predicting software qualities based on complexity metrics . By studying evaluation frameworks which are proposed until now, it seems that: a) there is not any appropriate framework for evaluating methodologies. b) there are frameworks which mostly based on feature- based analysis or simple case study methods. c) there are only small amount of frameworks based on quantitative approach.

Therefore presenting a proper framework containing both quantitative and qualitative methods and using feature based analysis, survey and case study methods could be really helpful.

5 Game Development Methodology Framework (GDMF)

The quantitative evaluation is an important part of the evaluation process because it is based on a fixed result for comparison purposes. The majority of the previous research has focused on the qualitative approach to enable a comparison to be made between methodologies.

Some difficulties were encountered during the literature review regarding the quantitative evaluation: a) the majority of the evaluations such as [14] [15] compared two methodologies by using a specific case study to determine their results. b) there were no standard attributes or metrics which had been used for evaluation.

To evaluate our framework, the following criteria will be used:

- Select the common criteria. Regarding qualitative evaluation, we decided to adopt [11] because his evaluation precision covered qualitative criteria and used feature based analysis methods and [13] since his evaluation is based on survey methods. Regarding quantitative evaluation criteria, criteria are divided into three sub-criteria: First by transferring the existing qualitative attribute values to the quantitative numbers; second, by dealing with Meta model metrics and use case evaluation methods, and finally by dealing with conversion of existing diagrams such as use case to numerical results.

- Transfer the qualitative attributes into quantitative values, then convert those values by using a proposed common scale for each metric as shown in the following:
  - Yes-No To 0-1 and the common scale 0-10
  - None-Low- Medium- High To 0-1-2-3 and the common scale 0-3-7-10

5.1 *Converting the Qualitative Results to Quantitative Evaluations*

In this section, we adopted [11] and [13] by converting their criteria to numerical results to facilitate the comparison between MaSE and Tropos.

Table 1 Comparison of MaSE and Tropos using criteria adopted from [11]

Criteria	Sub-Criteria	Tropos	MaSE	Comment
Concept and Properties	Autonomy	10	10	
	Mental Mechanism	10*	10	achieve goals and soft goals
	Reactivity	10	10	
	Pro-activeness.	10	10	
	Adaptation	10	0*	Needs to add iteration
	Concurrency	10	10	
	Agent interaction	10	0*	Needs to add agent interaction
	Collaboration	10	10	
	Teamwork	10	10	
	Agent-oriented	10	10	
Modeling techniques	Expressiveness	10	10	
	Modularity	10	0	
	Refinement	10	10	
	Traceability	10	10	
	Accessibility	10	10	
Process	Life-cycle Coverage	10	10	
	Architecture Design	10	10	
	Implementation Tools	10	10	
	Deployment.	0	10	
	Management	0	0*	Needs to add management
	Requirement capture	10	10	
Pragmatics	Tools Available	0	10	
	Modeling Suitability	0	10	
	Domain Applicability	10	10	
	Scalability	10	10	

Table 2 Comparison regarding steps and usability of Tropos and MaSE adopted from [13]

Steps	Tropos	MaSE
Identify system goal	10	10
Identify system tasks/behaviors	10	10
Specify use case scenario	10	0
Identify roles	10	0
Identify agent classes	10	10
Model domain conceptualization	0	0
Specify acquaintance between agent classes	10	7
Define interaction protocol	10	10
Define content of exchange message	7	7
Specify agent architecture	7	0
Define agent mental	0	7
Define agent behavior interface	0	0
Specify system architecture	0	0
Specify organizational structure	0	10
Model MAS environment	0	10
Specify agent environment interaction mechanism	0	0
Specify agent inheritance	0	0
Instantiate agent classes	10	0
Specify instance agent deployment	10	0

Initially we represent the summarized lists which have been provided from [11] in each of the main four criteria (i.e. Concept and properties, Modeling techniques and Process and pragmatics), as shown in table 1.

From the criteria defined [13], we selected the criteria of steps and usability, however around twenty steps were used to compare Tropos and MaSE as shown in table 2.

5.2 *Evaluating the Methodologies by Meta-Model Metrics*

This part of our framework deals with Meta-Model diagrams, using meta-modeling techniques for defining the abstract syntax of MAS modeling languages (MLs) which is a common practice today. We used a set of metrics in order to measure the meta-models. These metrics helped to quantify two features of the language: specificity and availability.

- The availability metric as shown in equation 1 measures how appropriate an ML is to model a particular problem domain. A higher value is better in the domain problem.

  The ncmm indicates the number of necessary meta-model elements;nc indicates the number of necessary elements; mc indicates the number of missing concepts.

$$Availability = nccm \div (nccm + mc) \quad (1)$$

- The specificity metric as shown in equation 2 measures the percentage of the modeling concepts that are actually used for modeling a particular problem domain. If the value of this metric is low, it means there are many ML concepts which are not being used for modeling the problem domain [16].

$$Specificity = nccm \div cmm \quad (2)$$

The term cmm represents the number of all the concepts in the meta-model.
Table 3 has been created by [16] whereby he has used four use cases and has selected six methodologies to find the availability and specificity for the purposes of comparison.

Table 3 Availability and specificity to compare between them. of Tropos and MaSE.

	Tropos		MaSE	
Case study	Availability	Specificity	Availability	Specificity
Cinema	75	75	77.7	60.9
Request example	91.7	45.8	100.0	52.2
Delphi	72.2	66.782.4	82.4	60.9
Crisis Management	63.6	58.3	77.7	60.9
Total Average	75.8	61.5	84.5	58.7

5.3 *Evaluation of the Methodologies by Diagrams*

The majority of the AOSE methodologies were delivered in some phase diagrams or tables especially during the analysis and design phase, such as UML diagrams and agent diagrams. The important point to consider in this evaluation is that we worked within an abstract level of methodology, which presented difficulties in finding artifacts that can be qualified. The important point in this evaluation is that it was based on a case study and the results depend on the case study itself [17]. Therefore in some case studies, the MaSE may obtain higher results than those associated with Tropos and vice versa. According to results found in the paper of Basseda et al [17],which compared by diagrams the complexity of dependency of modules for three AOSE, MaSEs has greater complexity of dependency modules than those of MESSAGE and Prometheus. This means that MaSE has more dependencies between its models; thus, using MaSE needs more time and effort. Although MaSE is suggested in critical systems which not only detailed designs are not considered overhead, but also are essential and worthwhile.

6 Critical Evaluation

In this section, the proposed framework has been applied to MaSE and Tropos as two case studies to demonstrate how the framework could be used to evaluate the methodologies. There are slight changes between the results of comparison. Both [18] [4] have evaluated MaSE to be better than Tropos. Moreover when we made the suggested change in MaSE, this would increase efficacy of MaSE. In [19], the author compared two popular reference works [4]and [18]and used profile analysis which is a multivariate statistical method. The majority of the results were found to be similar to those of our own evaluation results.

In Section 5.1, from table 1 we calculated the means of Tropos =0.84 and we obtained the same results as with MaSE, in case we applied the suggested enhancement to MaSE shown in comment column of Table 1, the means of MaSE after enhancement is=0.96.

The means for MaSE is=1.7894 from Table 2 which was greater than Tropos=1.2631.

As we noticed from Table 3, MaSE obtained a higher percentage in availability than Tropos; but Tropos had a higher percentage in terms of specificity. Availability is a more important parameter than specificity in game development because game development implements the modules based on the priority of modules. When we calculated the total percentage of both specificity and availability, MaSE obtained 71.6 and Tropos obtained 68.65. Thus, MaSE was considered to be better than Tropos with regards to the final results from that metrics.

From the previous measurement, we found that MaSE in the majority of measurements used in our frameworks obtained higher scores than Tropos by using feature base analysis, survey and case study evaluation methods .

We noticed the following weaknesses in most AOSE methodologies:

- All AOSE methodologies lacked industrial strength tools and standards. In addition, they did not seem to deal with team work since project management and time planning were not considered in AOSE methodologies found in the literature.
- There was a weakness during the implementation phase.
- There were no standard metrics for used in each phase to evaluate the phase output, the complete system and the most effective methodology for an application.

7 Conclusion

This study focused on comparing different AOSE methodology from the point of view of the game development domain. The results of the experiment were summarized to select the MaSE as a methodology to be adopted in game development methodology as future work for the three reasons. Firstly, the MaSE from the previous evaluation was better than Tropos. Secondly, many references such as [20]have used MaSE in creating methodology for robotics which is similar to this particular area of research. Finally, MaSE has defined the goal in the first stage and every goal has to be associated with its role which is an important feature in game development.

8 Future Works

An interesting future piece of work would be to use MaSE with a software engineering methodology such as agile in game development methodologies following the addition of the following improvements to enhance the MaSE methodologies: Firstly, it would be necessary to add iterations to the methodology to solve any problems from the previous stage; or to add new goals or requirement to the system and to obtain module prototypes. Secondly, it would be necessary to utilize project management from software engineering. Finally, it would be necessary to put more attention on the implementation and testing phases.

References

1. Akbari, O.: A survey of agent-oriented software engineering paradigm: Towards its industrial acceptance. Journal of Computer Engineering Research 1, 14–28 (2010)
2. Basseda, R., Alinaghi, T., Ghoroghi, C.: A dependency based framework for the evaluation of agent oriented methodologies. In: IEEE International Conference on System of Systems Engineering, SoSE 2009, pp. 1–9 (June 2009)
3. Dam, K.: Evaluating and comparing agent-oriented software engineering methodologies. PhD thesis, School of Computer Science and Information Technology, RMIT University, Australia (2003)
4. Dam, K.H., Winikoff, M.: Comparing agent-oriented methodologies. In: Giorgini, P., Henderson-Sellers, B., Winikoff, M. (eds.) AOIS 2003. LNCS (LNAI), vol. 3030, pp. 78–93. Springer, Heidelberg (2004)
5. DeLoach, S.: Multiagent systems engineering of organization-based multiagent systems. In: SELMAS 2005: Proceedings of the 4th International Workshop on Software Engineering for Large-Scale Multi-Agent Systems, pp. 1–7. ACM, New York (2005)
6. DeLoach, S.: Analysis and Design using MaSE and agentTool. In: Midwest Artificial Intelligence and Cognitive Science, pp. 1–7. Miami University Press (2001)
7. Bresciani, P., Perini, A., Giorgini, P., Giunchiglia, F., Mylopoulos, J.: Tropos: An Agent-Oriented Software Development Methodology. Autonomous Agents and Multi-Agent Systems 8, 203–236 (2004)
8. Mouratidis, H.: Secure Tropos: An Agent Oriented Software Engineering Methodology for the Development of Health and Social Care Information Systems. International Journal of Computer Science and Security 3(3), 241–271 (2009)
9. Kanode, C., Haddad, H.: Software Engineering Challenges in Game Development. In: 2009 Sixth International Conference on Information Technology: New Generations, pp. 260–265. IEEE Computer Society (2009)
10. Gomez-Rodriguez, A., Gonzalez-Moreno, J.C., Ramos-Valcarcel, D., Vazquez-Lopez, L.: Modeling serious games using AOSE methodologies. In: 11th International Conference on Intelligent Systems Design and Applications (ISDA), pp. 53–58 (2011)
11. Lin, C.-E., Kavi, K.M., Sheldon, F.T., Potok, T.E.: A methodology to evaluate agent oriented software engineering techniques. In: 40th Annual Hawaii International Conference on System Sciences, HICSS 2007, Island of Hawaii, USA, pp. 1–20. IEEE Computer Society (2007)
12. Akbari, O., Faraahi, A.: Evaluation Framework for Agent-Oriented Methodologies. In: Proceedings of World Academy of Science, Engineering and Technology, WCSET, Paris, France, vol. 35, pp. 419–424 (2008)

13. Tran, Q., Graham, C.: Comparison of ten agent-oriented methodologies, ch. XI, p. 341. Idea Group Inc. (2005)
14. Saremi, A., Esmaeili, M., Rahnama, M.: Evaluation complexity problem in agent based software development methodology. In: Second International Conference on Industrial and Information Systems (ICIIS 2007), pp. 577–584 (August 2007)
15. Cernuzzi, L.: On the evaluation of agent oriented modeling methods. In: The OOPSLA Workshop on Agent-Oriented Methodologies, Seattle (2002)
16. García-Magariño, I., Gómez-Sanz, J.J., Fuentes-Fernández, R.: An Evaluation Framework for MAS Modeling Languages Based on Metamodel Metrics. In: Luck, M., Gomez-Sanz, J.J. (eds.) AOSE 2008. LNCS, vol. 5386, pp. 101–115. Springer, Heidelberg (2009)
17. Basseda, R., Taghiyareh, F., Alinaghi, T., Ghoroghi, C., Moallem, A.: A framework for estimation of complexity in agent oriented methodologies. In: IEEE/ACS International Conference on Computer Systems and Applications, AICCSA 2009, pp. 645–652 (May 2009)
18. Sturm, A., Shehory, O.: A framework for evaluating agent-oriented methodologies. In: Giorgini, P., Henderson-Sellers, B., Winikoff, M. (eds.) AOIS 2003. LNCS (LNAI), vol. 3030, pp. 94–109. Springer, Heidelberg (2004)
19. Cernuzzi, L.: Profile based comparative analysis for AOSE methodologies evaluation. In: SAC 2008 Proceedings of the 2008 ACM Symposium on Applied Computing, pp.60–65 (2008)
20. DeLoach, S., Matson, E.T., Li, Y.: Applying agent oriented software engineering to cooperative robotics. In: The 15th International FLAIRS Conference (FLAIRS 2002), Pensacola, Florida, pp. 391–396 (May 2002)

Periodic Chemotherapy Dose Schedule Optimization Using Genetic Algorithm

Nadia Alam[1], Munira Sultana[1], M.S. Alam[1], M.A. Al-Mamun[2], and M.A. Hossain[2]

[1] Department of Applied Physics, Electronics and Communication Engineering, University of Dhaka, Dhaka-1000, Bangladesh
{nadia14,munirasultana17}@gmail.com, msalam@univdhaka.edu

[2] Computational Intelligence Group, Faculty of Engineering and Environment, University of Northumbria at Newcastle, UK
{mohammed.al-mamun,alamgir.hossain}@northumbria.ac.uk

Abstract. This paper presents a design method for optimal cancer chemotherapy schedules using genetic algorithm (GA). The main objective of chemotherapy is to reduce the number of cancer cells or eradicate completely, if possible, after a predefined time with minimum toxic side effects which is difficult to achieve using conventional clinical methods due to narrow therapeutic indices of chemotherapy drugs. Three drug scheduling schemes are proposed where GA is used to optimize the doses and schedules by satisfying several treatment constraints. Finally, a clinically relevant dose scheme with periodic nature is proposed. Here Martin's model is used to test the designed treatment schedules and observe cell population, drug concentration and toxicity during the treatment. The number of cancer cells is found zero at the end of the treatment for all three cases with acceptable toxicity. So the proposed design method clearly shows effectiveness in planning chemotherapy schedules.

Keywords: Cancer Chemotherapy, Drug Scheduling, Mathematical Model, Optimization, Genetic Algorithm.

1 Introduction

Cancer is a class of diseases characterized by an imbalance in the mechanisms of cellular proliferation and apoptosis [1-2]. There are four major approaches to cancer treatment: surgery and radiotherapy as local treatments, chemotherapy and the use of biological agents (such as hormones, antibodies and growth factors). Traditional chemotherapeutic agents act by killing cells that divide rapidly, one of the main properties of cancer cells. Chemotherapy also harms other cells that divide rapidly: cells in the bone marrow, digestive tract, hair follicles. This causes common side-effects: myelosuppression (decreased production of blood cells, hence

S. Omatu et al. (Eds.): *Distrib. Computing & Artificial Intelligence*, AISC 217, pp. 503–511.
DOI: 10.1007/978-3-319-00551-5_60

also immunosuppression), mucostisitis (inflammation of the lining of the digestive tract), and alopecia (hair loss). Chemotherapy treatment schedule, defined as dose amount and frequency is needed to be conveniently chosen to reduce the number of cancer cells after a number of fixed treatment cycles with acceptable/minimum toxic side effects. Researchers have designed optimal drug schedules of cancer chemotherapy and developed mathematical models to predict tumor growth. Evolutionary algorithms have been employed to design the chemotherapy drug scheduling for cancer treatment [3-6]. Considering clinical limitations in maintaining continuous treatment and giving emphasis on clinical relevance and patient's comfort, this paper presents a design method of optimal cancer chemotherapy treatment schedules where genetic algorithm (GA) is used to optimize drug doses and intervals by minimizing treatment main objective (cancer cells) and satisfying other key objectives.

2 Mathematical Model, Design Objective and Constraints

Here we consider a mathematical model (Equations 1 –7), originally developed by Martin and Teo in [1] that accounts for a tumor proliferating in Gompertzian fashion along with therapeutic and toxicity effects of intravenous administration of drug.

$$\dot{C}(t)=D(t)-C(t) \tag{1}$$

$$\dot{N}(t)=\frac{1}{\tau_g}\left[\frac{\ln\left(\frac{\rho_g}{N_0}\right)}{\ln\left(\frac{\rho_g}{2N_0}\right)}\right]N(t)\ln\left(\frac{\rho_g}{N(t)}\right)-k_{eff}C_{eff}(t)N(t) \tag{2}$$

$$C_{eff}(t)=(C(t)-C_{th})H(C(t)-C_{th}) \tag{3}$$

$$H(C(t)-C_{th})=\begin{cases}1, & if\ C(t)\geq C_{th}\\ 0, & if\ C(t)<C_{th}\end{cases} \tag{4}$$

$$C(0)=C_0=0 \tag{5}$$

$$N(0)=N_0 \tag{6}$$

$$\dot{T}(t)=C(t)-\eta T(t) \tag{7}$$

Eqn 1 gives the pharmacokinetics of drug. The plasma drug concentration, $C(t)$ increases with intravenous infusions of the drug, $D(t)$, and decreases according to first–order elimination kinetics at a rate λ. Equation 2 gives the number of cancer cells proliferating in a Gompertzian fashion and the therapeutic effect of the drug on the tumor is represented by adding a negative bilinear kill term to the tumor growth equation. Here ρ_g is the asymptotic plateau population, N_0 is the initial number of tumor cells, τ_g is the first doubling time of the tumor during exponential growth [1]. The bilinear term is proportional to both the current size of the tumor, $N(t)$ and the effective drug plasma concentration, $C_{eff}(t)$ with constant of proportionality k_{eff} [1]. k_{eff} is called the fractional kill term per day of the drug. $C_{eff}(t)$ is the drug concentration above the minimum therapeutic concentration, C_{th}, as given in equation 3. Equation 4 is a Heaviside step function that implies drugs may not become effective until a therapeutic plasma concentration is reached (C_{th}). The initial drug concentration and number of cancer cells are given by C_0 and N_0, respectively. Equation 7 gives the toxicity level in body after infusion of drugs where is η a constant [8]. The values of λ, τ_g, ρ_g, N_0, k_{eff}, C_{th} and η are respectively considered to be 0.27 days^{-1}, 150 days, 10^{12}, 10^{10}, 2.7×10^{-2} days^{-1}[D]$^{-1}$, 10[D] and 0.4 days^{-1} [1],[8]. Here [D] is a unit of dose concentration/mass. The model is implemented in Matlab/Simulink environment and used in following sections.

Here we considered three types of toxicity constraints. The drug concentration in plasma should not exceed maximum allowable level C_{max} and the measurement of toxicity must not exceed an acute level T_{max} stated by following inequalities 8 and 9

$$C(t) \le C_{\max}; \qquad \forall t \in [0,t_f] \tag{8}$$

$$T(t) \le T_{\max}; \qquad \forall t \in [0,t_f] \tag{9}$$

Total exposure of drugs in plasma is commonly calculated by integrating drug plasma concentration over the treatment must not exceed a value C_{cum} as in 10

$$\int_0^{t_f} C(t)dt \le C_{cum} \tag{10}$$

The values of C_{max}, T_{max} and C_{cum} are taken to be 50[D], 100[D] and 4.1×10^3[D].days [7,8]. Finally, the efficacy constraint limits the number of cancer cells not to surpass the initial condition, N_0. Which gives

$$N(t) \le N_0; \qquad \forall t \in [0,t_f] \tag{11}$$

Here we have used the optimal control problem considered by Martin and Teo in [1] to minimize cancer cell no. at a final time. It can be expressed as:

$$MIN_{D(t)} N(tf) \; s.t.(1-11) \tag{12}$$

In words, we have to design a chemotherapy schedule for 1 year to minimize the final number of cancer cell. The drug concentration should range between 10 and 50 and cumulative plasma drug concentration at the end of the treatment should be lower than a value 4.1×10^3. Finally, the cancer cell number should never exceed 10^{12}.

3 Optimal Chemotherapy Schedule Using GA

Genetic Algorithm (GA) is a stochastic global search method that replicates the metaphor of natural biological evolution [8]. Selection, crossover and mutation are its main operators. The fundamental element processed is a string formed by concatenating sub-strings, each of which is a numeric coding of a parameter. Each string stands for a solution in the search space. Performance of each solution is assessed through an objective function imposed by the problem and used in the selection process to lead the search towards the best individual. Crossover can cause to swap the properties of any two chromosomes via random decision in the mating pool and provides a means to produce the desirable qualities. Mutation is a random alternation of a bit in the string to keep diversity in the population. Here we propose three drug scheduling schemes, all planned for 364 days and GA is employed to find doses and intervals throughout the period.

Dose Pattern 1: Variable Interval Variable Dose (VIVD): In VIVD scheme, chemotherapy treatment will be administered to patients only first two days of each week depending on decision variable. For each week, decision variable is encoded with one bit; '1' to indicate that a patient will receive treatment on that week and '0' to indicate rest week, i.e, no drugs will be administered on that week. Giving clinical relevance, same drug doses are administered to patient treatment for first two days of any treatment week and one variable is required for each week. So, two variables are defined for each week; one for dose and one for decision. For a year (364 days = 52 weeks) long treatment plan, 52×2=104 variables are required and GA is used find an optimum solution set.

Dose Pattern 2: Fixed Interval Variable Dose (FIVD): In FIVD, interval between two consecutive treatments is fixed throughout the whole treatment period. Drugs are administered to patients on first two days of every 4th week following a rest period of 26 days. For any treatment week, same doses are administered on first two days. So, only one control variable is required to define the dose level of any treatment week. For a total period of one year, treatments are given only in 52/4=13 weeks and a total of 13 variables are required in designing this dose pattern. Aiming clinical relevance and to meet treatment efficacy, a high dose, called bowl (dose level of 50[D]) is administered to a patient at the beginning of the treatment.

Dose Pattern 3: Periodic Dose: Like FIVD, drugs are administered on first two days of every 4th week followed by a rest period of 26 days in this case. Unlike dose pattern 2, in any treatment week, different drug doses are administered on first two days and similar doses are followed in subsequent treatment weeks. As a result only two control variables are required to design treatment schedules for a year.

Encoding Scheme and GA Optimization Process: To design optimum dose pattern 1(VIVD), the GA optimization process begins with a randomly generated population called chromosome of size 50 × 676 where 50 is the number of individuals and 676((52×12)+(52×1)) is the length of the chromosome structure for 104 control variables. First 52 parameters are encoded as 12 bits binary strings which will define drug doses for each week while the remaining 52 parameters are encoded as 1 bit to define decision variables, i.e., whether treatment will be given to a patient. First 52 binary strings are converted into real numbers within a range of 10 to 50. Using each individual (solution), a chemotherapy drug schedule is designed for 1 year as discussed earlier and used as input $D(t)$ to the tumor model stated in Section 2. The model is simulated and several important output parameters: number of cells, drug concentration and toxicity are measured. The number of tumor cells at the end of treatment is used as objective function in GA optimization process. Before calculating fitness function, each individual is checked for constraints. If any of the constraint is not satisfied, that individual is penalized by adding a big penalty value so that it will have less chance to be selected for following generations. Once individuals are evaluated, fit individuals are selected through selection process to form the mating pool [8]. Genetic operators such as crossover, mutation and reinsertion are applied to form the new population for the next generation [8]. The crossover rate and mutation rate are set as 0.8 and 0.01 respectively. The maximum number of generations is set to 50. It is noted that, binary-coded GA is preferred and used in this optimization/design procedure because half of the control variable (=52) are binary-type decision variables represented by only single bit. GA with aforementioned parameters is run several times on the model. Table 1 gives a summary of the simulation results for five different runs.

Table 1 A summary of the simulation results of different runs for dose pattern 1

Run	Drug Dose		Drug Concentration		Toxicity		No. Cell at end	Cell Reduction
	Max	Avg	Max	Avg	Max	Avg		
1	32	10.7	49.4	11.2	83.4	27.7	≈ 0	≈ 100%
2	32	10.5	49.9	11.1	83.3	27.8	≈ 0	≈ 100%
3	32	10.6	49.5	11.2	81.3	27.9	≈ 0	≈ 100%
4	32	10.7	49.8	11.2	84.7	27.7	≈ 0	≈ 100%
5	32	10.7	49.9	11.2	82.2	27.7	≈ 0	≈ 100%

Table 2 Results of all three dose patterns

Scheme	Drug Dose		DrugConcentration		Toxicity		No. Cell at End	Cell Reduction
	Max	Avg	Max	Avg	Max	Avg		
VIVD (Run-1)	32	10.7	49.4	11.2	83.4	27.7	≈ 0	≈ 100%
FIVD	50	32.6	49.6	8.8	69.9	21.7	≈ 0	≈ 100%
Periodic	50	34.6	49.3	9.3	74.4	23	≈ 0	≈ 100%

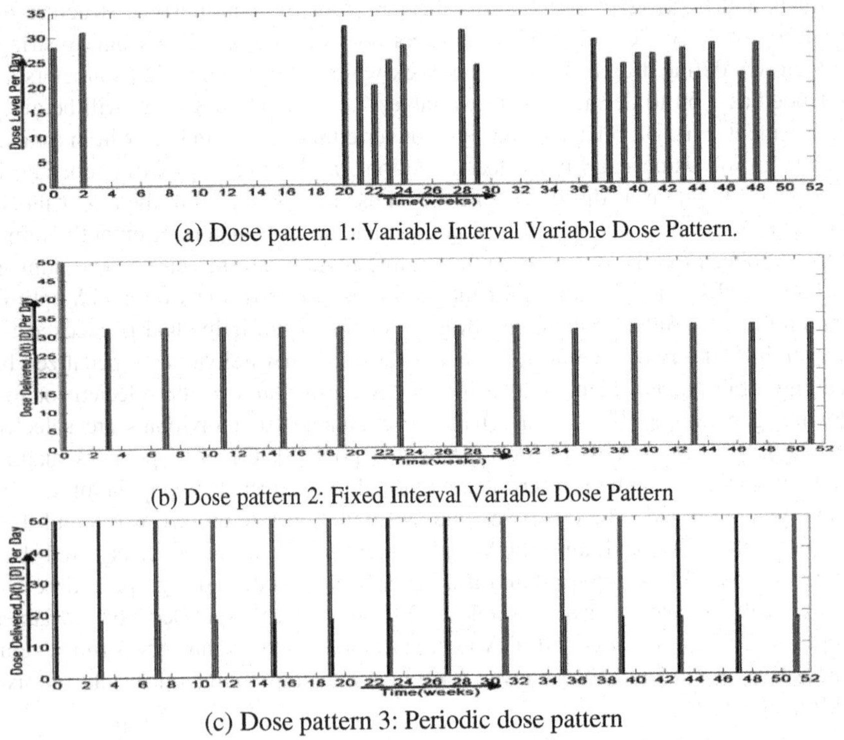

Fig. 1 Different dose patterns

The minimum values of objective function out of the five runs are approximately same (see Table 1). Moreover, other performance measures, recorded and presented in Table 1, are also very close to one another. All these show repeatability and consistency of GA optimization process and the overall design procedure as well. Similar GA optimization process but with different number of control variables are used to design optimum treatment schedules for dose pattern 2 and dose pattern 3. Due to space constraints, like dose pattern 1, results of all runs for dose pattern 2 and dose pattern 3 are not shown here. Instead, results of best runs for dose pattern 2 and dose pattern 3 are provided against dose pattern 1(run-1) in

Table 2. The drug schedules obtained with corresponding runs for three dose patterns are shown in Figure 1.

For dose pattern 3 (see Figure 1(c)), a pre-decided constant dose concentration/mass of 50[D] on the 1st day of the treatment is applied to guarantee efficacy constraint. For every 4th week the dosing can be clearly seen to be periodic. Each 4th week is dosed with 50[D] on the 1st day followed by 18[D] on the 2nd day. The remaining days of a week are kept as rest period for the patient to recover from toxic side effects, if occur or tend to occur. Fig. 2 shows how the number of tumor cells reduces and finally reaches to approximately zero. Toxicity profile due to the dosing is displayed in Fig. 3 where it can be seen that the limiting value Tmax is never exceeded during the whole course of treatment.

It is relevant to mention that dose pattern 1: VIVD, highly lacks clinical and logical acceptability. We can call the solution mathematically optimal. But the plan requires too much information to record. As the dosing is at random weeks, a patient may forget to visit the clinic for administration. On the other hand, dose pattern 2: FIVD, leads towards logical and clinical relevance with its fixed dosing days (1st 2 days every 4th week), same level of doses for the two consecutive days of a week, an efficient resting period and a defined starting dose of 50[D] to meet efficacy constraint. 13 weekly dose levels/parameters calculated by applying GA in this scheme vary from 30 to 32 (Fig. 1(b)).

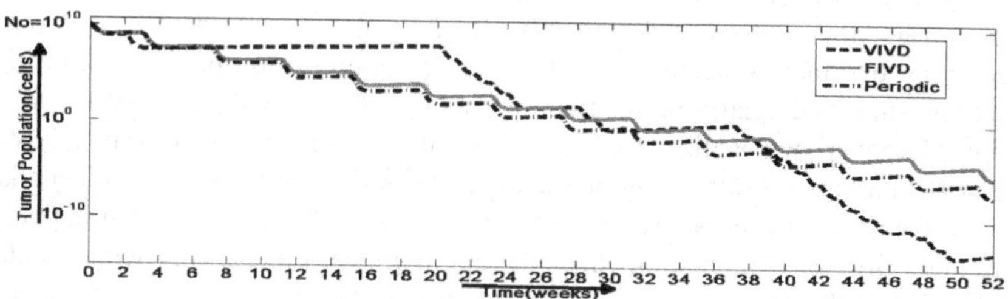

Fig. 2 Tumor cell population throughout the treatment period for all three dose patterns

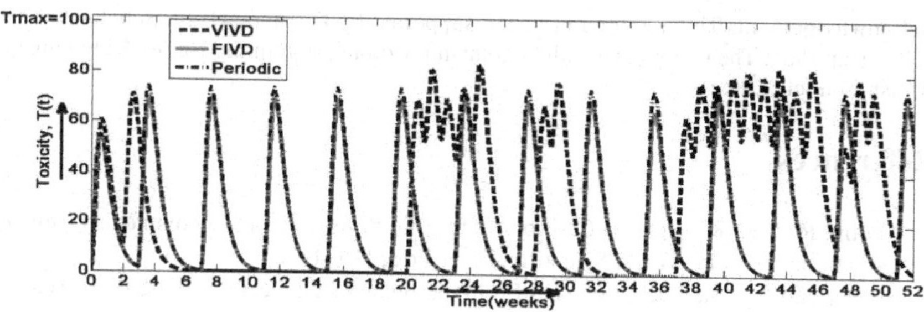

Fig. 3 Toxicity profile throughout the treatment period for all three dose patterns

If we compare FIVD and periodic pattern, both have same dosing days a, in the former, same level in two successive days are chosen, where in the latter we have used two separate levels. When FIVD is formulated, periodicity has not been imposed rather near periodicity is achieved from the 4th week in the optimal result (negligible variation of dose levels). So for more simplification and clinical relevance, in our final step we have approached periodicity and proposed periodic dose pattern. Having two separate levels is just an additional variation to get a lower value of the objective function. The main feature of periodic pattern is it is highly simplified, clinically appropriate, but still effective.

4 Conclusion

This paper presents a optimal cancer chemotherapy schedule mothod using GA. All of the dose plans have successfully converged resulting in 100% elimination of cancer tumors without violating treatment constraints. More importantly, the maximum toxicity levels during the whole period of treatment remain lower than the maximum allowable value as indicated earlier and suggested by other researchers [1],[3],[8]. So the periodic dose plan can be preferred for clinical implementation. The proposed technique clearly demonstrates that 'GA with reliable tumor growth model' can help oncologists/clinicians in planning optimum chemotherapy drug scheduling besides conventional methods. Although multi-objective evolutionary algorithms can design different drug doses by trading of different conflicting treatment objectives, single-objective optimization process with efficient encoding and clearly defined constraints can also provide very satisfactory result. Moreover, personalized treatment schedules can also be obtained by adjusting model parameters depending on the physiological condition of the patient and state of the tumor. Furthermore, the same design method can be extended in planning multi-drug or combination chemotherapy regimen. Future work will include verification of the proposed scheduling with clinical data and efforts are underway in that direction.

Acknowledgement. This research has been supported by Commonwealth Scholarship and Fellowship Plan, The Commonwealth Scholarship Commission in the United Kingdom and British Council, UK.

References

1. Martin, R., Teo, K.: Optimal Control of Drug Administration in Chemotherapy Tumour Growth, pp. 95–111. World Scientific, River Edge (1994)
2. Slingerland, M., Tannock, I.F.: Cell Proliferation and Cell Death. In: The Basic Science of Oncology, 3rd edn., pp. 134–165. McGraw-Hill (1998)
3. Algoul, S., Alam, M.S., Hossain, M.A., Majumder, M.A.A.: Multi-objective Optimal Chemotherapy Control Model for Cancer Treatment. Springer Journal on MBEC 49(1), 51–65 (2010)

4. Algoul, S., Alam, M.S., Sakib, K., Hossain, M.A., Majumder, M.A.A.: MOGA-based Multi-drug Optimization for Cancer Chemotherapy. In: Rocha, M.P., Rodríguez, J.M.C., Fdez-Riverola, F., Valencia, A. (eds.) PACBB 2011. AISC, vol. 93, pp. 133–140. Springer, Heidelberg (2011)
5. McCall, J., Petrovski, A., Shakya, A.: Evolutionary algorithms for cancer chemotherapy optimization. In: Fogel, G.B., Corne, D.W., Pan, Y. (eds.) Computational Intelligence in Bioinformatics, pp. 265–296. John Wiley & Sons, Inc., Hoboken (2007)
6. Kiran, K., Jayachandran, D., Lakshminarayanan, S.: Multi-objective Optimization of Cancer Immuno-Chemotherapy. In: Lim, C.T., Goh, J.C.H. (eds.) ICBME 2008. IFMBE Proceedings, vol. 23, pp. 1337–1340. Springer, Heidelberg (2008)
7. Harrold, J.M.: Model–based design of cancer chemotherapy treatment schedules. Ph.D. Thesis, University of Pittsburgh, USA (2005)
8. Algoul, S., Alam, M.S., Hossain, M.A., Majumder, M.A.A.: Feedback control of chemotherapy drug scheduling for phase specific cancer treatment. In: Proc. IEEE 5th International Conference on BIC-TA, Liverpool, United Kingdom (2010)
9. Chipperfield, A.J., Fleming, P.J., Pohlheim, H.: A Genetic Algorithm Toolbox for MATLAB. In: Proc. International Conference on Systems Engineering, Coventry, UK, pp. 200–207 (1994)

Mobile-Agent Based Delay-Tolerant Network Architecture for Non-critical Aeronautical Data Communications

Rubén Martínez-Vidal[1], Sergio Castillo-Pérez[1], Sergi Robles[1], Miguel Cordero[2], Antidio Viguria[2], and Nicolás Giuditta[3]

[1] Department of Information and Communication Engineering, Universitat Autònoma de Barcelona, Edifici Q. Bellaterra, Barcelona, Spain
rmartinez@deic.uab.cat

[2] Center for Advanced Aerospace Technologies (CATEC), Parque Tecnológico y Aeronáutico de Andalucía. La Rinconada, Sevilla, Spain

[3] Deimos Space, Ronda de Poniente 19, Tres Cantos, Madrid, Spain

Abstract. The future air transportation systems being developed under the NextGen (USA) and SESAR (European Commission) research initiatives will imply new levels of connectivity requirements between the concerned parties. Within this new aeronautical connectivity scenario, this paper proposes a new communication architecture for non-critical delay-tolerant communication (e.g. passenger data communications such as email and news services and non-critical telemetry data) based on mobile agents. Mobile agents carry both the user data and the routing algorithm used to decide the next hop in the path to the final destination. Therefore, mobile agents allow to increase the dynamism of the routing process. The proposed architecture constitutes an evolution of DTN (Delay Tolerant Networks), more flexible than the traditional layer-based approaches such as the Bundle Protocol and Licklider Transmission Protocol (LTP). This paper also presents the results obtained after network emulation and field experimentation of our proposed architecture.

Keywords: Mobile Agents, Delay Tolerant Networking, Aeronautical applications.

1 Introduction

The year 2020 will mark a turning point in the field of air traffic management (ATM) and control, as the next evolution in ATM is expected to become fully operational and deployed both in Europe and in North America. The Single European Sky (SES) initiative will unify the heterogeneous air traffic control models used by each country, transforming the European airspace into a single integrated air management scenario.

The correct deployment of these initiatives will depend on network connectivity able to support multiple applications at the same time. However, the mobility of the network elements and their heterogeneity (ranging from commercial passenger planes to small, general aviation users) introduce several obstacles in terms of maintaining constant connection, such as interoperability, disruption, security, delay

S. Omatu et al. (Eds.): *Distrib. Computing & Artificial Intelligence,* AISC 217, pp. 513–520.
DOI: 10.1007/978-3-319-00551-5_61

and so on. This renders the majority of current-day communication protocols invalid for their inclusion in such scenarios.

Hence, PROSES [1] was conceived to tackle these issues and offer a stable and reliable communication solution for the future ATM framework. The starting point of our research has been Delay Tolerant Networks, a scheme originally created [2, 3] to answer the difficulties of deep space communications, and currently being applied in other areas thanks to its possibilities. Our objective is to transfer this concept to ATM to design a protocol oriented to every possible participant.

PROSES introduces a new architecture, moving from prevailing store-carry-forward technologies used in previous DTN applications, to focus instead on the advantages offered by mobile agents (both data and routing algorithm carried together). This new architecture (see section 2) represents a low-cost solution suitable for diverse aeronautical applications which need diverse routing schemes (see section 3). A set of virtual emulations and field tests using radio-controlled aircraft is presented in sections 4-5. These experiments show the practical effects of the new architecture under different conditions, meant to reflect application scenarios derived from the future implementation of the new ATM framework.

2 Mobile-Agent Based DTN Architecture

Traditionally, computer networks have been designed using protocol stack schemes organised in several layers, with the theoretical OSI model being the most representative standard. The first ideas on DTN still use this scheme, posing it as an additional "overlay" layer providing the application layer with the support needed to withstand the problems derived from disrupted connectivity and delays in data exchange. However, the current proposals of DTN models, such as Bundle Protocol or LTP [8], do not effectively answer a number of issues, mainly regarding the routing problem and its underlying processes, preventing the formation of a distinct DTN standard.

PROSES introduces a layer-less DTN conception, doing away with the hierarchical model of tiers and instead proposing a model based on several modules interacting without rigid interactions. This novel communication model for DTN networks requires a basic communication service beneath, the so called Network Abstraction. This abstraction defines a set of minimum specifications in order to maintain an effective connection between the nodes.

Taking into account our previous assertion, the mobile code - and the mobile agents in particular - is a clear way to provide a solution of a multi-routing protocol scenario like in the DTN networks. PROSES is sustained through the use of mobile agents, software entities moving from host to host in a network, and able to carry data and code which can be autonomously run, such as routing schemes. In this sense, our proposal for the PROSES environment is to use the JADE framework [4] as the basis of our infrastructure. In fact, a modified implementation of the Inter-Platform Mobility Service (IPMS) module [5] of the JADE framework has been made in order to overcome the main problems of the DTN routing. This gives us the chance of introducing as many routing protocols as applications we want to deploy.

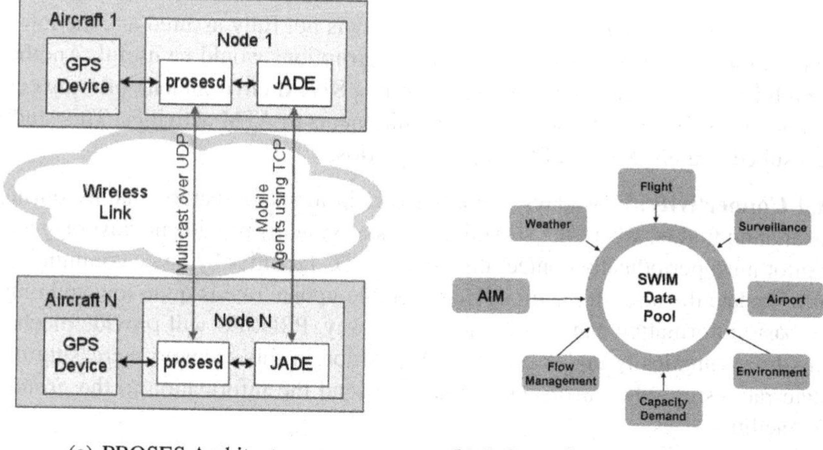

(a) PROSES Architecture

(b) Information sources present in SWIM

Fig. 1 Mobile-agent based DTN architecture

The PROSES architecture (see figure 1(a)) proposes the use of JADE mobile agents as carriers of data messages. Mobile agents allow the construction of active messages composed by data and routing algorithms. PROSES agents implement forward decision algorithms which use information about the network nodes to decide which node will be the next hop in the path towards the final destination of the data message. This routing information is gathered by an independent software daemon called "*prosesd*" running on each node and is stored in a memory array accessible by the JADE platform running in that node. Each *prosesd* daemon basically reads the current node's position from a GPS device attached to the node and sends multicast messages announcing its presence and position information to the neighbour nodes, which will store the received information in the routing arrays so it can be accessed by the PROSES agents.

3 Application Scenarios

One of the main features that characterize SESAR (SES ATM Research) is a massive increase in data communications between the participants in European ATM. The system that will provide the new networking functionalities is called SWIM (System Wide Information Management [6]). This infrastructure will integrate distributed and/or geographically sparse services, like third party services (e.g. weather information services), monitoring systems, by means of the different communication links (air-ground datalinks, satellite, grounded networks...), acting as an "intranet" for ATM users and services (see figure 1(b) [7]).

PROSES has identified a number of scenarios where it is possible to improve the connectivity and interoperability of SWIM using the DTN approach, as an intermediate step towards its full deployment or as a complementary system to augment its planned capabilities. In order to select the PROSES scenarios, the project team has focused on the situations where connectivity is not fully assured and the introduction of a new system able to withstand interruptions would be useful. Another research line is to act as a secondary system in a SWIM environment, offering certain low-cost services with non-critical requirements to ATM activities. This study has resulted in the selection of two main scenarios:

Total Connectivity: Nowadays, there are certain airspace sectors, mainly during transatlantic flights, where the surveillance radar systems provide no coverage and the pilot must periodically contact the ground ATC facilities by voice communication to update the aircraft position. However, the update rate is quite low and only very basic information can be exchanged this way. PROSES will provide mechanisms to automatically update the aircraft position in this scenario increasing the update rate using other airplanes as relays to send the information to the ground ATC facilities.

Non-critical Data Exchange: PROSES is aimed to support the interconnected scenario provided in SESAR by offering users a set of basic data communication services (e.g. e-mail, SMS) using air-ground communication capabilities without the need of costly, additional equipment onboard.

4 Emulation and Field Experimentation

The implementation of the proposed architecture was first tested and validated in a software emulated network environment. A commercial software network emulation and simulation called EXata was used for this purpose. Using its emulation mode, real applications can be run in virtual emulated networks which operate in real time. The emulated network behaves exactly as a real network allowing SIL (Software-in-the-Loop) tests to be done. This mode was used for testing real nodes running JADE platforms and prosesd daemons over an emulated wireless network. The SIL tests seek to evaluate the feasibility and suitability of using mobile agents as carriers of user data in DTN scenarios.

Following the emulation tests a set of field tests was performed. The field tests pursued a two-fold objective. Firstly, to validate some of the figures found out in the emulations. Secondly, to check the feasibility of deploying the architecture in a real scenario. These field tests evaluate if both the paradigm and the equipment are able to perform their task with an acceptable level of performance.

The field work was performed in an aerodrome near Seville at the end of 2011, and it took one day for equipment integration and two days for flying tests. The communication scenario was formed by three nodes: one on ground node installed inside a van (see figure 2(a)) and two aerial nodes installed onboard a RC fixed-wing aircraft and a RC helicopter (shown in figure 2(b)). Each node carried a mini computer, a GPS receiver and a WiFi transmitter. Agents where launched from the static node located in the van which was as well used as the monitoring node.

The experiments are divided into several categories based on node connectivity and routing criteria. The emulation experiments were to be similar to those of the

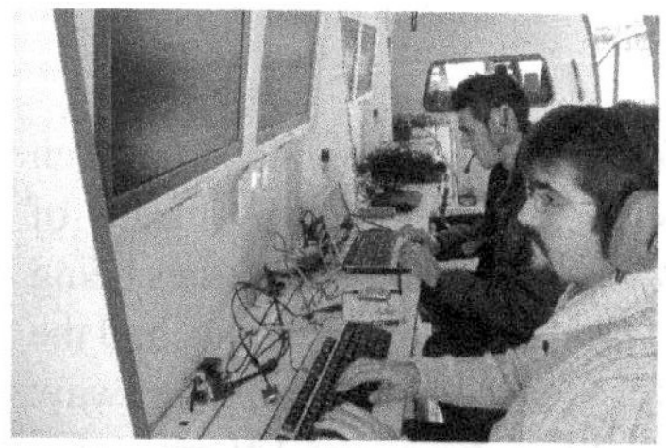

(a) Monitoring node

(b) Helicopter platform

Fig. 2 Nodes used in field experiments

field work. Only three nodes have been used in both cases as only two aerial vehicles and a ground station were available.

In each experiment all nodes are initially located close to each other on the center of the aerodrome runway. At the start, the aerial vehicles take off in opposite directions until they reach a separation of 300 meters. Then, they make a turn around and go back to the center of the runway, then again they separate themselves 300 meters in the opposite direction. In each experiment, the two aerial nodes moved using different mobility patterns but always keeping this motion of meeting and separating. For simplification purposes, from now on the nodes will be called *A*, *B*, and *C*. *A* corresponds to the static ground station, *B* and *C* are the mobile nodes (helicopter and aircraft, respectively).

First Experiment: Intermittent Connectivity with routing constrained to predefined itinerary. A single agent is launched from the static node with a predefined route (A-B-C-A). Hence, if the node defined as the next hop is not in range, that agent will wait until the node is close enough for the agent to jump. The path has been defined in a way that assures that the agent will traverse all the nodes in the network and then come back to the origin.

Second Experiment: Intermittent Connectivity with Dynamic Routing. Two simultaneous agents are used with a routing decision function that varies its criterion depending on the situation. This experiment seeks to analyze the coexistence of multiple routing schemes in a single network, by offering the same options to multiple agents and observe them taking different routing decisions.

Third Experiment: Intermittent Connectivity with Opportunistic Routing. Only two nodes (the static one and a mobile node) have been used. Several agents are launched simultaneously from the static node. The routing decision function makes the agent jump to whichever node is in range. This experiment seeks to observe the performance when the routing has no restrictions.

5 Results

This section contains the main conclusions made from the results obtained for each one of the experiments. The behavioural data has been represented by means of chronograms. The chronograms show the state of connectivity and the actions being performed by the several nodes of the experiment. The abscissae show time, and the state or action is represented in the ordinates. Possible states are numbered this way: **[0]** disconnected, **[1]** connected, **[2]** performing neighbor discovery, **[3]** executing routing code, **[4]** performing migration.

In the connected state, the several neighbors that are connected are listed. The location of the agent is shown by a crossed circle; the dotted arrows depict exchange of neighbor discovery messages between platforms; and finally, the solid lines show the exchange of migration data. For each chronogram we picked the most common behavior or the one that better shows the purpose of the experiment. Then we proceed to give a detailed description of such concrete case.

First Experiment : Predefined Itinerary. This experiment uses three nodes: A,B, and C. The agent starts its execution in node A, its routing function has been set to follow the predefined itinerary A-B-C-A. When the experiment starts, A performs neighbor discovery, and locates node C but not B, the routing code is executed and the agent decides to stay in the platform because B is not available. A periodically performs neighbor discovery until the mobility of nodes brings B in range. Then the routing code is executed again, B is chosen as the destination and the agent migrates. Once in B a neighbor discovery is performed again, both A and C are in range, so the agent immediately jumps to C. Once in C the discovery is repeated and A is in range, so the agent jumps there, finally it returns to A and the agent finishes its execution. This process can be seen in figure 3(c), which is a chronogram that follows the conventions explained at the beginning of this section.

Second Experiment: Dynamic Routing. This scenario uses three nodes: A,B,C. A remains static, B and C are mobile nodes. Two agents are used, using dynamic routing criteria, they start their execution in node A. First, A performs neighbor discovery, it finds connectivity with B and C, then both agents are notified, the routing code for each agent is executed, the first one chooses to stay in the platform while the other one chooses to migrate to C. Platform A performs neighbor discovery once again and notifies the agent. This time, it migrates to B. Simultaneously, the agent residing in node C is executed and selects to migrate to A. Node B becomes disconnected from the other two carrying one of the agents. The other two nodes exchange the remaining agent with some intermittent disconnections between themselves. Those agents do not have a predefined final destination, so the exchange continues until the experiment finishes. This process can be seen in figure 3(a).

Third Experiment: Opportunistic Routing. This experiment uses two nodes A,B, A remains static, B is a mobile node. We use several agents with a routing policy that decides to migrate every time it has the chance. Several agents have been consecutively sent in this experiment. The agents start their execution in platform A, they decide to migrate as soon as platform B is discovered. After this, the agents keep

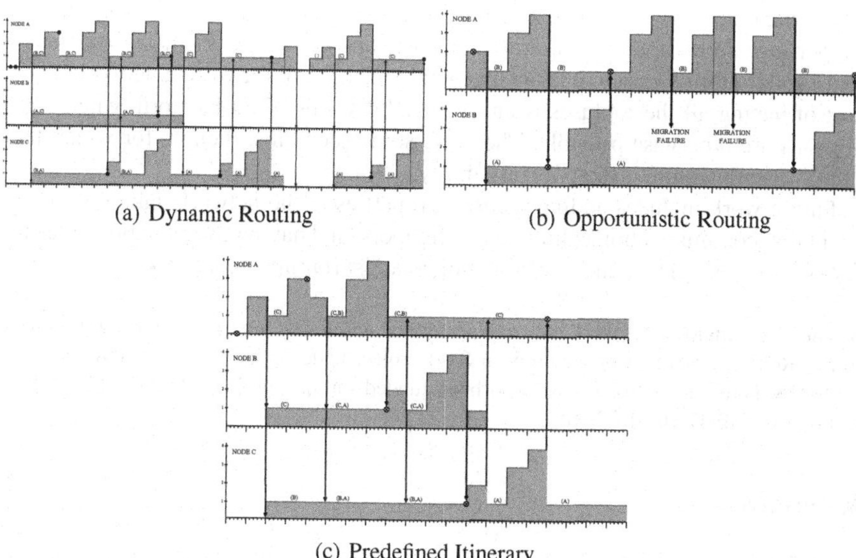

(a) Dynamic Routing

(b) Opportunistic Routing

(c) Predefined Itinerary

Fig. 3 Behavior Chronograms

going back and forth between the two platforms, several interruptions occur due to loss of connectivity when the mobile node goes out of range. The mutual exchange continues until the experiment is finished, see figure 3(b).

6 Conclusions and Future Work

This paper introduces a novel network architecture based on a new approach to Delay Tolerant Networking, using mobile agents as dynamic messages. The resulting communication network sought to take advantage of the new communication landscape, and represents an evolution of current networks in order to adapt to challenged scenarios.

The presented communication model is quite different from traditional communication networks, for delay, bandwidth, and throughput are second to data delivery. In scenarios were there is not a contemporaneous path between the communicating ends, most network solutions are just outstripped. Other networks can deal with these constraints, but the novelty of the proposal in this paper dwells in the possibility of simultaneously having different messages using different routing algorithms by using dynamical processes. Furthermore, the scheme also allows to introduce new routing algorithms, or removing them, without uploading new information to node-routers. It is the message itself which carries its own routing algorithm.

The system has been successfully designed and implemented. Emulations were run in laboratory to verify the feasibility of the proposal. The system was subsequently uploaded aboard unmanned drones to perform test flights to validate the results and check feasibility. Albeit the proposed network has a number of advantages over other networks, such as being resistant to long connectivity disruptions or supporting several applications simultaneously and yet having the best expediency routing algorithm for each, it is not a general replacement for current networks. The cost of having all these pluses is an overhead that might not be worth when connections are otherwise possible. There are some scenarios, though, for which the PROSES architecture is the only possibility.

Future work includes adding security properties to the network, taking into account that communication is limited to neighbors, and having dynamic processes in nodes for exchanging context information to assist routing algorithms.

Acknowledgement. This work has been performed within the context of the PROSES (network PROtocol for the Single European Sky) project under grant AVANZA TSI-020100-2009-115. Parts of this work have also been funded through project TIN2010-15764 and grant (UAB PIF 472-01-1/E2010).

References

1. PROSES, network PROtocol for the Single European Sky, Spanish Ministerio de Indústria, Turismo y Comercio project, ref. TSI-020100-2009-115
2. Cerf, V., et al.: Delay-Tolerant Network Architecture, IETF RFC 4838, informational (April 2007)
3. Warthman, R.: Delay-Tolerant Networks (DTNs) - A Tutorial. Warthman Associates (2003)
4. Bellifermine, F., Poggi, A., Rimassa, G.: JADE - A FIPA-compliant Agent Framework. In: Proceedings of the Practical Applications of Intelligent Agents (1999)
5. Cucurull, J., Marti, R., Navarro, G., Robles, S., Overeinder, B., Borrell, J.: Agent Mobility Architecture based on IEEE-FIPA Standards. Computer Communications 32(4), 712–729 (2009)
6. SESAR Joint Undertaking, WP 14 - SWIM Technical Architecture, Statement of Work (2008)
7. Wilson, S.: Information and Service Models - SWIM Technical Briefing. In: ATC Global 2011, Amsterdam (March 2011)
8. Farrell, S., Cahill, V.: Delay-and Disruption-Tolerant Networking. Artech House (2006)

A Preliminary Study on Early Diagnosis of Illnesses Based on Activity Disturbances

Silvia González[1], José R. Villar[2], Javier Sedano[1], and Camelia Chira[1]

[1] Instituto Tecnológico de Castilla y León (ITCL), Spain
{silvia.gonzalez,javier.sedano,camelia.chira}@itcl.es
[2] University of Oviedo, Spain
villarjose@uniovi.es

Abstract. Recently, the human stroke is gathering the focus as one of the diseases with higher mortality and social impact. In addition, it has a long-term treatment and high rehabilitation costs. Therefore, the early diagnosis of stroke can take advantage in avoiding the stroke itself or highly reducing its effects. Up to our knowledge, no previous study on stroke early diagnosis has been published in the literature. This study deals with the early detection of the stroke based on accelerometers and mobile devices. First, a discussion on the problem is presented and the design of the approach is outlined. In a first stage, it is necessary to determine what is the subject doing at any moment; thus, human activity recognition is performed. Afterwards, once the current activity is estimated, the detection of anomalous movements is proposed. Nevertheless, as there is no data available to learn the problem, a realistic proposal for simulating stroke episodes is presented, which lead us to draw the conclusions.

Keywords: Ambient Assisted Living, Stroke Early Diagnosis, Genetic Algorithms, Fuzzy State Machines.

1 Introduction

Stroke is a cerebrovascular abnormal blood circulation upheaval that disrupts the normal function of a region in the brain. Stroke is considered one of the main causes for mortality and a leading cause of disability throughout the world [1, 2]. In case of disability, the rehabilitation is a long term process that does not guarantee reaching the patient normality.

Physical activity helps in the reduction of the stroke risks [3], but it is not enough. Current methods for diagnosing the stroke are based on the observation of the patient's behaviour when passing through a simple positional test in which the patient is told to keep his/her arms in parallel or to walk through a line [3, 1]. It has been found that the sooner the diagnosis the lower the stroke consequences [1]. Consequently, detecting the abnormality in the patient movements represents a challenge in the diagnosing of the stroke.

S. Omatu et al. (Eds.): *Distrib. Computing & Artificial Intelligence,* AISC 217, pp. 521–527.
DOI: 10.1007/978-3-319-00551-5_62

It is well known that human beings have very characteristic movements. Actually, the most well-known and characterized movement is walking, which has been completely defined [4]. The main part of the studies have been carried out in video based motion analysis system [5, 1]. Recently, the use of accelerometer-based methods are reaching the focus due to several reasons [6, 7]: the reduction of the diagnosing costs, the possibility of further data analysis and the ubiquity among others. The possibility for automatically detecting several human movements and its validity are pushing the balance towards the use of accelerometers as main sensors for human motion detection and classification [8].

This study is focused on using accelerometers for the early detection of the stroke, which would provide the method for reducing its effects in the society and in the individual: an early detection would reduce the patient recovery period. An analysis of the problem and the solution design lead towards a two stage method: the former stage includes the Human Activity Recognition (HAR), while the latter is the responsible of detecting abnormalities in the activities and the alarm generation. This is a preliminary study, so we adapt a HAR technique and then evaluate the data from the accelerometer and propose how the diagnosing of the stroke can be afforded for one of the analyzed activities.

The organization of the manuscript is as follows. Next section deals with the Stroke characterization and the design of the solution. In addition, this section will briefly introduce the hypothesis for stroke episode detection. Afterwards, the HAR method is outlined and its adaptation to the specific stroke problem is detailed. In Sect. 4, the full experimentation of the HAR is included, as well as a further discussion on detecting abnormal movement episodes when the subject is walking. Subsequently, the conclusions drawn and the future work are depicted.

2 Analysing the Stroke Early Detection

The most common and well-known symptoms of stroke episode include mainly one side only numbness of the face, arm, hand or leg; dizziness and trouble walking; collapse; loss of balance and coordination; and speech disorders among others. As stated before, the first neurological exam test carried out is requesting the subject to carry on some specific movements that will clearly be far from normal.

The characterization of the human movement, specially when walking, has been well documented in the literature [4]. Figure 1 shows the schematic representation of a normal walk extracted from the previously referred study. As known, stroke highly influences the way patients move [9], particularly the patient's gait is clearly affected. Due to this reason, and also because walking is one of the most sensible parameters in human dependence, the study of the gait and the patient kinematics has been discussed so far in the literature [10, 9].

As stated in [10], there are several abnormalities in stroke patients, as the reduction of both the length of the stride and the cadence of the limbs, or the "abnormal movements of the upper extremity, the trunk, the pelvis, and the lower extremity on the unaffected side in an effort to compensate for the decreased velocity on the hemiplegic side." Moreover,Manaf et al [1] found an evidence of disability in the

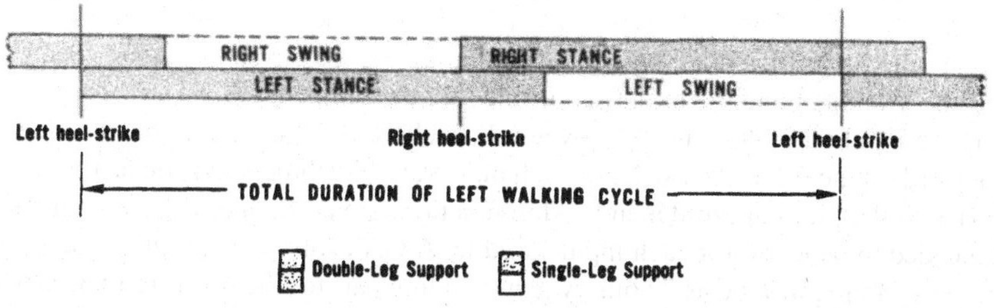

Fig. 1 Schematic representation of left walking cycle extracted from [4].

gait when turning, and a difference in the pattern of the number of steps, time, gait velocity, step width and step and stride length has been observed.

Interesting readers should notice that there is no information about the very moment of a stroke episode. The only available information is, on the one hand, the experiences with stroke survivors and, on the other hand, the neurological exam test itself. Therefore, there are several hypothesis of how a patient behaves when an ictus episode occurs. The most relevant hypothesis are: i) the individual collapses and no movements are make for a relative extent, and ii) the individual behaves anomalously. The former hypothesis means that the subject keeps quiet for far too long, while the second considers the findings from [10, 1].

In this study, the problem of detecting stroke episodes is faced, so both hypothesis should be analyzed. Consequently, one of the very first issues to obtain is to detect the current human activity the subject is doing. HAR needs the number of states to be defined before the learning phase starts. Thus, for the stroke early diagnosis we need establishing which are the most relevant states or activities to detect. As mentioned before, accelerometer-based HAR methods have been found successful, thus they will be used in this study.

Besides, the abnormalities expected to occur in stroke episodes are those mentioned before: i) there is an extended period of time without any movement, and ii) the dissimilarities of the movements with respect to those considered normal get increased. For identifying both kind of events a set of sensors is also need. Accelerometers are proposed for this aim because i) they are already deployed for HAR and ii) the lower the number of issues the subjects need to wear the lower the costs and the better the subjects would accept to participate. In order to measure such differences, we propose placing a pair of sensors in both wrists. This configuration would help in detecting abnormalities in all of the possible states. Unfortunately, this design decision forces having HAR using those sensors instead of the most common solution of placing only one accelerometer on the hip or in the central part of the body.

3 Human Activity Recognition

In this study, we adopt the solution proposed by [11], where a Genetic Algorithm (GA) evolves the Fuzzy Finite State Machine for detecting human activity

$GFFSM = \{Q,U,f,Y,g\}$, learning both the rules and the partitions. From a predefined Finite State Machine, which is depicted in Fig. 2, the set of states {Seated, Upright, Walking} and the initial set of rules are determined.

Three input variables -the dorso-ventral acceleration, the amount of movement and the tilt of the body- are used. For each input variable, three linguistic labels ($n_i = 3, \forall i$) with Ruspini trapezoid membership functions are used, thus $n_i + 1$ parameters are needed to be learnt for each input variable. A GA evolves the partitions and the rules in a Michigan style as 72 binary genes coding part for the rules; 12 real-coded genes for the membership function parameters.

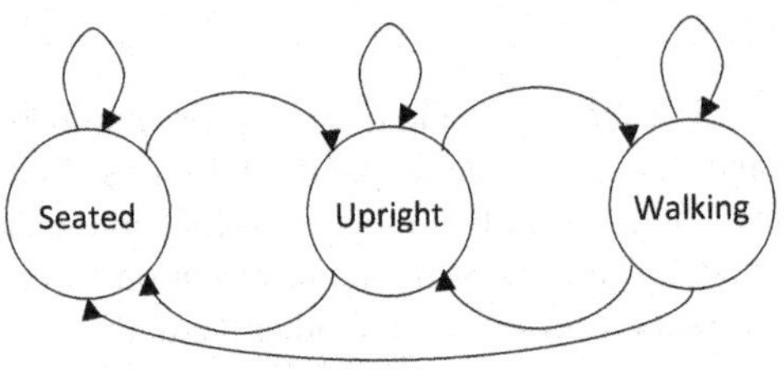

Fig. 2 The Fuzzy Finite State Machine proposed for this preliminary study

The fitness function is the mean absolute error (MAE), calculated as $MAE = \frac{1}{N}\frac{1}{T}\sum_{i=1}^{N}\sum_{j=0}^{T} abs(s_i[j] - s_i^*[j])$, where T is the number of examples in the data set, $s_i[t]$ and $s_i^*[t]$ are the degree of activation and the expected degree of activation, respectively, of state q_i at time $t = j$.

3.1 Adapting GFFSM Approach to a Different Sensor Placing

In this study, we consider the previous outlined method but with a different set of variables as far as we propose the use of accelerometers in both wrists instead of being in the hip. For the HAR, the dominant wrist acceleration data is used. Let a_i^x, a_i^y and a_i^z be the raw acceleration data, let g_i^x, g_i^y and g_i^z be the acceleration components due to the gravity G, extracted from the accelerometer raw data using low-pass filters [12]. Let b_i^x, b_i^y and b_i^z the body acceleration (BA) components, calculated as $a_i^{\{x,y,z\}} - g_i^{\{x,y,z\}}$ [12, 13, 14]. Then, this study proposes the use of a sliding window of size $w = 10$ samples and, for each one, the following input variables: i) *Signal Magnitude Area* (SMA), a well-known measure for discriminating between G and BA, which is calculated as $SMA_t = \frac{1}{w}\sum_{i=t+1}^{t+w}(|b_i^x| + |b_i^y| + |b_i^z|)$ [13, 14], ii) the sensor vibration $\Delta_t = \frac{1}{w}\sum_{i=t+1}^{t+w} \mid a_i^{x2} + b_i^{y2} + c_i^{z2} - g_i^2 \mid$ [13] and iii) the amount of movement, calculated as $\Delta_t = \sum_{v=\{x,y,z\}} \mid max_{i=t+1}^{t+w}(a_i^v) - min_{i=1}^{w}(a_i^v) \mid$ [11].

4 Experimentation in HAR

To test this prototype a well-known stroke patients rehabilitation test (for short, SRT) [5] will be carried out. Two bracelets will be given to a subject, each one with a tri-axial accelerometer with sampling frequency 16 Hz. Firstly, ten runs will be registered for a normal subject. All the data will be segmented and classified according to the activity the subject is owe to do. The data for these runs will be used for training and testing the HAR in a leave-one-folder-out manner, in order to obtain statistics results.

Table 1 shows the results obtained from this experimentation. The GA parameters employed in this study are those indicated in [11]: population size of 100 individuals in the population, crossover probability 0.8, the crossover α parameter is set to 0.3, the mutation probability 0.02, the maximum number of generations set to 200 and the maximum number with MAE unchanged fixed to 50.

From results it is shown that the current deployment of the GFSSM correctly classifies some of the activities while the walking activity recognition certainly performs worse. Comparing with the results presented in [11], we obtained remarkable similar MAE's results for walking. We have tested different GA parameters, but the obtained results are pretty close to those shown. Nevertheless, It is clear that the walking activity detection should be improved. Certainly, a PCA feature subset and different sliding window schema could help in this task.

Table 1 Obtained results for the HAR: mMAE and sMAE stand for mean and standard deviation of the MAE, respectively. On the right hand, the box plot of the right hand MAE over the leave-one-fold-out cross validation.

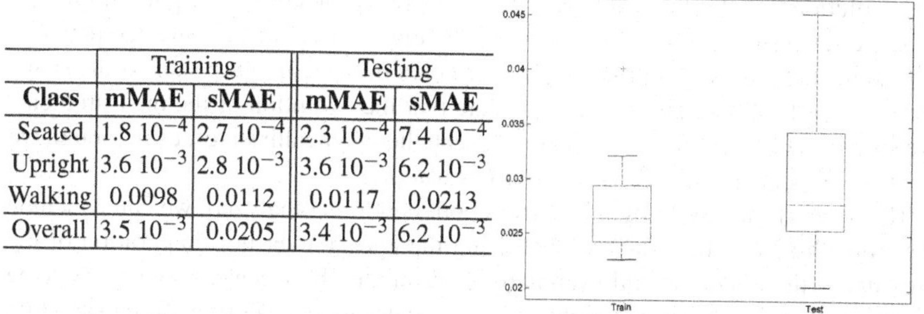

	Training		Testing	
Class	**mMAE**	**sMAE**	**mMAE**	**sMAE**
Seated	$1.8\ 10^{-4}$	$2.7\ 10^{-4}$	$2.3\ 10^{-4}$	$7.4\ 10^{-4}$
Upright	$3.6\ 10^{-3}$	$2.8\ 10^{-3}$	$3.6\ 10^{-3}$	$6.2\ 10^{-3}$
Walking	0.0098	0.0112	0.0117	0.0213
Overall	$3.5\ 10^{-3}$	0.0205	$3.4\ 10^{-3}$	$6.2\ 10^{-3}$

4.1 A Discussion on the Results and the Stroke Detection

As explained before, several hypothesis are addressed in analyzing what do occur when an ictus episode arises. One of such hypothesis is that, in certain activities, the subject suffering the episode tends to make the corresponding movements asymmetrically. Let us focus in the activity of walking, where the subject has more difficulties making the movements of the affected part of the body (upper and lower limbs) than in the non-affected part and, thus, the signals from the sensors should be rather different. To evaluate the performance of ictus episode we develop an SRT in which the subject simulates an stroke episode by carrying a weight in one hand [5]. Four runs with the weight plus four runs without it were registered and all the data segmented and classified according to the activity the subject is owe to do.

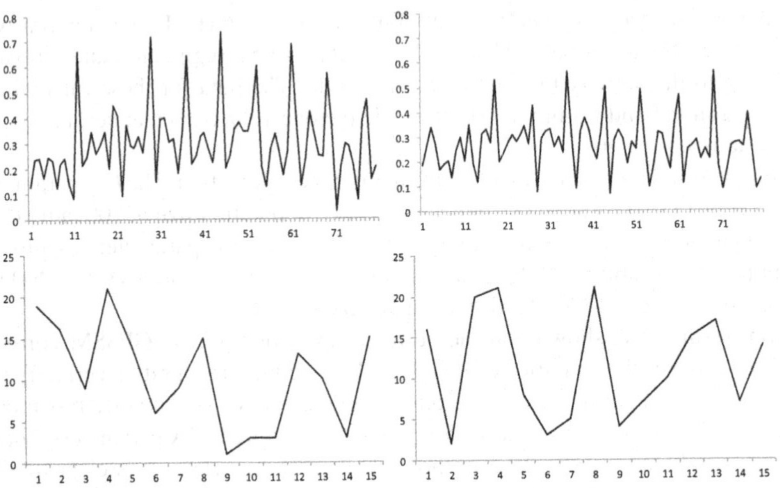

Fig. 3 Top: Evolution of the sum of the absolute values of the BA $|b_i^x| + |b_i^y| + |b_i^z|$ obtained from normal walking (left part) and walking with a weight (right part). Bottom: evolution of the SMA for normal behavior (left) and for an simulated ictus (right).

Fig. 3 shows the evolutions of two variables with and without the weight. Thought the evolutions of the left side seem different from those to the right, simple approaches as those presented in [15] weren't able to discriminate between patterns. The same happens with well-known pattern matching techniques, such as the shape index factor [16]. Clearly, we can not state that the evolution of the measurements are so different to easily discriminate between the two health states, even though we a priori know the subject current activity is walking.

However, if the BA components from both hands are analyzed, we can conclude that something can be done and that using the energy measure -integration of the area under the curve- would eventually discriminate between the two stages (see Figure 4). This is why we claim that if we can determine the current human activity, then specific and simple techniques would allow us to generate suitable ictus episode alarms. Provided the methods are kept simple, the solutions found would eventually be deployed in mobile devices.

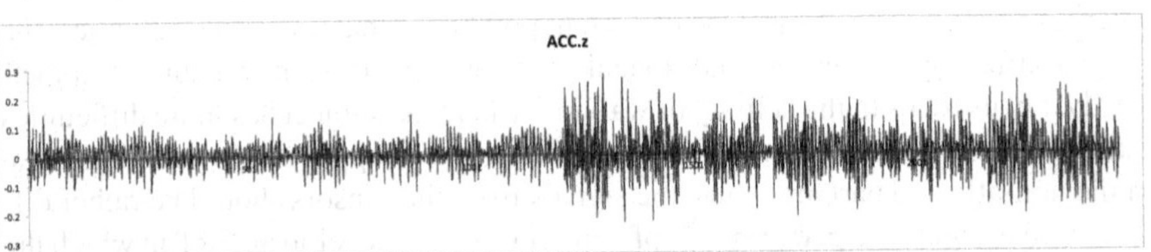

Fig. 4 SRT runs overall BA components: right hand b_i^z signal would eventually allow the discrimination of a stroke episode

5 Conclusions and Future Work

This study faces the problem of early ictus diagnosis, which is a very challenging task. This study proposes a two step method that firstly estimates the human activity and then analyses the data for detecting anomalous movements. This hypothesis has been analyzed for the walking activity. Results show that although the approach should be highly improved in both stages, the high rate of success encourage to continue this research line. Consequently, future work includes developing several different techniques for human activity recognition and a rather sophisticated approach for the ictus diagnosis.

Acknowledgement. This research has been partially supported through the projects of the Spanish Ministry of Science and Innovation PID 560300-2009-11 and TIN2011-24302, Junta de Castilla y Leon CCTT/10/BU/0002 and Agencia de Inversiones y Servicios de Castilla y Leon (record CCTT/10/LE/0001).

References

1. Manaf, H., Justine, M., Omar, M., Isa, K.A.M., Salleh, Z.: ISRN Rehabilitation (in press)
2. Murray, C.J., Lopez, A.: Lancet 349(9061), 1269 (1997)
3. Gallanagh, S., Quinn, T.J., Alexander, J., Walters, M.R.: ISRN Neurology (2011), doi:10.5402/2011/953818
4. Murray, M.P., Drought, A.B., Kory, R.C.: Journal of Bone and Joint Surgery 46(2), 335 (1964)
5. Hollands, K.L.: Whole body coordination during turning while walking in stroke survivors. Ph.D. thesis, School of Health and Population Sciences, University of Birmingham (2010)
6. Rand, D., Eng, J.J., Tang, P.F., Jeng, J.S., Hung, C.: Stroke 40, 163 (2009)
7. Roy, S.H., Cheng, M.S., Chang, S.S., Moore, J., Luca, G.D., Nawab, S.H., Luca, C.J.D.: IEEE Transactions on Neural System Rehabilitiation Engineering 17(6), 585 (2009)
8. Rothney, M.P., Schaefer, E.V., Neumann, M.M., Choi, L., Chen, K.Y.: Obesity 16(8), 1946 (2008)
9. Ahmed, M., Ahmed, S.: Annals of King Edward Medical University 14(4), 143 (2008)
10. de Quervain, I., Simon, S., Leurgans, S., Pease, W., McAllister, D.: The Journal of Bone and Joint Surgery. American Volume 78(10), 1506 (1996)
11. Ávarez-Álvarez, A., Triviño, G., Cordón, O.: In: IEEE 5th International Workshop on Genetic and Evolutionary Fuzzy Systems (GEFS), pp. 60–65 (2011)
12. Allen, F.R., Ambikairajah, E., Lovell, N.H., Celler, B.G.: Physiological Measurement 27, 935 (2006)
13. Wang, S., Yang, J., Chen, N., Chen, X., Zhang, Q.: In: Proceedings of the International Conference on Neural Networks and Brain ICNN&B 2005, vol. 2, pp. 1212–1217. IEEE Conference Publications (2005)
14. Yang, J.Y., Wang, J.S., Chen, Y.P.: Patter Recognition Letters 29, 2213 (2008)
15. Burchfield, T.R., Venkatesan, S.: In: Proceedings of the 1st ACM SIGMOBILE International Workshop on Systems and Networking Support for Healthcare and Assisted Living Environments HealthNet 2007, pp. 67–69 (2007)
16. Phan, S., Famili, F., Tang, Z., Pan, Y., Liu, Z., Ouyang, J., Lenferink, A., McCourt-O'Connor, M.: International Journal of Computer Mathematics 84(5), 585 (2007)

Towards an Adaptable Mobile Agents' Management System

Mohammad Bashir Uddin Khan, Ghada Abaza, and Peter Göhner

Institut für Automatisierungs- und Softwaretechnik,
Universität Stuttgart, Stuttgart, Germany
bashir_jubd@yahoo.com,
{ghada.abaza,peter.goehner}@ias.uni-stuttgart.de

Abstract. Mobile agent is the software paradigm that has been widely employed from comparatively small system to complex industrial system to realize their activities. The key features of a mobile agents' management system are the agent management, communication management, agent mobility and monitoring management. The real challenge arises when developing such a system from scratch for each application domain. Such a process is effort and time consuming. System developers have to spend more time and effort to realize these features for each application domain. If system developers were to be provided by a software system approach that can adapt the system key features for different application domains, the development time and effort will be reduced considerably. In this paper, a mobile agents' management system approach is developed that can be adapted by different application domains and is proved by implementing a case study traffic management system.

Keywords: Adaptation, Mobile agent, Multi agent system, Adaptable mobile agents' management system.

1 Introduction

During the last few years, the internet has grown considerably fast, increasing by that the available information and services online. To access these information and services, traditional distributed systems practice a static configuration of the environment where they execute. A set of hosts are communicated by enabling a set of physical links whose configuration is fixed and statically determined. Similarly the distributed applications that run on the hosts of the system are typically bound to such hosts for their whole life. Therefore, the topology is essentially assumed as fixed both at the physical and logical level of the system.

This view is being challenged by developing the mobile agent technology. It is the most prominent technology for developing distributed and open system applications [5]. It introduces the degree of mobility. In many distributed system applications mobile agents have revealed a significant degree of exploitation. Applications benefit form mobile agents are: E-commerce, telecommunication,

S. Omatu et al. (Eds.): *Distrib. Computing & Artificial Intelligence*, AISC 217, pp. 529–536.
DOI: 10.1007/978-3-319-00551-5_63

network services, network management, manufacturing, secure brokering, monitoring and notification, parallel processing, workflow application, personal assistant, etc.

The purpose of this work is to develop an adaptable mobile agents' management system approach using an existing mobile-agent platform. The main focus of the work is to develop and analyzed the key features of a mobile agents' management system and modeled each feature separately. This approach also introduces a module called 'Adaptation Module' that makes relationship between the configurations of these models. Finally it generates the agent stab code according to their configurations. So it will help the agent developers to find out optimal behaviours and functionalities whilst designing their system and the developers can save their development time and effort.

The organization of the paper is as follows: Section 2 reviews the related work in adaptive mobile agent and mobile agents' management systems. Section 3 explains the proposed approach. Section 4 describes the results and discussion. Finally Section 5 concludes this work and provides an insight to the future work.

2 Related Works

Complex systems are very difficult to understand and hence the primary requirements of such systems suffer from errors. Therefore, to handle such complex system, adaptability is the crucial property. To solve this problem scientific community works in the area of adaptive multi agent system [1][2]. Some research on adaptable mobile agent and multi agent systems are given below.

Thomas Ledoux and Bouraqadi-Saâdani proposed an approach for adaptability in mobile agent systems using reflection [6]. They considered runtime adaptability of agent mechanism. Puppeteer is a component based adaptation system in mobile environments and is proposed by Eyal de lara [7]. Nejla Amara-Hachmi proposed a framework for building adaptive mobile agents [8]. For adaptability it emphases on mobile agent fundamental services for example Own control, agent profile, mobility, messages center, actions and knowledge base.

Though the management of mobile agent is highly complex and cannot be dealt with one paper, our proposed approach has the ability to adapt itself with different application domains (in real world) with changing requirements. For adaptability it only focuses on the key features of mobile agents' management system. Existing agent framework 'Java Agent Development Framework' (JADE) [10] is used to develop our adaptive mobile agents' management system.

3 Description of the Proposed Approach

The proposed approach has two parts: The Application Domain and the System Domain. The Application Domain is divided into 'Real World' and 'Mapping Module'. The 'Real World' extracts the entities, their behaviours and relationships for an application and the 'Mapping Module' maps those to mobile agents. This mapping works as the foundation of the system domain.

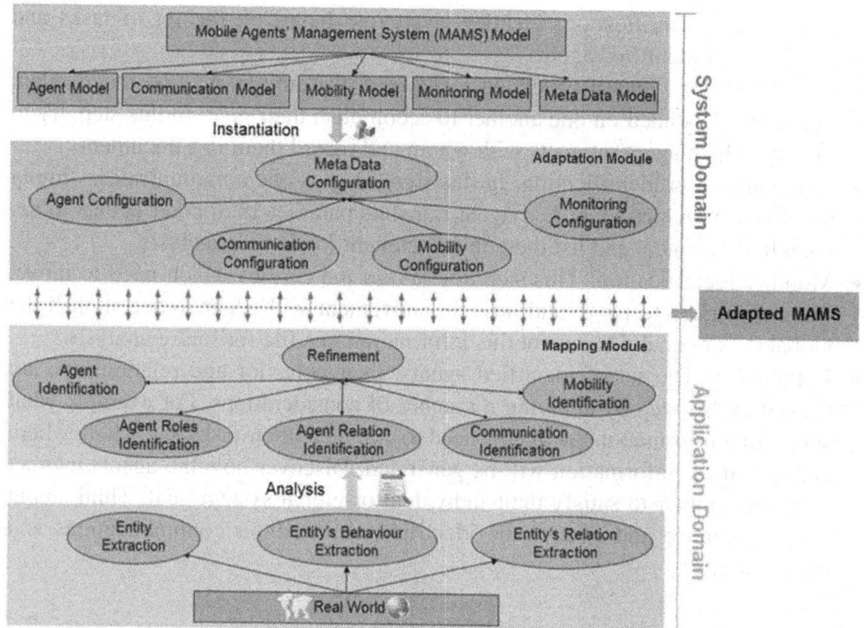

Fig. 1 An overview of the proposed approach: an adapted mobile agents' management system

The system domain contains the mobile agents' management system (MAMS) model and 'Adaptation Module'. MAMS provides several models for mobile agents' execution, communication, migration and monitoring. The necessary services to manage the mobile agents and to maintain the mobile agents' life cycle are provided by the selected mobile agent platform. The user of the system will configure the services and the properties according to the mobile agents found in mapping module. Finally the 'Adaptation Module' generates stab codes for each mobile agent. An overview of the proposed approach is shown in the Fig. 1

3.1 Application Domain

The user of the approach is responsible to accomplish this part. In this part the user will analyses the real world applications and maps them to the system domain (as shown in Fig. 1) by mapping module pursuing the following steps

- Agent Identification: This step identifies the agents of a specific application. The subsequent rules should be useful to identify agents from the real world. 1) Choose an agent for each user/device 2) Choose an agent for each resource (that includes legacy software). The agents need a vast knowledge about their behavior to accomplish their task.

- Agent Role Identification: Identify agent role based on the set of tasks and properties of identified agent.
- Agent Relations Identification: In real world, the entities may be independent or may be depended on one another to accomplish their goal. In this step, try to identify which agent interacts with whom and record them in a document.
- Communication Identification: In this step, identify the communication simply by observing which identified agent transfers parameters to other agents to accomplish their goal and list them in a document for further analysis.
- Mobility Identification: This step determines the entities which need to move, when they need to move and how the movement will occur to accomplish the global system goal. Document this information in a file for future analysis.
- Refinement: The set of identified agents, their behavior and relationships are refined in this step by applying a number of considerations. For example what supporting information the agents need to complete their tasks and when, where and how these information will be generated. Discover how the agents interact with one another to satisfy their individual or global system goal. Think about how to monitor and control the identified agents, their communications and their mobility.

3.2 System Domain

In system domain the user will configure the MAMS models and the adaptation module dynamically configures the instances of these models through the adaptation process.

Mobile Agent Management System Model (MAMS)

The details of MAMS models are described below

- Agent Model: The mobile agent has a set of properties and tasks. It must have the ability to complete a number of concurrent tasks in response to different external events [3]. The tasks may be simple (atomic task, cyclic task) or complex (sequential task, concurrent task; tasks performed in the state of a finite state machine) [9]. Therefore each agent is modeled with its set of tasks. In the proposed approach, the agents will be modeled according to their properties and behaviour. There are 2 types of properties: System and user properties. Therefore properties are configured as 1) System-configured properties, such as the agent identifier and 2) User-configured properties, such as the agents local names, agents class name, agents locations, agents roles etc (shown in Fig. 2 a). Every mobile agent has its own behaviour that is implemented by one or more tasks to accomplish its goal. The behaviour is configured as 1) Execution related details that include the white page services, yellow page services and the agent life cycle services. 2) User-configured tasks and services include the task type (simple or complex), the task execution manner (atomic, cyclic, sequential, concurrent etc.), the task execution location, the task source code, the service name, the service type, the service language and the DF agent description language.

- Communication Model: Agent communication is the most fundamental feature and is implemented according to the FIPA specifications [4]. Agents may communicate with identical or different platforms and this communication is followed by the Message Transport Protocol (MTP). In the proposed approach, the communication between agents is modeled into services and Agent Communication Language (ACL) message (shown in Fig. 2 b). The services are configured by the system and called the system-configured communication services. The ACL message is configured as 1) System-configured communication details: it provides message transport services for identical or different agent platforms. 2) System-configured message details: configures the envelop (message transport information) and the payload (encoding/decoding of message) part of FIPA compliant message whereas in 3) User-configured message details: configures the message content (static/dynamic) and message parameters (for example sender, receiver, communicative act, content language, reply with and reply by) of the message.

- Mobility Model: Mobile agents have the ability to suspend their execution in the source location, transfer their code, data and state to another location; and resume their execution in the remote location. Therefore, mobility can occur either on identical or between different platforms in a guided or an autonomous way. In both platforms mobile agent can transfer by migration or cloning process. In the proposed approach the mobility model is divided into two parts 1) System-implemented details: It provides the necessary mobility services to transfer the agent from the source to the destination location as much as the selected agent platform provides. 2) User-configured properties: The user of this approach also has the ability to configure mobility properties according to the involved platform (identical or different), mobility process (migration or cloning) and mobility invocation (guided or autonomous). Shown in Fig. 2 d.

- Monitoring Model: The monitoring model monitors and controls the existing mobile agents, their communications and mobility. It is modeled according to the display and control in our proposed approach (Fig. 2 c). Both display and control is related to the agent role (provided by the user of the approach). The display part of the monitoring model views agent, and agent communication information. The control part controls the agent, agent communication and agent mobility.

- Meta-Data Model: The Meta-data model controls the overall system as well as the other models of the system. It also provides the security of the system by applying proper authentication, authorization and performing administrative tasks. In the proposed approach the Meta-data is modeled according to the system meta-data and the system security (Fig. 2 e). 1) System Meta-Data: System meta-data is related to the number of agents in the overall system, number of hosts, number of containers, number of agent groups, number of agents in the container, number of agents in the agent group etc. 2) System-Security: The user will configure the system security to prevent the system from unnecessary incidents like attacks from malicious agents/ hosts and to provide needed fault tolerance strategy.

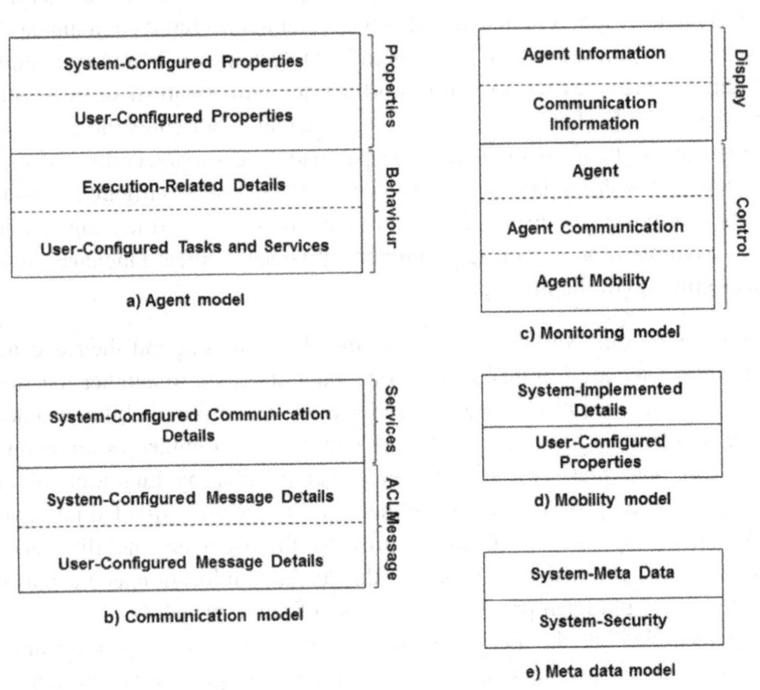

Fig. 2 Models of adaptive mobile agents' management system

Adaptation Module

The adaptation module takes the instance of each model, configures through their corresponding configuration and generates or updates the stab codes in java file. The main objective of this module is to adapt the mobile agents' management system model for the considered real world application described in application domain and the process is shown in Fig. 3.

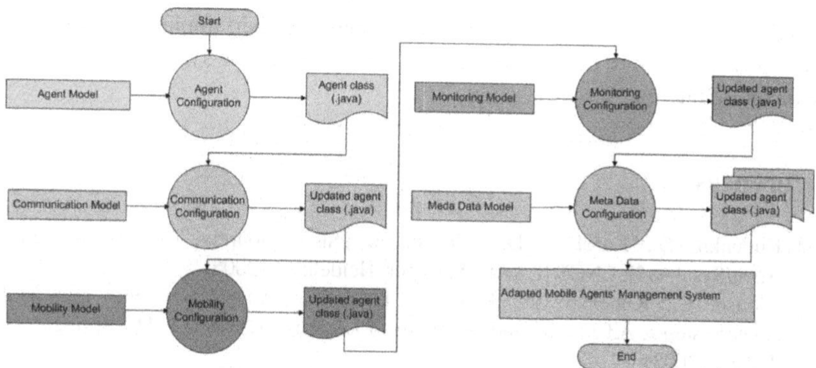

Fig. 3 Adaptation process

4 Results and Discussion

To prove the applicability of the proposed approach, "agent based traffic management system" is implemented as a case study. By analyzing the case study, several agents, their behaviours and functionalities are identified in 'Application Domain' part. After that these agents are mapped to the 'System Domain', configured them and finally the 'Adaptation Module' generates the necessary stab codes according to their configuration. The time, cost and effort to implement the case study using this approach is compared to the same case study same agent platform (JADE) that is built from scratch. As a result this approach reduces approximately 20% of the total cost and effort as well as time. The calculation is given below

- According to the COCOMO method, 1) software without interface to the existing system uses PM = 2.4 x $(KLOC)^{1.05}$ where PM stands for person month and KLOC means thousand source lines. 2) Software with a couple of interface uses PM = 3.0 x $(KLOC)^{1.12}$ and 3) software that has to fit in existing system uses PM = 3.6 x $(KLOC)^{1.20}$. By applying the third method in the case study that is developed from scratch uses 213 PM for 30 KLOC. Our approach reduces the line of code to 25 KLOC by generating different agent files and the cost reduces to 171 PM. As a result this approach reduced approximately 20% of the total cost and effort.

5 Conclusion and Future Work

For adaptable mobile agents' management system different key features such as agent model, communication model, mobility model, monitoring model and metadata model has been developed and analyzed. Each model is guided by a set of constraints for adaptability. Finally adaptation process is used to adapt different application domains in the proposed approach. Proposed approach may be extended by considering system security issues. Further implementation of the developed approach to mobile devices and movement of the system to the "Cloud" are of future interest.

References

[1] Kudenko, D., Kazakov, D., Alonso, E. (eds.): Adaptive Agents and Multi-AgentSystems II. LNCS (LNAI). Springer, Heidelberg (2005)
[2] Alonso, E., Kudenko, D., Kazakov, D. (eds.): Adaptive Agents and Multi-AgentSystems: Adaptation and Multi-Agent Learning. LNCS (LNAI). Springer, Heidelberg (2003)
[3] Bellifemine, F., Caire, G., Trucco, T. (eds.): JADE Programmer's Guide. Telecom Italia S.p.A. (2008)
[4] FIPA. FIPA agent message transport service specification (2002)
[5] Hacini, S., Cheribi, H., Boufaïda, Z.: Dynamic Adaptability using Reflexivity for Mobile Agent Protection. World Academy of Science, Engineering and Technology 23 (2008)
[6] Ledoux, T., Bouraqadi-Saâdani, N.: Adaptability in mobile agent systems using reflection. In: ECOOP (2000)
[7] de Lara, E., Wallach, D.S., Zwaenepoel, W.: Puppeteer:Component based adaptation for mobile computing. Proceedings of the 3rd USENIX Symposium on Internet Technologies and Systems, 159–170 (March 2001)
[8] Hachmi, N.A.: A Framework for Building Adaptive Mobile Agents. In: AAMAS 2005, Utrecht, NL, July 25-29 (2005)
[9] Bellifemine, F., Caire, G., Trucco, T.: JADE Programmer's Guide. Copyright (C) 2008 Telecom Italia S.p.A. (2008)
[10] Bellifemine, F., Poggi, A., Rimassa, G.: JADE - A FIPA-compliant AgentFramework. In: Proceedings of Practical Application of Intelligent Agents and Multi-Agents (PAAM 1999), London, UK, pp. 97–108 (April 1999)

Glove-Based Input for Reusing Everyday Objects as Interfaces in Smart Environments

Ionuţ-Alexandru Zaiţi and Ştefan-Gheorghe Pentiuc

University Ştefan cel Mare of Suceava, 13, Universitatii, Suceava, 720229, Romania
ionutzaiti@yahoo.com, pentiuc@eed.usv.ro

Abstract. Gestural interfaces can be naturally adopted for ambient intelligence applications given the fact that most interactions we have with our environment are performed through our hands. Human hands can be extremely expressive and can perform very precise actions which in turn could benefit ambient interactions and enhance user experience. However, today's common gesture-sensing technologies make heavily use of motion in the detriment of fine hand posture and finger movements. We propose in this work a glove-based input technique for inferring properties about grasped objects by using measurements of finger flexure alone to explore new interaction opportunities in intelligent environments.

1 Introduction

Hands play an essential part in the process of human communication while they support practically all the interactions we have within our environment [11]. Therefore, the interest of researchers and practitioners towards capturing and using hand posture for interactive purposes has been growing in the last decades [4]. However, even in the age of ambient intelligence adorned with gesture-sensitive interfaces [6, 8, 23], hand postures and fine finger movements are still being little tapped in contrast with motion gestures. For example, today's most popular gesture sensing devices[1] make heavily use of body motion but leave hand postures little employed. However, the information gained from monitoring finger movements and hand poses while the hand is grasping and manipulating objects can be incredibly rich and offer many interaction opportunities [15].

In this work we reverse the process of hand pose and gesture recognition by extracting information about grasped objects rather than recognizing the gesture being performed. A data glove is used to measure the flexion of fingers in order to inform on the approximate shape and size of the objects being grasped and manipulated.

[1] Video game sensors: Nintendo Wii (http://www.wii.com/), PlayStation MoveMe (http://us.playstation.com/ps3/playstation-move/move-me/), and Microsoft Kinect (http://www.xbox.com/en-GB/kinect).

S. Omatu et al. (Eds.): *Distrib. Computing & Artificial Intelligence*, AISC 217, pp. 537–544.
DOI: 10.1007/978-3-319-00551-5_64

The information is exploited in order to turn everyday non-interactive objects into practical working interfaces for ambient intelligence applications.

2 Research Premises and Related Work

Jones and Lederman [11] provided a comprehensive overview of the functions performed by the human hand acting as motor element and sensor alike. In the first case, the hand manipulates the environment with simple actions such as pushing a button or grasping a ball while it also performs complex activities such as playing a musical instrument. In its second function, the hand acts as a sensor collecting information about the environment through tactile, active haptic, prehensile and nonprehensile mechanisms (e.g., sense temperature or texture roughness of grasped objects) [11]. Therefore, the hand can be modeled as a two-way device delivering input and output alike. In this context, Buchholz et al. [2] were concerned with proposing a model for the human hand in order to noninvasively predict postures as the hand grasps different objects and Paulson et al. [15] showed that hand posture can be used to distinguish between several office activities.

In order to infer such information, precise measurements must be performed on the grasping hand and, to date, numeric gloves are the only devices that can deliver such fine measurement accuracy. Dipietro et al. [4] provide a comprehensive review on the use of glove-based hand posture for input with gloves being frequently used by ambient intelligence researchers [8, 22]. Vision-based techniques have also been employed for hand gesture recognition [5, 17] however they come with low sensing resolution, demand high processing for running complex vision algorithms, and present line of sight and occlusion problems (e.g., fingers may overlap one another). The accuracy of the acquired data must be especially taken into consideration as it may cause fine finger movements not being registered correctly.

In this context, we focus on glove-based capture of hand posture in order to infer object properties. Discrimination between object geometries and sizes allows us to use any real-world object as a potential interface for AmI applications. We must note that specialized object identification systems exist such as the use of RFID technology [22] which has the obvious advantage of precise object identification. However, placing tags in the environment may prove impractical in terms of cost while restricting interactions to only those objects that have been tagged in advance. Also, Tanenbaum et al. noted how participants wearing an RFID glove tended not to actively engage in manipulating objects but simply touched the attached tags [22]. We argue that such interactions reduce the need for physical manipulation in the detriment of a realistic and stimulating experience. Using hand measurements alone allows us to work in any environment (with the glove as part of a wearable computing system) with considerable more freedom and interaction opportunities in terms of user behavior and possible applications.

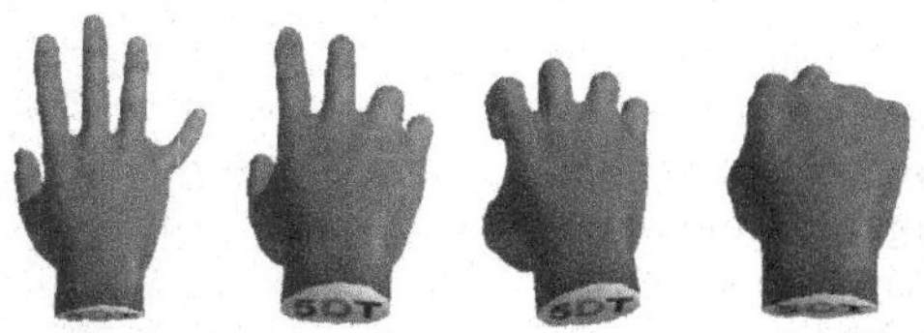

Fig. 1 Closing the fist

3 Object Recognition Using Hand Postures

We built a prototype employing the 5DT Data Glove[2] which includes 14 sensors: two for each finger (the knuckle and first joint) and four sensors that measure proximity between each pair of successive fingers. A recognizer was implemented in order to identify and differentiate both simple and complex hand gestures. We consider a simple gesture to be a static one in which the hand posture does not change for a period of time such as having a stable grip on an object. A complex gesture involves the movement of fingers and changes in hand posture such as closing the fist or tapping with the fingers (Figure 1). Such complex gestures are composed of multiple consecutive simple gestures. Simple hand postures are classified based on the Nearest Neighbor rule [15] and using the Euclidean distance working on postures represented as 14-dimensional vectors:

$$p = \{p_1, p_2, ..., p_{14}\} \in [0,1]^{14} \tag{1}$$

$$\|p - q\| = \left(\sum_{i=1}^{14} (p_i - q_i)^2 \right)^{\frac{1}{2}} \tag{2}$$

We limit our discussion on the recognition techniques here as they represent common practices of the domain and instead refer the reader to similar approaches [8, 15, 9]. We instead focus on the application opportunities of using hand posture as a tool for inferring object data, as described next in the paper. As a performance note, we found recognition rates over 95% when discriminating between a set of patterns representing 18 objects with different sizes and shapes with data acquired from 12 participants [25]. Our results are in agreement with those reported by existing research [15].

4 Application for Games

The field of gaming is stepping into the age of ubiquitous computing by no longer being solely a means of entertainment but accomplishing more complex roles such as connecting the player to the physical world and to other players as well [1, 14]. The examples we present in this article serve the goal of increasing the user's experience by encouraging him to explore his environment and pick his own tools [24]

[2] `http://www.5dt.com/products/pdataglove14.html`

Fig. 2 Using real objects to interact in a first person shooter: (a) a toy as a gun and (b) a cup as a grenade for a first person shooter

in the form of common object for the available actions fitting in the augmented reality subgenre [14] of pervasive games along with others such as ARQuake [16], an outdoor implementation of the classic first person shooter, Quake.

Ambient games are now designed to tackle essential issues of the human community such as health [13], education [7, 19] or social behavior [1, 18], while engaging and entertaining the players through what may sometimes be a very simple set of rules. Can You See Me Now (CYSMN) and Uncle Roy All Around You (URAAY) are simple games of catch in which some of the players are in a virtual environment and some are in the real world (mapped to the virtual one) [1, 14]. The two games are designed to encourage communication even with strangers in what could be a socially new environment.

We illustrate our techniques for video games due to their popularity [3] and recently adoption of gesture input which allows us to reach a large number of potential users. We also chose games due to the wide palette of tasks and activities they can simulate and which can be naturally performed using hand gestures. Many video games require dedicated accessories that enhance the sensing capabilities and the form factor of the primary controller such as driving wheels or golf clubs for the Wii console[3]. We argue that hand postures can inform on the object the user is grasping and holding and therefore the game can adapt accordingly. The result is transforming everyday objects into practical interfaces instead of using specialized equipment.

As a first example we discuss Counter Strike, a first person shooter in which the gamer must defeat opponents using various weapons (knife, pistol, grenade, etc.) and that makes heavy use of the PC keyboard or the controller keypad[4]. The data glove can sense when objects are grasped and the grasping shape can inform on object shape and size. For example, a simple toy pistol can be grasped to simulate the pistol in the game (static gesture) while firing works naturally by pulling the trigger (dynamic gesture). Changing the weapon is done by simply changing the hand posture (see Figure 2a and 2b). Each weapon has a corresponding static posture that allows the system to identify it and a corresponding dynamic gesture for triggering the action: pulling the trigger for a gun or throwing a grenade.

As a second example, in role playing games the user takes on a persona and accomplishes various missions (Figure 3a and 3b). This is currently done by use

[3] http://www.nintendo.co.uk/NOE/en_GB/systems/accessories_1243.html

[4] http://counterstrike.wikia.com/wiki/Controls

Fig. 3 Using real objects to interact in a role playing game: (a) a stapler as a sword and (b) a stapler as a fishing rod for role playing games

of mouse/keyboard or game controller through various combinations of keys and/or buttons. One could say that this defeats the purpose of the role playing game and the experience of submersing into the game is not complete.

5 Discussion

While the applications in gaming may present the opportunity of improving the user's experience in particular scenarios they do not however illustrate the potential impact that using an object based interaction could have in the day to day life, specifically in the context of ambient intelligence. By working with a gesture-based system the user gets the opportunity to engage familiar objects from his surrounding to interact with the virtual environment, a concept similar to tangible bits [10]. The tangible bits system put forward the idea of connecting the real world with the digital world by using physical objects which the user can manipulate and thus interact with or control the digital world.

Given that in a glove-based system the association between digital information and real objects is done entirely through the user's hand gestures this also means that the system is no longer restricted to a controlled and physically limited environment. As part of a wearable computing system the concept can be extended and applied to any environment in which the user finds himself thus granting a considerable worth to a glove-based interaction system.

The concept of physical tokens is found in multiple approaches [10, 21] but the possibilities offered by those systems are confined to a single limited and controlled environment. These physical tokens are usually dumb devices [12] with a unique assigned identity or to which various sensors have been added in order to make them a part of the smart environment. In the case of Sluis et al. [21], a smart environment in which the interaction is based on dedicated physical tokens, the initial tokens were associated by test subjects to a TV remote and thus seen as separate from the environment. The tokens were then redesigned to appear more as decorative objects and less as technological devices so that the interaction would seem more natural to the user. What this creates is a distinction between how the interaction takes place when the user moves from one space to another.

While it is viable in an in-home environment this system raises some problems in moving for example between two different environments, a scenario in which carrying the physical tokens could be inconvenient and would not much differ from how we currently carry data (using for example an USB flash memory). However, using a gesture based system allows the user to associate a process or data to a gesture which can further be supported by an object. The user can create his own menu of gestures associated with objects which, assisted by a wearable computing system, can be carried anywhere.

So far we have described the idea that devices and objects in an environment can act as a support for preexisting gesture commands (as in the Counter-Strike application the cup was acting as a replacement for the grenade which had its own established gesture). However gesture based interaction can work the other way around as well.

Considering an established communication between wearable computing systems and smart environments available commands can be discovered by the user through the simple physical manipulation of components in the environment. The user can explore the environment and interact with objects in it and then receive specific information. The greatest advantage of this system is that there is no need to embed or attach additional sensors to the objects themselves (which could turn to be quite costly given the variety of objects that can be found in a common house hold). This of course does not apply only to discovering the smart environment but to the objects themselves as the user can get relevant data either on the way they can be used individually or how they can be used to interact with and control the environment.

6 Conclusion and Future Work

Despite many application opportunities for ambient intelligence, hand posture is little utilized by today's common gesture acquisition technologies (e.g, sensors of video consoles). For example, Microsoft's Kinect sensor delivers low-cost full-body tracking but lacks resolution accuracy for detecting high fidelity finger movements. While ambient intelligence researchers have started using such popular technology [23] on top of data glove investigations[8], we estimate great benefit from having a mixed system that would take advantage of both approaches.

Hand posture and fine finger movements can offer a higher support for gesture-enhanced ambient interactions making them resemblant to our every day actions. Moreover, common non-interactive objects can become part of the interface as the hands naturally grasp and manipulate them. Users can then create their own customized interfaces by reusing familiar objects from the environment as command triggers leading to a more realistic and engaging experience.

In this work we elaborated on one particular scenario in which hand grasping can inform on the characteristics of the objects being grasped and manipulated. As future work we are considering applications for learning as the learning process seems to be influenced by gestures [26]. The same technique could be used in order to improve children experience while providing appropriate feedback. Haptic displays [20] are another field we feel would benefit making use of hand postures due to the accuracy of the data acquired while exploring a surface or manipulating an object.

Acknowledgement. This paper was supported by the project InteractEdu (Interactive gesture-based system for the educational development of school-age children: applications in education, tourism and discovery of patrimony) 588/2012, co-funded by UEFISCDI, Romania and WBI, Belgium.

References

1. Benford, S., Magerkurth, C., Ljungstrand, P.: Bridging the physical and digital in pervasive gaming. Commun. ACM 48(3), 54–57 (2005), http://doi.acm.org/10.1145/1047671.1047704, doi:10.1145/1047671.1047704
2. Buchholz, B., Armstrong, T., Goldstein, S.: Anthropometric data for describing the kinematics of the human hand. Ergonomics 35(3), 261–273 (1992)
3. Chatfield, T., Fun Inc.: Why Gaming will Dominate the Twenty-First Century. Pegasus Communications, Inc. (2011)
4. Dipietro, L., Sabatini, A.M., Dario, P.: A Survey of Glove-Based Systems and Their Applications. Trans. Sys. Man Cyber Part C 38(4), 461–482 (2008)
5. Erol, A., Bebis, G., Nicolescu, M., Boyle, R.D., Twombly, X.: Vision-based hand pose estimation: A review. Comput. Vis. Image Underst. 108(1-2), 52–73 (2007)
6. Farella, E., O'Modhrain, S., Benini, L., Riccó, B.: Gesture signature for ambient intelligence applications: A feasibility study. In: Fishkin, K.P., Schiele, B., Nixon, P., Quigley, A. (eds.) PERVASIVE 2006. LNCS, vol. 3968, pp. 288–304. Springer, Heidelberg (2006)
7. Hjert-Bernardi, K., Hernndez-Leo, D., Melero, J., Blat, J.: Do Different Hint Techniques Embedded in a Digital Game-Based Learning Tool have an effect on students' behavior? In: Proceedings of the International Joint Conference on Ambient Intelligence - AmI 2011, Amsterdam, Netherlands (2011)
8. Huang, Y., Monekosso, D., Wang, H., Augusto, J.C.: A Concept Grounding Approach for Glove-Based Gesture Recognition. In: Proc. of the 7th Int. Conf. on Intelligent Environments (IE 2011), pp. 358–361. IEEE Computer Society, Washington, DC (2011)
9. Huang, Y., Monekosso, D., Wang, H., Augusto, J.C.: A Hybrid Method for Hand Gesture Recognition. In: Proceedings of the 8th International Conference on Intelligent Environments (IE), pp. 297–300 (2012), http://dx.doi.org/10.1109/IE.2012.30
10. Ishii, H., Ullmer, B.: Tangible bits: towards seamless interfaces between people, bits and atoms. In: Proc. of the SIGCHI Conference on Human factors in Computing Systems (CHI 1997), pp. 234–241. ACM, New York (1997)
11. Jones, L.A., Lederman, S.J.: Human Hand Function. Oxford University Press, Inc., New York (2006)
12. van Loenen, E.: On the role of Graspable Objects in the Ambient Intelligence Paradigm. In: Proceedings of the Smart Objects Conference, Grenoble, France, pp. 15–17 (May 2003)
13. Madeira, R.N., Postolache, O., Correia, N.: Gaming for Therapy in a Healthcare Smart Ambient. In: Proceedings of the First Workshop on Ambient Gaming (AmGam 2011), Amsterdam, The Netherlands (2011)
14. Magerkurth, C., Cheok, A.D., Mandryk, R.L., Nilsen, T.: Pervasive games: bringing computer entertainment back to the real world. Comput. Entertain. 3(3), 4 (2005), http://doi.acm.org/10.1145/1077246.1077257, doi:10.1145/1077246.1077257
15. Paulson, B., Cummings, D., Hammond, T.: Object Interaction Detection using Hand Posture Cues in an Office Setting. Int. Journal of Human-Computer Studies 69(1-2), 19–29 (2011)

16. Piekarski, W., Thomas, B.: ARQuake: the outdoor augmented reality gaming system. Commun. ACM 45(1), 36–38 (2002), `http://doi.acm.org/10.1145/502269.502291`, doi:10.1145/502269.502291
17. Prodan, R.-C., Pentiuc, S.-G., Vatavu, R.-D.: An Efficient Solution for Hand Gesture Recognition from Video Sequence. Advances in Electrical and Computer Engineering 12(3), 85–88 (2012), doi:10.4316/AECE.2012.03013
18. Rijnbout, P., de Valk, L., de Graaf, M., Bekker, T., Schouten, B., Eggen, B.: i-PE: A Decentralized Approach for Designing Adaptive and Persuasive Intelligent Play Environments. In: Proceedings of the First Workshop on Ambient Gaming (AmGam 2011), Amsterdam, The Netherlands (2011)
19. Rosales, A., Arroyo, E., Blat, J.: Evocative Experiences in the Design of Objects to Encourage Free-Play. In: Wichert, R., Van Laerhoven, K., Gelissen, J. (eds.) AmI 2011. CCIS, vol. 277, pp. 229–232. Springer, Heidelberg (2012)
20. Ruspini, D.C., Kolarov, K., Khatib, O.: The haptic display of complex graphical environments. In: Proc. of SIGGRAPH 1997, pp. 345–352. ACM Press/Addison-Wesley Publishing Co., New York (1997)
21. van de Sluis, R., Eggen, J.H., Jansen, J., Kohar, H.: User Interface for an In-Home Environment. In: Hirose, M. (ed.) Human Computer Interaction, INTERACT 2001, Tokyo, pp. 383–390 (July 2001)
22. Tanenbaum, K., Tanenbaum, J., Antle, A.N., Bizzocchi, J., el-Nasr, M.S., Hatala, M.: Experiencing the reading glove. In: Proc. of the 5th Int. Conf. on Tangible, Embedded, and Embodied interaction (TEI 2011), pp. 137–144. ACM, New York (2010)
23. Vatavu, R.-D.: Nomadic Gestures: A Technique for Reusing Gesture Commands for Frequent Ambient Interactions. Journal of Ambient Intelligence and Smart Environments 4(2), 79–93 (2012)
24. Vatavu, R.-D., Zaiţi, I.-A.: An Investigation of Extrinsic-Oriented Ambient Exploration for Gaming Applications. In: Wichert, R., Van Laerhoven, K., Gelissen, J. (eds.) AmI 2011. CCIS, vol. 277, pp. 245–248. Springer, Heidelberg (2012)
25. Vatavu, R.-D., Zaiti, I.-A.: Automatic recognition of object size and shape via User-dependent measurements of the grasping hand. International Journal of Human-Computer Studies (2013), doi:10.1016/j.ijhcs.2013.01.002
26. Wagner Cook, S., Mitchell, Z., Goldin-Meadow, S.: Gesturing makes learning last. Cognition 106(2), 1047–1058 (2007)

Suboptimal Restraint Use as an Emergent Norm via Social Influence

Felicitas Mokom and Ziad Kobti

University of Windsor, School of Computer Science, Windsor, Canada N9B-3P4
{mokom, kobti}@uwindsor.ca

Abstract. Suboptimal restraint use is a prevalent problem worldwide. In developed countries injuries and deaths related to vehicle accidents persist despite increases in restraint use. In this study we investigate the emergence of patterns of restraint use in groups of agents and the population at large. Using age as an influential factor we simulate random encounters between group members where dominant individuals repeatedly alter the knowledge of less influential individuals. Belief spaces implemented as part of a cultural algorithm are used to preserve prevalent patterns of restraint use both at the group and population levels. The objective is to demonstrate restraint selection and use patterns emerging within a population and to determine whether a focus on influential members might have a positive effect towards optimal restraint use. We demonstrate that prominent patterns of behavior similar to the influential members of the groups do emerge both in the presence of social and cultural influence.

1 Introduction

Motor vehicle accidents and their associated injuries are a social and economic burden to households and society as a whole. Recognizing the significant threat they pose to the global community, the World Health Organization (WHO) simultaneously assigned "road safety" as the world health day theme in April 2004 and released a comprehensive report on road traffic injury prevention. Direct costs associated with roadway collisions were estimated at US$ 518 billion with costs in low income countries at US$65 billion surpassing amount received in aid [13]. These accidents are globally considered to be the leading cause of death in children aged 15-19 years and the second leading cause of death in children aged 5-14 years [12].

The increasing use of Child Restraint Systems (CRS) is noted as one of the major contributors to improvements in child road safety [13]. When used correctly CRS have been reported to reduce fatalities in children by 71% and serious injury by 67% in the event of a crash [7]. Reported rates of CRS misuse have been as high as 80% [4]. Research has shown that suboptimally restrained children are not only injured more often but are at a higher risk of more serious injuries [3]. Factors contributing

S. Omatu et al. (Eds.): *Distrib. Computing & Artificial Intelligence,* AISC 217, pp. 545–552.
DOI: 10.1007/978-3-319-00551-5_65

to suboptimal restraint use include lack of education, parenting style, short travel distance, ease of placement, CRS complexity and social influence [2, 4].

In this study we strive to answer the question: can suboptimal restraint use emerge as a norm in the presence of social and cultural influence? To address this we revert to a prior implemented model of learning artifact capabilities [9]. In the study artifacts were defined as physical objects in the environment that provide some functionality useful to an agent for an objective. Artifact capabilities refer to knowledge acquired by agents for artifact use. We simulate a population of drivers or artifact-capable agents, each with the goal of applying and evolving their capability for a restraint. The drivers are organized in hierarchical groups where age is used as a social influential factor within each group giving older agents a greater influence over younger ones. To simulate the effects of culture we utilize a cultural algorithm to extract prominent patterns of restraint use from the population and use them to further influence driver behavior. One of our objectives is to observe whether suboptimal restraint use can emerge as a norm as agents adapt their knowledge. Another objective is to demonstrate the effects on restraint use when agents are influenced socially by other agents or beliefs in the community as a whole.

The next section provides some background on related work. We then present our restraint use model. Details on experiments conducted and results obtained are provided. The final section provides conclusions deduced and future work.

2 Related Work

The subject of restraint use has gained recent popularity particularly in the social sciences [2, 3, 4]. Closely related to our work is the cultural model for learning to apply child safety restraint in [8]. Agents operating in a cultural framework learned to select a CRS and its location in the vehicle depending on the age, weight and height of the child. In our work CRS selection, its placement within the vehicle as well as all the steps that would constitute use of the CRS are considered.

Norms have been defined as expected behaviors in given situations [11]. Four ways in which norms could emerge have been identified: explicit statements by the group leader or other influential members, relevant events in the group's past, the primary behavior becoming the expected one, and inherited behavior from past experiences [6]. A comprehensive summary of the varying agent-based simulation models of norms is provided in [10].

3 Restraint Selection and Use Model

3.1 Environment Description and Approach

The environment consists of a set of agents with varying ages organized into groups of varying sizes. The idea is to simulate families of drivers living within a community. Normalized agent ages are used as influence rates. Influence rates are comparable such that for agent a and agent b, if $inf(a) > inf(b)$ then agent a has influence over agent b with a degree of influence: $inf(a) - inf(b)$. Agents are also randomly assigned retention rates specifying the probability that the agent keeps its knowledge when an attempt is made to influence it.

Each agent has the same objective: to apply its knowledge for selecting and using a restraint. During this process the agent may adapt its knowledge via influence from another group member or beliefs in its community.

3.2 *Knowledge Structures*

An agents' knowledge captures what it knows about selecting and using a restraint. Selection knowledge is maintained for four predefined age ranges obtained from standard knowledge [8]. The five possible selections are "rear facing", "forward facing", "booster", "seatbelt" and "none" represented in binary strings R_1 to R_5 respectively. "None" is used to represent the child being unrestrained in the vehicle. If a bit is set to '1' then the agent deems the respective restraint as a possible selection for the associated range. An agent's restraint selection knowledge is as follows:

SelKn(SK) = {
$SK_1 = <00, 13, [R_1R_2R_3R_4R_5]>$,
$SK_2 = <13, 49, [R_1R_2R_3R_4R_5]>$,
$SK_3 = <49, 97, [R_1R_2R_3R_4R_5]>$,
$SK_4 = <97, 145, [R_1R_2R_3R_4R_5]>$
}

For example an agent with partial knowledge (13, 49, [01100]) knows that a child aged between 13 and 49 months should be placed in a forward facing or booster seat. Knowledge maintained for restraint use involves the operations that the agent performs in order to use the restraint. In our model the number of operations is fixed for each type of restraint. Restraints have parts which have attributes. Restraints themselves can have attributes. All attributes have a fixed set of possible values $\{0,1,2\}$ representing *no use*, *suboptimal use* and *optimal use* respectively. Knowledge for each operation is represented in a binary string as a vector of selected values for each attribute. The agent's use knowledge is as follows:

UseKn(UK) = {
$UK_1 = <V_1 \ldots V_n>$,
$UK_2 = <V_1 \ldots V_n>$,
|
$UK_k = <V_1 \ldots V_n>$
}

for n total attributes and k operations. For example an operation knowledge for a restraint with 3 total attributes might be (100100) representing the chosen values $<2, 1, 0>$. The operation involves optimal use of the first attribute, suboptimal use of the second and no use for the third.

Belief spaces are used to capture patterns of selection and use within a group or the entire population. At any time these spaces can be probed for the current "belief" of the group or population, that is, the prevalent knowledge that exists. Belief space knowledge is represented as follows:

$$
\begin{aligned}
&\text{BeliefSpace(B)} = \{ \\
&\quad B(SK_i) = \quad <P_1, \ldots, P_s>, i = 1 \ldots 4 \\
&\quad B(UK_j) = \quad <Q_1, \ldots, Q_t>, j = 1 \ldots k \\
&\quad \}
\end{aligned}
$$

where s (fixed at 32 in the model) is the number of possible patterns used to represent selection knowledge for each of the 4 age ranges. Each P_l is a tuple representing a counter and a binary string: $\langle c, x \rangle$ where x ranges from [00000] to [11111] and c represents the number of agents with x as their selection knowledge for an age range.

For use knowledge t is the number of possible patterns for each of the k operations. If r is the total number of attributes then $t = 2^{r*2}$. Each Q_v is a tuple representing a counter and a binary string: $\langle d, y \rangle$ where d represents the number of agents with the binary string y as their use knowledge for a particular operation.

3.3 Simulating Social and Cultural Influence

Age is used as the social influential factor in the model. At the start of the simulation each agent is given its own knowledge on selecting and using the same restraint. All agents contribute their knowledge to their group's belief space and the population's belief space. At each simulation step two randomly selected agents in each group are paired. For each pair of agents the more influential agent injects its knowledge into the less influential agent by copying each bit of the selection knowledge and each attribute value of the use knowledge at a probability of the degree of influence. The adapted knowledge is contributed to the belief spaces.

To simulate the effects of culture, the most prominent pattern in the population space is used in each simulation step as an additional influence factor on the population. To implement this aspect we utilize a cultural algorithm [14]. Cultural algorithms were introduced to facilitate modeling the effects of culture on a population. In the algorithm selected individuals from a population space contribute knowledge to a global belief space via an acceptance function. Knowledge maintained in the belief space affects the population via an influence function. Different knowledge types may be maintained in the belief space including situational, normative, topographic, historical and domain. In our implementation the belief spaces maintain patterns of restraint selection and use in the form of normative knowledge contributed by all agents. The framework for our cultural algorithm is shown in Figure 1.

Adjusting knowledge contributed to belief spaces involves sorting it according to the prominent patterns of restraint selection and use within the group or the population respectively.

We do not model any form of influence from the group belief spaces. In our implementation cultural influence is modeled as influence from the population belief space on the population at large in order to simulate individuals in the population

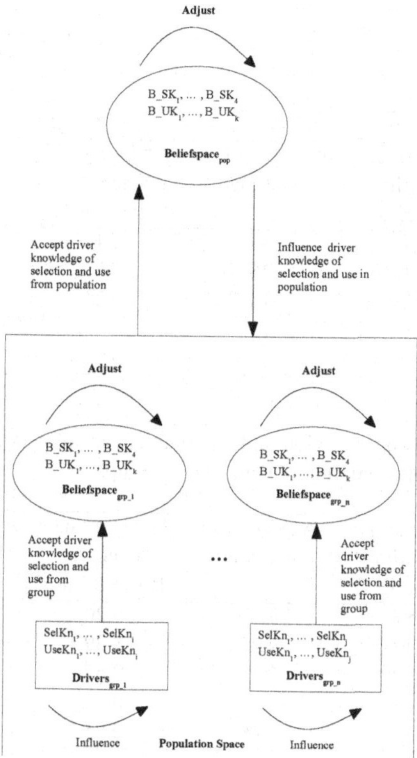

Fig. 1 Cultural algorithm applied to n group of drivers with group level and population level influence for i drivers in group 1 through j drivers in group n. Belief spaces maintain restraint selection knowledge for 4 child age ranges and restraint use knowledge for k operations.

being affected by beliefs in the community rather than beliefs in their particular group. To simulate cultural influence, in addition to the degree of influence the most prominent pattern in the population belief space is used as an additional influence factor on the influenced agent's knowledge. The agent copies each bit of the prominent selection knowledge and each attribute value of the prominent use knowledge at a probability of its retention rate.

4 Experiments and Results

Our test environment consists of a population of 50 agents randomly organized in 6 groups of varying sizes $[4, 12]$ with each group representing an extended family of drivers. We believe groups of varying sizes gives credibility to results observed at population level and and is a more accurate representation of reality. Agents are assigned random ages, random selection and use knowledge for a restraint: *booster*. The *booster* is a artifact with 2 artifact attributes: *location* and *attach* specifying

Table 1 Patterns of restraint selection and use with social and cultural influence extracted from population belief space for 50 agents randomly organized in 6 groups

	Social Influence		Cultural Influence	
Time Step	# Distinct Selection Patterns	# Distinct Use Patterns	# Distinct Selection Patterns	# Distinct Use Patterns
1	25	34	24	36
100	23	25	16	19
500	17	13	12	13
1000	15	12	10	12
5000	9	5	10	11

Table 2 Patterns of restraint selection and use with social influence extracted from group belief spaces for 50 agents randomly organized in 6 groups

Time Step	Avg Distinct Selection Patterns	Avg Distinct Use Patterns	Avg Dominant Selection Pattern Ratio	Avg Dominant Use Pattern Ratio
1	6.83	7.83	0.20	0.13
100	5.17	4.33	0.39	0.44
500	3.67	2.17	0.57	0.71
1000	3.33	2.00	0.66	0.75
5000	1.83	1.00	0.85	0.97

the where the booster is placed in the vehicle and its attachment to the vehicle. The *booster* has 2 artifact parts: *shoulder_belt* and *lap_belt*. Each artifact part has a single attribute: *attach* specifying attachment of the respective belt with the child in the seat. The *booster* therefore has a total of 4 attributes. As explained in the model possible values for all attributes are: $\{0,1,2\}$ representing no use, suboptimal use and optimal use respectively. Using guidelines on minimal recommendations for appropriate booster use [5] we define 3 operations for booster use. The first operation involves selecting a location for the booster, the second involves attaching the booster to the vehicle and in the final operation the shoulder and lap belts are attached.

In the first test case we investigate norm emergence in the presence of social influence only. The second and third columns of Table 1 show the number of distinct selection and use patterns that emerge from the population at large from 1 through 5000 time steps. Emergent patterns of selection and use at group level are shown in Table 2. The average distinct selection and use patterns across groups are computed. The last two columns show the mean of the number of agents using the most dominant patterns for groups relative to the group size.

In the second test case we investigate norm emergence in the presence of social and cultural influence. The results are shown in the last two columns of Table 1.

As demonstrated in the second and third columns of Table 1 norms do emerge in the population as a whole even though social influence occurs at group level. The number of distinct patterns decreases from 25 to 9 for selection patterns and from 34 to 5 for use patterns. It should be noted that there are only 6 groups from which 6 distinct patterns could emerge. Norm emergence via social influence is further demonstrated in Table 2 where average distinct patterns of selection and use across groups approaches a single pattern. Additionally the last two columns in the table demonstrate that irrespective of varying group sizes the majority of agents in each group converge towards the most dominant pattern in the group. Further investigation of the dominant patterns revealed that they always either belonged to or were similar to the patterns of the oldest agent. The results also demonstrate norm emergence in the presence of cultural influence. The last two columns of Table 1 demonstrate a reduction in the number of distinct patterns in the population when cultural influence is enabled. This is interesting because these agents are influenced both socially and by simulated cultural beliefs in the entire population. We believe this demonstrates that when the most influential agents in the group possess suboptimal knowledge that knowledge will propagate to the rest of the group members.

5 Conclusions and Future Work

In this study we implemented a multi agent simulation model to study the emergence of restraint misuse within a group of agents in the presence of social and cultural influences. Aspects of a prior model of artifact capabilities [9] are utilized to define restraints as artifacts and agents with capabilities for their use. Using age as an influential factor, agents are paired repeatedly to enable social influence. The dominating agent alters the knowledge of the less influential agent depending on its degree of influence. The agents "live" in a cultural environment where dominant patterns of behavior are maintained in a shared belief space. This knowledge is also used to influence the population at large in an effort to simulate the effects of cultural beliefs on restraint use.

Results obtained from conducted experiments show the emergence of norms in restraint selection restraint use in the presence of social and cultural influence. The results suggest that a focus on the more influential agents in a group may assist in the reduction of suboptimal restraint use.

In future work we would like to investigate the benefits of interventions during the emergence of norms and conduct experiments with much larger population sizes. It would also be useful to explore cultural influence via affinity relationships [1].

Acknowledgement. This work was supported in part by a grant from Auto21 and partial grants from CIHR and NSERC Discovery.

References

1. Axelrod, R.: The dissemination of culture: a model with local convergence and global polarization. Journal of Conflict Resolution 41(2), 203–226 (1997)
2. Blair, J., Perdios, A., Babul, S., Young, K., Beckles, J., Pike, I., Cripton, P., Sasgese, D., Mulpuri, K., Desapriya, E.: The appropriate and inappropriate use of child restraint seats in Manitoba. International Journal of Injury Control and Safety Promotion 15(3), 151–156 (2008)
3. Brown, J., McCaskill, M.E., Henderson, M., Bilston, L.E.: Serous injury is associated with suboptimal restraint use in child motor vehicle occupants. Journal of Paediatrics 42, 345–349 (2006)
4. Bruckner, R., Rocker, J.: Car safety. Pediatrics in Review 30, 463–469 (2009)
5. Bulger, E.M., Kaufman, R., Mock, C.: Childhood crash injury patterns associated with restraint misuse: implications for field triage. Prehospital and Disaster Medicine 23(1), 9–15 (2008)
6. Feldman, D.C.: The development and enforcement of group norms. The Academy of Management Review 9(1), 47–53 (1984)
7. Howard, A.: Automobile restraints for children: a review for clinicians. Canadian Medical Association Journal 167, 769–773 (2002)
8. Kobti, Z., Snowdon, A., Rahaman, S., Dunlop, T., Kent, R.: A cultural algorithm to guide driver learning in applying child vehicle safety restraint. In: Proceedings of the IEEE Congress on Evolutionary Computation, pp. 1111–1118 (2006)
9. Mokom, F., Kobti, Z.: Evolution of artifact capabilities. In: Proceedings of the IEEE Congress on Evolutionary Computation, pp. 476–483. IEEE Press, New Orleans (2011)
10. Neumann, M.: Homo socionicus: a case study of simulation models of norms. Journal of Artificial Societies and Social Simulation 11(4-6) (2008), `http://jass.soc.surrey.ac.uk/11/4/6.html`
11. Opp, K.D.: How do norms emerge? an outline of a theory. Mind & Society 2, 101–128 (2001)
12. Peden, M., Oyegbite, K., Ozanne-Smith, J., Hyder, A.A., Branche, C., Rahman, A.F., Rivara, F., Bartolomeos, K. (eds.): World report on child injury prevention. World Health Organization (2008)
13. Peden, M., Scurfield, R., Sleet, D., Mohan, D., Hyder, A.A., Jarawan, E., Mathers, C. (eds.): World report on road traffic injury prevention. World Health Organization (2004)
14. Reynolds, R.G.: An adaptive computer model of the evolution of agriculture for hunter-gatherers in the valley of oaxaca. Ph.D. thesis, Dept. of Computer Science, University of Michigan (1979)

Implementing MAS Agreement Processes Based on Consensus Networks*

Alberto Palomares, Carlos Carrascosa, Miguel Rebollo, and Yolanda Gómez

Universitat Politècnica de València,
Camino de Vera s/n 46022 Valencia, Spain
{apalomares,carrasco,mrebollo,ygomez}@dsic.upv.es

Abstract. Consensus is a negotiation process where agents need to agree upon certain quantities of interest. The theoretical framework for solving consensus problems in dynamic networks of agents was formally introduced by Olfati-Saber and Murray, and is based on algebraic graph theory, matrix theory and control theory. Consensus problems are usually simulated using mathematical frameworks. However, implementation using multi-agent system platforms is a very difficult task due to problems such as synchronization, distributed finalization, and monitorization among others. The aim of this paper is to propose a protocol for the consensus agreement process in MAS in order to check the correctness of the algorithm and validate the protocol.

1 Introduction

In agent-based networks, 'consensus' is referred to reach an agreement about a certain quantity of interest or distribution function that depends on the state of all agents [6]. Consensus algorithms can be modeled as iterative processes in which autonomous agents work in a distributed fashion, without the necessity of sending information to a central node that acts as coordinator.

When it is implemented in a real system, some considerations about how the communication process will take place must be taken into account. The set of rules that specifies the information exchange between an agent and all of its neighbors on the network are specified in a consensus protocol.

Distributed Constraint Optimization Problems (DCOP) [5,9] are a model for representing multiagent systems (MAS) in which agents cooperate to optimize a global objective, which could be used to deal with this type of problem too. Nonetheless, as [7] details, there are some limitations in the way DCOP algorithms approaches the problem. Firstly, they assume that the environment is deterministic and fully-observable, meaning that agents have complete information about the utility of the

* This work is supported by ww and PROMETEO/2008/051 projects of the Spanish government, CONSOLIDER-INGENIO 2010 under grant CSD2007-00022, TIN2012-36586-C03-01 and PAID-06-11-2084.

S. Omatu et al. (Eds.): *Distrib. Computing & Artificial Intelligence,* AISC 217, pp. 553–560.
DOI: 10.1007/978-3-319-00551-5_66

Algorithm 1. Consensus algorithm. Power iteration method for the matricial form

$D = \sum_{i \neq j} a_{ij}$
$L = D - A$
assign a $\varepsilon < 1/\Delta$
$P = I - \varepsilon L$
init x with random values
repeat
  $x = P * x$
until the system converges or maxiter reached

outcomes of their possible decisions and the system remains in a static state while the decision is been taken.

This paper propose a protocol for the consensus agreement process in MAS. This protocol has been implemented it in the MAGENTIX2 MAS platform, addressing the problems that a real implementation introduces in a theoretical mathematical model. To check its correctness, the results of the MATLAB simulations are compared with the results obtained by the real execution of the MAS in MAGENTIX2.

2 Consensus Networks

The theoretical framework for solving consensus problems in dynamic networks of agents was formally introduced by Olfati-Saber and Murray [6]. The interaction topology of the agents is represented by a graph and *consensus* means to reach an agreement regarding a certain quantity of interest that depends on the state of all agents in the network. This value represents the variable of interest in our problem *(agreement term)*, and might be for example a physical quantity, a control parameter or a price among others. The value of a node might represent physical quantities measured in a distributed network of sensors (temperatures, voltages, velocities, prices, qualities,...).

A consensus algorithm is an interaction rule that specifies the information exchange between the agents and all of its neighbors on the network in order to reach the agreement (see Algorithm 1). Distributed solutions of consensus problems in which no node is connected to all nodes are especially interesting.

Solving consensus is a cooperative task because if a single agent decides not to cooperate and keep its state unchanged, all others asymptotically agree with them. However, if there are multiple non cooperative agents then no consensus can be asymptotically reached. When an agent does not cooperate with the rest its links in the network might be disconnected in order all others will asymptotically agree. In this case it is impossible for all agents to reach an agreement but is possible to reach an agreement for the rest of the agents.

3 Implementation of Consensus Protocol for MAS

Usually, consensus algorithms are checked by means of simulation executions, where the multi-agent system is modeled as a matrix. Using a power iteration method, the matrix is applied over the solution vector until the changes between successive iterations are under a threshold. This kind of solutions are easier to implement and to test than real executions, because they avoid to deal with problems such as:

- Synchronization: how the agents coordinate to deal with the consensus process. To get this, a new consensus protocol is needed.
- Distributed finalization: how any agent detects when the consensus is reached, or when the process has to end.
- Monitorization: how can be observed the dynamic of the consensus process.
- Bottlenecks: are there any bottleneck in the process? how it is dealt with?
- Communication: how to deal with the overload of messages, and the ordering of the different messages from the different agents any agent is connected to?
- Scalability: is it possible to scale the MAS trying to reach the consensus? how it affects to the above mentioned problems?
- Lack of MAS design toolkits with connection networks, allowing the automatization in the developing of different scenarios.

Nevertheless, the use of a matricial system assumes a complete knowledge scenario in which all agents know the entire network structure and the values. This approach can be useful in centralized and relatively small systems, but if the size of the network grows or participants play in conditions of bounded rationality, a centralized approach is not feasible.

3.1 MATLAB *Implementation*

Algorithm 2 contains the code that corresponds to the MATLAB implementation of the consensus process.

The main limitation of this method is the memory needed to store the adjacency matrix and the Perron matrix. But this can be solved using sparse matrices and adapting the corresponding functions to deal with them. With this change, a model

Algorithm 2. MATLAB Code

```
n = size(A,1)
L = spdiag(sum(A)) - A
eps = 1 / max(sum(A))
P = speyes(n) - eps * L
x = rand(n,1)
for i = 1:maxiter
  x = P * x
end
```

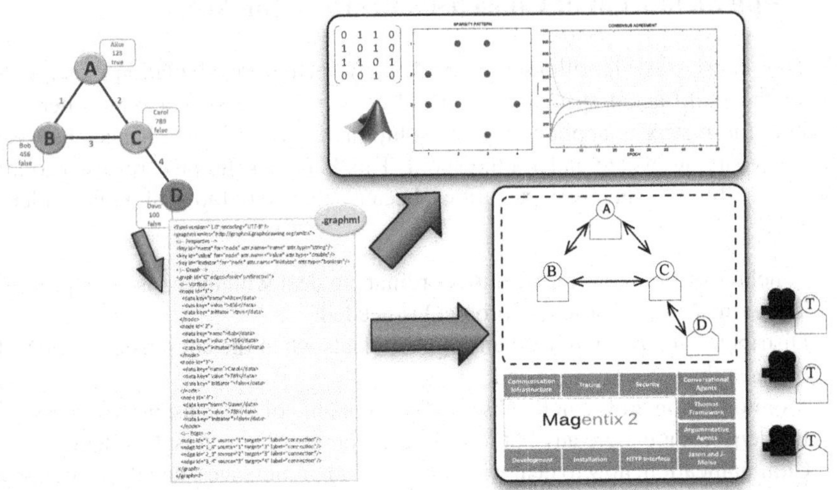

Fig. 1 Implementation process. The network is described in a GraphML file that feeds the MATLAB simulation and the initial configuration of the real MAS. A set of tracer agents can monitor the evolution of the process.

can be easily created with a maximum of 10^7 agents, which is the bound for the 32-bit version of MATLAB.

Nonetheless, the numerical method is a valid tool to (i) simulate the behavior of the system, (ii) check the correctness of the consensus algorithm in different situations other than the original Olfati-Saber and Murray proposal, and (iii) validate the implementation.

3.2 MAGENTIX2 *Implementation*

MAGENTIX2 is an agent platform that follows the levels proposed in [4], providing support for organization (according to the THOMAS framework specification [1]), interaction (by means of FIPA-ACL messages) and agent levels (through a Java API).

Moreover, all the generated agents that participate in the consensus process make use of the MAGENTIX2 tracing services [2] to generate events with all the needed information about the consensus process. So, any number of monitoring agents can be generated to connect to the events generated by such agents.

It has been generated a system that has as input a GraphML file that describes the connection network of a MAS, that is, nodes represent agents, and edges represent the connections among them. From this file, it generates a real MAS composed of MAGENTIX2 agents.

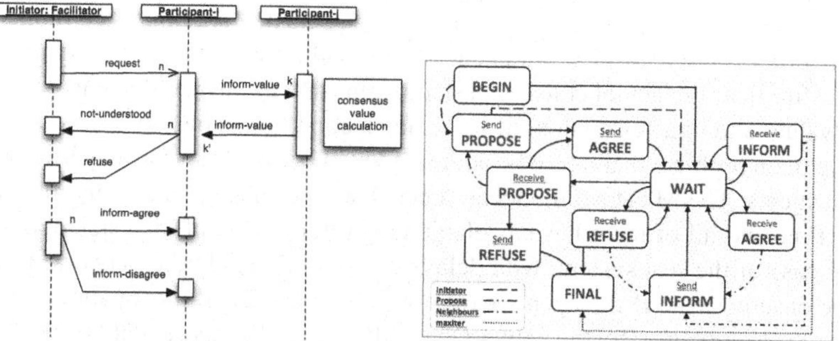

Fig. 2 (Left) General Consensus Protocol. An Initiator decides to start a consensus process and sends a proposal to its neighbors, that is propagated to the complete network. After that, agents exchange their values until the consensus is reached. (Right) State Automaton of the Consensus Communication Protocol in MAGENTIX2

3.2.1 Consensus Protocol

A general consensus protocol has been designed in order to allow a set of agents connected in a network to achieve an agreement exchanging information exclusively with their direct neighbors. Agents do not have information about the size of the network (how many agents are participating) nor how they are connected. The protocol is characterized for the absence of any kind of controller agent and all decisions are taken autonomously and in a decentralized way.

When an agent decides to start a consensus process, it sends a consensus proposal to all its neighbors (Figure 2). Each one of these neighbors decide to participate or not. If it participates, confirms its participation to the sender and re–sends the proposal to its corresponding neighbors. The process continues until all agents in the network have received the proposal. Therefore, all agents play two roles: initiator and participant in the consensus.

Once agents have received the confirmation of the neighbors desiring to participate in the consensus, the next step begins. Agents exchange their values with the neighbors and, after each iteration, update their internal values. In this step, all agents act as *participant* and they just send its value and recalculate it from the values received from their neighbors until a maximum number of iterations are executed or the difference with the value of the previous iteration $|x_i(k) - x_i(k-1)|$ is below a threshold (see termination problems for a more detailed analysis).

3.2.2 Agent Implementation

The protocol has been adapted to match the requirements of MAGENTIX2 platform. A *conversation factory* has been created that follows the finite-state machine shown in Figure 2. States are divided into three types: *send* states (in which the result is a message sent to the system), *receive* states (in which the agent is automatically awake to treat a concrete message) and general states.

The protocol has two steps. In the first one, agents that are interested in the consensus process are identified (an agent can individually decide if it participates or not). After that, the actual consensus step begins and all involved agents exchange their values and updates them until some finish condition is met.

The main problem that has to be addressed by the protocol is the synchronization of the agents. The MATLAB model assumes that all agents update their value at the same time and all of them have available the current value of their neighbors. But in the case of the real system, with delays or even failures in the network, this can not be ensured. And if the agents have not properly updated information, the final average value obtained by the network can differ from the theoretical average value.

When the platform starts, all agents instantiate a factory for the consensus protocol that stay in a wait state until some agent decides to launch a consensus process. Eventually, some agent will decide to begin a consensus. It sends a `propose` message to all its neighbors and it waits until all its neighbors have answered. Each agent decides if it wants to participate, in which case an `agree` answer is sent, or not, replying with a `refuse` message. Any agent accepting the consensus, for its part, sends a propose to its neighbors until the complete network acknowledges the propose. In order to let the second part of the consensus start, each agent must receive an answer from its neighbors.

In a second step, all participating agents inform their neighbors with their current values. The agent waits until all agents that have answered with their corresponding values. When all values have been received, each agent calculates their new updated value and informs its neighbor about the new value. This behavior makes the agent to be **synchronized** and evolve step by step at the same time.

As stated above, the messages may have a delay. Furthermore, the agents are free to inform when their direct neighbors have answered. These facts could cause that each agent receive messages belonging to different iterations. In order to solve this problem, a buffer has been created, in which the information of different iterations are saved, ensuring the agents to have a properly updated information.

The **distributed finalization** problem refers to how an individual agent can detect that the calculated value corresponds to the true consensus value or it is the process which has an slow dynamics. The considered possibilities are

1. to introduce a timeout
2. to specify a maximum number of iterations
3. to establish a threshold for two consecutive iterations
4. to include other dynamic conditions (for example, agents leaving the consensus)

In this work, static networks are considered only, so the last option is not applicable. The limit by timeout or number of iterations can be easily implemented by exchanging the corresponding value as a consensus parameter in the initial `propose` message. Each agent will stop itself when the specified condition is met. The third case is more complicated and requires the modification of the finite-state-machine to include additional states to handle a new set of messages. Basically, any agent can propose to stop the process because it has reached a possible consensus value and this message is propagated throughout the entire network. The consensus will stop when all the agents answer affirmatively to this question.[1]

MAGENTIX2 platform has a trace system available for developers to supervise the evolution of the system. It is implemented based on an event subscription scheme. All the agents can generate events to make notice diverse changes: internal state, activation, deactivation, message incoming, message sending, etc. Any other agent can subscribe to a set of events, so it will eventually receive any change in the system related with the event it is interested in. Using this mechanism, **monitoring** agents can be created and subscribed to changes in the consensus value of all the agents. When an agent changes its internal value, it triggers an event. The monitoring agent catches and treats it. An event is lighter than a message, so the overload introduced in the system is tractable.

As the proposed implementation is a completely decentralized system, there is no functional **bottlenecks**. Nonetheless, regarding with the network topology, it is possible for a reduce number of agents to concentrate the responsibility of the consensus process. This fact is related with the convergence speed of the network and some spectral properties of the adjacency matrix.

Regarding with the cost of the **communications**, the number of exchanged messages needed to complete the consensus process depends on the connectivity of the network. In each iteration, each agent sends exactly one message to each one of its neighbors. If we assume the network to be undirected, each link is used twice. In the case of a complete network, the number of messages in each step is $n(n-1)$ (where n is the number of agents in the network), that belongs to $O(n^2)$. Nevertheless, communication networks tend to be very sparse (with a low density: only a small fraction of nodes are connected). In that case, the cost of the messages can be considered as $\Theta(n\bar{d})$, where $\bar{d}$ is the average degree of the network.

Finally, **scalability** of the system is provided by the MAGENTIX2 platform. It is supported by AMQP standard for the communication, which is widely used in the industry as messaging middleware (see [2, 3, 8]).

4 Conclusions

This paper shows the application of consensus networks in a distributed and self-organized fashion, implemented in an agent platform with real agents.

The contribution of the paper is the proposal of a protocol that allows a network of agent achieve agreements using the consensus algorithm. This protocol has been implemented in the MAGENTIX2 MAS platform and tested, comparing the obtained results with the MATLAB simulation data. Both implementations arrive to the same numerical result, validating the correctness of the proposed protocol.

As future work, the integration with a development tool is planned, in order to create virtual organizations that follow a concrete network topology. After that, is

[1] The variation has not been explained because it does not introduce any change in the consensus process but complicates the comprehension of the protocol. The complete version has been implemented and it is available to download.

possible to automatically generate templates for the agents forming a network that follows the links specified at design time. Different network topologies can be easily generated just providing some basic configuration parameters, such as the topology itself (random, preferential attachment, small-world, growing), degree distribution, clustering, or assortativity index. Furthermore, more complicated scenarios can be tested, as the effect of time delays, changes in the dynamics when agents are allowed to enter or leave the system, multi–variable consensus and consensus in coupled systems, generated when several consensus processes take place at the same time in different groups with some common participants.

References

1. Argente, E.: et al: An Abstract Architecture for Virtual Organizations: The THOMAS approach. Knowledge and Information Systems 29(2), 379–403 (2011)
2. Búrdalo, L.: et al: TRAMMAS: A tracing model for multiagent systems. Eng. Appl. Artif. Intel. 24(7), 1110–1119 (2011)
3. Fogués, R.L., et al.: Towards Dynamic Agent Interaction Support in Open Multiagent Systems. In: Proc. of the 13th CCIA, vol. 220, pp. 89–98. IOS Press (2010)
4. Luck, M., et al.: Agent technology: Computing as interaction (a roadmap for agent based computing). Eng. Appl. Artif. Intel. (2005)
5. Mailler, R., Lesser, V.: Solving distributed constraint optimization problems using cooperative mediation. In: AAMAS 2004, pp. 438–445 (2004)
6. Olfati-Saber, R., Fax, J.A., Murray, R.M.: Consensus and cooperation in networked multi-agent systems. Proceedings of the IEEE 95(1), 215–233 (2007)
7. Pujol-Gonzalez, M.: Multi-agent coordination: Dcops and beyond. In: Proc. of IJCAI, pp. 2838–2839 (2011)
8. Such, J.: et al: Magentix2: A privacy-enhancing agent platform. Eng. Appl. Artif. Intel. 26(1), 96–109 (2013)
9. Vinyals, M., et al.: Constructing a unifying theory of dynamic programming dcop algorithms via the generalized distributive law. Autonomous Agents and Multi-Agent Systems 22, 439–464 (2011)

Agent-Based Interoperability for e-Government

Fábio Marques[1], Gonçalo Paiva Dias[2], and André Zúquete[3]

[1] ESTGA-UA/IEETA, Campus Univ. de Santiago 3810 – 193 Aveiro, Portugal
fabio@ua.pt
[2] ESTGA-UA/GOVCOPP, Campus Univ. de Santiago 3810 – 193 Aveiro, Portugal
gpd@ua.pt
[3] DETI-UA/IEETA, Campus Univ. de Santiago 3810 – 193 Aveiro, Portugal
andre.zuquete@ua.pt

Abstract. The provision of valuable e-government services depends upon the capacity to integrate the disperse provision of services by the public administration and thus upon the availability of interoperability platforms. These platforms are commonly built according to the principles of service oriented architectures, which raise the question of how to dynamically orchestrate services while preserving information security. Recently, it was presented an e-government interoperability model that preserves privacy during the dynamic orchestration of services. In this paper we present a prototype that implements that model using software agents. The model and the prototype are briefly described; an illustrative use case is presented; and the advantages of using software agents to implement the model are discussed.

1 Introduction

E-Government is defined as the use of ICT by government agencies to transform the relations with their clients (citizens, businesses and other government agencies) [14]. The need to deliver services and in a more effective way was a catalyst towards presenting solutions for transversally integrating business processes and, consequently, to the interoperability problem [11, 12]. This process often triggers the re-engineering of existing business processes in the Public Administration (PA), which adds extra complexity. Finally, several other aspects (legal, social and technical) that must be as well taken into consideration (see, for instance, [2, 8]).

Security is one of the main issues regarding interoperability. Trust, authentication and access control are issues that must be dealt carefully, mainly to keep sensitive information private. In [4, 5] it has been proposed a model that addresses the dynamic orchestration of services (ensuring interoperability) and security, in particular trust, authentication and privacy, to provide government services. According to the model, the PA consists of a set of heterogeneous entities with heterogeneous systems, which may deliver overlapping services. Each entity is autonomous, reactive, proactive and able to communicate with other entities. These are the same characteristics that are identified by Luck et al. in [3] for a software agent (SA). SAs also support distributed computing efficiently and allow the dynamic composition of services [10], two key characteristics of the model. Based on these reasons, and

S. Omatu et al. (Eds.): *Distrib. Computing & Artificial Intelligence,* AISC 217, pp. 561–568.
DOI: 10.1007/978-3-319-00551-5_67

assuming that an SA represents an entity, we believe that the SA is the appropriate technology for implementing a prototype for validating the referred e-government interoperability model.

However, besides verifying the accordance between the characteristics of the technology and of the entities comprising the PA, a set of questions remain: are SAs a solution to implement the proposed model? What are the problems that are raised by using this technology? With this paper it is our purpose to present an agent-based prototype that implements the referred interoperability model and to evaluate the advantages and pitfalls of using SAs to accomplish that task.

Our paper is organized as follows. The related work is presented in Sect. 2. In Sect. 3 we present the basic principles of the model and an illustrative use case. The agent based prototype is presented in Sect. 4. In Sect. 5 we discuss the utility of agents and we conclude the paper in Sect. 6.

2 Related Work

There are a few examples of the use of software agents for improving some e-government functionalities. In this section we will describe them briefly as a base for comparison with our proposal.

In [9] it is addressed the use of mobile software agents as a facilitator for people migration. In this platform, the software agent represents the user interacting with the PA. The remaining components allow interoperability and a way to communicate with the service repositories and with legacy systems of the PA.

In [7] it is proposed an Intelligent Agent Technology to support decision making on government agencies. This system consists on 3 types of agents: User agents; Service Manager agents; and PA agents. The User agents support the access to services. The Service Manager agents allow searching for the best suited service for the user. PA agents give support for the decision making of a manager.

In [13] it is presented a web-based intermediary for the management of workflows that integrate services. The approach used by the authors combine workflow engines with agents' technologies. In this model the workflow controls the logic of the service delivery and selects the appropriate agent to continue with the service, achieving, this way, the interoperability.

In [6] it is briefly described an agent-based architecture, implemented with Web Services, which has four components: Citizen agent; Portal agent; Ministries and agencies agents; and the Service Agents. The Citizen agent represents the citizen and communicates with the Portal agent. The Portal agent collaborates with the Ministry and agencies agents to search the appropriate service information. The Service agents participate in the delivery of the service.

In [16] it is proposed a structure based on multi-agent technology. Agents are grouped according to their duty and forming a hierarchical structure. There are three types of agents: Organization agents are responsible for decision making and for intermediary management; the Function agents can set or invoke tasks; and Resource agents manage the information resources and provide the services.

These models stress the suitability of agent technology to implement service orchestration and interoperability in e-government. The majority of the models use agents as a representation of clients (people) and of PA agencies. Our approach uses agents only to model the agencies that provide services, which simplifies service orchestration. In fact, agents, besides providing services, may also request services from other agents, since they have an active role in the orchestration process.

3 e-Government Orchestration and Interoperability

To understand the prototype it is important to describe the basic principles of its orchestration and interoperability model. The basic concept is fairly simple: the PA is composed by entities that deliver simple services and that work together to build complex services suitable to be requested by citizens. The services that each entity is able to deliver are registered in service repositories. The following principles apply:

- All services (simple or complex) are provided by entities (public or private). Entities that have Local Information Systems (LIS) provide simple services;
- Interoperability is achieved through the orchestration services. The complex services are delivered by Intermediary entities (which may or may not be real entities) that compose services, which are consumed by other government agencies.
- The management of the delivery of a complex service is not centralized. Entities may delegate the management of the workflow (the sequence of steps for delivering a service) or part of it to other entities. So, workflows are established dynamically and there is no prior knowledge of what entities participate in the service being provided. Thus, results of service requests might be produced without knowing to whom they will be delivered, causing confidentiality or privacy issues;
- Consequently, a result is kept by the entity that produces it until requested by another entity, to be consumed. Thus, the result is provided only when its addressee is known and confidentiality or privacy measures can be enforced. The knowledge about the existence of the result is transmitted through all entities that participate in the service delivery, allowing the result to be requested when needed. This principle solves the confidentiality and privacy issues raised before.

3.1 Illustrative Use Case – Trip to a Foreign Country

This use case is very simple: a citizen needs to travel to another country and he uses a travel agency to book the tickets and the accommodation and to deal with the visa request (see Fig. 1). Only the visa acquiring and usage is addressed in the use case. It was chosen to illustrate various characteristics of the model and, therefore of the prototype. Although this use case is based on a real scenario, the sequence of services defined is the result of re-engineering the actual service to take advantage of the model and thus does not directly corresponds to how it is delivered today.

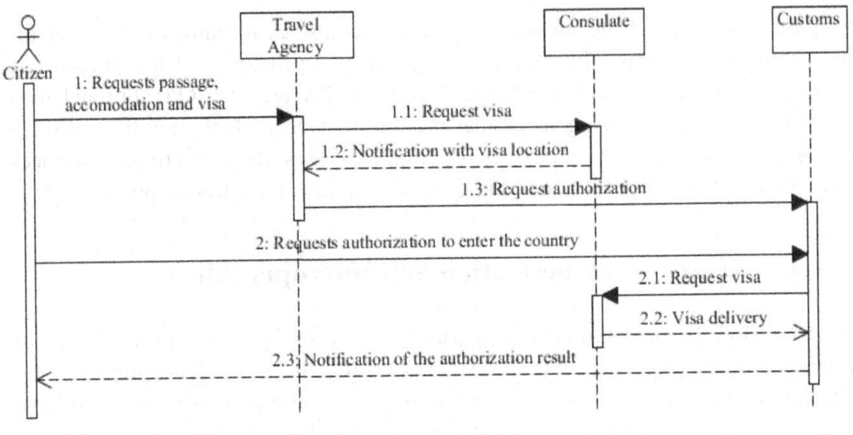

Fig. 1 Representation of the acquisition and presentation of a visa

After receiving the request from its client (1), the travel agency sends a request for a visa to the consulate of the country to which the client plans to travel (1.1). The consulate produces the visa and informs the travel agency that it is available in a given location (1.2). Then, the travel agency sends the information to the Customs of the country of destination as a request to admit its client in the country (1.3). Finally, when the client arrives at the country of destination (2), the Customs already has the information that enables it to acquire the visa from the consulate, which it does (2.1; 2.2). After receiving the visa, the Customs has the information it needs to assess if the person is allowed to enter the country or not (2.3).

In this use case there are private entities interacting with public entities which are from a foreign country. The process is asynchronous: the production of a visa is not instantaneous and the trip is booked in advance. The use case shows the utility of the basic characteristics of software agents: entities communicate amongst them (communication between agents); each entity does isolated work autonomously (autonomy); each entity reacts when a request arrives, producing some result (reactivity); and entities autonomously anticipate actions (pro-activity), an example being the sending of information from the travel agency to the Customs.

4 Prototype Implementation

The implementation of the prototype follows the model presented in [4]. The model is composed by: support services, which allow Certification and Service publication; support data structures, which define the data structures that are used to store the information for service provisioning; and by the messages that are used to communicate between entities. For implementing the architecture we used the Java Agent Development Framework (JADE).

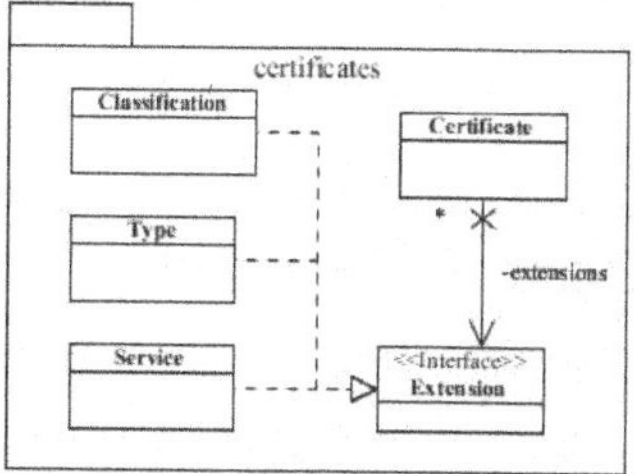

Fig. 2 Class diagram of the Certificate data structure

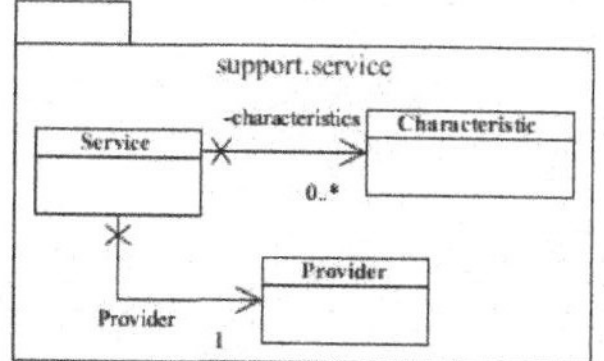

Fig. 3 Class diagram of the Service data structure

Support Services

There are three support services in the model: Time-stamping services; Certification services; and Directory services. The time-stamping services allow defining the date and time of message creation and sending, which allow the chronological assortment of messages exchanged between entities. The directory services allow the search, publication and removal of services. Finally, the certification services allow establishing, managing and issuing the certificates that certify what services an entity is able to provide and the services an entity is authorized to request.

Although these services are essential to support interoperability in the model, specially its dynamism and its cryptography functions, they were not deployed since they are fully validated technologies and are not essential for the proof of concept we desire.

Support Structures

There are two support structures in the model, both being based on open standards. The Certificate data structure (see Fig. 2), represents a certificate that is issued to an entity, is based on version 3 of the X.509 standard [1], and supports the identification, accreditation (by creating the extensions Classification, Type and Service) and authorization (by implementing the extensions Classification and Type) of entities with respect to services.

The Service Description data structure (see Fig. 3) is also based in an open standard: the Web Services Description Language (WSDL) [15]. This data structure allows the description of a service that is active and that is provided by an entity.

Both these structures were implemented and are used by the prototype. So, although the services repositories are not available in the prototype, the description of the services, described by the instantiation of these structures, are available to the agents that need it.

Messages

The model defines four types of messages (see Fig. 4). Each type has its own function: Request Service; Notify; Request Result; and Deliver Result. All the types were implemented in the prototype allowing the communication between agents and ensuring the correct use of the guiding principles of the model.

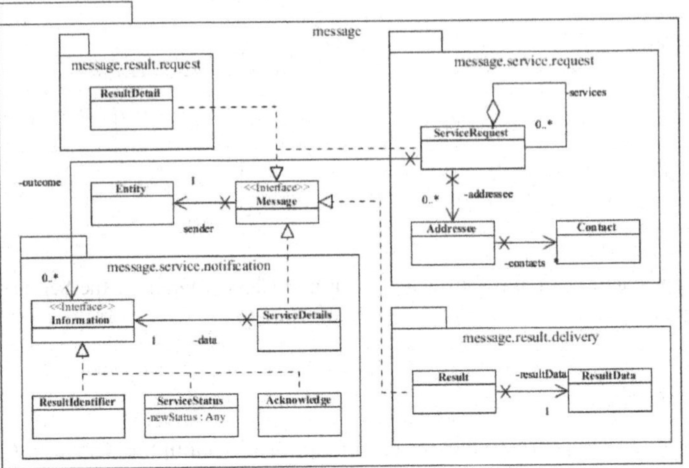

Fig. 4 Class diagram of the Message data structure

The messages also make use of the data structures defined in Sect. 4. The Message structure defines the language which is used by the agents to communicate. This structure is the common point between all messages types: the requested entity is identified and the timestamp is included. The only requirement is that agents must recognize and interpret messages with this structure.

Agent

One of the main components of the prototype is the agent. In the implementation we used the WADE extension of the JADE platform. The agent implementation was divided over four packages: the **workflow** package - includes the classes that define the tasks that an agent must execute and knows how to execute, it is used during the internal execution of the agent while the service is being provided; the **com** package - includes the classes that are responsible to build and to prepare the messages to send, likewise it includes all the necessary features to receive and validate the messages upon reception; the **security** package - contains the classes with the cryptographic functions and the assessment of accreditation and authorization; the **servicedeliverers** package - the class within this package represents a generic agent, each entity that is represented in the prototype is a subclass of this class.

Implementation of the Use Case

The use case presented (see Sect. 3.1) was deployed using three agents. Each agent (despite of the entity it represents) is an instantiation of a class that inherits the behavior of the class *BaseAgentGov*, which belongs to the *servicedeliverers* package. The differences between agents are materialized in the workflow, in the set of certificates and in the description of the services it offers.

5 Discussion

All the basic characteristics of software agents (see Sect. 4) are used in the prototype and proved to be important for the model implementation. The illustrative use case described in 3.1 take advantage of all these characteristics.

The characteristic that relates to the ability to communicate with other software agents is critical not only for the model itself, but even in a more basic way: to achieve interoperability. Only with this characteristic it is possible to delegate the management of the workflow and to require and provide services, which are crucial for service orchestration and for the presented interoperability model.

The reactive characteristic is an important feature with respect to the service provisioning since after the reception of the service request the agent knows how to deal with it and perform the necessary tasks to complete its execution.

The remaining two characteristics, autonomy and pro-activity, are essential to implement two of the main features of the model – its dynamism and concurrency. The concurrency is a feature that can only be achieved if there are two or more entities able to provide the same service. The dynamism of the model is only achieved if it is possible to know that a new entity is available or that an old entity was removed. The autonomy and the pro-activity allow the agent to be aware of its surroundings, being able to identify new entities and to identify services that are provided by different entities but with similar outcomes.

6 Conclusions

In this paper we presented an agent-based prototype that implements a model for e-government interoperability, an illustrative use case that demonstrate its deployment; and the advantages of using agents to implement the referred model.

We conclude that agent technology proves to be an appropriated choice to implement the tested e-government interoperability model, namely because of its key characteristics: interoperability, service orchestration, dynamism and concurrency. No major pitfalls were identified.

References

1. Cooper, D., Santesson, S., Farrell, S., Boyeen, S., Housley, R., Polk, W.: Internet X.509 Public Key Infrastructure Certificate and Certificate Revocation List (CRL) Profile. RFC 5280. Internet Society (2008), `http://tools.ietf.org/html/rfc5280`
2. Gugliotta, A., Cabral, L., Domingue, J.: Knowledge Modelling for Integrating e-Government Applications and Semantic Web Services. In: AAAI Spring Symposium: Semantic Web Meets eGovernment, pp. 21–32 (2006)
3. Luck, M., Ashri, R., D'Inverno, M.: Agent-Based Software Development. Artech House, Norwood (2004)
4. Marques, F., Dias, G.P., Zòquete, A.: A General Interoperability Architecture for e-Government based on Agents and Web Services. In: 6th Iberian Conf. of Inf. Sys. and Tech., pp. 338–343 (2011)

5. Marques, F., Dias, G.P., Zòquete, A.: Modelo de Segurança para a Composição Dinâmica de Workflows em Arquiteturas de e-Government. Iberian J. of Inf. Sys. and Tech. 9, 15–26 (2012)
6. Mellouli, S., Bouslama, F.: Issues in Mult-agent Systems for e-Government Applications. In: Current Trends in Information Technology (CTIT), pp. 53–56 (2011)
7. De Meo, P., Quattrone, G., Terracina, G., Ursino, D.: Utilization of intelligent agents for supporting citizens in their access to e-government services. Web Intelligence and Agent Systems: An International Journal 5, 273–310 (2007)
8. Muthaiyah, S., Kerschberg, L.: Achieving Interoperability in e-Government Services with two Modes of Semantic Bridging: SRS and SWRL. Journal of Theoretical and Applied Electronic Commerce Research 3 (2008)
9. Piotrowski, Z.: Perspectives for using Software Agents in e-Government Applications. Annales UMCS, Informatica 8, 203–212 (2008)
10. Rishi, O.P., Sharma, A., Bhatnagar, A., Gupta, A.: Service Oriented Architecture for Business Dynamics: an Agent-based Approach. Emerging Technologies in E-Government, 19–28 (2008)
11. Sabucedo, L.M.A., Rifon, L.E.A., Corradini, F., Polzonetti, A., Re, B.: Knowledge-based Platform for eGovernment Agents: A Web-based Solution using Semantic Technologies. Expert Systems with Applications 37, 3647–3656 (2010)
12. The European Communities: Linking up Europe: The Importance of Interoperability for eGovernment Services. Commission staff working paper (2003)
13. Verginadis, Y., Mentzas, G.: Agents and Workflow Engines for Inter-Organizational Workflows in e-government cases. Business Process Management Journal 14, 188–203 (2008)
14. World Bank, http://go.worldbank.org/JKO5DVDMQ0
15. World Wide Web Consortium (W3C): Web Services Description Language (WSDL), version 1.1, W3C Recommendation (2001), http://www.w3.org/TR/wsdl
16. Zhang, P., Wang, Y., Wang, X.: Research on the Integration in E-government Based on Multi-agent. In: Web Intelligence and Intelligent Agent Technology, vol. 3 (2006)

Context- and Social-Aware User Profiling for Audiovisual Recommender Systems

César A. Mantilla, Víctor Torres-Padrosa, and Ramón Fabregat

Institute of Informatics and Applications (IIiA), University of Girona,
Girona 17071, Spain
{cesar.mantilla,victor.torres,ramon.fabregat}@udg.edu

Abstract. User profiles are the base to obtain knowledge about users of recommender systems. We propose a context- and social-aware user profiling for audiovisual recommender systems that combines explicit preferences, implicit preferences and stereotypes modeling, taking advantage of information available in social networks and the current user context. We examine how the user profile is represented, acquired, built and updated; and how the profile information is exploited by an audiovisual recommender system that uses both collaborative filtering and the content-based method.

1 Introduction

Recommender systems are aimed to provide users a more proactive and intelligent information service [1-3]. Currently, they combine user profiling, information filtering and machine learning among other techniques, to provide services or products that match the learned user preferences. In general, recommender systems use either content-based filtering [4] or collaborative filtering [5]. The content-based approach recommends items whose content is similar to content that the user has previously viewed. Collaborative filtering collects information about users or even asks them to rate items and makes recommendations based on their views or on highly rated items corresponding to users with similar characteristics, i.e., it relies on user profiles. Some proposals combine both content and collaborative filtering approaches [6].

User profiling [7] for recommender systems is one of most challenging tasks because it is generally difficult to elicit this information. There are many approaches and proposals to model user preferences for audiovisual content recommendation [8-11] and some which also consider the context where consumption is completed [12-14]. The user context variables can be used to improve both the user model and the recommender system, and to differentiate the recommendations according to the current conditions and situations of users. However, none of the aforementioned approaches deal with the new paradigm arising from the exploitation of social networking. Social information can be used to extend or to initialize user profiles to be able to generate more reliable recommendations.

S. Omatu et al. (Eds.): *Distrib. Computing & Artificial Intelligence*, AISC 217, pp. 569–577.
DOI: 10.1007/978-3-319-00551-5_68

The main contribution of this paper relies on the adoption of a user model influenced by the context and that is based on implicit, explicit and stereotypical user preferences, which derive not only from user behaviour and settings in the system, but also from the information extracted from social networks. We give a special attention to implicit preferences and how they are updated according to user actions, considering content consumption as well as social actions over content.

The paper is structured as follows. In section 2 we describe the high-level architecture and components of the recommendation system, i.e., the user model and the recommendation engine. In section 3 we describe for each part of the user model how preferences can be obtained from different sources and how they are updated. Section 4 presents the conclusions and the envisaged future work.

2 Recommender System Architecture

In this section we present the high level architecture of the recommender system (Figure 1), which is mainly composed of a user model, a context model and the recommendation engine. The user model contains the user preferences, as described in section 3. These preferences will come from different sources (explicit, implicit and stereotypical), will be updated according to the user consumption and social actions regarding content, and will be influenced by the user session context.

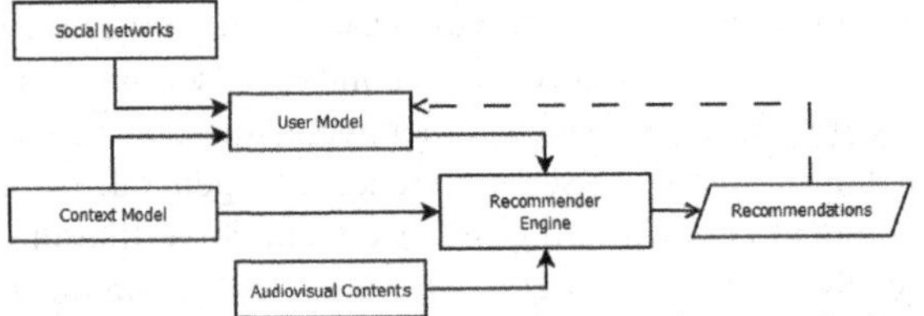

Fig. 1 Architecture of the Recommender System

The context model represents external factors surrounding the user such as the device, location and weather, among others, which will be relevant to improve recommendation. Finally, the recommendation engine combines different elements from the user model, context model, contents and social networks in order to output the best possible recommendations, which feedback the user model.

3 User Model

The user model (Figure 2) contains the characteristics, preferences and data of the user, which are further processed in the recommendation engine to generate suitable recommendations. When new users are registered, the system needs to obtain some information about them to model their preferences. In our case, the user model initialization and update is carried out of three forms:

- Data provided directly by the user through web forms or questionnaires, or extracted from social networks (explicit preferences).
- Data observed from users behavior in the system within a specific context or from comments or actions related to audiovisual contents available in social networks (implicit preferences).
- Data bundled to user stereotypes (stereotypical preferences).

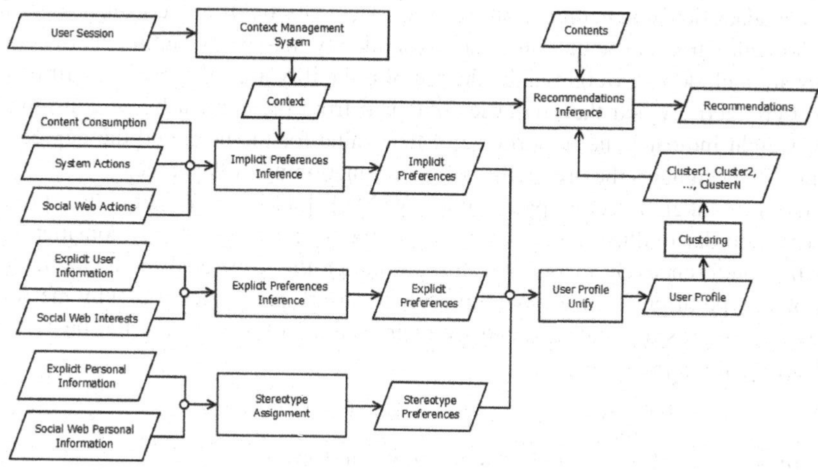

Fig. 2 User Model

Contents will help to model the user preferences. To achieve this objective, contents have to be classified into categories and characterized by metadata and their respective values. Each user preference will have a weight according to its importance, and a degree of reliability depending on the number of times it appears in the content in which the user has interacted (implicit preferences), the source of obtaining it (explicit preferences) or the matching of users with the stereotypes.

3.1 *Explicit Preferences*

Explicit preferences are those related to types or other attributes of audiovisual content (e.g., comedy, action, preferred actors or directors, etc.) and general preferences (e.g., cars, hobbies, etc.). This information is either explicitly entered by the user or obtained from the data extracted from social networks in which the user participates. Through web forms presented to the user, a set of items will be weighted or ranked based on a scale. If users allow access to their data in social networks, those cases where they explicitly show interest in certain products or services will be considered as being explicit information. For instance, from Facebook the system can access the user's "likes", preferred movies, books or music, while from YouTube the list of favorite or published videos can be determined.

3.2 *Implicit Preferences*

Implicit preferences are those related to specific content preferences. Preferences will be inferred according to users' activities in the system (videos viewed, tagged actions, opinions, etc.) or in social networks (whenever they can be accessed). Implicit preferences are classified in preferences by metadata values, preferences for metadata fields and content preferences. Metadata fields and content types are predefined, while metadata values are dynamically defined at runtime as the contents are included. A weight and a degree of reliability are calculated according to the user's activity and the properties obtained from the content being consumed. The weight indicates the importance of that value for the user, and the degree of reliability designates the probability of good functioning of the value.

We have taken a vector space model (VSM) [9], [15] to model the user preferences. In order to illustrate how it works, in the formulation we only consider the implicit preferences by metadata values, although the same analysis is applicable for other type of preferences. We represent the user's preferences (UP) by metadata values (m_i) as a vector ordered by their weights (w_i), and also the degree of reliability of the preference (r_i):

$$UP = ((m_1,w_1,r_1),(m_2,w_2,r_2),\dots,(m_n,w_n,r_n)) \quad (1)$$

where $w_i \geq w_{i+1}$, $1 \leq i \leq n$ and r_i ranges between 0 and 1.

Preferences Weight Update

Updating user preferences is carried out according to quantification of the user actions and analysis of the content consumption. Metadata values and their respective weights and degree of reliability will be updated and then rearranged according to their current weight. When the user has viewed contents partially over a period of time, or has carried out some actions, the user profile is recalculated and updated. An initial way to take into account the consumption of contents is based on the percentage of the total time of the viewed content, where the viewing ratio is calculated as:

$$\text{Viewing ratio} = Tr / Tt \quad (2)$$

where *Tr* is the real time viewing by the user and *Tt* the total time of the content. This ratio will directly influence the weighting of metadata that conform the content.

When users perform actions over the contents, which may be executed in the proprietary system or in the social networks (Facebook, Twitter, YouTube, etc.), we need to add a positive or negative quantification (W_{action}) of these actions according to their relevance to calculate the new weight of each metadata value. The users actions taken into account are detailed in Table 1.

Semantic analysis is necessary to infer negative or positive qualifications about contents in the case of tweets or comments. Thus, the updated weight of a specific metadata value in the preferences vector is defined as:

$$w_{i\_updated} = (1-\eta) * w_{i\_current} + \eta * (K_{visualization} * W_{visualization} + K_{action} * W_{action}) \quad (3)$$

where:

- η is the constant learning rate that establishes how quickly the system forgets the previous preferences and taken into account the new ones.
- $w_{i\_current}$ is the weight of the metadata value located at the i_{th} position of the vector.
- $K_{visualization}$ and K_{action} are constants according to the importance given respectively by the system to viewing contents and to the actions being executed. $K_{visualization} + K_{action} = 1$ and $0 \leq K_{visualization} \leq 1$, $0 \leq K_{action} \leq 1$
- $W_{visualization}$ is (Tr / Tt) x ε, where ε is a calculated value according to the importance of the metadata (i.e., actor, director) for the user.
- W_{action} is the value of the executed actions over the contents.

Table 1 Actions taken into account

Actions	W_{action}
"I like", "+1", positive vote, sharing or recommending contents	+2
"Dislike", negative vote	-2
Users' tweets (Twitter), making comments	+1 or -1
Content Rating	-2 to +2
Viewing content later	+1

Next we illustrate the case of viewing content and executing actions. We define $K_{visualization} = 0.7$, $K_{action} = 0.3$ and $\eta = 0.75$. Users view a film by Steven Spielberg during 90 minutes out of 100 minutes, and they perform the action "Recommend". We assume that the value "Steven Spielberg" of the metadata field "director" already exists in the user's preferences at the position 5 with a weight of 0.85. The updated weight will be:

$$w_{5\_updated} = 0.25 * 0.85 + 0.75 * (0.7 * (90/100) + 0.3 * (2)) = 1.135$$

We can include any other variable that may influence the user model for obtaining the preferences. For instance, if social networks and the user context directly affect user preferences, we would be adding two new terms into the formula:

$$w_{i\_updated} = (1-\eta) * w_{i\_current} + \eta * (K_{visualization} * W_{visualization} + K_{action} * W_{action} + K_{social} * W_{social} + K_{context} * W_{context}) \quad (4)$$

Reliability of Preferences Update

The degree of reliability indicates the probability of good functioning of the preferences. The elements that have to influence its calculation are the consumption, the actions carried out and the number of occurrences of each of the items. The degree of reliability generated from consumption will be left for future work. If we have a predefined value *n,* such that after executing a certain number of actions, the grade of reliability is maximum, a value *A* that quantifies each action, as was described previously, and a value *B* that represents the viewing ratio, the degree of reliability of any metadata is:

$$r_{i\_updated} = r_{i\_current} + (|A| + B)/n \tag{5}$$

If a metadata did not previously exist in the preferences, the initial value of $r_{i\_current}$ is equal to 0. The calculation is carried out as long as $r_{i\_current}$ is less than 1, which is the highest reliability value.

3.3 *Stereotypical Preferences*

Stereotypical preferences refer to the preferences associated to some specific user stereotypes, where a user stereotype stands for a group of users with a common set of characteristics that makes them have similar audiovisual consumption preferences. The main objective of having stereotypical preferences is to be able to classify users in stereotypes, according to their personal information and profile, and solve the major drawback of collaborative filtering and content based approaches referred to as the cold start problem.

This mechanism is based on a previous demographic knowledge and content influence statistics in the field of interest, which in this case is audiovisual content, so that stereotypes and their preferences can be predetermined. This approach is different from dynamic clustering, since the latter relies on finding groups of users with similar preferences without any previous knowledge of the users other than their behavior in the system. Starting from the personal information of a user, the user will be mapped to one or more stereotypes with a certain degree of coincidence, so that actions affecting implicit user preferences should also affect the stereotype profile proportionally. In our proposal we identify attributes like age, gender, education level and profession to classify the stereotypes and their content types preferences according to surveys carried out by experts [16]. Thus, we use vector similarity between stereotype groups and user personal information to classify the user in one or several stereotypes.

As people interact further with the system, the implicit preferences will be more important than the stereotypical preferences and collaborative filtering and content-based approaches will be taken into account to carry out the recommendations.

3.4 Context Model

The context model represents external factors surrounding the user such as the device, location and weather, among others, which are relevant to improve recommendation. It is an input to the recommendation engine together with the user model that is used to generate the best possible recommendations. The considered context data is the information about the device properties from which the user connects (type and screen resolution, speed and cost of connection), location and time (expressed explicitly by the user or inferred from its IP address), the place (home, work, on the move), the companionship (single, family, children, friends, etc.). Although this information is external to the user, it can help to contextualize their actions and their environment to perform better recommendations.

Thus, depending on the user's device capabilities, the recommender will decide to recommend a high or low definition audiovisual content, or to suggest contents whose properties are related to the current city or weekday. On the other hand, these context variables will be used to know the people preferences according to their context. Thus, the recommender will know that when users are at home they prefer to view television series and when they are at work, they prefer to view economy news. For these reasons the context is considered and bundled to the implicit preferences.

3.5 User Profile Unification and Recommendations Inference

The recommender engine receives as input a set of clusters obtained from the unification of the preferences of each user, audiovisual contents specified by their metadata and the current user session context.

The unified user preferences, given that each preference has a weight and a degree of reliability, can be obtained as a new ordered vector with an overall weight as follows:

$$W_{i\_overall} = (w_{i\_explicit} * r_{i\_explicit}) + (w_{i\_implicit} * r_{i\_implicit}) + (w_{i\_stereotype} * r_{i\_stereotype}) \quad (6)$$

Collaborative filtering will be applied to cluster these new preference vectors, taking into account the context and the user's relationships in the social networks to filter the adequate and relevant contents to carry out the recommendations.

4 Conclusions and Future Work

In this paper we have presented an enhanced user model for audiovisual recommender systems that goes beyond some previous proposals. Our approach uses different types of preferences (i.e. implicit, explicit and stereotypical), taking into account the context and the social network information.

We have proposed to update implicit preferences according to the user actions and behavior in the system, among which social actions over content (e.g. like, recommend, rate, etc.) are considered as a novel contribution. A context model is also presented and linked to the implicit preferences. We have seen how to take advantage of the user profile present in social networks (e.g. personal details, likes) to improve the knowledge of the user and initialize their explicit profile. Finally, we have proposed to use predefined stereotypes based on the knowledge content consumption habits in order to avoid cold start issues when applying clustering techniques.

Ongoing work involves the inclusion of the user and context models in the recommendation process and the implementation of the full framework in the MIREIA project [17], which will have visible outputs about the middle of 2013. Recommendation accuracy, usability and user satisfaction will be evaluated through the development of targeted pilots.

Acknowledgements. This work has been partly funded by the Spanish Ministry MINECO through Project MIREIA (IPT-2011-2015-430000) “Distribution of Contents through Intelligent Recommendation Engines adapted to Social Networks” and Juan de la Cierva subprogram grant JCI-2010-08322 and by the Generalitat de Catalunya through the research support program project SGR-1202.

References

1. Adomavicius, G., Tuzhilin, A.: Towards the Next Generation of Recommender Systems: A Survey of the State-of-the-Art and Possible Extensions (2005)
2. O'Donovan, J., Smyth, B.: Trust in recommender systems. In: Proceedings of the 2005 International Conference on Intelligent User Interfaces, pp. 167–174. ACM (2005)
3. Pascual-Miguel, F., Chaparro-Peláez, J., Fumero-Reverón, A.: Presente y futuro de los sistemas recomendadores en la web 2.0. El Profesional de la Información 20(6), 645–651 (2011)
4. Pazzani, M.J., Billsus, D.: Content-Based Recommendation Systems. In: Brusilovsky, P., Kobsa, A., Nejdl, W. (eds.) Adaptive Web 2007. LNCS, vol. 4321, pp. 325–341. Springer, Heidelberg (2007)
5. Schafer, J.B., Frankowski, D., Herlocker, J., Sen, S.: Collaborative Filtering Recommender Systems. In: Brusilovsky, P., Kobsa, A., Nejdl, W. (eds.) Adaptive Web 2007. LNCS, vol. 4321, pp. 291–324. Springer, Heidelberg (2007)
6. Burke, R.: Hybrid Web Recommender Systems. In: Brusilovsky, P., Kobsa, A., Nejdl, W. (eds.) Adaptive Web 2007. LNCS, vol. 4321, pp. 377–408. Springer, Heidelberg (2007)
7. Kobsa, A.: Generic User Modeling Systems. User Modeling and User Adapted Interaction 11, 49–63 (2001)
8. Ardissono, L., Gena, C., Torasso, P., Bellifemine, F., Chiarotto, A., Difino, A., Negro, B.: Personalized recommendation of TV programs. In: Cappelli, A., Turini, F. (eds.) AI*IA 2003. LNCS (LNAI), vol. 2829, pp. 474–486. Springer, Heidelberg (2003)
9. Zhiwen, Y., Xingshe, Z.: TV3P: An Adaptive Assistant for Personalized TV. IEEE Transactions on Consumer Electronics 50(1), 393–399 (2004)

10. Lee, W.P., Yang, T.H.: Personalizing Information Appliances: a Multi-agent Framework for TV Programme Recommendations. Expert Systems with Applications 25(3), 331–341 (2003)
11. Blanco-Fernández, Y., Pazos-Arias, J., López-Nores, M., Gil-Solla, A., RamosCabrer, M.: AVATAR: An improved solution for personalized TV based on semantic inference. IEEE Transactions on Consumer Electronics 52(1), 223–231 (2006)
12. Park, W.I., Park, J.H., Kim, Y.K., Kang, J.H.: An Efficient Context-Aware Personalization Technique in Ubiquitous Environments. In: Proceedings of the 3rd International Conference on Ubiquitous Information Management and Communication, pp. 415–421. ACM (2010)
13. Palmisano, C., Tuzhilin, A., Gorgoglione, M.: Using context to improve predictive modeling of customers in personalization applications. IEEE Transactions on Knowledge and Data Engineering 20(11), 1535–1549 (2008)
14. Adomavicius, G., Tuzhilin, A.: Context-Aware Recommender Systems. In: Recommender Systems Handbook, pp. 217–253 (2011)
15. Salton, G., Wong, A., Yamg, C.S.: A Vector Space Model for Automatic Indexing. Communications of the ACM 18(11), 613–620 (1975)
16. Redondo, I., Holbrook, M.B.: Modeling the appeal of movie features to demographic segments of theatrical demand. Journal of Cultural Economics 34, 299–315 (2010)
17. MIREIA project, `http://mireia.laviniainteractiva.com/`

Current Trends in Bio-Ontologies and Data Integration

Rafael Pereira and Rui Mendes

University of Minho, Campus of Gualtar, Braga, Portugal
rafatp@di.uminho.pt, rcm@di.uminho.pt

Abstract. Biological data integration is currently one of the major challenges in the field of Bioinformatics. Several studies show that biological knowledge is growing at a continuous rate and is usually distributed among many databases. This paper presents an overview of data integration and the existing approaches that aim to solve this problem. The paper will also review the different ontology approaches that were introduced by researchers for representing biological data.

Keywords: Data integration, ontologies, bioinformatics, databases.

1 Introduction

Nowadays one of the main issues addressed in the bioinformatics field is understanding the structure and behaviour of complex molecular interaction networks that model genetic/metabolic regulation. The vast amount and complexity of biological data retrieved in recent years requires an integrated approach, thus forcing scientists to look for novel approaches in order to address this issue. This task is difficult because researchers often need to retrieve information from several databases and work with different data types at the same time.

Integrating data from different data sources is a very difficult task within the bioinformatics domain. The reason for the complexity of this area is because of the term ambiguity in available databases However, there are usually several differences between both the structure and data types in bioinformatics databases. One way to solve the data problem of integration is to use ontologies to create relationships between different sources.

In this context, ontologies provide a good solution to this problem. They are crucial for maintaining the coherence between grouped data within a complex collection of related concepts. According to Grubber [15] an ontology is an explicit specification of a conceptualization. Whilst controlled vocabularies only restrict words that will be used in a particular domain, ontologies extend this characteristic and allow a formal specification of terms and their relationships. A specification of a conceptualization is a written, formal description of a set of concepts and relationships in a domain of interest [19].

S. Omatu et al. (Eds.): *Distrib. Computing & Artificial Intelligence*, AISC 217, pp. 579–586.
DOI: 10.1007/978-3-319-00551-5_69

2 Data Integration

Currently, a large part of the work of scientists involves querying multiple heterogeneous data sources, manually retrieving and integrating data and manipulating it with advanced data analysis and visualization tools [22]. Thus, different profiles of professionals are necessary to gather and process the data, as depicted in Figure 1.

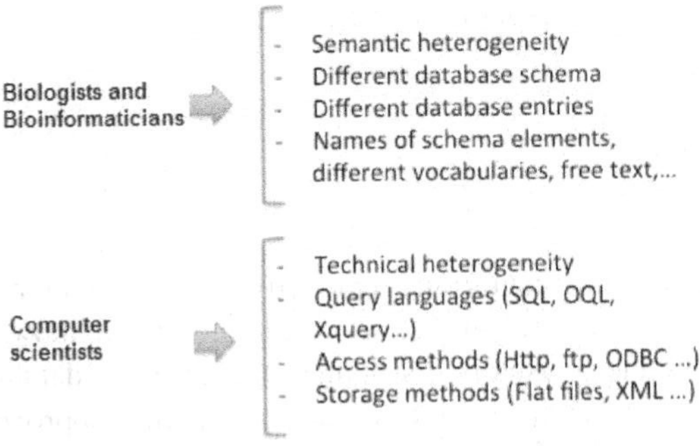

Fig. 1 Differences between biologists, bioinformaticians and computer scientists and their perspectives

2.1 Data Integration Approaches

NCBI (National Center for Biotechnology Information) attempts to solve the data integration issue by creating database that incorporates several types of biological information. In NCBI it is possible to find a large number of tools and sources to explore the bioinformatics field, as well as search engines to find papers on PubMed, DNA sequences on GenBank, tools to analyse sequence alignment, data mining and many others.

Other approaches aim to solve the integration data problem by using ontologies, like INDUS (Intelligent Data Understanding System), a system for information integration and knowledge acquisition from semantically heterogeneous distributed data [10] and DAS (Distributed System Annotation) a widely adopted protocol for the integration of biological data types in user-driven contexts [17].

The main objective of the INDUS system is to provide knowledge acquisition for semantically heterogeneous distributed data sources. INDUS uses methods ontology based methods in order to query multiple repositories in a flexible manner in such a way that each data source can be regarded as a single table. This system can help answer queries to several sources without the need of a centralized database or a common global ontology [11]. When the user performs a query, it is dispatched to a part of system, called *query answer engine*, that operates as an intermediate step between clients and servers. This engine divides the query into sub-queries and directs them to the corresponding data sources then it organizes the results of the sub-queries into a final view and sends the reply back to the user [10].

DAS was initially developed as a system for sharing, visualizing and aggregating genomic annotation. The rationale is that information is not concentrated in a single centralized database, but distributed over multiple sources. This protocol may be regarded as a standard for sharing biological information because it defines how data should be shared and visualized. The system architecture follows the client-server model [17]. The DAS system works with a set of annotations, divided into several layers where each contains a particular kind of annotation produced by a specific data source. The information exchange is performed using a XML file. The DAS servers are able to provide annotation layers in real-time and a DAS client can overlay them to produce a single integrated view [23]. DAS may also be used to annotate different types of data. It employs a coordinate system that is used to describe several supported reference data types, where each coordinate can be seen as a common information model used by bioinformaticians to describe biological entities.

2.2 *Difference between Data Integration Approaches*

The difference between the INDUS and DAS approaches, is the way they perform the data integration. The main features of INDUS are: a clear distinction between data and their semantics which help define mappings from data sources ontologies to user ontologies; Each user can specify their ontologies; It has a user-friendly ontology and mappings editor. DAS uses a decentralized structure, its best advantage is when integrating frequently changing data. It also allows more complex integration strategies to be inserted, but is not simple to use because it needs more involvement in the setup process and textual display.

3 Biological Ontologies

According to Grubber[15], an ontology is an explicit specification of a conceptualization. Whilst controlled vocabularies only restrict words that will be used in a particular domain, ontologies extend these characteristics and describe a formal specification of terms and their relationships. A specification of a conceptualization is a written, formal description of a set of concepts and relationships in a domain of interest [19].

Nowadays, it is possible to find a large variety of biological ontologies. They differ in the type of biological knowledge they describe, their intended use, the adopted level of abstraction and the knowledge representation language [16].

One of the major challenges faced by the biomedical research community is how to access, analyse, and visualize heterogeneous data in ways that lead to novel insights into biological processes or that lead to the formulation hypothesis that can be tested experimentally [9]. An advantage of using ontologies is their ability to define relationships between concepts that distinguishes the integration and interoperability approaches. 'Bio-ontologies' are formal representations of areas of knowledge in which the essential terms are combined with structuring rules that describe the relationships between the terms [6].

An ontology of biological terminology provides a model of biological concepts that can be used to form a semantic framework for many data storage, retrieval and analysis tasks [4]. Nowadays, is possible to find a large number of bio-ontologies that were developed to address the biological data integration problem. Gene Ontology (GO) was initially developped in 1998 with the intent of creating a consistent way of representing the cross-species vocabularies that are used by multiple databases. It was a collaboration of the following organism databases: FlyBase, the genome database for *Drosophila* [29]; the *Saccharomyces* Genome Database (SGD) [5]; and integrated Mouse Genome Informatics Database, Mouse Genome Database, MGD [8] and Gene Expression Database, GXD [24]. GO is currently the most successful ontology in this endeavor, measured both in number of users and by reach across species [26] and also combines several model organism database groups by providing a shared annotation system for describing some primary aspects of organismal biology [30].

3.1 *Examples of Biological Ontologies*

The use of biological ontologies emerged recently as an important topic [3]. The proliferation of ontologies can be a problem and create obstacles for data integration. The Open Biomedical Ontologies (OBO) consortium aims to overcome this problem by performing a coordinated effort where each ontology shares a set of principles governing their development [26]. The aim is to design ontologies that can be interoperable and follow a logical structure that incorporates accurate representations of biological reality. Many biological ontologies are available via OBO; currently, it provides 112 ontologies for shared use across different biomedical domains. OBO is open to everyone who works with ontologies within the biomedicine domain.

There are other initiatives using biological ontologies: EcoCyc [20], ONDEX [2], RIBOWEB [27], LinkBase [13], Molecular Biology Ontology (MBO) [25], Gene Regulation Ontology (GRO) [12]. These approaches are summarized below:

- **EcoCyc:** it is the most popular database to study *Escherichia coli*, that describes all known genes of this bacteria, the enzyme of small-molecule metabo lism that are encoded by these genes, the reactions catalysed by each enzyme, and the organization of these reactions in metabolic pathways [19]. EcoCyc, uses an ontology to describe the richness and complexity of the domain and the constraints acting within that domain, to specify the database [18]. Table 3.1 summarizes approximately the current of content stored in EcoCyc [20].
  The advantage of using ontologies in this case is the definition of a database schema to cope necessary changes in biological information.
- **ONDEX:** the main idea here is to represent ontologies and databases using the same structures to assist in linking, integrating and relating data from different data sources through algorithms and methods that can be applied in the same way both for ontologies and for databases. The ontology used allows the user to import data using a graph template. It uses a mechanism to ensure the correct

Table 1 Number and types of data stored on EcoCyc

Data Type	Number
Genes	4489
Genes products covered by a mini-review	3666
Enzymes	1450
Metabolic Reactions	1446
Compounds	2105
Transporters	252
Transport reactions	292
Transported substrates	207
Transcription Factors	175
Transcription units	3409
With experimental evidence	1043
Regulatory Interactions	5345
Transcription initiation	2746
Transcription attenuation	18
Enzyme modulation	2473
Other	108
Literature citations	20 284

semantics of the edges, as well as the links that represent relationships between the entities. Thus, different networks from different levels of the biological hierarchy can be integrated into the same database structure [21].

- **RiboWeb:** this approach implements four types of ontologies. The first, called *physical-thing ontology*, is used to specify the molecular components and cofactors that are present in ribosome, and to determine the objects and relations for representing data about ribosomal structure. The second ontology, called *data ontology*, provides a specification of data types that are gathered from biological experiments. The third one, called *reference ontology*, determines the types of publications. The key contribution here is to recognize/associate scientific data with a particular publication, enabling users to review and validate the data introduced. The fourth one is called *methods ontology*. It specifies the types of actions the system can perform and declares the key attributes required to execute these actions [1]. The main idea of RiboWeb, is to use these four ontologies to provide a integrated system and make it intuitive to the users, thus facilitating the reconstruction of 3D models of ribosomal components and to compare the experimental results with existing studies [28].
- **MBO:** The Ontology for Molecular Biology is a general ontology whose aim is to collect all relevant concepts that are required to describe biological objects, experimental procedures and computational aspects of molecular biology [14]. The reason for the creation of this ontology was to avoid a frequent semantic confusion that happens when working with several biological data sources. Thus, by adhering to an ontology that is strongly supported in concepts, the frequent uncertainty and misunderstandings about semantic relations between

database entries from different repositories can be eliminated [28]. The main objective of this ontology is to collect all types of entities that molecular biologists include in their work and placing those concepts in a appropriate hierarchical system and annotating them with additional properties [25].

- **GRO:** Gene Regulation Ontology is a conceptual model that was developed as a novel approach to design complex events that are part of the gene regulation process. The ontology defines gene regulation events in terms of ontological classes and imposes constraints on them by specifying the participants involved [7]. GRO can be represented by a direct acyclic graph structure where the classes are represented by a vertex of the graph and the relationships between these classes are represented by edges [7].

The ontologies reviewed here are summarized in Table 2, according to their knowledge domain and application.

Table 2 Biological Ontologies

Ontology	Knowledge domain	Application
GO	Cellular components, molecular function and biological process	Controlled vocabulary of biological terms
EcoCyc	*E.coli* genome, transcriptional regulation, transporters, and metabolic pathways	Database schema
ONDEX	General	Data integration
RiboWeb	Information related ribosome and their components	Database schema
MBO	Molecular biology and experimental procedures	Community reference
GRO	Regulation process	Semantic annotation

4 Conclusion

The traditional approaches in biological data representation suffer from a large proliferation of databases of heterogeneous data, each database using different structures and formats, some of which using poor structure or integrity. All these factors along with the dispersion of information entail the fact that it is very difficult to perform data integration.

Ontologies help establish standards for the concepts, they help provide a way of representing the structure and relationship between concepts and are easily extendible. They can also help to integrate data from several sources by establishing an ontology for each source and a way of incorporation all the data obtained into a single body of knowledge. Because of all these characteristics, this field is currently experiencing a steady growth with several approaches being created in parallel. The main challenges right now are the definition of standards that will allow researchers to be able to share information from different organisms.

Acknowledgement. The work is partially funded by ERDF - European Regional Development Fund through the COMPETE Programme (operational programme for competitiveness) and by National Funds through the FCT (Portuguese Foundation for Science and Technology) within projects ref. COMPETE FCOMP-01-0124-FEDER-015079 and PEst-OE/EEI/UI0752/2011 and by CNPq, National Council for Scientific and Technological Development - Brazil.

References

1. Altman, R.B., Bada, M., Carillo, M.W., et al.: RiboWeb: An Ontology-Based System for Collaborative Molecular Biology. IEEE Inteligent Systems and Their Applications 14(5), 68–76 (1999)
2. Ananiadou, S., Canevet, C., et al.: Ondex Data Integration and Visualization (2008), http://www.ondex.org/
3. Baher, J.O.C., Kei-Hoi, C.: Semantic Web: Revolutionizing Knowledge Discovery in the Life Sciences. Springer Science+Business Media (2007)
4. Baker, P., Goble, C., Bechhofer, S., et al.: An ontology for bioinformatics applications. Bioinformatics 15(6), 510–520 (1999)
5. Ball, C.A., Dolinski, K., Dwight, S.S., et al.: Integrating functional genomic information into the Saccharomyces genome database. Nucleic Acids Research 28(1), 77–80 (2000)
6. Bard, J.B.L., Rhee, S.Y.: Ontologies in biology: design, applications and future challenges. Nature Reviews. Genetics 5(3), 213–222 (2004)
7. Beisswanger, E., Lee, V., et al.: Gene Regulation Ontology (GRO): design principles and use cases. EHealth Beyond the Horizon: Get IT There 136, 9–14 (2008)
8. Blake, J., Eppig, J.: The Mouse Genome Database (MGD): expanding genetic and genomic resources for the laboratory mouse. Nucleic Acids Research 28(1), 108–111 (2000)
9. Blake, J.A., Bult, C.J.: Beyond the data deluge: data integration and bio-ontologies. Journal of Biomedical Informatics 39(3), 314–320 (2006)
10. Caragea, D., Pathak, J., Bao, J., Silvescu, A., Andorf, C., Dobbs, D., Honavar, V.G.: Information integration and knowledge acquisition from semantically heterogeneous biological data sources. In: Ludäscher, B., Raschid, L. (eds.) DILS 2005. LNB, vol. 3615, pp. 175–190. Springer, Heidelberg (2005)
11. Caragea, D., Pathak, J., Honavar, V.: Learning classifiers from semantically heterogeneous data. In: Proceedings of the International Conference on Ontologies, Databases, and Applications of Semantics (2004)
12. European Bioinformatics Institute: Gene Regularion Ontology, http://www.ebi.ac.uk/Rebholz-srv/GRO/GRO.html
13. Flett, A., dos Santos, M.C., Ceusters, W.: Some ontology engineering processes and their supporting technologies. In: Gómez-Pérez, A., Benjamins, V.R. (eds.) EKAW 2002. LNCS (LNAI), vol. 2473, pp. 154–165. Springer, Heidelberg (2002)
14. Gomez, S.M., Krauthammer, M., Kaplan, S.H., et al.: A knowledge model for analyses and simulation of regulatory networks. Bioinformatics Ontology 16(12), 1120–1128 (2000)
15. Gruber, T.: Toward principles for the design of ontologies used for knowledge sharing. In: Formal Ontology in Conceptual Analysis and Knowledge Reresentation. Kluwer Academic Publishers (1995)
16. Jakoniene, V., Lambrix, P.: Ontology-based integration for bioinformatics. In: VLDB 2005: Proceedings of the 31st International Conference on Very Large Data Bases. VLDB Endowment, Trondheim (2005)

17. Jenkinson, A.M., Albrecht, M., Birney, E., et al.: Integrating biological data–the Distributed Annotation System. BMC Bioinformatics 9(suppl. 8), S3 (2008)
18. Karp, P.D., Paley, S.: Integrated access to metabolic and genomic data. Journal of Computational Biology: a Journal of Computational Molecular Cell Biology 3, 191–212 (1996)
19. Karp, P.: An ontology for biological function based on molecular interactions. Bioinformatics 16(3), 269–285 (2000)
20. Keseler, I.M., Collado-Vides, J., Santos-Zavaleta, A., et al.: EcoCyc: a comprehensive database of Escherichia coli biology. Nucleic Acids Research 39(Database Issue), D583–D590 (2011)
21. Koehler, J., Rawlings, C., Verrier, P., et al.: Linking experimental results, biological networks and sequence analysis methods using Ontologies and Generalised Data Structures. Silico Biology 5(1), 33–44 (2005)
22. Lacroix, Z.: Biological data integration: wrapping data and tools. IEEE Transactions on Information Technology in Biomedicine 6(2), 123–128 (2002)
23. Macías, J.R., Jiménez-Lozano, N., Carazo, J.M.: Integrating electron microscopy information into existing Distributed Annotation Systems. Journal of Structural Biology 158(2), 205–213 (2007)
24. Ringwald, M., Eppig, J.: GXD: a Gene Expression Database for the laboratory mouse: current status and recent enhancements. Nucleic Acids Research 28(1), 115–119 (2000)
25. Schulze-kremer, S.: Ontologies for molecular biology. In: Proceedings of the Third Pacific Symposium on Biocomputing, pp. 693–704. AAAI Press (1998)
26. Smith, B., Ashburner, M., Rosse, C., et al.: The OBO Foundry: coordinated evolution of ontologies to support biomedical data integration. Nature Biotechnology 25(11), 1251–1255 (2007)
27. Stevens, R.: The RiboWeb Ontology (2001), `http://www.cs.man.ac.uk/~stevensr/onto/node6.html/`
28. Stevens, R., Baker, P., Bechhofer, S., et al.: TAMBIS: Transparent Access to Multiple. Bioinformatics Application Notes 16(2), 184–185 (2000)
29. The Flybase Consortium: The FlyBase database of the Drosophila genome projects and community literature. Nucleic Acids Research 31(1), 172–175 (January 2003)
30. The Gene Ontology Consortium: Creating the gene ontology resource: design and implementation. Genome research 11(8), 1425–33 (August 2001)

Challenges in Development of Real Time Multi-Robot System Using Behaviour Based Agents

Aleksis Liekna, Egons Lavendelis, and Agris Nikitenko

Riga Technical University,
Faculty of Computer Science and Information Technology,
Department of Systems Theory and Design,
1 Kalku Street, Riga, LV 1658, Latvia
{aleksis.liekna,egons.lavendelis,agris.nikitenko}@rtu.lv

Abstract. This paper presents a case-study regarding development challenges of multi-agent system for multi-robot system management based on our previous research of the given topic. During the development and implementation of multi-agent system prototype using JADE platform, several implementation challenges regarding messaging system were faced. These challenges may negatively impact system maintenance, burden system evolution and also cause performance issues. The latter is of special importance in the context of multi-robot systems that operate under real-time constraints. In this paper we adopt our previous research as a case study and share challenges faced during prototype multi-robot system development. We believe that potential drawbacks and pitfalls of multi-agent system development such as challenges identified in this paper should be considered with great care especially when applying multi-agent systems to real-time constrained applications such as multi-robot systems.

Keywords: Multi-Robot Systems, Behaviour Based Agents, Real Time Agent Applications, Development Challenges.

1 Introduction

During our previous research we have proposed a multi-agent based approach for multi robot management system development [1]. Currently we have implemented a simulator where the multi-agent system based robot software manages a virtual multi-robot system [1]. The simulator was used for practical experiments with task allocation protocols. It was concluded that a widely used Contract Net [2] task allocation protocol does not guarantee optimal task allocation and thus some more specific protocol must be developed for task allocation [3]. Behaviour oriented agent development platform JADE was chosen for our practical implementation. By term "behaviour" we mean the implementation behaviour used in JADE platform to specify actions performed by the agents.

Despite successful implementation of the simulator that works with virtual environment and its successful usage in our experiments, the development process

S. Omatu et al. (Eds.): *Distrib. Computing & Artificial Intelligence*, AISC 217, pp. 587–595.
DOI: 10.1007/978-3-319-00551-5_70

and practical experiments revealed that the behaviour based agent approach and the messaging system used in the JADE platform may lead to some practical implementation challenges.

In this paper we use previously implemented simulator as a case study and share our experiences and lessons learned. In particular, we analyse the design and implementation challenges of messaging system using behaviour-based approach.

The remainder of the paper is organized as follows. Section 2 describes our previous work regarding the given topic. Section 3 is dedicated to the case study of the development challenges. A short conclusion is given in Section 4.

2 Previous Work

In this section we describe our previous work regarding multi-robot system development using multi-agent system for multi-robot system management. This section serves as a context for our contribution to analysis of implementation challenges introduced in Section 3.

2.1 Multi-Robot System and Real-Time Constraints

The multi-robot system under consideration is being developed for indoor applications like floor washing, area painting, vacuum cleaning and others that are common in their essence and known by the robotics community as “area coverage” or “coverage planning” [4]. Solutions exist that allow single robots (such as iRobot Roomba) to process (e.g. clean) limited-size areas (such as hotel-rooms). However, joining multiple autonomous robots in a multi-robot system provides opportunities to process areas that cannot be processed by single robots in reasonable time (e.g. large-scale warehouses). This is exactly the topic of our previous research [1].

While the robots operate in physical world they are bound by real-time constraints [5]. Within the paper’s context these constraints are defined as critical time intervals for particular data processing. One of the most important tasks to be performed by each of the robots is reaching some particular point in space (see Figure 1). Keeping in mind robot positioning errors, it is possible to define some interval ΔS, within which the robot is considered having reached the goal position.

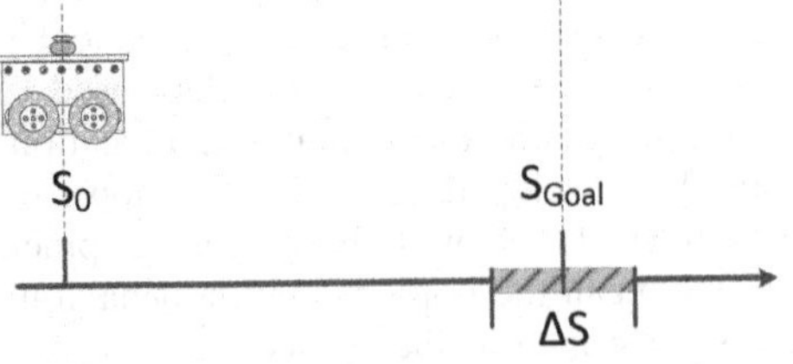

Fig. 1 Processed data availability critical area.

To check if the goal conditions are met it is necessary to process feedback data that in our case includes both odometry data and visual landmark data, which requires to process frames sensed by the on-board camera. The feedback data processing includes frame data filtering and landmark recognition [6, 7, 8], Kalman filtering [9], sensor data fusion [10, 11] and final position estimation, which requires some time $t_{processing}$. Having this time estimation it is possible to define one of the real time constraints:

$$t_{processing} \le \frac{\Delta S}{v} \text{, where} \tag{1}$$

v is speed of the robot. If the constraint (1) is not met then the feedback data will be processed too late and the robot will run over the ΔS, causing robot alternation around ΔS. Similarly we define real-time constraints for inter-agent communications, global map update and other data processing procedures thereby defining a set of real-time constraints T:

$$T = \{\ t_i\ :\ \ t_i = f(\Delta S_i, v)\ \}, \text{where} \tag{2}$$

t_i is i-th constraint, ΔS_i is critical area where the processed data are relevant and v is robot's speed. In case the specific processed data is not bound to area, then function $f(\Delta S_i, v)$ is replaced by a threshold constant $t_{critical}$.

2.2 *Multi-Agent System for Multi-Robot System Management*

Multi-robot system should be supported by the appropriate management system allowing the integration of multiple robots into a single multi-robot system. In our previous research [1] we developed architecture for multi-robot system that is managed by a multi-agent system (see Fig. 2). System architecture is composed of three layers: robot layer, interface layer and multi-agent system layer.

Robot layer consists of physical robots that operate in real world environment. This layer represents software running on-board robots and is responsible for real-time reactive control, such as motor control, while higher-level functions, such as planning and task-allocation, are handled by multi-agent system.

Multi-agent system serves as the management component. There are four types of agents: robot agents that represent physical robots in the multi-agent system, manager agent that is responsible for task decomposition and task allocation, gateway agent that communicates with physical robots through interface layer, and user interface agent representing the user in the multi-agent system. Each robot agent in the multi-agent system represents a physical robot in the robot layer. Exactly one robot agent per physical robot exists in the system. One user interface agent exists per user in the system, while manager agent and gateway agent are both singletons. Gateway agent is introduced to avoid direct interaction between robot agents and the interface layer thus leaving out the need to implement interface-layer communication specific features in robot agents. Although gateway agent is meant to be a singleton, it may be possible to create multiple instances of it to reduce potential communication bottlenecks.

Interface layer contains an interface that provides mapping between multi-agent system and physical robots. This layer is introduced to prevent tangling robot-specific data structures and functionality in the multi-agent system and vice versa.

The principle of operation is as follows. User issues a command via user interface agent. User interface agent converts user input into agent-specific task and communicates the task to manager agent. Manager agent provides task decomposition into sub-tasks and assigns each subtask to the particular robot agent. For example, if the task is to clean a specified area, manager agent performs task decomposition in such way that each subtask is a sub-area of the initial and can be processed by a single robot. Task allocation can be done in various ways and during our previous research we did experiment with Contract NET protocol [3]. After task is allocated to particular robot agent, it takes partial control of the robot and communicates high-level commands to the software running at respective robot. Communication goes through gateway agent and the interface layer to the appropriate robot. A two-way communication is present to provide the appropriate feedback. When the task is completed, robot agent informs the manager agent. When all sub-tasks of the particular task are completed, the manager agent informs the user interface agent.

During our previous research, a prototype system has been implemented for experimental purposes. The prototype system consists of multi-agent system implemented in JADE managing virtual (simulated) robots. Experiments were conducted to determine the applicability of Contract NET protocol to task allocation among robots. It was concluded that sequentially allocating tasks using Contract Net protocol does not guarantee the optimal result [3].

Despite the successful application of JADE to development of multi-agent system layer, a number of challenges regarding design and implementation of messaging system were observed.

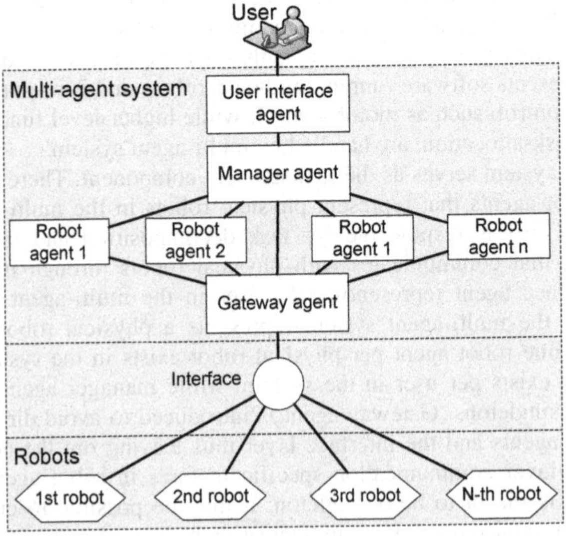

Fig. 2 System architecture

3 Development Challenges

In this section we analyse challenges that arise in design and implementation of the multi-agent system introduced in Section 2.

3.1 *Messaging System*

To design a messaging system for the multi-agent system layer depicted in Fig. 2, possible types of messages, their senders and receivers must be identified. Let us look at possible message types assuming that FIPA predefined communication protocols are used.

User interface agent uses FIPA REQUEST [12] protocol to request the manager agent to perform the task. Manager agent uses FIPA REQUEST protocol to accept tasks from user interface agent and FIPA Contract NET [2] protocol to allocate subtasks to individual robot agents. Robot agent participates in Contract NET protocol issued by manager agent, as well as it uses FIPA QUERY [13] protocol to query gateway agent for robot status information and also uses FIPA REQUEST protocol to execute robot commands through gateway agent. Gateway agent participates in FIPA REQUEST and FIPA QUERY protocols initiated by robot agents. Given that FIPA QUERY and FIPA REQUEST protocols result in minimum 2 messages (request or query as initiating message and inform as reply) and Contract NET protocol results in 2 or 4 messages (2 messages in the case of refusal or 4 messages in the case when participant wins the contract), the total number of message types for each agent are the following. User interface agent: 2; manager agent: 2 + 4 = 6; robot agent: 2 + 2 + 4 = 8. This gives a total of 2 + 6 + 8 = 16 message types to be sent and received by the agents in the system.

Considering the total number of message types in the system it is reasonable to think about the way they will be handled. In the context of JADE all actions that agent performs should be defined within the appropriate behaviours. This pattern is referred to as "behaviour-per-action" in further text. This means that to send a message, a dedicated behaviour "MessageSendingBehavior" is created in a separate class rather than a method "sendMessage()" in the agent class. Similarly, to receive a message, a behaviour "ReceiveMessage" should be created as a separate class rather than a while-loop in the agent class. In the following subsections sending and receiving of messages is analysed more closely

3.2 *Receiving Messages*

There are two alternatives to receive a particular message. In the first case a single behaviour for message receiving is created in each agent. This behaviour receives all possible messages and dispatches them accordingly. In this case the behaviour has a method like "handleMessage" that contains a (potentially large) number of if-then statements (e.g. if message type is X1 and content is Y1 then do A1, else if message type is X1 and content is Y2 then do A2 etc.). This may lead to potential difficulties introducing new messages or modifying existing messaging logic.

The second alternative is to use message filters – a built-in feature of JADE. In this case, multiple behaviours for message receiving are created in each agent. Each behaviour receives only specific messages, filtered by type or other parameters. To completely escape if-then statements mentioned in the previous alternative, a behaviour-per-message-type approach can be used. This way the number of behaviour classes in the agent matches the number of message types an agent may receive. This may not be a problem if the number of message types in the system is relatively small (i.e. less than 3 per agent). But the robot agent in our case would require a total of 16 behaviour classes to receive all possible messages. Creating the behaviour classes themselves is not the problem. The problem is to determine how this affects system maintenance and evolution. It is not uncommon during development of a prototype system to change existing or introduce new message types as well as change or replace the ontology or communication language used by the agents in the system. Introducing such changes leads to changes in the message receiving system as well. That means some or all of message-receiving behaviour classes have to be modified. This burdens system maintenance and evolution, as it was observed during development of prototype system described in previous section.

3.3 Sending Messages

Sending a message in JADE involves specifying message type, communication language, communication protocol and other parameters before the actual "sending" action is performed [14]. To reduce the tangling of messaging-specific code in all modules where the message is about to be sent a dedicated behaviour can be introduced for message sending. This way messaging-specific code (such as specifying the communication language and other relevant parameters) is kept in a single place. This also matches the JADE behaviour-per-action pattern described earlier. So, to send a message, a corresponding behaviour is instantiated, specifying the message to be sent. This behaviour then executes and the message sending is performed.

The described messaging pattern is suitable for one-way messages. However, FIPA protocols (REQUEST, QUERY and Contract NET) used in the messaging system are asynchronous and are based on two-way communication. For a request, a response is expected. This means that the behaviour responsible for message sending should not only send the specified message, but also wait for the appropriate reply, introducing the problem of waiting for an asynchronous reply. In JADE there are predefined behaviours for asynchronous message sending and receiving for REQUEST, QUERY and Contact NET protocols [14] that used in implementation of the particular system. These behaviours help to solve the issue of asynchronous waiting, allowing user to define behaviour or method that execute when the appropriate reply message is received. Nevertheless, the asynchronous waiting mechanism together with message receiving pattern in JADE introduces potential real-time issues, described in the following subsection.

3.4 Messaging and Real-Time Issues

In JADE, whenever a behaviour is waiting for a message to arrive, it is in a blocked (waiting) state [14]. When a message arrives, all blocked behaviours are notified. Each behaviour checks if the message is of particular interest of that behaviour. If the result is positive, the appropriate action is performed. If the result is negative, the behaviour returns to the blocked state. The problem is that the total time required for all behaviours to check the message contents increases according to the number of behaviours waiting for the messages. It is worth noting that increasing the number of waiting behaviours reduces the chance that a particular message is of interest of a particular behaviour, while increasing the chance of "worthless" activation of particular behaviour and thus – wasting processing time.

While experimenting with the developed simulator, it was noted that increasing the number of agents in the system (and thus message count and the number of behaviours waiting for messages), the system response time increased. No clear measurements indicating system response time according to the number of messages in the system were taken however, because the system response time in our particular case depends also on other factors discussion about which is beyond the scope of this paper. Nevertheless, the overall tendency was observed – using debugging tools provided by JADE it was noted that system spends a great deal of time activating and then suspending behaviours waiting for messages.

The observed increase of system response time must be taken into account considering that the multi-agent system is meant to be a management layer of multi-robot system. Delays due to the messaging system issues may lead to undesirable behaviour of a multi-robot system that operate under real-time constraints, described in Section 2.1.

4 Conclusion

During our previous research regarding multi-agent system development for multi-robot system management, two main development challenges were identified during system prototype design using JADE platform. First, there were issues regarding message receiving pattern design that would not burden system maintenance and evolution. It was determined that whether Second, the way JADE handles waiting for messages may cause performance degradation in systems with large number of message interchange. This may cause issues in systems operating under real-time constraints, such as multi-robot systems.

Identified challenges are subject for further investigation, and "clean" supporting measurements are required to support or deny a design decision. Nevertheless we believe that the identified challenges serve as a good starting point for further investigation. While trying to apply multi-agent system paradigm to increasingly large number of problems, potential drawbacks and pitfalls, such as challenges identified in this paper must be considered with great care.

Supporting measurements, their analysis and deeper investigation of this topic is the subject of further research.

Acknowledgement. The work has been partly supported by ERDF European Regional Development Fund project 2010/0258/2DP/2.1.1.1.0/10/APIA/VIAA/005 Development of Intelligent Multiagent Robotics System Technology.

References

1. Lavendelis, E., et al.: Multi-Agent Robotic System Architecture for Effective Task Allocation and Management. In: Recent Researches in Communications, Electronics, Signal Processing & Automatic: Proceedings of the 11th WSEAS International Conference on Signal Processing, Robotics and Automation (ISPRA 2012), UK, Cambridge, February 22-24, pp. 167–174 (2012)
2. FIPA Contract Net Interaction Protocol Specification. Foundation for Intelligent Physical Agents, `http://www.fipa.org/specs/fipa00029/` (last visited: February 21, 2002)
3. Liekna, A., Lavendelis, E., Grabovskis, A.: Experimental Analysis of Contract NET Protocol in Multi-Robot Task Allocation. Scientific Journal of RTU. 5th Series. Computer Science 213, 6–14 (2012)
4. Choset, H.: Coverage for robotics – a survey of recent results. Annals of Mathematics and Artificial Intelligence 31, 113–126 (2001)
5. Ben-Ari, M.: Principles of Concurrent and Distributed Programming, IInd edn., 384 p. Prentice-Hall (2006)
6. Zitov, B., Flusser, J.: Landmark recognition using invariant features. Pattern Recognition Letters 20, 541–547 (1999)
7. Jang, G., et al.: Metric Localization Using a Single Artificial Landmark for Indoor Mobile Robots. In: Proceedings of IEEE International Conference on IROS, pp. 2857–2862 (2005)
8. Open source augmented reality project GRAFT, `http://www.aforgenet.com/articles/glyph_recognition/` (cited: November 16, 2012)
9. Mitchell, H.B.: Data Fusion: Concepts and Ideas, 348 p. Springer (2010)
10. Raol, J.R.: Multi-Sensor Data fusion with MATLAB, 534 p. CRC Press (2010)
11. Hall, D.L., Llinas, J.: Handbook of multisensory data fusion, 537 p. CRC Press (2001)

12. FIPA Request Interaction Protocol Specification. Foundation for Intelligent Physical Agents (2002), http://www.fipa.org/specs/fipa00026/ (last visited: February 21, 2013)
13. FIPA Query Interaction Protocol Specification. Foundation for Intelligent Physical Agents (2002), http://www.fipa.org/specs/fipa00027/ (last visited: February 21, 2013)
14. Bellifemine, F.L., Caire, G., Greenwood, D.: Developing Multi-Agent Systems with JADE, 300 p. John Wiley & Sons (2007)

Application of Hybrid Agents to Smart Energy Management of a Prosumer Node

Pasquale Caianiello, Stefania Costantini, Giovanni De Gasperis, Niva Florio, and Federico Gobbo

Department of Information Engineering, Computer Science and Mathematics, University of L'Aquila, Italy
{pasquale.caianiello,stefania.costantini, giovanni.degasperis,niva.florio,federico.gobbo}@univaq.it

Abstract. In this paper we propose an intelligent control scheme based on a multi-agent system composed of two main components: a logical planner and forecaster obtained by machine learning techniques. The chosen benchmark application lays in the field of smart energy management, were we consider the concept of a prosumer node. We discuss a case-study where a simplified multi-agent system is applied to the supervision of an air conditioner, leading to an energy saving of about 17%.

Keywords: hybrid agents, energy management, prosumer node.

1 Introduction

Energy management of modern smart buildings requires distributed intelligent control in order to save energy. An energy efficient building has its passive intrinsic energy efficiency which is embedded by design in its structure. However, if a control methodology inspired to intelligent agents is adopted, a further improvement in energy efficiency can be achieved. Such intelligent control should be dynamic by nature, including the real-time requirement as the building has its own dynamical thermo-physical behaviour and it is immersed in a dynamical environment where weather events change its energy footprint in function of time.

A smart building with energy management capability is a node which is part of a smart grid of energy, such as a packet router in a packet-switched network. The main difference is that energy packets can hardly be addressed to a specific destination, but a local smart energy real-time management policy could be beneficial to the overall energy management of the overall smart grid. This kind of energy node can be considered a prosumer, i.e. producer and consumer of energy in different forms.

In the case of thermal regulation of a building, it can account for the 80% of the total energy footprint, especially during the summer period in the temperate zones of the planet when air conditioning energy spending is at its top, or at the equator for the whole year.

In this paper, we assume that all controllable energy forms are electric in nature. Fossil fuel can also be controlled, but would imply the introduction of other

S. Omatu et al. (Eds.): *Distrib. Computing & Artificial Intelligence,* AISC 217, pp. 597–607.
DOI: 10.1007/978-3-319-00551-5_71

variables into the complexity of the system that we intentionally would not consider here.

2 Our Solution

Our agent architecture includes a perception layer – composed of the set of prosumer forecasters – and the reasoning layer – made by the set of prosumer planners. The perception layer has the job to transform the sensors array signals into abstract predicates, i.e. fluents, that can be handled by the reasoning layer. The application that we are now considering falls into the realms of Distributed Intelligent Control Systems that can be described in terms of Complex Event Processing (CEP) [3], where the need for some degree of autonomy is crucial in order to enable components to respond dynamically to ever-changing circumstances while trying to achieve overarching objectives. In the application that we are considering, many events should be properly handled. In fact, each definite area of the building – e.g., the dining room, the bathroom, the hall, etc. – can be managed by a hybrid agent which plays the role of the prosumer, in communication with its siblings of the other areas, in order to accomplish the performance goals. Therefore, the energy management system of the building can be represented in terms of a Multi-Agent Systems (MAS), where each agent deals with the others in terms of planning and forecasting, as in [2].

The trends extracted from cooling load predictions, i.e. energy consumption, made by the forecaster feeds the symbolic interpretation needed for a rule-based logical agent, described in the DALI language. From the planner point of view, this abstraction becomes a fluent in an event calculus context, i.e. events to be handled proactively by the DALI MAS. The latter should also take into account external requirements and user preferences, in order to achieve the performance goals, which should be defined a priori.

3 Related Works

In the field of Active Logic Programming, we have chosen to use DALI [6, 7] for its unique features in respect of the other development frameworks that are available. The DALI semantics is fully defined in [4], while the operational semantics of the interpreter is described in [8]. A DALI agent is a logic program that contains a particular kind of rules, reactive rules, aimed at interacting with an external environment. The environment is perceived in the form of external events, that can be exogenous events, observations, or messages by other agents. In response, the agent can perform actions, send messages, adopt goals, etc. The reactive and proactive behavior of the agent is triggered by several kinds of events: external events, internal, present and past events. When an event arrives to the agent from its "external world", the agent can perceive it and decide to react. However, when an agent perceives an event from the "external world", it doesn't necessarily react to it immediately: it has the possibility of reasoning about the event, before (or instead of) triggering a reaction. Furthermore, internal events make a DALI agent proactive.

In particular, [5] proposes an extension so to manage complex reaction in rule-based logical agents using a DALI-like syntax which associates a set of preference rules, that can express the performance goals directly into the language.

The literature in MAS is wide and large; see [14] for a survey. However, there is no specific application of MASs to the smart energy management of a building, as far as the authors know. However, there are some points in contact with different solutions, both at the domain and at a design and development levels.

3APL [9, 10] is a programming language and a platform specifically tailored to implement MAS, where the agents are designed in terms of "data structure" or "mental attitudes" (beliefs, goals, plans and reasoning rules) and "deliberation process" – implemented as programming instructions. Based on its beliefs, an agent can reach its goals by planning its actions thanks to the deliberation cycle, that can leads also to the check and revision of its mental attitudes. Although 3APL is based on the concepts of rules and planning, it lacks to implement the idea of event, unlike for instance DALI. 3APL has been used mostly to realize application for virtual training (e.g. [17]).

A similar approach is known as KGP (Knowledge, Goals and Plans), which is based on logic programming with priorities, taking beliefs, desires and intentions as a starting point, but adding reasoning capabilities, state transitions and control – see [15]. Furthermore, it is implemented directly in computational logic. KGP could be a valid alternative candidate compared to DALI.

METATEM (and its extension Current MERATEM) [1] is a language based on first-order linear temporal logic, and thus it is suitable for temporal planning and temporal knowledge representation. At the basis it has concepts such as beliefs, intentions, goals and plans, but the very fundamental rules are of the form "past and present formula implies present and future formula" [16]. So, like DALI, at the center of METATEM there are the concept of time and the idea that the past determines the present and the future, but it has no different classes of events and METATEM agents are just reactive agents, unlike DALI. METATEM can have a wide range of applications (e.g. patient monitoring, fault tolerance system, process control etc.) [13, 12], but asfar as the authors know, it has not been used for the energy management of a building.

4 The Prosumer Node

In Figure 1 an average daily power consumption profile of a prosumer node is shown, where energy load can be obtained by integration. There are times of the day when energy is in surplus, and times were it is in deficiency. An ideal energy manager should level its energy of the given goals, limiting peaks and avoiding surplus periods, and founding its decision on an accurate dynamical behavioral model of the underlying building infrastructure in its environment.

A smart building connected to a smart energy grid can be considered at the same time a consumer and a producer of energy. So, a energy prosumer node, as defined by [11], should have local energy sources which are independent from the grid. Usually, these sources include sustainable energy types – e.g. solar thermal

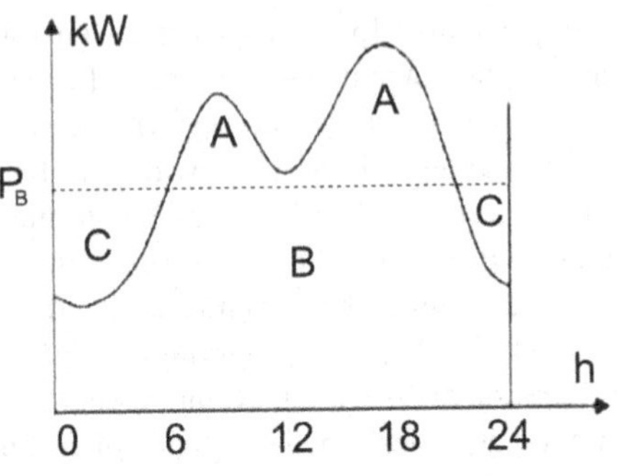

Fig. 1 A typical daily load diagram of a prosumer node: P_B is the designed power level of the node. A: load peaks, B: basic load, C: load surplus.

and PV, wind, geothermic, etc. – but also conventional fossil fuel based sources, as for instance continuity groups and diesel generators, that need to be managed appropriately. The energy prosumer definition should also take into account the local consumption profile and the local energy metering tools available. An intelligent energy manager should also be present in order to improve the energy efficiency of the prosumer node. Such an intelligent prosumer manager should derive its behavior from the capability to dynamically predict the energy consumption and production given the intrinsic time constant of the building itself – e.g. 15 minutes as a typical thermal inertia of a modern energy efficiency building.

In Figure 2, a general block diagram of a prosumer node is shown. Let's start analyzing it from the physical point of view: the **prosumer physics** block, which describes the actual behaviour of the building, given its physical concrete walls, window areas, constituent materials and given the actuators energy output – e.g. conditioners air flow in the summer, or heat flow from heaters in the winter. The building energy status is given at each time instant by: external weather conditions, internal user behaviour, internal signals and settings in terms of comfort, i.e. internal temperature and reference point. Also the internal sensor set should include what that a modern smart building can be equipped with **energy metering**, which are electrical load sensors at the power plugs of electrical equipments or at the electrical cabinet for each floor. In this way, a complete energy consumption profile of the building can be obtained at each sampling time. The **prosumer controller** will then implement a first local control loop that should keep at equilibrium the goals given by the **prosumer planner**, having an impact on the energy flows by means of the actuators, i.e. heaters, water chillers, air conditioners, windows controls, etc..., together with other controllable energy generators or accumulators. The external sensors array should return the state of the most important environmental signals with an impact to the energy profile of the building, i.e. external temperature, solar irradiance, humidity, etc.. The resolution and the distribution of the external sensors at which such signals should be sampled depends on the structural configuration of the building. Diffused solar irradiance has a major impact on the energy demand of the prosumer node, also depending on neighborhood buildings for the shading account. The **prosumer planner** gets the whole set of signals, internal and external, so to feed an important part of itself made by the **prosumer forecaster**.

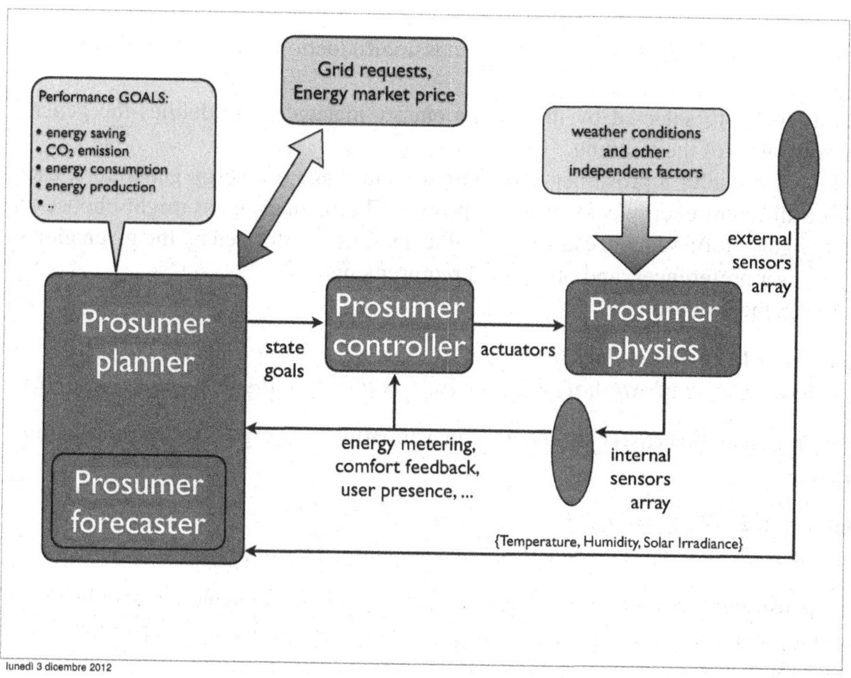

Fig. 2 The Prosumer Node block diagram

Then, the prosumer forecaster dynamically produces an estimate of the near future energy requirements (i.e. at the next time sample) of the prosumer node, needed for plan generation. In order to predict energy consumption data in the next time interval we used resilient backpropagation to train a perceptron with 4 layers and 33 sigmoid neurons. The training set was constructed with raw data coming from 30 sensors recording several metereological variables, indoor temperature at different spots of the experimentation room, and actual energy consumption. The target learning output was average consumption in the following hour. Data recorded over a few days in summer 2012 allowed a quick convergence of the learning procedure to an irrelevant mean square error over less than two thousand epochs. The experimentation supports the feasibility of a real-time local adaptation to obtain the optimal predictive power of the forecaster.

5 The Energy Management Problem

The energy management problem at a hand could be summarized as in the following. Planning goals could be targeted to possibly conflicting functions to minimise: energy, cost, and CO_2 emission. An example of possible global goal set is:

$G = \{$ EnergySaving, CostReduction, EmissionReduction, HighestComfort $\}$

These goals are selected by the human energy manager who defines the general energy policy of the building.

Let us consider a prosumer node with a wide range of energy generators, each with a different energy/cost/emission profile. The planner agent might choose an energy generation sequence among feasible ones, as constrained by the given global goals, user preferences, and other local requirements.

Given the input set:

Definition 1. *The input data set* $I = I_e \cup I_i$
where I_e *is the set of external sensor array,* I_i *is the set of the internal sensor array.*

The **prosumer forecaster** should give an estimation of energy demand at each time interval:

Definition 2. *The prosumer forecaster agent* $F = f(I_e, I_i)$
where f is the function implemented by the machine learning module.

The **prosumer planner** should then generate a plan as a sequence of actions taken from a pre-defined set of possible energy related actions, given the prosumer forecaster estimates:

Definition 3. *The prosumer planner possible actions* $P = p(F)$
where $p = \{a(q_a)\}$ *sequence of actions with at least one action,* $a \in A$, $q_a \in \mathbb{R}$, *taken from* $A = \{E_p, E_c, E_a, E_b, E_s\}$ *the action set*

where:

- E_p stands for *produce energy*
- E_c stands for *consume energy*
- E_a stands for *accumulate energy*
- E_b stands for *buy energy*
- E_s stands for *sell energy*

q_a is the quantity argument relative to the given action a. Then, for each action to be taken, there should be a function that associates an internal state goal, also depending on the action argument q_a.

The building energy management goals are processed by the overall planner, whose results are broadcasted to every active prosumer node (supervised by a local agent) managed by the MAS. The agents are then in charge of adapting the planner results to the specific situation they are responsible for. As an example, let's consider the global goal to be selected at a given time as: **EnergySaving**, and let it be as state goals for the prosumer controller, i.e. the air conditioner, the set point at which the thermal zone temperature should be oscillating around, not too far from the comfort user requirements. This implies that a specific domotic technology should be adopted so that the planner agent can send control signals to the air conditioner by the local area network.

Definition 4. *The DALI knowledge base with preferences of the prosumer planner agent*

energy_savingE :>
  increaseAirConditionerSetPointA |
  decreaseAirConditionerSetPointA |
  holdA ::
  increaseAirConditionerSetPointA > *decreaseAirConditionerSetPointA*
  :- *externalTemperatureIrradianceHaveSameTrend*.
  holdA > *decreaseAirConditionerSetPointA*

increaseAirConditionerSetPointA :< *userComfortIsMaintained*, *plannerHasEnabled*

holdA :< *userComfortIsMaintained*, *plannerHasEnabled*

decreaseAirConditionerSetPointA :< *userComfortIsMaintained*, *plannerHasEnabled*.

This very concise planner agent knowledge base is interpreted as follows. The local planner, relative to a single thermal zone, has three possible actions: increase the temperature set point of the air conditioner, decrease it or standby (hold). There are two preferences, expressed by the rules with connective '>'. Each rule states that the leftmost action is preferred. The preference rule applies whenever its body (the part after the ' :- ') holds, and preferences apply to feasible actions. An action is feasible if its preconditions (if any) are verified. Preconditions are expressed by rules with connective ' :< '. A basic precondition for an action to be feasible is that the action has been enabled by the overall planner (e.g., the planner might not enable temperature decrease if too much energy is being consumed at the moment). If the perception layer, i.e. the prosumer forecaster agent, detects that the external environmental signals of temperature and solar irradiance shows the same variation trend, then the planner agent would prefer to increase the temperature set point so to save energy waiting for a better thermal equilibrium. Also, in order to save energy, it would prefer not doing anything rather than reducing the set point, i.e., decreasing the room temperature, unless user comfort requirements are no more satisfied. In this latter case, being this a precondition to all action, it would definitively reduce the temperature set point until the system is in equilibrium and user comfort requirement is met. An other important pre-condition is the *plannerHasEnabled* predicate. It allows to configure a hierarchy of planner agents: a global planner agent associated to the whole building and many other sub-agents associated to each thermal zone, or rooms. In this way the knowledge base of the global planner can enable or disable the possible actions of the local planners, depending on global energy reasoning goals, like energy availability or special energy needs.

The above rule is re-evaluated periodically, at a certain (customizable) frequency. This mechanism (which is a generalization of the DALI internal event construct) makes the agent proactive, i.e., capable of autonomously operating on its environment. The frequency will be customized according to the kind of appliances that are being controlled and to the granularity of results that one needs to obtain about temperature and energy consumption.

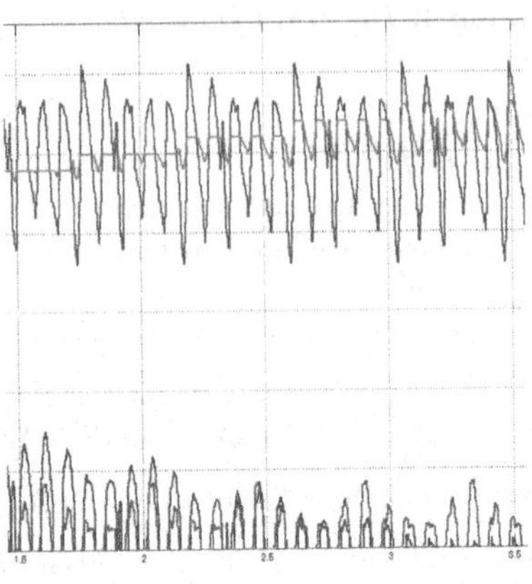

Fig. 3 Cooling load prediction of the prosumer forecaster depending on AC reference points and external weather conditions measured by the sensor array. GREEN: external temperature, RED: solar diffused irradiance, MAGENTA: internal temperature, BLU: predicted cooling load.

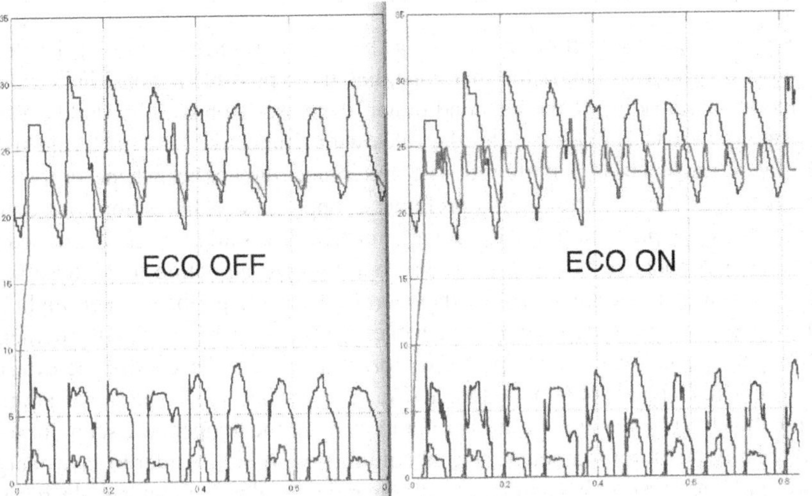

Fig. 4 LEFT: Temperature time series when ECO mode is OFF. RIGHT: ECO mode is ON. The net energy saving is 42kWh on a maximum of 252kWh, about 17%. Legend: GREEN: external temperature, RED: solar diffused irradiance, MAGENTA: internal temperature, BLU: predicted cooling load.

This kind of multi-agent system based control allows the temperature to be kept within user preferences, overcoming the control scheme of a simple thermostat, which is based on a infinite energy supply hypothesis in order to keep the temperature constant and that are no longer feasible in modern buildings.

6 Case Study: Air Conditioning Economizer

Let us consider as a case-study a simple prosumer physical model which does not consider any local energy source. The environment at hand consists of a small 20msq room with a door and a window, with a inverter-class air conditioner as the prosumer controller. The prosumer forecaster has been trained to correlate the measures of external temperature, solar irradiance, and internal temperature with the air conditioner energy consumption at each next sampling time interval of 15 minutes. Here the prosumer planner just applies the simple rule of increasing the air conditioner set point when external temperature and solar irradiance have the same trend, and reducing it during hold periods. Matlab simulation results for the proposed solution are shown in Figures 3 and 4.

Simulation shows that when the energy saving goal is activated, the hybrid MAS is capable to achieve about 17% gain in respect to a pure thermostat control setup. In this way, external weather conditions have been exploited to forecast energy demand and regulate the energy spending of the building so to save energy, which implies in general to reduce CO_2 emissions proportionally as well.

7 Conclusions

In this paper, we have proposed a MAS-based solution for managing energy consumption of a modern smart building. The overall objective is that of combining user comfort and energy saving. The architecture includes a multi-agent system comprising logical planners and machine learning trained estimators. This solution is particularly flexible, as the overall planning strategy can be integrated by local strategies implemented by single agents which are responsible for sub-parts of the building, and are aware of the particular constraints to be applied on that part. Notice in fact that different parts of the building might have different requirements, e.g., food preservation w.r.t. human presence. Experiments performed using as benchmark a prosumer building have shown that we are able to obtain significant energy saving in the air conditioning. The use of the posterior predictive model obtained by the machine learning module avoids the expensive and complex a priori modelling of the prosumer physics for the building-system, necessary for correct planning. The versatile extended DALI framework that we have adopted has been shown to be usefully applicable to real world applications with real-time constraint related with thermal inertia of a typical energy efficient building. In the future, we plan to extend the solution and the experiments to the integrated management of several energy sources.

Acknowledgement. Giovanni De Gasperis and Pasquale Caianiello dealt with the domain problem, the case study and the prosumer forecaster, while Federico Gobbo and Niva Florio helped definining the energy problem and revising the work. Stefania Costantini leads the research group devoted to logical agents and contributed defining the DALI rule set to handle user preferences. The final logical agent-based approach is the result of the work of the whole team.

We also wish to thank Prof. Francesco Muzi from the Department of Industrial and Information Engineering and Economics at the University of L'Aquila, for the insightful discussions about the smart grid concept and the requirements of the prosumer node.

References

1. Barringer, H., Fisher, M., Gabbay, D., Gough, G., Owens, R.: Metatem: An introduction. Formal Aspects of Computing 7(5), 533–549 (1995)
2. De Gasperis, G., Bevar, V., Costantini, S., Tocchio, A., Paolucci, A.: Demonstrator of a multi-agent system for industrial fault detection and repair. In: Demazeau, Y., Müller, J.P., Rodríguez, J.M.C., Pérez, J.B. (eds.) Advances on PAAMS. AISC, vol. 155, pp. 237–240. Springer, Heidelberg (2012)
3. Chandy, M.K., Etzion, O.O., von Ammon, R.: 10201 Executive Summary and Manifesto – Event Processing. In: Event Processing, Dagstuhl, Germany. Dagstuhl Seminar Proceedings, vol. 10201. Schloss Dagstuhl - Leibniz-Zentrum fuer Informatik, Germany (2011)
4. Costantini, S., Tocchio, A.: About declarative semantics of logic-based agent languages. In: Baldoni, M., Endriss, U., Omicini, A., Torroni, P. (eds.) DALT 2005. LNCS (LNAI), vol. 3904, pp. 106–123. Springer, Heidelberg (2006)
5. Costantini, S., De Gasperis, G.: Complex reactivity with preferences in rule-based agents. In: Bikakis, A., Giurca, A. (eds.) RuleML 2012. LNCS, vol. 7438, pp. 167–181. Springer, Heidelberg (2012)
6. Costantini, S., Tocchio, A.: A logic programming language for multi-agent systems. In: Flesca, S., Greco, S., Leone, N., Ianni, G. (eds.) JELIA 2002. LNCS (LNAI), vol. 2424, pp. 1–13. Springer, Heidelberg (2002)
7. Costantini, S., Tocchio, A.: The DALI logic programming agent-Oriented language. In: Alferes, J.J., Leite, J. (eds.) JELIA 2004. LNCS (LNAI), vol. 3229, pp. 685–688. Springer, Heidelberg (2004)
8. Costantini, S., Tocchio, A.: A dialogue games framework for the operational semantics of logic agent-oriented languages. In: Dix, J., Leite, J., Governatori, G., Jamroga, W. (eds.) CLIMA XI. LNCS, vol. 6245, pp. 238–255. Springer, Heidelberg (2010)
9. Dastani, M., Birna Riemsdijk, M., Meyer, J.-J.: Programming multi-agent systems in 3apl. Multi-agent Programming, 39–67 (2005)
10. Dastani, M., De Boer, F., Dignum, F., Meyer, J.-J.: Programming agent deliberation: an approach illustrated using the 3apl language. In: Proceedings of the Second International Joint Conference on Autonomous Agents and Multiagent Systems, pp. 97–104. ACM (2003)
11. De Gasperis, G., De Lorenzo, M.G., Muzi, M.: Intelligence improvement of a prosumer node through the predictive concept. In: Proceedings of the Sixth UKSim European Symposium on Computer Modeling and Simulation. IEEE Computer Society, Malta (2012)
12. Fisher, M.: A survey of concurrent metatem-the language and its applications. Temporal Logic, 480–505 (1994)

13. Fisher, M.: Metatem: The story so far. Programming Multi-Agent Systems, 3–22 (2006)
14. Fisher, M., Bordini, R.H., Hirsch, B., Torroni, P.: Computational logics and agents. a roadmap of current technologies and future trends. Computational Intelligence (2007)
15. Kakas, A.C., Mancarella, P., Sadri, F., Stathis, K., Toni, F.: Computational logic foundations of kgp agents. J. Artif. Intell. Res. (JAIR) 33, 285–348 (2008)
16. Mulder, M., Treur, J., Fisher, M.: Agent modelling in metatem and desire. In: Intelligent Agents IV Agent Theories, Architectures, and Languages, pp. 193–207 (1998)
17. Narayanasamy, G., Cecil, J., Son, T.C.: A collaborative framework to realize virtual enterprises using 3apl. In: Baldoni, M., Endriss, U. (eds.) DALT 2006. LNCS (LNAI), vol. 4327, pp. 191–206. Springer, Heidelberg (2006)

Structuring and Exploring the Biomedical Literature Using Latent Semantics

Sérgio Matos, Hugo Araújo, and José Luís Oliveira

DETI/IEETA, University of Aveiro, 3810-193 Aveiro, Portugal
{aleixomatos,hugo.rafael,jlo}@ua.pt

Abstract. The fast increasing amount of articles published in the biomedical field is creating difficulties in the way this wealth of information can be efficiently exploited by researchers. As a way of overcoming these limitations and potentiating a more efficient use of the literature, we propose an approach for structuring the results of a literature search based on the latent semantic information extracted from a corpus. Moreover, we show how the results of the Latent Semantic Analysis method can be adapted so as to evidence differences between results of different searches. We also propose different visualization techniques that can be applied to explore these results. Used in combination, these techniques could empower users with tools for literature guided knowledge exploration and discovery.

1 Introduction

Being able to conduct a systematic literature search is an essential skill for researchers in any field. In a thriving and evolving research area such as biomedicine, where the scientific literature is the main source of information, containing the outcomes of the most recent studies, this becomes even more important. However, the fast increasing amount of articles published in this field is creating difficulties in the way information can be efficiently searched and used by researchers [8, 6].

Another important aspect is the inherent interrelations between concepts. Additionally, researchers may be interested in studying a given idea or concept from a particular perspective. Given a disease, for example, they may be interested on different aspects, from the underlying genetics, to previous studies using a particular laboratory technique or experiment, to more clinically oriented information.

Although many literature retrieval tools have been developed for this particular domain, many limitations are still present, specially in the form the results are presented, forcing users to continually reformulate their queries in view of information they gather at each point, looking for more specific or more relevant information [4].

In this work, we evaluate the use of Latent Semantic Analysis (LSA) for structuring the results of a literature search into high-level semantic divisions, or themes. LSA is a natural language processing technique that allows analysing the relations

S. Omatu et al. (Eds.): *Distrib. Computing & Artificial Intelligence,* AISC 217, pp. 609–616.
DOI: 10.1007/978-3-319-00551-5_72

between a set of documents and the terms that belong to those documents, by representing them in a multi-dimensional semantic space [5]. Each dimension in this semantic space is represented as a linear combination of words from a fixed vocabulary (the words that compose the documents in the collection), and is usually represented by the list of words with highest value for that dimension. Since each dimension can be regarded as a different view of the results, looking at a given dimension corresponds to exploring the documents from a different perspective. This analysis allows organizing the documents according to the themes they include, providing an intuitive way for exploring the document collection.

The next sections are organized as follows: related works are presented in Section 2, Section 3 describes the proposed methodology, Section 4 presents and discusses the results obtained. Final conclusions are made in Section 5.

2 Related Work

PubMed is the most popular and widely used biomedical literature retrieval system. It combines boolean and vector space models for document retrieval with expert assigned Medical Subject Headings (MeSH) categories, giving researchers access to over 20 million citations [6] . However, as most information retrieval (IR) systems, PubMed uses query proximity models to search documents matching a user's query terms, returning results in the form of a list. Similarly, several other IR tools based on the MEDLINE literature database have been developed (see [6] for a comprehensive list of tools).

More recently, the focus has been on the use of Latent Semantic Analysis (LSA) [5, 2] and probabilistic topic models such as Latent Dirichlet Allocation (LDA) [1] . These models allow identifying the relevant themes or concepts associated to a document. Zheng et al. [11] and Jahiruddin et al. [3] have proposed document conceptualization and clustering frameworks based on LSA and domain ontologies. Zheng et al. base their methods on a user-defined ontology, matching the terms that compose this ontology to phrase chunks extracted from the documents in a collection. LSA is then applied to the term-document matrix constructed from these matches. The authors demonstrated that the application of LSA considerably improves document conceptualization. Jahiruddin et al. integrate natural language processing (NLP) and semantic analysis to identify key concepts and relations between those concepts. Their method starts by selecting candidate terms from the noun phrases in the document collection. LSA is then applied to the matrix constructed from these terms in order to identify the most important ones. Relation extraction is also performed, by identifying relational verbs in the vicinity of biomedical entities and concepts. Validated concepts and interactions are then used to construct a semantic network, which can be used to navigate through the information extracted from the documents. In this work, we use LSA to identify the latent semantics within a corpus, and borrow the term topic to refer to the underlying theme(s) for a given LSA dimension. However, this should not be confused with the meaning of this term within (probabilistic) topic models.

3 Methods

As mentioned before, our aim is to structure the results of a literature search into high-level themes, or topics, in order to help researchers search and explore the information enclosed in the scientific biomedical literature. We apply our method to a corpus related to neurodegenerative disorders, containing around 135 thousand Medline documents composed by the title and abstract of the publication. The PubMed query used to obtain the documents was: "Neurodegenerative Diseases"[MeSH Terms] OR "Heredodegenerative Disorders, Nervous System"[MeSH Terms]. Articles in languages other than English or not containing an abstract were discarded. The list of MeSH term assigned to each document was also obtained.

Our approach consists of an offline phase followed by two online steps. In the offline phase we calculate the LSA transformation matrix and transform the corpus to the LSA space. This operation is performed once for the complete corpus, and the transformation matrix is kept for transforming the user queries into the semantic space. Given a query, the two online steps consist of identifying and ranking the relevant documents within each topic and obtaining a list of representative MeSH terms for each topic. These steps are described in the next sections.

3.1 Corpus Processing and LSA

Before applying LSA, the corpus was processed in order to identify terms from a fixed vocabulary. This vocabulary contains terms from the biomedical domain, and was created based on the UMLS Metathesaurus [9]. The documents are therefore represented by the set of domain terms occurring in them, instead of through the common bag-of-words approach, and the term-document matrix used for calculating the LSA is constructed from this representation. The Gensim framework [7] was used for calculating LSA.

3.2 Ranking Relevant Documents

This step starts by selecting only the most relevant LSA dimensions (topics) for the query, given the query representation in the LSA space. A threshold is applied to eliminate those dimensions to which the query is less related, i.e. has a smaller coefficient. Next, we proceed to ranking the relevant documents within each of the selected topics. For this, and given the selected dimensions, we want to consider two aspects: the similarity between the query and the document, and the association of the document to each given topic. To reflect these two aspects, we propose the following score:

$$Score(D_k, T_j) = |Sim(D_k, Q) \times V(D_k, T_j)|, \quad (1)$$

where $V(D_k, T_j)$ is the LSA coefficient for document D_k in dimension T_j and $Sim(D_k, Q)$ is the cosine similarity between document D_k and the query Q, in the LSA space. Finally, we use a second threshold to filter these scores, obtaining the most similar documents for the query regarding each of the considered topics.

3.3 Identifying Representative Terms

In order to represent each identified topic and facilitate the exploration of results by users we make use of the MeSH terms, which represent the major concepts in each Medline article, selected by expert annotators. The important aspect to consider is that we expect that each cluster of results represents a distinct topic or theme. Therefore, the documents assigned to each dimension, after the previous step, should not only be different but should also be focused on different themes. In order to evaluate this, we can compare the set of MeSH terms assigned to the documents in different topics to see how different they are. In order to do so, we first calculate an association score for each MeSH term in each topic, creating a vector representation that we then use to compare the topics. For each MeSH terms and each topic, this score is calculated as the sum of the coefficients of the documents containing the MeSH term, normalized by the corresponding rank of that document in the topic, as shown in Eq. 2.

$$Score(M_i, T_j) = \sum_{k=1}^{N_j} \frac{V(D_k, T_j)}{Rank(D_k, T_j)}, \quad M_i \in D_k,\ D_k \in T_j, \tag{2}$$

Table 1 Top results in topics 12 and 25 for the query term "Dopamine"

topic 12	
20926973	Intense dopamine innervation of the subventricular zone in Huntington's disease.
10838590	Neuronal cell death in Huntington's disease: a potential role for dopamine.
9822765	Dopamine modulates the susceptibility of striatal neurons to 3-nitropropionic acid in the rat model of Huntington's disease.
10829080	Severe deficiencies in dopamine signaling in presymptomatic Huntington's disease mice.
17065224	Dopamine enhances motor and neuropathological consequences of polyglutamine expanded huntingtin.
topic 25	
9620058	Polymorphisms of dopamine receptor and transporter genes and Parkinson's disease.
17290452	Higher nigrostriatal dopamine neuron loss in early than late onset Parkinson's disease?
8464534	Brain dopamine receptors: 20 years of progress.
8848171	Involvement of ventrolateral striatal dopamine in movement initiation and execution.
20188048	Recent discoveries on the function and plasticity of central dopamine pathways.

where N_j is the number of documents assigned to dimension T_j, $V(D_k, T_j)$ is the LSA coefficient for document D_k in dimension T_j, and $Rank(D_k, T_j)$ is the ranking position of document D_k in topic T_j.

4 Results

Using the methods proposed in the previous section, it is possible to organize the literature search results into separate lists, each associated to a certain theme. The documents retrieved by LSA similarity will be distributed across these topics, allowing an easier navigation. Also, although a given document may occur in more than one topic, which is expected since articles discuss interrelated subjects, it will appear in different ranking positions in each result list. Therefore, users looking at two different topics will see two different sets of results. This also justifies using the document ranking when calculating the score for each MeSH term and topic pair, since the most important results for the users are the top ones in each result list.

Table 1 shows the first five results for the query "Dopamine", in topics 12 and 25. As can be noticed, the results lists are significantly different between topics, illustrating how the retrieved results are organized around separate themes.

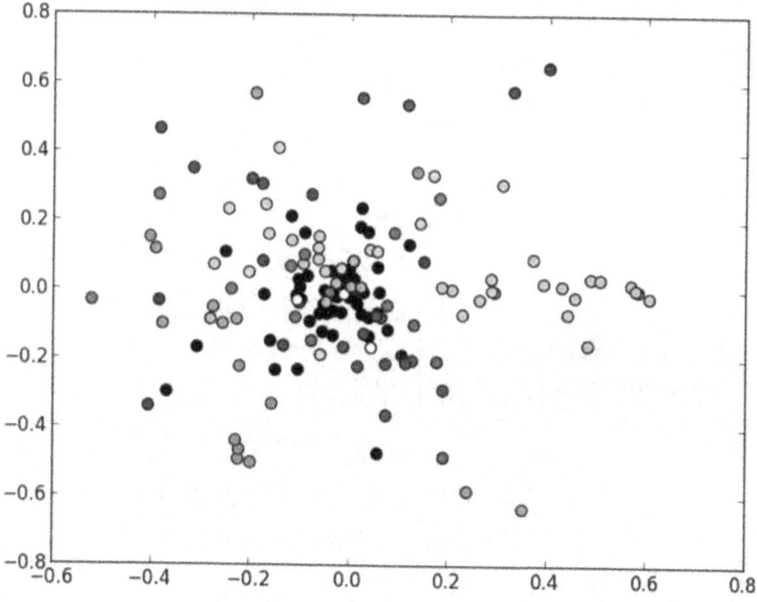

Fig. 1 Distribution of documents in the geometrical space created by MDS. The top 20 documents for each topic relevant for the query "Dopamine" are shown. The colour of the circle indicates the topic to which the document is assigned; black circles indicate documents assigned to more than one topic.

4.1 Multidimensional Scaling

An intuitive way to represent the different topics in the results is by using multidimensional scaling (MDS), an exploratory technique used to visualize proximities in a low dimensional space [10]. Using MDS, the documents or the topics resulting from a search can be displayed on a two-dimensional space, where they appear distributed according to their similarities. Figure 1 shows the result of MDS for the query "Dopamine", using the LSA cosine similarity between each pair of documents. Only the top 20 documents in each topic were considered. Each document is represented by a circle, coloured according to the topic that document belongs to; black circles represent documents assigned to more than one topic. This representation, used within a literature retrieval system, would allow users to navigate the results while visualizing the relations or similarities between the resulting documents.

Fig. 2 MeSH term cloud for topics 12 and 25 showing the most relevant terms in these topics for the query "Dopamine". The size of the font is proportional to the score of that term for the topic and query combination. The cloud was created in the Wordle website (http://www.wordle.net/)

4.2 Word Clouds

Another way of visualizing the results of LSA is by representing the topics by word clouds. In this case, we want the word cloud for a topic to reflect the most important terms for that topic given the specific query. Therefore, we use the most significant MeSH terms for resulting documents assigned to that topic. Figure 2 illustrates the MeSH term cloud corresponding to topics 12 and 25 for the query "Dopamine". From the most prominent terms, one can identify that topic 12 is about physiology and physiopathology in Huntington's disease, while topic 25 is mostly about receptors, transport and metabolism of Dopamine. It is important to emphasize that, although the LSA dimensions are defined for the entire corpus and are kept constant, the word cloud for this same topic would be different for a different query, as given by the score defined in Eq. 2.

5 Conclusions

We have described an approach for structuring the results of a literature search based on the latent semantic information extracted from the documents in a corpus, as expressed by LSA. Moreover, we show how the results of LSA can be adapted so as to evidence differences between results of different queries and propose several visualization techniques that can be applied to explore these results.

Further work is required for evaluating how users would benefit from the proposed solutions. Although objective evaluation of methods such as the one proposed here is usually very difficult, the results presented indicate that methods for structuring literature search results, used in combination within a literature retrieval system, could empower users with tools for literature guided knowledge exploration and discovery.

Acknowledgement. This research work was partially funded by FEDER through the COMPETE programme and by national funds through FCT - "Fundação para a Ciência e a Tecnologia" under project number PTDC/EIA-CCO/100541/2008 (FCOMP-01-0124-FEDER-010029). Sérgio Matos is funded by FCT under the Ciência2007 programme.

References

1. Blei, D.M., Ng, A.Y., Jordan, M.I.: Latent Dirichlet Allocation. Journal of Machine Learning Research 3, 993–1022 (2003)
2. Deerwester, S., Dumais, S.T., Furnas, G.W., Landauer, T.K., Harshman, R.: Indexing by Latent Semantic Analysis. Journal of the American Society for Information Science 41, 391–407 (1990)
3. Jahiruddin, Abulaish, M., Dey, L.: A concept-driven biomedical knowledge extraction and visualization framework for conceptualization of text corpora. Journal of Biomedical Informatics 43, 1020–1035 (2010)
4. Kim, J.J., Rebholz-Schuhmann, D.: Categorization of services for seeking information in biomedical literature: a typology for improvement of practice. Briefings in Bioinformatics 9(6), 452–465 (2008)

5. Landauer, T.K., Foltz, P.W., Laham, D.: An introduction to latent semantic analysis. Discourse Processes 25(2-3), 259–284 (1998)
6. Lu, Z.: PubMed and beyond: a survey of web tools for searching biomedical literature. Database 2011, baq036 (2011)
7. Řehůřek, R., Sojka, P.: Software Framework for Topic Modelling with Large Corpora. In: Proceedings of LREC 2010 Workshop New Challenges for NLP Frameworks, pp. 46–50. LREC (2010)
8. Shatkay, H.: Hairpins in bookstacks: information retrieval from biomedical text. Briefings in Bioinformatics 6(3), 222–238 (2005)
9. UMLS Metathesaurus Fact Sheet, `http://www.nlm.nih.gov/pubs/factsheets/umlsmeta.html`
10. Van Deun, K., Heiser, W.J., Delbeke, L.: Multidimensional unfolding by nonmetric multi-dimensional scaling of Spearman distances in the extended permutation polytope. Multivariate Behavioral Research 42(1), 103–132 (2007)
11. Zheng, H.-T., Borchert, C., Jiang, Y.: A knowledge-driven approach to biomedical document conceptualization. Artificial Intelligence in Medicine 49, 67–78 (2010)

Upper Ontology for Multi-Agent Energy Systems' Applications

Gabriel Santos, Tiago Pinto, Zita Vale, Hugo Morais, and Isabel Praça

GECAD Knowledge Engineering and Decision-Support Research Center, Institute of Engineering - Politechnic of Porto (ISEP/IPP), Porto, Portugal
{gajls,tmcfp,zav,hugvm,icp}@isep.ipp.pt

Abstract. Energy systems worldwide are complex and challenging environments. Multi-agent based simulation platforms are increasing at a high rate, as they show to be a good option to study many issues related to these systems, as well as the involved players at act in this domain. In this scope the authors research group has developed three multi-agent systems: MASCEM, which simulates the electricity markets; ALBidS that works as a decision support system for market players; and MASGriP, which simulates the internal operations of smart grids. To take better advantage of these systems, their integration is mandatory. For this reason, is proposed the development of an upper-ontology which allows an easier cooperation and adequate communication between them. Additionally, the concepts and rules defined by this ontology can be expanded and complemented by the needs of other simulation and real systems in the same areas as the mentioned systems. Each system's particular ontology must be extended from this top-level ontology.

1 Introduction

Electricity markets worldwide are complex and challenging environments, involving a considerable number of participating entities, operating dynamically trying to obtain the best possible advantages and profits [20]. Market players and regulators are very interested in foreseeing market behavior by understanding its principles and learning how to evaluate their investiments in such a competitive environment [13]. A clear understanding of the impact of power systems physics on market dynamics and vice-versa is required.

The development of simulation platforms based in multi-agent systems (MAS) is increasing as a good option to simulate real systems in which stakeholders have different and often conflicting objectives. The use of MAS in energy systems is a reality, particularly the simulation of electricity markets [9, 10, 18] and hierarchical decision making, as smart grid (SG) and microgrids (MG) [4, 15]. Several modeling tools can be fruitfully applied to study and explore restructured power markets, such as AMES [10] and EMCAS [9].

MASCEM (Multi-Agent Simulator for Electricity Markets) [18, 21] is a simulation tool to study and explore restructured electricity markets. It aims to simulate

S. Omatu et al. (Eds.): *Distrib. Computing & Artificial Intelligence,* AISC 217, pp. 617–624.
DOI: 10.1007/978-3-319-00551-5_73

as many market models and player types as possible, enabling it to be used as a simulation and decision-support tool for short/medium term purposes but also as a tool to support long-term decisions, such as the ones taken by regulators.

The strategic definition in MASCEM is undertaken using a connection with another MAS: ALBidS (Adaptive Learning strategic Bidding System) [17]. It provides decision support to electricity markets negotiating players, allowing them to analyze different contexts of negotiation and automatically adapt their strategic behavior according to the current situation.

Another relevant tool to our research group is MASGriP (Multi-Agents Smart Grid Simulation Platform) [15], a MAS that models the internal operation of SG and considers all the typical involved players. Modeling SG as MAS allows exploring both the individual and internal performance of each player, as well as the global and specific interactions between the involved players.

Since these three MAS are independent and heterogeneous, it is important to define common language and semantics for the communication between them to be coherent. In order to provide full interoperability and prepare the basis for an adequate integration with external MAS, open standards are needed [3, 12].

This paper describes the development of an ontology divided by layers, allowing communication between simulation platforms of electricity markets, SG and MG. Section 2 presents an overview on MASCEM, ALBidS and MASGriP. Section 3 describes the multi-agent open standards on interoperability and the proposed approach to integrate these three systems. Finally, section 4 features the final conclusions of the presented work.

2 MASCEM, ALBidS and MASGriP Overview

The new paradigm of power system operation with a large number of players and with decentralized operation of electric network and the energy resources, impose new methodologies in decision support tools to help the players to opt to best strategy in the operation and in the negotiation process between players and in the electricity market participation. The use of MAS system allows the simulation of this complex environment and the interaction between players.

2.1 MASCEM

MASCEM [18] simulates the electricity market, considering the most important entities, allowing the definition of their offers and strategies, granting them competitive advantage in the market. It modulates the complexity of dynamic market players, their interaction and medium/long-term gathering of information.

MASCEM main entities include: a market operator agent, a system operator agent, a market facilitator agent, buyer agents, seller agents, Virtual Power Player (VPP) agents, and VPP facilitators. The market operator is an independent entity that regulates market negotiations. The system operator is responsible for the system's security, by examining the technical feasability, and assures that all conditions are met within the system. The market facilitator coordinates and assures the proper operation of the market by regulating all existing communications.

The key elements of the market are buyer and seller agents. The buyer agents represent the electricity consumers and distribution companies. Electricity producers or other entities able to sell energy in the market are represented by seller agents.

VPPs [16, 19] represent alliances of small/medium players, manly based on distributed generation (DG) and renewable sources, providing means to adequately support their increasing use and its participation in the competitive electricity markets. VPP agents are implemented as a coalition of players, each one acting as an independent MAS, maintaining high performance and allowing agents to be installed on separate machines.

MASCEM allows the simulation of the main market models: day-ahead pool (symmetric or asymmetric, with or without complex conditions), bilateral contracts, balancing market, forward markets and ancillary services.

2.2 ALBidS

The way prices are predicted for each market can be approached in several ways, through the use of statistical methods, data mining techniques [2], artificial neural networks (ANN) [1], support vector machines, among others [8, 10]. To take advantage of the best characteristics of each technique, it was developed ALBidS, which integrates several distinct methodologies and approaches.

ALBidS uses reinforcement learning algorithms to choose the players most adequate action, from a set of different proposals provided by the several algorithms that use distinct approaches. To this end, it considers the past experience of the actions responses and the present characteristics of each context, such as the week day, the period, and the particular market that the algorithms are being asked to forecast. One of the main strategic approaches considered by ALBidS is based on an error analysis, considering the forecasting errors tendencies over time.

ALBidS is implemented as a MAS itself where each algorithm is under the responsibility of an agent. Thus, the system executes the algorithms simultaneously, increasing the performance of the system. As each agent gets its answer, sends it to the main agent, which is responsible for choosing the most appropriate answer among all.

To guarantee the minimum degradation of the processing time, a methodology to manage the efficiency/effectiveness *(2E)* balance of ALBidS has been developed [17].

2.3 MASGriP

MASGriP [15] is a MAS that proposes a set of possible coalitions that facilitate the management of SG and MG. It models the distribution network and the involved players, such as Domestic Customers, Small, Medium and Large Commerce, Small, Medium and Large Industry and Rural Consumers, all of them may consider Demand Response (DR) and/or micro/mini-generation and/or Electric Vehicles (EVs), as well as different sizes of DG. Each player is represented by an agent.

The players establish contracts with two types of aggregators: the VPP or the Curtailment Service Provider (CSP). A CSP can be defined as a special player aggregating consumers DR participation, enabling small and medium consumers to participate in DR events. Its main tasks are to identify curtailable loads, to enroll customers, to manage curtailment events, and to calculate payments or penalties for its customers [14].

The MG operator can be a player of a SG, a CSP or a VPP, allowing the MG members to benefit of the advantages of being in the coalition as well. SG can also be a part of a VPP. A VPP can have several SGs and MGs alongside large players, mixing traditional and renewable forms of generation, large consumers with great demands and medium and small consumers with good possibilities of load curtailment. A large player can be connected to a coalition, like the CSP, and at the same time participate in the market by itself to sell or buy energy.

The integration between MASGriP and MASCEM provides the means for simulating appropriately the resources management in the scope of SG, including all the most important features it requires, such as the internal management of SG and MG, the use of DR, the management by VPPs, and the actual market negotiations.

3 Upper Ontology to Multi-Agent Systems Interoperability

The integration of MAS raises inherent issues to the inter-operation of those systems, particularly the ones involving the use of different ontologies. To disseminate the development of interoperable multi-agent systems, especially in the power industry, these issues need to be addressed [3, 11, 12]. In order to take full advantage of the functionalities of those systems, there is a growing need for knowledge exchange between them. Open standards are needed to provide full interoperability.

3.1 Multi-agent Open Standards on Interoperability

When developing MAS, the use of standards is important to allow the integration of separate systems. Within power engineering, the increasing application of multi-agent technology promotes the adoption of standards that enable the communication between heterogeneous systems, bringing future advantages [11, 12].

The Foundation for Intelligent Physical Agents (FIPA) is devoted to develop and promote open specifications that support interoperability among agents and agent-based applications [6]. MAS using FIPAs standards should be able to interoperate but it doesn't mean that the agents are able to share useful information due to the employment of different ontologies.

FIPA proposes the Agent Communication Language (ACL) as a standard for communications between agents. The content of the message includes the content language and the ontology. The former specifies the syntax, while the latter provides the semantics of the message [7]. This way the correct interpretation of the meaning of the message is assured, removing the ambiguity about the content. The FIPA-SL content language is the only one that reached a stable standard. Ontologies are used by agents for exchanging information, ask questions, and request the execution of actions related to their specific domain.

Presently, MAS in the power systems domain are developed with their own specific ontologies. These systems share common concepts, which are differently represented between ontologies. Translating these concepts automatically is not as straightforward as it seems. To solve the problem of multiple ontologies, FIPA proposes the use of an ontology agent that provides some related services [5]. This is still an experimental standard and mappings between ontologies still must be performed by ontologies' designers, which increases the human effort required and costs of implementation.

3.2 Upper Ontology

Alternatively, the use of an upper ontology representing general concepts of the domain is proposed in [3]. This approach ensures a common basic representation for the same concepts between systems, and their relations while reducing the complexity of ontology mapping.

For the integration of our MAS, this latter approach was adopted. Initially a top-level ontology is defined with concepts and relationships suitable for our systems. From this ontology specific ontologies are extended for each of our platforms. Fig. 1 illustrates the agents and existing communications between platforms.

The systems communicate with each other according to their needs and trying to make the best possible negotiations. MASCEM's players interact with ALBidS' Main Agent, sending him information about the market in which they are presenting

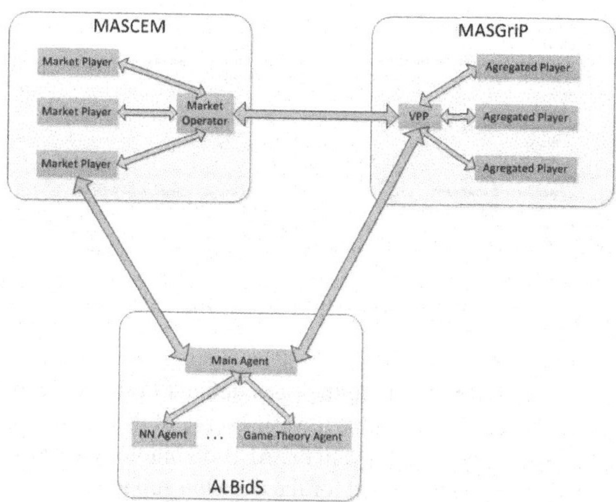

Fig. 1 Main communications between MASCEM, ALBidS & MASGriP

bids. The Main Agent spreads this information among its agents that process data from the past accordingly to this information and present a proposal for the market. The best of all proposals is returned to the player. Similarly, VPPs in MASGriP can also communicate with the ALBidS to support the decision on which is the best offer to present in the market. Communication between MASCEM and MASGriP happens when the VPPs in MASGriP decide to present offers in the market in order to buy or sell energy. To such, VPPs communicate with the Market Operator.

Each of the MAS has an internal ontology shared among its agents. However, sharing their ontology is not enough to allow proper communication between systems. For defining an upper-ontology abstract enough to allow inter-communication of our systems, an analysis must be performed to each platforms ontology in order to extract the concepts and relationships common to all. And then, the ontology of each system must be redesigned so as to extend the top-level ontology. In Fig. 2 a small fraction of our Electricity Markets Upper Ontology is shown, where one can see a part of MASCEM and ALBidS ontologies.

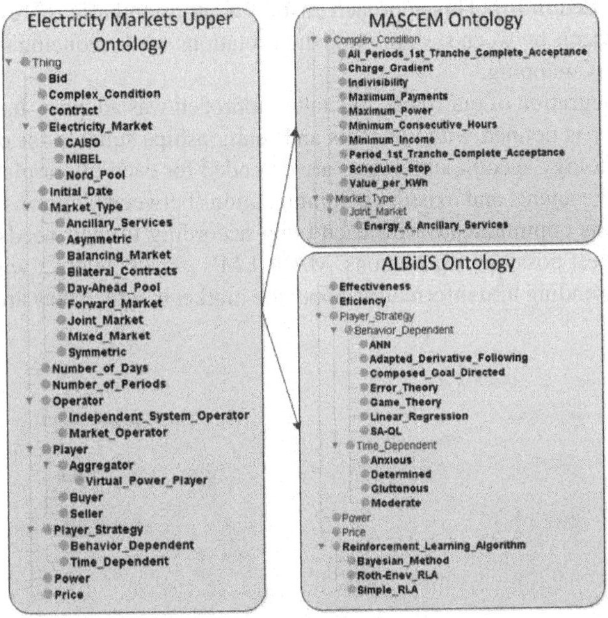

Fig. 2 Excerpt of Electricity Markets, MASCEM and ALBidS ontologies

The classes represented in bold are the ones defined in the respective ontology. In the MASCEM ontology it is possible to verify some of the extensions made from the Electricity Markets upper ontology. In the ALBidS ontology are also represented some of the extended concepts needed for the decision support.

The gain obtained with this inter-operability brings added value to our research, allowing us to take better advantage of the various platforms developed, in the study of competitive electricity markets. This also provides the basis for the interconnection of our systems with other systems, allowing our agents to participate in different types of markets, different perspectives of SG, and vice versa.

4 Conclusions

The use of MAS for the simulation and study of competitive energy markets is increasingly the main option. These platforms have proved a great asset in the area of Power Systems.

This paper presents three MAS developed by the authors research group, and their importance for the study and better comprehension of complex energy systems. Although these systems are independent platforms, to achieve better results in the study of these systems and from the interaction between the involved agents, the need to connect them arises. For this it is necessary that the agents involved are able to interpret messages from other platforms.

To achieve systems interoperability an upper-ontology was developed, from which the ontologies of each platform must be extended. Although this approach does not avoid the need for mapping, it significantly reduces the effort expended for this purpose. It also aims to enable communication from external systems with ours, allowing a much more complete study of this domain.

Acknowledgement. This work is supported by FEDER Funds through COMPETE program and by National Funds through FCT under the projects FCOMP-01-0124-FEDER: PEst-OE/EEI/UI0760/2011, PTDC/EEA-EEL/099832/2008, PTDC/SEN-ENR/099844/2008, and PTDC/SEN-ENR/122174/2010.

References

1. Amjady, N., Daraeepour, A., Keynia, F.: Day-ahead electricity price forecasting by modified relief algorithm and hybrid neural network. IET Generation Transmission Distribution (2010)
2. Azevedo, F., Vale, Z., Oliveira, P.: A decision-support system based on particle swarm optimization for multiperiod hedging in electricity markets. IEEE Transactions on Power Systems (2007)
3. Catterson, V., Davidson, E., McArthur, S.: Issues in integrating existing multi-agent systems for power engineering applications. In: Proceedings of the 13th International Conference on Intelligent Systems Application to Power Systems (2005)
4. Dimeas, A., Hatziargyriou, N.: A mas architecture for microgrids control. In: Proceedings of the 13th International Conference on Intelligent Systems Application to Power Systems (2005)
5. F. for Intelligent Physical Agents (FIPA). Fipa ontology service specification (2001)
6. F. for Intelligent Physical Agents (FIPA). Agent management specification (2002)
7. F. for Intelligent Physical Agents (FIPA). Fipa acl message structure specification (2002)
8. Greenwald, A., Kephart, J.O.: Shopbots and pricebots. In: IJCAI (1999)
9. Koritarov, V.: Real-world market representation with agents. IEEE Power and Energy Magazine (2004)

10. Li, H., Member, S., Tesfatsion, L.: Development of open source software for power market research: The ames test bed (2009)
11. McArthur, S., Davidson, E., Catterson, V.: Building multi-agent systems for power engineering applications. In: IEEE Power Engineering Society General Meeting (2006)
12. McArthur, S., Davidson, E., Catterson, V., Dimeas, A., Hatziargyriou, N., Ponci, F., Funabashi, T.: Multi-agent systems for power engineering applications - part ii: Technologies, standards, and tools for building multi-agent systems. IEEE Transactions on Power Systems (2007)
13. Meeus, L., Purchalaa, K., Belmans, R.: Development of the Internal Electricity Market in Europe. The Electricity Journal (2005)
14. Moran, D., Suzuki, J.: Curtailment service providers: They bring the horse to water–do we care if it drinks? In: 16th Biennial ACEEE Summer Study on Energy Efficiency in Buildings (2010)
15. Oliveira, P., Pinto, T., Morais, H., Vale, Z.: Masgrip: A multi-agent smart grid simulation platform. In: 2012 IEEE Power and Energy Society General Meeting (2012)
16. Oliveira, P., Pinto, T., Morais, H., Vale, Z., Praça, I.: Mascem - an electricity market simulator providing coalition support for virtual power players. In: 15th International Conference on Intelligent System Applications to Power Systems, ISAP 2009 (2009)
17. Pinto, T., Vale, Z., Rodrigues, F., Morais, H., Praça, I.: Bid definition method for electricity markets based on an adaptive multiagent system. In: Demazeau, Y., Pĕchoucĕk, M., Corchado, J.M., Pérez, J.B. (eds.) Advances on Practical Applications of Agents and Multiagent Systems. AISC, vol. 88, pp. 309–316. Springer, Heidelberg (2011)
18. Praça, I., Ramos, C., Vale, Z., Cordeiro, M.: Mascem: a multiagent system that simulates competitive electricity markets. IEEE Intelligent Systems (2003)
19. Rahwan, T., Jennings, N.: Coalition structure generation: Dynamic programming meets anytime optimisation. In: Proc 23rd Conference on AI, AAAI (2008)
20. Shahidehpour, M., Yamin, H., Li, Z.: Market Operations in Electric Power Systems: Forecasting, Scheduling, and Risk Management. Wiley-IEEE Press (2002)
21. Vale, Z., Pinto, T., Praça, I., Morais, H.: Mascem - electricity markets simulation with strategic agents. IEEE Intelligent Systems (2011)

A Practical Mobile Robot Agent Implementation Based on a Google Android Smartphone

Dani Martínez, Javier Moreno, Davinia Font, Marcel Tresanchez, Tomàs Pallejà, Mercè Teixidó, and Jordi Palacín

Computer Science and Industrial Engineering Department,
University of Lleida, 25001 Lleida, Spain
{dmartinez,jmoreno,dfont,mtresanchez,tpalleja,
mteixido,palacin}@diei.udl.cat

Abstract. This paper proposes a practical methodology to implement a mobile robot agent based on a Google Android Smartphone. The main computational unit of the robot agent is a Smartphone connected through USB to a control motor board that drives two motors and one stick. The agent program structure is implemented using multi-threading methods with shared memory instances. The agent uses the Smartphone camera to obtain images and to apply image processing algorithms in order to obtain profitable information of its environment. Moreover, the robot can use the sensors embedded in the Smartphone to gather more information of the environment. This paper describes the methodology used and the advantages of developing a robot agent based on a Smartphone.

Keywords: Mobile robot, robot agent, Google Android Smartphone.

1 Introduction

Agents are considered a reference for many robotic systems and applications. An agent can be defined as an autonomous system in which is capable to sense the environment, react to it, develop collaborative task, and take an initiative to complete a task [1]. According to this, agent objectives are in tight correlation with the objectives of artificial intelligence processes and algorithms. Usually, the implementation of an agent application requires several complex and heavy algorithms to make decisions in order to achieve its objective. For example, computer vision methods combined with other sensors usually exploit most of the computational resources delivered by the agent to obtain information of the environment. All this embedded features can raise considerably the economic cost of a robot and so, can make the project be non-viable. In this work we propose the

S. Omatu et al. (Eds.): *Distrib. Computing & Artificial Intelligence,* AISC 217, pp. 625–632.
DOI: 10.1007/978-3-319-00551-5_74

development of an agent system by using the computational power, the sensors and actuators, and the communication capabilities of an Android Smartphone.

The popularity and computational power of Smartphones are increasing significantly from the last years. This evolution has fostered many research initiatives [2,3] focused on the relatively new Google Android operating system for mobile devices. This user-friendly operating system is designed for low power consumption while having constant connectivity. The Android powered devices give access to its integrated peripherals such as cameras, wireless connectivity modules, embedded sensors and touch screen. Such devices usually require a high memory capacity and high computational power that is currently delivered by powerful multi-core processors. The computational resources offered by such devices can be directly applied in the development of new amazing applications and also to drive small mobile robots. The main advantage of a Smartphone based mobile robot is that the vision sense can be based on the onboard cameras of the Smartphones without wasting time in connections and in procedures to transfer the image of the camera to the Smartphone. Thus, the computational power of the Smartphone can be focused in the agent implementation required to drive the mobile robot.

This paper proposes a methodology to implement robot agents using the onboard Smartphone resources (Figure 1) accessible through by Google Android Software Development Kit [4]. The motivation of this research is the evaluation of the performances of a mobile robot agent based in a Google Android Smartphone. In this paper, the objective of the developed mobile robots will be playing a game inspired in the soccer competition but without following any specific standard [5] and without a centralized command host. The soccer game requires very specific and well developed agents and also enables the future development of collaborative agent methodologies [6] and strategies when playing with teams of several mobile robots.

Fig. 1 Soccer robot with a Google Android Smartphone as the main computational unit

2 Materials and Methods

The materials used in this paper are the set of physical components which form the soccer robot agent: the Smartphone, the mobile robot structure, and its internal control devices. The method used in this paper is mainly the vision sense required for gathering environment information in order to play the proposed game.

2.1 Central Processing Unit

The central processing unit used in this paper is an Android HTC Sensation Smartphone device which is powered by a dual-core 1.2GHz processor, 768MB of RAM memory, and Android 4.0.3. The Smartphone also integrates WIFI and Bluetooth connectivity modules and other embedded sensors such as GPS, ambient light sensor, digital compass, three-axial accelerometer, gyroscope, multi-touch capacitive touch screen, proximity sensor, microphone, and a frontal and a rear camera. In this case, the Android SDK [4] provides an easy method to manage such sensors and also to implement multi-threading applications.

2.2 Mobile Robot

The soccer mobile robot is composed by an external case made in fused ABS plastic material that is also very resistant. The plastic case has been colored in red and in blue to distinguish the soccer team and allow the differentiation of the team by the different agents. A feature that will be specially needed in future team implementations. This external case is designed to support the motion of the mobile robot and also a kicking mechanism with a motor to push a small ball. The Smartphone is installed horizontally in a support on the top of the mobile robot case to ensure an adequate angular view of its rear camera. Figures 1 and 2 show the external aspect of the complete mobile robot based on a Google Android Smartphone.

The robot mobility is accomplished with a motor control board plugged to the Smartphone through an USB interface. The board controls two small DC motors

Fig. 2 Soccer robot agent in its playfield environment

and has a microcontroller which implements the methods required to establish a communication with the Smartphone. In addition, the board is powered by an auxiliary 5.000 mAh battery which also powers and charges the Smartphone.

2.3 Vision Sense

The application of the soccer robot agent uses some image processing algorithms to process the rear image of the Smartphone that is pointed to the playfield. The vision sense has to generate profitable information of the environment from the images acquired that will be used by the agent to play the game. The methodology used to extract the information of the images is explained in [7]. The images acquired are converted to the H and V color space layers to perform a color indexing by an established look up table. Once a new image is acquired, the pixels are indexed to reference the objects which represent the game environment elements such as the ball, the goals, the field, the lines, and the teammates and adversaries. Then, the different objects (or layers) of the classified image are analyzed to extract profitable information for the agent such as relative orientation and distance to the ball, to the goals, to the lines, and to the other mobile robots. Figure 2 shows the screen of the Smartphone of the mobile robot that has windows showing the image acquired, the indexed classified image were each object has an identifying color, and the estimated position of the mobile robot in the playfield.

3 Implementation

The software implementation of the agent in the Google Android Smartphone has been performed in Java language and executed internally in Dalvik virtual machine which is capable to execute several instances simultaneously with an optimization in process management and in use of memory. Such applications can use several context instances called activities which manage the lifecycle of an application. An Android activity is executed in a main execution thread in which can access and manage the user interface resources such as buttons, images, or touch input data.

The programming architecture proposed in this paper to implement the agent, requires several simultaneously execution threads. The Android-level programming enables the initialization and start of different thread instances and services from the main thread. Figure 3 shows our methodology, is structured as a distribution of distinguished tasks, and processed in separate threads using shared memory protocols for data demands.

3.1 Main Thread

The main thread has lifecycle activity methods such as *onCreate()* and *onDestroy()* that are called when the activity starts, when the activity goes to background to focus the Smartphone resources on new activity created, or when is destroyed to free memory and resources. When the activity is initializing, the other threads and services that define the agent are also created and registered. The main thread firstly creates and configures the camera service that is associated with a surface class and could be configured with different parameters such as camera resolution, frame rate, and several image filters. However, the final effect of the parameters over the camera depends on the Smartphone used and not all Android devices support all parameter configurations.

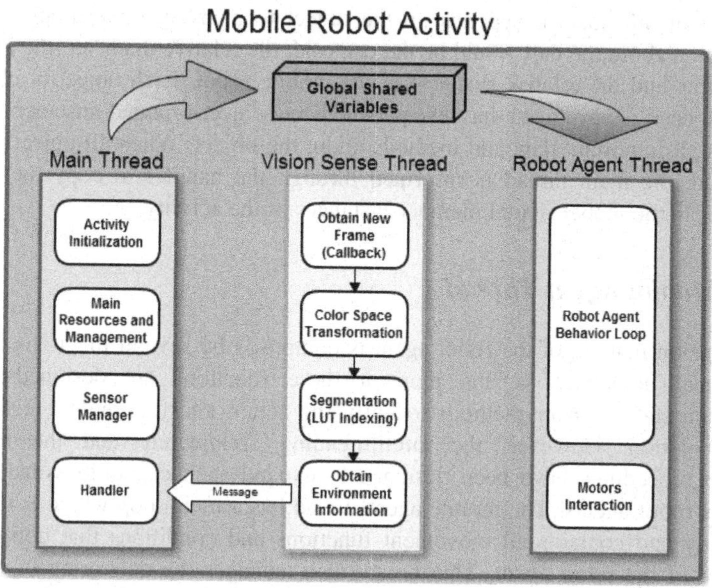

Fig. 3 Threading and processes structure of the Android activity

Next, the robot agent thread is also initialized and started. When all execution threads of the agent are fully operative, the main thread manages a handler which is listening for asynchronous messages which will be sent between the threads to establish an inter-threading communication. The handler, thus, is designed to copy the profitable data processed from the vision sense thread to global shared memory in the activity, then, such variables from global shared memory are updated when new data are just processed in the vision sense thread. Moreover, this thread also initialize the USB interface in order to send motion orders to the motor control board, and also registers the services to obtain data from the available embedded sensors.

3.2 Vision Sense Thread

Android provides to developers the possibility of accessing to services offered by the integrated cameras of the device. In order to obtain an image from the camera, the camera service should be initialized and configured, then, it requires implementing a callback which calls a method each time the hardware delivers a frame image to the application. The callback method onPreviewFrame(...) is internally executed in a separate thread in order to avoid a block in the main thread and consequent performance loose. This separate thread is the vision sense thread of the mobile robot.

The color segmentation and indexing methods used in this thread are explained in [7]. After identifying and labeling the game elements, all objects of the image are analyzed to extract profitable information which would be useful for the robot agent. Such information depends on the object to analyze, for example, useful information about the ball would be the centroid, the relative diameter, the relative orientation, and the relative distance to the mobile robot. Each object or element of the soccer environment has his personalized object-oriented instance which contains all important data and methods about the object. When all object data is processed, the main thread is informed through the handler to copy the object instances in the global shared memory instances of the activity.

3.3 *Mobile Agent Thread*

The decision making of the robot agent is composed by several functions and by conditional branches. At this moment these functions are designed as an automaton state transfer methods to unify the action and to simplify this initial implementation. However, the multithreading architecture and the internal function procedures have been designed to expand such capabilities and create complex robot agents. The mobile agent thread has a main loop which is running constantly and contains all movement functions and conditions that defines the behavior of the robot agent. The thread, also manages the delivering of motion orders to the motor control board through the USB interface.

An example of basic functions implemented in the mobile agent thread is the "search_ball" procedure which retains mobile robot control until the ball has been detected in the image acquired by the camera of the Smartphone. Other example is the "approach_ball" procedure which retains mobile robot control until touching the ball but or until the ball goes out from the image, for example, when kicked by another player. The method will return true or false depending on the result of the robot movement. Figure 4 shows a code of a loop in the thread to perform a simple task:

The functions of this thread are considered as blocking methods until the objective of the function is completed or interrupted by external causes. The implementation of such functions are structured as methods in which a determined action or objective is done. However, these functions require environmental information to achieve its objective and have to perform data queries to global shared memory instances in order to obtain the profitable data processed in the vision sense thread.

```
while(!thread_stop)
{
  if(motors.connected)
  {
            found = search_ball();
            if(!found)continue;

            aimed = aim_ball();
            if(!aimed)continue;

            caught = approach_ball();
            if(!caught)continue;

            kick_ball();
            wait(2000);
  }
}
```

Fig. 4 Example code of a simple loop in which the robot searches the ball, goes to it and kicks it

Many agent functions generate relative order displacements for the robot which, in certain cases, uses the orientation sensors embedded in the Smartphone. Then, the mobile robot can execute certain tasks such as rotate a number of degrees based only on the orientation sensor and, alternatively, based on the information of the encoders in the wheels of the mobile robot. All motion functions can be altered by some defined situations such as detecting a collision or a whistle sound (used to start and to end the game). The detection of collisions is performed through the analysis of the information of the accelerometer of the Smartphone [8]. The detection of a starting whistle sound requires a more complex approach because it requires the implementation of a media recorder for requesting the sound and analyzing the amplitude and frequency of the sound captured by using the class *MediaRecorder* of Android SDK. For example, this implementation can be used to capture the amplitude of the ambient sound and applying a threshold intensity level to detect a large sound signal. The agent of the mobile robot is always aware of the results of these detections in order to adapt the strategy or to control the evolution of the game.

4 Conclusions and Future Work

This paper presents a practical methodology to implement a mobile robot agent based on a Google Android Smartphone device to take advantage of its mobile features and its computational power, as well as its integrated sensors and peripherals. This paper explains all main parts that conforms the mobile robot agent that is based on a multi-threading methodology to manage all the processes required to implement an effective robot agent. The agent uses shared memory instances to asynchronously communicate the different threads which compound the agent activity while maintain its own execution rhythm. The main conclusion of this paper is that a Smartphone offers extended possibilities in order to develop and implement mobile robot agent.

Future work will be focused in the improvement of the vision sense thread to include some spatial memory of the elements and objects detected in the playfield. The inclusion of spatial memory will affect the behavior of the agent and the evolutions of the mobile robot that are currently based only on what the agent sees in the image. Other future improvement will be the use of wireless communication to intercommunicate different mobile robot agents to develop collaborative team behaviors.

References

1. Wooldridge, M., Jennings, N.: Intelligent Agents: Theory and Practice. The Knowledge Engineering Review 10(2), 115–152 (1995)
2. Paul, K., Kundu, T.K.: Android on Mobile Devices: An Energy Perspective. In: IEEE 10th International Conference on Computer and Information Technology (CIT), pp. 2421–2426. IEEE Press, Bradford (2010)
3. Son, K., Lee, J.: The method of Android application speed up by using NDK. In: 3rd International Conference on Awareness Science and Technology (iCAST), pp. 382–385. IEEE Press, Dalian (2011)
4. Android Developers website, `http://developer.android.com` (accessed November 2012)
5. Robocup website, `http://www.robocup.org/` (accessed November 2012)
6. Veloso, M., Stone, P.: Individual and Collaborative Behaviors in a Team of Homogeneous Robotic Soccer Agents. In: Proceedings of the Third International Conference on Multi-Agent Systems, pp. 309–316. IEEE Computer Society, Paris (1998)
7. Martínez, D., Moreno, J., Tresanchez, M., Font, D., Teixidó, M., Pallejà, T., Palacín, J.: Evaluation of the color-based image segmentation capabilities of a compact mobile robot based on Google Android Smartphone. In: International Conference on Practical Applications of Agents and Multi-Agent Systems, Special Session in Agents and Mobility (accepted, 2013)
8. Yazdi, N., Ayazi, F., Najafi, K.: Micromachined inertial sensors. Proceedings of the IEEE 86(8), 1640–1659 (1998)

Cloud-Based Platform to Labor Integration of Deaf People

Amparo Jiménez[1], Amparo Casado[1], Javier Bajo[2], Fernando De la Prieta[3], and Juan Francisco De Paz[3]

[1] Universidad Pontificia de Salamanca, Salamanca, Spain
{ajimenezvi,acasadome}@upsa.es
[2] Department of Computer Science and Automation Control, Universidad de Salamanca, Plaza de la Merced s/n, 37007, Salamanca, Spain
{fer,fcofds}@usal.es
[3] Departamento de Inteligencia Artificial, Universidad Politécnica de Madrid
javier.bajo@upm.es

Abstract. The new model of labor relations established by the Spanish Royal Decree-Law (3/2012) on urgent measures for labor reform has among its objectives the promotion of inclusion in the labor market of more advantaged groups, including the people with disabilities. This paper presents a cloud-based platform aimed at obtaining an on-line workspace to provide facilities to inform, train and evaluate the competencies of disabled people, and more specifically those skills required to facilitate the labor integration of individuals with auditory disabilities. This platform presented in this paper has been tested in a real environment and the results obtained are promising.

Keywords: disabled people, auditory disability, competence, intelligent systems, learning and training processes.

1 Introduction

Nowadays, the context of education and training for disabled people has acquired a growing relevance, especially for labor integration. Information and communication technologies play a very important role in this evolution. This study presents a research project, carried out during past year that was focused on two different realities: professional training and proper professional performance with the special needs, such as auditory disabled people, of some people with difficulties to access to employment.

Within this study, the target term auditory disabled refers to a person on with hearing difficulty that can be alleviated with technical aids (FESORCV) [2], as well as Prieto indicates [1]. In other words, people with a degree of disability (now disabled) greater than 33% by deafness or hearing limitations that encounter communication barriers (Spanish Law 27/2007, of 23 October, recognizing the

S. Omatu et al. (Eds.): *Distrib. Computing & Artificial Intelligence*, AISC 217, pp. 633–640.
DOI: 10.1007/978-3-319-00551-5_75

Spanish sign languages and regulates the means of support for oral communication of the deaf, hearing impaired and deaf-blind, 2007) [3], or as a term currently used for the Confederation of Deaf People (CNSE) [4] or the Spanish Confederation of Deaf Families [5].

The study started with the selection of a target group of deaf people in a specific job profile and its performance: Auxiliary Operations and General Administrative Services. Subsequently, a technology-based training tool had developed which allows the disabled people to effectively develop their professional performance, as well as to improve the professional training previous to the integration into the labor environment. In order to perform a proper design of this professional guidance, it is necessary to detail and analysis the specific characteristics of the position, profile and skills associated with their good work performance.

These skills have to be acquired by the worker, by means of professional qualification, in order to achieve the goals of the position. As Spanish Law 5/2002 on Qualifications and Vocational Training [6] states professional qualification is the "*set of skills with significance in employment that can be acquired through training or other types of modular training, and through work experience*". From a formal point of view, the qualification is the set of professional competencies (knowledge, skills, abilities, motivations) that allow us to perform occupations and jobs with a valuable labor market impact and that can be acquired through training or work experience.

Thus, we propose a cloud-based platform that focuses on obtaining on-line workspace for exchanging digital contents in an easy, intuitive and accessible manner. The main objective of the platform is to provide facilities to inform, train and evaluate the competencies of disabled people, and more specifically those skills required to facilitate the labor integration of individuals with auditory disabilities. This process may take place in the workplace or in the place of address via television, computer and mobile phone.

The rest of the paper is structured as follows: section 3 presents the problem formalization, section 4 describes the developed technological platform and, finally, section 5 presents the preliminary results and the conclusions obtained.

2 Problem Formalization

Our aim is not to find a professional qualification which corresponds to the Initial Professional Qualification Programmes (PCPI) but, based on the characteristics and requirements related to the position as identified in the Spanish Royal Decree 229/2008 of 1 February (BOE, No. 44 of February 20, 2008) [7], the objective is to identify some actions, strategies and more appropriate training resources, technologically updated and valid for the training and evaluation of the disabled individuals.The formalization of the problem as relied on the Spanish Catalogue of Professional Qualifications and professional qualifications and an Auxiliary Operations and Administrative Services for the Family General Administration and Management Professional with Level 1 were selected.

From our point of view, it is essential to follow the determination of the legal requirements and current proposals in the employment context. This allow us to train competent workers taking into account the parameters required in our socio-labor context, as well as the parameters shared by any worker (with or without disabilities) to develop such activities.

We define, therefore, and employment and social integration strategy for people with different skills but that can afford with guarantees the demands of the position. Therefore, we respect the design of general competencies, skills units and professional achievements with performance criteria proposed in the Spanish National Catalogue of Professional Qualifications, as well as the different existing guidelines in Spain and those proposed by various international organizations.

Taking as starting point the document from the Spanish National Institute of Vocational Qualifications, it is defined a structure of the professional qualifications that will serve to design programs, resources, methodologies and educational interventions. In this sense, we have made a major effort to assign to each qualification a general competence. This competence includes the roles and functions of the position and defines the specific skills or competency units. It is described also the professional environment in which you can develop the skills, relevant productive sectors and occupations or jobs relevant to access it.

Furthermore, in a complementary manner, we analyzed the professional achievements for each unit of competence along with their performance criteria.

The process started with the following situation:

- **General competence** is to distribute, reproduce and transmit the required information and documentation in the administrative and management task, internal and external, as well as to perform basic verification procedures on data and documents when senior technicians require it. These tasks are carried out in accordance with the existing instructions or procedures.
- **Competence units** are able to o provide support for basic administrative operations, to transmit and receive operational information to external agents to the organization and, finally, to perform auxiliary operations for reproduction and archiving data on conventional computational support.
- **Professional field:** This individual operates as an employee in any company or private/public entity, mainly in offices or departments oriented to administrative or general services.
- **Productive Sectors:** it appears in all the productive sectors, as well as public administration. It is necessary to remark the high degree of inter-sectoriality.
- **Relevant occupations and positions are office assistant**, general services assistant, file assistant, mail classifier and/or message, ordinance, information assistant, telephonist and ticket clerk.

However, looking for a more specific training support, it is necessary to complete this information with the detailed description of the most common tasks that arise in professional performance. Thus, describing the specific tasks, we have established the type of support that this group of disabled people requires to carry out an effective performance of the assigned tasks.

Finally, we have established the most appropriate training strategies. Thus, we have described the most common tasks related to the professional profile and professional qualification presented in the previous table. The following example in Table 1 illustrates our proposal.

Table 1 Example of Competence Unit and Professional development

Competence Unit: To provide support for basic administrative operations.

Professional Development 1: To periodically register the Information updates of the organization, department, areas, personnel, according to the instructions previously received, with the aim of obtaining key Information to improve the existing services.

1. Make a list of phone and fax references of the various members of the company.
2. Update the directory of people.
3. To register the physical location of people and areas within the company.
4. To update the physical location of people and areas within the company.
5. Safe-keeping of keys.
6. Opening and closing the workplace and departments.
7. Bring documentation to other centers in the city (unions, Delegation, City Council, County Council, etc.).
8. Turn off and turn on the lights.
9. Opening and closing windows.
10. Open and lock any room.
11. To register the inputs and outputs of the employees.
12. To register a list for people who want to take the annual medical review.

3 Technological Platform

Based on the problem formalized in section 2 we obtained a technological platform,shown in Figure 1, which is based on Cloud Computing paradigm and it is specifically designed to create intelligent environments [8] oriented to facilitate the labor integration of people with auditory disabilities. From one side, the main objective of the Ambient Intelligence (AmI) is to achieve transparent and ubiquitous interaction of the user with the underlying technology [8]. From the other side, Cloud Computing is a model for enabling ubiquitous, convenient, on-demand network access to a shared pool of configurable computing resources [11].

In this paper we use AmI to design a software technology specialized on determining the professional qualification, and providing on-line tools focused on transmitting signed orders that are easily accessed via mobile devices. Meanwhile, the platform is deployed in +Cloud [11], which is a cloud computing platform. This platform offers services as Platform as a Service (PaaS) and Software as a Service (SaaS) level. The project developed in this study is deployed at SaaS level and it uses for data storage the services provides by the platform at PaaS, concretely the service OSS (Object Service Storage) which makes use a non-relational database. And all educational resources are stored in the service FSS (File System Storage) provided by the platform.

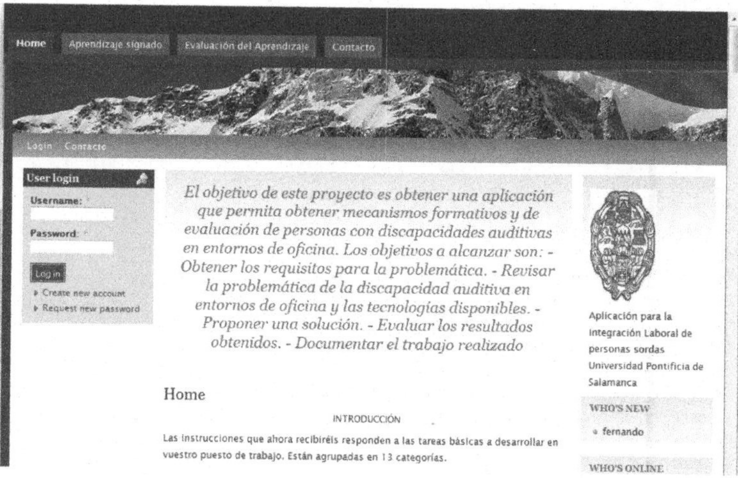

Fig. 1 Cloud-based platform

The functionality of the platform consists of a training web-based tool and a communication tool to send signed orders via mobile phone. Following it is described the main elements of the application:

- **Order signing.** Once the competences to evaluate were identified, and the related professional developments were defined, we proceeded to signing the actions and tasks that can be performed by the disabled person.To make the signing we counted on the cooperation of the Federation of the Deaf of Castile and Leon, who have participated in the signing process. The process followed consisted on recording a series of videos in which the sign interpreters transmit specific orders for each of the actions to be carried out by the disabled person. The recording was done in blocks, taking into account the professional developments taken into consideration.Once the recording process finished, we proceeded to edit the videos obtained by separating each action individually and including subtitles in Spanish.
- **Web platform.** In this task we obtained the design and development of a web platform that allows us to transmit work orders to the auditory disabled person using sign language format. The orders are transmitted via the Internet, television or mobile devices. The appearance of the platform is simple, trying to facilitate the accessibility and usability. The navigation through menus and contents is easy and intuitive. All the pages have been designed with the same structure, trying to facilitate a familiar environment and similar interaction patterns independently of the page or section in the platform.

Once the user is in the learning section (see Figure 2), the learning process is started, displaying the videos for the different blocks of accomplishments that can occur in the office environment:

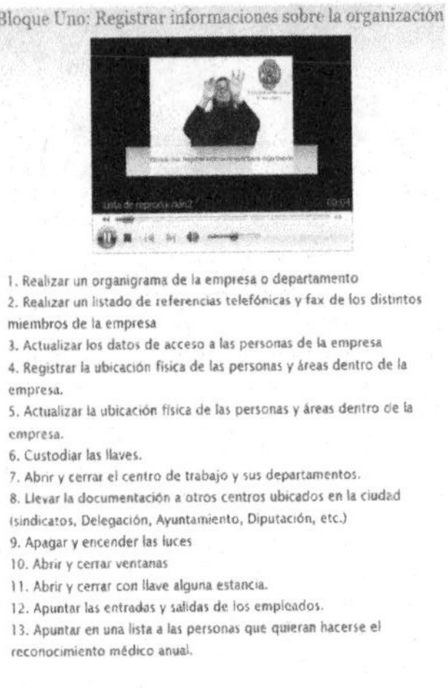

Fig. 2 Learning section

- **Mobile application.** In this work a mobile application had been developed for the platform that allows quick transmission of orders in the office workplace. The application includes voice recognition [9], so that a person at work may transmit voice instruction. These instructions will be recognized by the mobile device, which accesses a remote server and display the video corresponding to the order in sign language.

  The related work and the existing technologies were revised in order to choose the best option for the mobile module. An analysis was made of all mobile platforms on the market to see which is more suited to our requirements. The module was developed for iOS, and can be installed on a device like iPad iPhone, as long as it has the same operating system version iOS 5 or above. This module uses an XML file containing the structure of the data to be displayed. This XML file is stored in the cloud, and is parsed by our application. When the application starts, it parses the file and inserts into a table all the blocks, so that the user can choose one of them. Once the user clicks on a block, a screen containing an explicative video will be shown. The videos are also stored in the cloud. The advantages of using cloud storage are that the content can be updated very easily and without jeopardizing the proper functioning of the application. Below, in Figure 3, some screenshots for the developed application are presented, showing its operation.

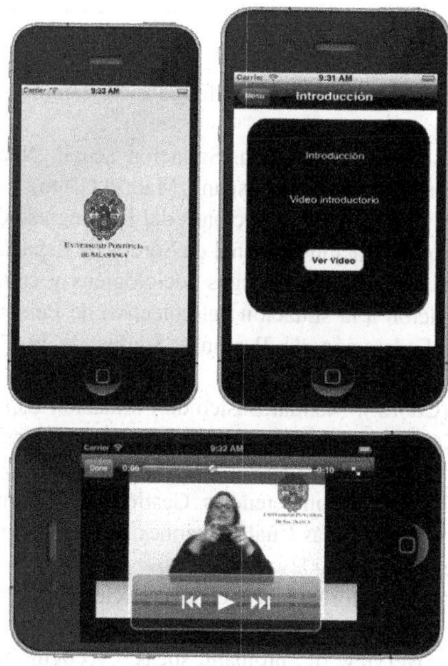

Fig. 3 Mobile application overview. Up-left: Main screen; Up-right: Block detail; Down: Signed video.

4 Conclusions

Our aim is to contribute to the goal of labor integration by means of a technological platform specifically designed to facilitate labor insertion in office environment of people with auditory impairment. The developed cloud-based platform has a web interface and an interface for mobile devices, and is based on pre-recorded videos that contain instructions on actions to be performed by the disabled person in the office environment. The web interface was successfully tested in teaching through television, in collaboration with the company CSA and the results have been promising. Moreover, the mobile application was tested in an office environment. Users and FAPSCyL specialists have highlighted the utility and advantages of the application. A test was designed with 10 basic tasks performed by 3 disabled people before and after the platform presented in this paper was installed. The platform provided a new tool that contributed to increase the percentage of completed tasks up to 85%, when the initial percentage (without the platform) was 42%. The disabled users have remarked the ease of understanding of instructions they receive from their supervisors and ease of use of the system.

Acknowledgments. This research has been supported by the project SOCIEDADES HUMANO-AGENTE: INMERSION, ADAPTACION Y SIMULACION. TIN2012-36586-C03-03funded by the Spanish Ministry of Science and Innovation.

References

1. Calvo Prieto, J.C. (ed.): La sordera. Un enfoque socio-familiar. Amarú Ediciones, Salamanca (1999)
2. CNSE. Las personas sordas en España. Situación actual. Necesidades y demandas. Confederación Nacional de Sordos de España, Madrid (1996)
3. CNSE. Retos para el siglo XXI: Resoluciones del II Congreso de la Confederación de Sordos de España. Confederación Nacional de Sordos de España, Madrid (1998)
4. FESORCV. Minguet, A. (Coord.): Rasgos sociológicos y culturales de las personas sordas: una aproximación a la situación del colectivo de Personas Sordas en la Comunidad Valenciana. Federación de Personas Sordas de la Comunidad Valenciana (FESORD C.V.), Valencia (2001)
5. FIAPAS. Jáudenes (Coord.): Manual Básico de Formación Especializada sobre Discapacidad Auditiva. Confederación Española de Padres y Amigos de los Sordos, Madrid (2004)
6. Levy-Leboner, C.: Gestión de competencias. Gestión 2000, Barcelona (1997)
7. Ley 5/2002, de 19 de junio de las Cualificaciones y de la Formación Profesional. Madrid: BOE del 20 de junio de (2002)
8. Weiser, M.: The computer for the 21st century. Scientific American 265(3), 94–104 (1991)
9. Reynolds, D.A.: An overview of automatic speaker recognition technology. In: 2002 IEEE International Conference on Acoustics, Speech, and Signal Processing (ICASSP), vol. 4, pp. 4072–4075 (2002)
10. Mell, P., Grance, T.: The NIST definition of Cloud Computing. NIST Special Publication 800-145 (September 2011)
11. Heras, S., De la Prieta, F., Julian, V., Rodríguez, S., Botti, V., Bajo, J., Corchado, J.M.: Agreement technologies and their use in cloud computing environments. Progress in Artificial Intelligence 1(4), 277–290 (2012)

Erratum: Mobile-Agent Based Delay-Tolerant Network Architecture for Non-critical Aeronautical Data Communications

Rubén Martínez-Vidal[1], Sergio Castillo-Pérez[1], Sergi Robles[1], Adrián Sánchez-Carmona[1], Joan Borrell[1], Miguel Cordero[2], Antidio Viguria[2], and Nicolás Giuditta[3]

[1] Department of Information and Communication Engineering, Universitat Autònoma de Barcelona, Edifici Q. Bellaterra, Barcelona, Spain
rmartinez@deic.uab.cat
[2] Center for Advanced Aerospace Technologies (CATEC), Parque Tecnológico y Aeronáutico de Andalucía. La Rinconada, Sevilla, Spain
[3] Deimos Space, Ronda de Poniente 19, Tres Cantos, Madrid, Spain

S. Omatu et al. (Eds.): *Distrib. Computing & Artificial Intelligence,* AISC 217, pp. 513–520.
DOI: 10.1007/978-3-319-00551-5_61

DOI 10.1007/978-3-319-00551-5_76

In the original version, the fourth and fifth author names were missed in this chapter. The names are given below:

Adrián Sánchez-Carmona[1] and Joan Borrell[1]

[1] Department of Information and Communication Engineering, Universitat Autònoma de Barcelona, Edifici Q. Bellaterra, Barcelona, Spain

---

The original online version for this chapter can be found at
http://dx.doi.org/10.1007/978-3-319-00551-5_61

Author Index